SOLUBILIZATION IN SURFACTANT AGGREGATES

SURFACTANT SCIENCE SERIES

1. Nonionic Surfactants, *edited by Martin J. Schick* (see also Volumes 19 and 23)
2. Solvent Properties of Surfactant Solutions, *edited by Kozo Shinoda* (see Volume 55)
3. Surfactant Biodegradation, *R. D. Swisher* (see Volume 18)
4. Cationic Surfactants, *edited by Eric Jungermann* (see also Volumes 34, 37, and 53)
5. Detergency: Theory and Test Methods (in three parts), *edited by W. G. Cutler and R. C. Davis* (see also Volume 20)
6. Emulsions and Emulsion Technology (in three parts), *edited by Kenneth J. Lissant*
7. Anionic Surfactants (in two parts), *edited by Warner M. Linfield* (out of print)
8. Anionic Surfactants: Chemical Analysis, *edited by John Cross* (out of print)
9. Stabilization of Colloidal Dispersions by Polymer Adsorption, *Tatsuo Sato and Richard Ruch*
10. Anionic Surfactants: Biochemistry, Toxicology, Dermatology, *edited by Christian Gloxhuber* (see Volume 43)
11. Anionic Surfactants: Physical Chemistry of Surfactant Action, *edited by E. H. Lucassen-Reynders* (out of print)
12. Amphoteric Surfactants, *edited by B. R. Bluestein and Clifford L. Hilton* (out of print)
13. Demulsification: Industrial Applications, *Kenneth J. Lissant*
14. Surfactants in Textile Processing, *Arved Datyner*
15. Electrical Phenomena at Interfaces: Fundamentals, Measurements, and Applications, *edited by Ayao Kitahara and Akira Watanabe*
16. Surfactants in Cosmetics, *edited by Martin M. Rieger*
17. Interfacial Phenomena: Equilibrium and Dynamic Effects, *Clarence A. Miller and P. Neogi*
18. Surfactant Biodegradation: Second Edition, Revised and Expanded, *R. D. Swisher*

19. Nonionic Surfactants: Chemical Analysis, *edited by John Cross*
20. Detergency: Theory and Technology, *edited by W. Gale Cutler and Erik Kissa*
21. Interfacial Phenomena in Apolar Media, *edited by Hans-Friedrich Eicke and Geoffrey D. Parfitt*
22. Surfactant Solutions: New Methods of Investigation, *edited by Raoul Zana*
23. Nonionic Surfactants: Physical Chemistry, *edited by Martin J. Schick*
24. Microemulsion Systems, *edited by Henri L. Rosano and Marc Clausse*
25. Biosurfactants and Biotechnology, *edited by Naim Kosaric, W. L. Cairns, and Neil C. C. Gray*
26. Surfactants in Emerging Technologies, *edited by Milton J. Rosen*
27. Reagents in Mineral Technology, *edited by P. Somasundaran and Brij M. Moudgil*
28. Surfactants in Chemical/Process Engineering, *edited by Darsh T. Wasan, Martin E. Ginn, and Dinesh O. Shah*
29. Thin Liquid Films, *edited by I. B. Ivanov*
30. Microemulsions and Related Systems: Formulation, Solvency, and Physical Properties, *edited by Maurice Bourrel and Robert S. Schecter*
31. Crystallization and Polymorphism of Fats and Fatty Acids, *edited by Nissim Garti and Kiyotaka Sato*
32. Interfacial Phenomena in Coal Technology, *edited by Gregory D. Botsaris and Yuli M. Glazman*
33. Surfactant-Based Separation Processes, *edited by John F. Scamehorn and Jeffrey H. Harwell*
34. Cationic Surfactants: Organic Chemistry, *edited by James M. Richmond*
35. Alkylene Oxides and Their Polymers, *F. E. Bailey, Jr., and Joseph V. Koleske*
36. Interfacial Phenomena in Petroleum Recovery, *edited by Norman R. Morrow*
37. Cationic Surfactants: Physical Chemistry, *edited by Donn N. Rubingh and Paul M. Holland*
38. Kinetics and Catalysis in Microheterogeneous Systems, *edited by M. Grätzel and K. Kalyanasundaram*
39. Interfacial Phenomena in Biological Systems, *edited by Max Bender*
40. Analysis of Surfactants, *Thomas M. Schmitt*
41. Light Scattering by Liquid Surfaces and Complementary Techniques, *edited by Dominique Langevin*
42. Polymeric Surfactants, *Irja Piirma*
43. Anionic Surfactants: Biochemistry, Toxicology, Dermatology. Second Edition, Revised and Expanded, *edited by Christian Gloxhuber and Klaus Künstler*
44. Organized Solutions: Surfactants in Science and Technology, *edited by Stig E. Friberg and Björn Lindman*
45. Defoaming: Theory and Industrial Applications, *edited by P. R. Garrett*

46. Mixed Surfactant Systems, *edited by Keizo Ogino and Masahiko Abe*
47. Coagulation and Flocculation: Theory and Applications, *edited by Bohuslav Dobiáš*
48. Biosurfactants: Production · Properties · Applications, *edited by Naim Kosaric*
49. Wettability, *edited by John C. Berg*
50. Fluorinated Surfactants: Synthesis · Properties · Applications, *Erik Kissa*
51. Surface and Colloid Chemistry in Advanced Ceramics Processing, *edited by Robert J. Pugh and Lennart Bergström*
52. Technological Applications of Dispersions, *edited by Robert B. McKay*
53. Cationic Surfactants: Analytical and Biological Evaluation, *edited by John Cross and Edward J. Singer*
54. Surfactants in Agrochemicals, *Tharwat F. Tadros*
55. Solubilization in Surfactant Aggregates, *edited by Sherril D. Christian and John F. Scamehorn*

ADDITIONAL VOLUMES IN PREPARATION

Anionic Surfactants: Organic Chemistry, *edited by Helmut Stache*

Foams: Theory, Measurements, and Applications, *edited by Robert K. Prud'homme and Saad A. Khan*

SOLUBILIZATION IN SURFACTANT AGGREGATES

edited by

Sherril D. Christian
John F. Scamehorn
University of Oklahoma
Norman, Oklahoma

CRC Press
Taylor & Francis Group
Boca Raton London New York

CRC Press is an imprint of the
Taylor & Francis Group, an **informa** business

CRC Press
Taylor & Francis Group
6000 Broken Sound Parkway NW, Suite 300
Boca Raton, FL 33487-2742

First issued in paperback 2019

ISBN-13: 978-0-8247-9099-8 (hbk)
ISBN-13: 978-0-367-40179-5 (pbk)

Library of Congress Cataloging-in-Publication Data

Solubilization in surfactant aggregates / edited by Sherrill D. Christian, John F. Scamehorn.
p. cm. – (Surfactant science series ; v. 55)
Includes bibliographical references and index.
ISBN 0-8247-9099-5 (acid-free)
1. Solubilization. 2. Surface active agents. 3. Organic compounds–Solubility. I. Christian, Sherril Duane. II. Scamehorn, John F. III. Series.
QD543.S66296 1995
547.1'342–dc20 95-7011
CIP

Visit the Taylor & Francis Web site at
http://www.taylorandfrancis.com

and the CRC Press Web site at
http://www.crcpress.com

To Dee with thanks for her cooperation, patience, and love

Sherril Christian

Preface

Solubilization of organic compounds in surfactant aggregates is a phenomenon of great theoretical and practical significance. For example, physicochemical studies of solubilization have been very influential in improving our understanding of the forces involved in micelle formation and the structure and character of micelles. Studies of solubilization in other types of surfactant aggregates, such as admicelles and vesicles, are also providing information about surfactant properties. Applications of solubilization are numerous; examples include pharmaceuticals, detergency, agricultural products, surfactant-based separations, food products, emulsion polymerization, surface modification, biomedical products, and analytical chemistry.

This volume includes chapters that discuss the solubilization of organics of various types in single-surfactant and mixed-surfactant micelles, as well as studies of the forces involved and mathematical descriptions of these effects. Solubilization in admicelles and vesicles is also addressed in several chapters. Methods of measuring solubilization, both classical and new instrumental techniques, are thoroughly covered by several authors. Lastly, a chapter on solubilization in detergency describes one practical application of the phenomenon.

This volume brings together both current knowledge and information about the newly developing research areas where advances are most rapid. Solubilization is one of the earliest observed aspects of surfactant solution behavior and continues to be an important and exciting field to explore.

Sherril D. Christian
John F. Scamehorn

Contents

Contributors

Masahiko Abe Faculty of Science and Technology, Science University of Tokyo, Noda, Chiba, Japan

Elsa Abuin Departamento de Quimica, Facultad de Ciencia, Universidad de Santiago de Chile, Santiago, Chile

Paschalis Alexandridis Department of Chemical Engineering, Massachusetts Institute of Technology, Cambridge, Massachusetts

Guy Broze Advanced Technology, Colgate-Palmolive Research & Development, Inc., Milmort, Belgium

Sherril D. Christian Department of Chemistry and Institute for Applied Surfactant Research, University of Oklahoma, Norman, Oklahoma

Rosario De Lisi Department of Physical Chemistry, University of Palermo, Palermo, Italy

Connie S. Dunaway Department of Chemistry and Institute for Applied Surfactant Research, University of Oklahoma, Norman, Oklahoma

Jeffrey H. Harwell School of Chemical Engineering and Materials Science and Institute for Applied Surfactant Research, University of Oklahoma, Norman, Oklahoma

T. Alan Hatton Department of Chemical Engineering, Massachusetts Institute of Technology, Cambridge, Massachusetts

Patricia N. Hurter Department of Chemical Engineering, Massachusetts Institute of Technology, Cambridge, Massachusetts

Bengt Jönsson Division of Physical Chemistry 1, Chemical Center, University of Lund, Lund, Sweden

Allen D. King, Jr. Department of Chemistry, University of Georgia, Athens, Georgia

Jan C. T. Kwak Department of Chemistry, Dalhousie University, Halifax, Nova Scotia, Canada

Mikael Landgren Division of Physical Chemistry 1, Chemical Center, University of Lund, Lund, Sweden

Eduardo Lissi Departamento de Quimica, Facultad de Ciencia, Universidad de Santiago de Chile, Santiago, Chile

Lance L. Lobban School of Chemical Engineering and Materials Science, University of Oklahoma, Norman, Oklahoma

D. Gerrard Marangoni Department of Chemistry, Saint Francis Xavier University, Antigonish, Nova Scotia, Canada

Stefania Milioto Department of Physical Chemistry, University of Palermo, Palermo, Italy

Nagamune Nishikido Faculty of Science, Fukuoka University, Jonan-ku, Fukuoka, Japan

John H. O'Haver School of Chemical Engineering and Materials Science, University of Oklahoma, Norman, Oklahoma

Edgar A. O'Rear III School of Chemical Engineering and Materials Science and Institute for Applied Surfactant Research, University of Oklahoma, Norman, Oklahoma

Keizo Ogino Faculty of Science and Technology, Science University of Tokyo, Noda, Chiba, Japan

Gerd Olofsson Division of Thermochemistry, Chemical Center, University of Lund, Lund, Sweden

John F. Scamehorn School of Chemical Engineering and Materials Science and Institute for Applied Surfactant Research, University of Oklahoma, Norman, Oklahoma

Peter Stilbs Department of Physical Chemistry, The Royal Institute of Technology, Stockholm, Sweden

C. Treiner Laboratoire d'Electrochimie, Université Pierre et Marie Curie, URA CNRS 430, Paris, France

Edwin E. Tucker Department of Chemistry and Biochemistry, University of Oklahoma, Norman, Oklahoma

Anthony J. Ward Department of Chemistry, Clarkson University, Potsdam, New York

Timothy J. Ward Department of Chemistry, Millsaps College, Jackson, Mississippi

Karen D. Ward Department of Chemistry, Millsaps College, Jackson, Mississippi

Hitoshi Yamauchi Developmental Research Laboratories, Daiichi Pharmaceutical Co., Ltd., Edogawa-ku, Tokyo, Japan

I
Overview

1

Overview and History of the Study of Solubilization

CONNIE S. DUNAWAY and SHERRIL D. CHRISTIAN Department of Chemistry and Institute for Applied Surfactant Research, University of Oklahoma, Norman, Oklahoma

JOHN F. SCAMEHORN School of Chemical Engineering and Materials Science and Institute for Applied Surfactant Research, University of Oklahoma, Norman, Oklahoma

SYNOPSIS

In this introductory chapter, a historical background of solubilization is presented by referring to various review articles, books, and monograph chapters that trace the development of the present understanding of the solubilization of organic molecules and ions by organized surfactant assemblies. A general description of the typical solubilization behavior of ionic micelle/solubilizate systems is included. In addition, methods for reporting and representing solubilization results are described. The solubilization and activity coefficient isotherms for three prototype solutes (hexane, 2,3-dichlorophenol, and benzene) in CPC (*N*-hexadecylpyridinium chloride or cetylpyridinium chloride) are summarized. Solubilization in other types of surfactant aggregates besides ionic micelles is briefly discussed. The last section of this chapter provides a guide to the organization and contents of the chapters included in this monograph.

I. INTRODUCTION

To introduce the papers in this monograph on solubilization in micelles and other surfactant aggregates, we shall refer to previous reviews [1–4] of the topic, including those by a number of scientists who have played important roles in the development of our present understanding of micellar and related systems. From the work of McBain [5–17], Hartley [18–22], Langmuir [23–26], Debye [27], and others, during the teens through the forties, knowledge gradually accumulated about the probable geometry, size, shape, and other characteristics of surfactant micelles, vesicles, monolayers, bilayers, and even more complicated structures. Part of our goal will be to acquaint the reader with the numerous review articles, books, and monograph chapters that trace the development of our understanding of the solubilization of organic molecules and ions by organized surfactant assemblies.

The articles included in the present monograph will deal with solubilization in micelles, in vesicles, and in adsorbed surfactant layers. Particular emphasis will be given to novel techniques for inferring the extent of solubilization, measuring spectral and thermodynamic properties related to the states of solubilized species in surfactant aggregates, and determining properties of systems in which the relative concentrations of solubilizate and surfactant are varied throughout considerable ranges. Applications of solubilization in areas such as detergency and analytical chemistry will also be discussed in several of the articles. Little attention will be given to solubilization phenomena occurring in emulsions and microemulsions. Such topics have been reviewed [28–31] extensively, particularly in relation to the use of microemulsions in the tertiary recovery of oil, caused by the solubilization and hence mobilization of hydrocarbons trapped in reservoirs after primary and secondary recovery treatments have been completed. Many of the surface and solution phenomena involved in solubilization in emulsions are similar to those discussed in the articles in this monograph, but the molecular interpretations of such effects would lead us far afield from our major goal: providing clear descriptions of systems in which organic solutes are solubilized in surfactant aggregates of various types.

A general description of solubilization results for the ubiquitous ionic micelle/solubilizate systems will be included in this introductory chapter. Methods for reporting and plotting solubilization results will be described, and the behavior of solubilization isotherms for typical systems will also be summarized. From having examined solubilization results obtained for many hundreds of systems, we have become convinced that it is unreasonable to assume that solubilization equilibrium or partition constants will remain constant for a given micellar system if the "loading" of solubilizate is varied throughout any sizable range. Thus it becomes important to develop methods for studying the variation in thermodynamic and physical properties of micelle/solubilizate systems throughout wide ranges of concentration of both the surfactant and the organic solute. The procedure for describing and interpreting such results will be described in some detail.

II. HISTORICAL BACKGROUND

Along with the accumulation of information about hydrophilic colloids in aqueous solutions, there was an important parallel development of knowledge about amphiphilic compounds adsorbed or spread at liquid/vapor and liquid/liquid interfaces. Such results will not be reviewed here, but it should be noted that information about monolayers of carboxylic acids, their insoluble salts, and related compounds, studied initially by Langmuir-Blodgett techniques, have provided and continue to provide fundamental knowledge that can help us understand the properties of surfactant assemblies and species solubilized in these aggregates [23–26, 32–37]. Quantitative results obtained from careful physicochemical studies of x-ray diffraction patterns [38–50], light scattering [27, 51–54], gas adsorption [55–62], surface and interfacial tension [63, 64], surface activity (in the Gibbs sense) [65], and properties of matter in small droplets or clusters [66–68], have provided a framework for interpreting properties of the "wet" colloidal systems of direct interest in most studies of solubilization by surfactant aggregates. Excellent treatises, textbooks, and monographs summarize the surface chemistry and physics that form the basis for many modern interpretations of the properties of aqueous and nonaqueous micellar solutions [34, 65, 69–72]. Developments in the theory of solutions have led to the use of models involving solubility parameters [73, 74], group interaction parameters [75, 76], and polymer interaction coefficients [77–79] to describe intramicellar "solutions." Modern theories must account for hydrophobic effects (hydrophobic association and hydrophobic interaction) in describing the formation of clusters such as micelles and their solubilizates from monomers in aqueous solution. Electrical effects [80–83], including properties of the electric double layer, counterion binding, Debye-Hückel formalism, and advanced statistical treatments [84, 85] of ionic surfactant solutions are also accounted for in various empirical and theoretical treatments of the properties of micelles and micelle/solubilizate systems.

The early studies of solubilization have been reviewed in articles by McBain [1] and by Klevens [2], as well as in the books by McBain and Hutchinson [3] and by Elworthy, Florence, and Macfarlane [4]. The various effects of such factors as the nature and structure of the solubilizer and solubilizate, temperature, and the presence of additives such as electrolytes on solubilization are discussed in detail in the reviews. McBain and Hutchinson have traced the development of the concept of solubilization in solutions that would have contained micelles back to at least 1846, in which year Persoz noted that soap solutions are able to enhance the solubility of slightly soluble substances [3]. During the last half of the nineteenth century, there were numerous studies of the effects of colloids in increasing the solubility of organic compounds and in enhancing the miscibility of liquids that are only sparingly soluble in each other. By 1900, there had been several studies of the solubilization of cholesterol and fats by soaps and biological surfactants, as well as reports of the use of fat-soluble dyes to label fats incorporated in such systems. In 1942 Merrill and McBain [10] defined solubilization as a process in which otherwise insoluble matter is brought into solution in a particular manner, namely by colloidal matter, specifically by micelles. Solubilization, according to McBain, "consists in the spontaneous passage of molecules of a substance insoluble in a given solvent, to form a thermodynamically stable solution" [13]. Most workers since that time have applied similar definitions, with the proviso that "otherwise insoluble matter" may include the increment in solubility caused by micelles (or other aggregates), even for substances that may already have appreciable solubility in the pure medium.

Following the introduction of the concept of micelles by McBain in 1913 [86], scientists began to speculate about the size and shapes of these species, and to relate these ideas to the possible mechanisms of solubilization of organic solutes in micelles. Progress in describing the mixed micelle/solubilizate systems awaited resolution of the controversy between those like McBain [1, 3, 10, 11, 16, 17, 45], who interpreted results in terms of lamellar structures, and those with Hartley [21, 87–89], who proposed spherical, oblate or prolate spheroid, or relatively small rodlike structures for micelles. Various references give historical accounts, as well as interpretations, of x-ray diffraction experiments which were utilized to investigate the structure and organization of micelles [1–3, 16, 17, 48, 50, 65, 90, 91]. By the 1950s and 1960s, a general molecular picture of micelles and their solubilizates had begun to emerge, particularly as a result of light scattering studies and other physical studies which led to estimates of the sizes and shapes of micelles [51–54, 92–98]. More recently, physical experiments such as small angle neutron scattering (SANS) [99–106] and fluorescence probe [107–119] methods have added to our knowledge about the surfactant assemblies that are capable of solubilizing neutral and ionic organic species, the location of solutes within micelles, the changes in micellar structure and size caused by the presence of dissolved solutes, and the physical and chemical forces responsible for the formation of aggregates containing dissolved solutes. Theories and model descriptions have been useful in predicting (or at least correlating) solubilization results for particular types of micellar systems, including those formed from the common ionic and nonionic surfactants [84, 120], as well as polymeric systems containing blocks of polyethylene oxide (PEO) and polypropylene oxide (PPO) [77–79]. Applications of solubilization have been found and continue to be found in numerous areas, including pharmaceuticals [4, 10, 78, 121–131], biology [4, 132], detergency [3, 4, 10, 122, 124, 127, 133–138], cosmetics [4, 139–142], textile processing [3, 4, 143], engineering [3, 127], agricultural products [3, 4, 127, 144, 145], surfactant-based separations [146–157], analytical chemistry [158–160], emulsion polymerization [4, 127, 161–163], micellar catalysis [124, 136, 164–174], and environmental processes [175–183].

III. TYPICAL BEHAVIOR OF IONIC MICELLE/ SOLUBILIZATE SYSTEMS

It may be informative to describe the behavior of aqueous surfactant/ organic solubilizate systems, with emphasis on solubilization by the typical aqueous ionic micellar systems, which have been by far the most extensively studied. There seems to be general agreement that micelles of many of the common ionic surfactants are spherical at concentrations ranging from the cmc (critical micelle concentration) to at least 10 times the cmc, in the absence of added electrolyte [83, 184, 185]. Within the micellar sphere, the hydrocarbon core region consists of intertwined, randomly oriented hydrocarbon groups [115, 186, 187], forming a liquid-like region having a viscosity approximately an order of magnitude greater than that of liquid hydrocarbons of similar chain length [188]. At higher concentrations, or with the addition of salts, spherical micelles convert to ellipsoidal, rodlike, or other nonspherical forms, in which the surfactant head groups are packed more tightly than at low concentrations and in the absence of added electrolytes [51, 97, 189–200]. Transitions between the spherical (or only slightly deformed spherical) structures and more-elongated micelles have been investigated with several types of experiments, including light-scattering [51, 189, 191, 194, 196], SANS [101, 106], ultrafiltration [201], fluorescence-probe [110], small-angle x-ray scattering [97, 98], electric birefringence [195, 202], Raman spectroscopy [186], conductance [203], calorimetry [204–206], and other methods. At still higher concentrations of surfactant, or in the presence of certain additives, there is ample evidence that lamellar forms and other quite complicated surfactant/solvent structures exist [207].

An important concept that has guided the modeling of surfactant aggregate structures for many decades is the principle that the length of the extended chain of the surfactant molecule is an important natural dimension in micellar and related systems [12, 17, 21, 26, 36, 208–212]. Thus dense monolayers and bilayers of soaps and ionic surfactants often form at interfaces with a near-parallel arrangement of the hydrocarbon chains; the thickness of these layers can be estimated quite accurately from simple models of the sizes and shapes of the surfactant molecules themselves. This idea can be applied in particularly simple form to spherical micelles, where it leads to good estimates of the micellar radius and volume, the number of molecules per micelle, and the effective area of the headgroup [184, 190, 208, 210–214]. In the case of spheroidal or rodlike aggregates, one of the dimensions can again be equated to the extended chain length, although simple geometrical considerations are of no use in estimating the lengths of rods or the diameters of disklike structures. Consideration of electrostatic effects (headgroup repulsion and the effects of counterion binding) makes it possible to estimate approximately when a transition from sphere to rod may be expected to occur [190, 211]. Thus either the addition of a 1:1 electrolyte or increasing the concentration of an ionic surfactant will increase the ionic strength, partially screening repulsive interactions between the headgroup anions or cations, and cause the more densely packed nonspherical micellar forms to become thermodynamically more stable than spheres.

There has been a consensus for several decades that solutes like hexane or heptane tend to solubilize in the core region of spherical or rodlike micelles, although simple geometry dictates that at least some of the methyl and methylene groups of most of the solubilized molecules will be within a few angstroms of the micellar headgroup region [215]. The alkane molecules are thought to mix with the hydrocarbon tails of the surfactants in a liquid-like solution, having a viscosity perhaps an order of magnitude greater than that of a corresponding liquid alkane [188]. There is no oil drop of pure solubilized hydrocarbon in the interior of the spherical micelle, as some earlier representations of alkane solubilization seemed to imply. Of course, addition of an alkane to a spherical micelle will be expected to cause an increase in volume and ordinarily to increase the micelle aggregation number [123, 136, 206, 216–219]. The transition from spheres to rods usually causes moderate increases in the extent of solubilization of alkanes, and it has long been argued that this reflects the fact that elongated micelles will have a larger volume of core (per surfactant molecule) than do spherical forms [220, 221].

It is also commonly agreed that amphiphiles which contain highly polar groups attached to aliphatic or aromatic moieties tend to solubilize with their polar groups anchored in the micellar headgroup region [2, 14, 46, 83, 91, 132, 136, 222–227]. If the hydrocarbon group of an amphiphile is sufficiently large, this group will presumably extend downward through the so-called palisade layer and into the micellar core. The addition of salts like NaCl or KCl to ionic surfactant/solubilizate systems, or an increase in the concentration of the surfactant tends to reduce the solubilization of the amphiphilic solutes [3, 14, 123, 228]. This may be attributed to the screening effect of the surfactant counterions and ions from the added electrolyte, which allows the ionic headgroups in the micellar surface region to pack more densely [136, 206]. Consequently, the amphiphile polar headgroups will bind or adsorb less strongly in the headgroup region. Ordinarily, the transition from spheres to rodlike micelles will cause a decrease in the tendency of phenols, aliphatic alcohols, and related solutes to solubilize, for a given total number of surfactant molecules [198, 221, 229].

There is still no general agreement regarding the location and binding of benzene and other aromatic molecules within micelles. Arguments have been presented to show that benzene solubilizes mainly within the micellar core, in the palisade layer or near the micellar surface, or in all of these regions [75, 230–239]. Certainly, if the hydrocarbon core region resembles a not-too-viscous hydrocarbon liquid, one would expect considerable solubilization of benzene and related compounds in this region; on the other hand, the possibility that aromatic π-electron systems could interact (albeit weakly) with ionic charges and dipoles in the outer regions of the micelle might account for the fact that aromatics interact somewhat more strongly with ionic micelles than do the aliphatic hydrocarbons [75, 238]. Thermodynamic data (*vide infra*) do in fact indicate that aromatics are intermediate between the alkanes and numerous types of amphiphiles in their tendency to transfer into micelles, using the pure component standard state for all of the liquid compounds compared [227, 240, 241].

IV. REPRESENTATION OF SOLUBILIZATION RESULTS

There has been no uniformity in the definitions of partition coefficients or equilibrium constants that may be used to represent the solubilization of solution components by surfactant micelles. One of the commonest procedures is to calculate a partition coefficient representing the ratio of the solubility of an organic compound in the micelle to its solubility in the solvent outside the micelle. Thus, treating the micelle as a pseudophase in which the surfactant and solubilized substance(s) reside, one may define a dimensionless partition ratio by

$$P = \frac{X_{\text{micelle}}}{X_{\text{bulk}}} \tag{1}$$

where X_{micelle} represents the mole fraction of the solubilizate in the micelle and X_{bulk} is the mole fraction of the solute in the extramicellar bulk solvent phase (the so-called monomer phase of the pseudophase separation model) [242–245]. Other concentration units may be used in place of mole fraction, including the solute molarity in both the micelle and the bulk aqueous phase, although knowledge of molarity in the micelle requires detailed partial molar volume results [85, 246–249].

Alternatively, and more in line with the mass action model, it is possible to relate solubilization to equilibrium constants for the reaction

$$\begin{aligned}&\text{surfactant (micelle)} + \text{organic solute (aqueous)}\\ &\qquad = \text{surfactant}\cdot\text{organic solute (micelle)}\end{aligned}$$

where the transfer of the solute into the micelle is treated as a binding phenomenon, for which the equilibrium constant is

$$K_s = \frac{[\text{organic solute in micelle}]}{[\text{surfactant in micelle}][\text{organic solute in bulk}]} \tag{2}$$

where the brackets indicate molar concentrations measured with respect to the entire volume of solution [147, 168, 207, 240, 250, 251]. Thus, assuming a method is available for inferring the partitioning of a solute between the bulk and micellar "phases," it is possible to express the equilibrium constant, K_s, as

$$K_s = \frac{X}{[(1 - X)c_o]} \tag{3}$$

where X is the mole fraction of organic solute in the micelle (the same as X_{micelle} in Eq. (1)) and c_o is the molar concentration of unsolubilized organic solute in the bulk aqueous phase. The value of c_o is of course readily calculated from X_{bulk} in Eq. (1), and for dilute solutions of solutes in water at 25°C, $X_{\text{bulk}} = c_o/55.34$, where c_o is expressed as a molarity. Thus conversion of K_s from Eq. (3) into P from Eq. (1) is a trivial problem, at least for solutes that are present in dilute solution in the bulk aqueous phase. K_s is of course defined without regard to the aggregation number

(n) of surfactant molecules in the micelle. If n is known, K_s can be multiplied by n to obtain a modified equilibrium constant for the transfer of organic solute molecules into the micelle, which is considered to be a molecular aggregate of known mass. In the limit as $X \to 0$, this modified equilibrium constant represents the binding constant for a single molecule of the organic solute transferred from the bulk aqueous phase into a single micelle.

In recent studies, we have found it useful to define yet another equilibrium constant, which preserves the M^{-1} units of the binding constant (K_s), while being more closely related to the partition ratio (P from Eq. (1) [127, 198, 199, 226, 240, 252–254]. Thus, in numerous studies of solubilization equilibria throughout a wide range of mole fractions of organic solute in the micelle (X), we have studied the variation of the equilibrium constant

$$K = \frac{X}{c_o} \tag{4}$$

where K is easily converted into both P and K_s. Thus

$$K = K_s(1 - X) = \frac{PX_{\text{bulk}}}{c_o} \tag{5}$$

and in sufficiently dilute solution at 25°C, $K = K_s = P/55.34$.

A minor correction is required to convert any of the constants (K, K_s, or P) into thermodynamic constants *if* the organic solute is not ideal in the Henry's law sense in the aqueous solution phase. For example, solutions of pentanol in water deviate significantly from ideality in the concentration range in which solubilization in ionic micelles is usually investigated; therefore, in place of bulk mole fractions or molarities of the organic solute (in Eqs. (1)–(5)), one should strictly speaking use activities based on ideal dilute solution standard states in calculating and presenting K, K_s, or P values as a function of X [255]. This has rarely been done in the literature, although as the practice of determining and interpreting detailed solubilization isotherms (the solubilization equilibrium constant, K, plotted as a function of X) becomes more widespread, it may become more important to account for nonideal behavior in the bulk aqueous phase.

Finally, in relating properties of micelle/solubilizate systems to various solution theories, it is convenient to define and calculate activity coefficients for organic solutes and surfactants in the "intramicellar solution." Treating the micelle as a binary solution of surfactant and solubilizate, one may define the activity coefficient of the organic solute as

$$\gamma_o = \frac{a_o}{X} \tag{6}$$

where a_o is the activity of the organic solute [199, 221, 241, 256]. In reports from our laboratory, a_o is based on the pure organic liquid or supercooled liquid standard state; ordinarily a_o can be calculated with sufficient accuracy from p_o/p_o°, where p_o is the partial pressure of the organic solute above the intramicellar solution and p_o° is the vapor pressure of the pure liquid organic compound at the given temperature. Values of the activity coefficient of a solute (γ_o) in a micelle, based on the definition given in Eq. (6), indicate the relative volatility of a compound from the intramicellar solution, compared to the pure liquid standard state for that component [198]. Thus, if the activity coefficient of hexane in a dilute solution of hexane in micelles of CPC (*N*-hexadecylpyridinium chloride or cetylpyridinium chloride) is 3, this indicates that at the given mole fraction in the CPC micelle, hexane will have three times the partial pressure (or escaping tendency) that it has in an ideal solution (one obeying Raoult's law) at the same mole fraction [221]. In the case of solutes which obey Henry's law at all concentrations up to saturation, there is a particularly simple relation between K (defined by Eq. (4)) and γ_o:

$$\gamma_o = \frac{1}{Kc_o^\circ} \tag{7}$$

where c_o° is the solubility of the organic solute in pure water [226, 227, 257].

One consequence of the definition of activity coefficient introduced in the previous paragraph is that γ_o will ordinarily tend toward unity as the mole fraction of the solute in the micelle (X) approaches 1 [199, 221, 241]. Of course, X cannot actually reach 1 in any micelle/solubilizate system, but many polar organic solutes which interact strongly with ionic micelles will have limiting values of X as large as 0.8 or 0.9; in such systems, γ_o increases quite rapidly as saturation is approached, and at saturation γ_o must actually exceed unity.

V. THE SOLUBILIZATION ISOTHERM

In many recent reports from this laboratory, we have emphasized that it is rarely justifiable to assume that the solubilization equilibrium constant (K) or partition ratio (P) will remain constant as the intramicellar composition (X) varies throughout wide ranges [127, 199, 240, 252]. Unfortunately, most methods for inferring the extent of solubilization only provide information about limit points on the solubilization isotherm. For example, the solubility or maximum additive concentration (MAC) method is only capable of determining the partitioning of a solute into the micelle under conditions where the micelle contains the maximum mole fraction (X) of solute [124, 172, 258–261]. Other methods lead to the determination of the partition constant or solubilization equilibrium constant in the opposite region, namely in the Henry's law limit, where $X \rightarrow 0$ [242, 262]. Thus the extent of our current knowledge about thermodynamic and other properties of most surfactant/solubilizate systems is much more restricted than the information available for binary mixtures of organic compounds, which are typically studied throughout the entire miscibility range (from zero mole fraction to the saturation limit). In our laboratory, a major fundamental and practical goal in recent solubilization studies has been to determine solubilization equilibrium constants as a function of X, throughout as much of the region of miscibility as possible [127, 236, 263–265].

There is of course another dimension to the problem of reporting solubilization data at different concentrations of surfactant and organic solubilizate: that is the effect of variation in total concentration of surfactant, which can ultimately cause the size and shape of micelles to vary, independent of any effects caused by incorporating the solute in micelles [98, 195]. Thus, as the total concentration of surfactant increases from a few times the cmc to more than 100 times the cmc, many ionic micelles will at some point undergo a transition from near-spherical to rodlike or other elongated forms. This may be expected to modify the extent of solubilization of organic solutes, normally causing alkane solubilization to increase and the solubilization of highly polar amphiphiles to decrease [198, 221, 229, 256]. Fortunately, in the case of many ionic surfactants like CPC, solubilization results are usually consistent for total surfactant concentrations varying from just above the cmc to at least 100 times the cmc. That is, the intramicellar composition (X) is a function of the thermodynamic activity of the organic solubilizate, but not the total concentration of surfactant.

Figures 1–3 show solubilization isotherms (K in reciprocal molarity units, represented as a function of X) for three prototypical solutes in CPC micelles, under conditions where the micelles are assumed to be nearly spherical, at least in the absence of added solubilizate. In Fig. 1, data for hexane [220, 221] are plotted in the form K vs. X, for X varying from near zero to near saturation; these results were obtained with an accurate vapor pressure method (see Chapter 13). Similar plots are shown in Fig. 2 for 2,3-dichlorophenol in CPC micelles, using data obtained with the semiequilibrium dialysis method [226], and in Fig. 3 for benzene in CPC micelles, using vapor pressure results [221, 238].

The monotonic increase in K values with increasing X shown in Fig. 1 supports the general belief that hexane solubilizes primarily in the interior hydrocarbon-core region of the CPC micelle. Thus, incorporation of hex-

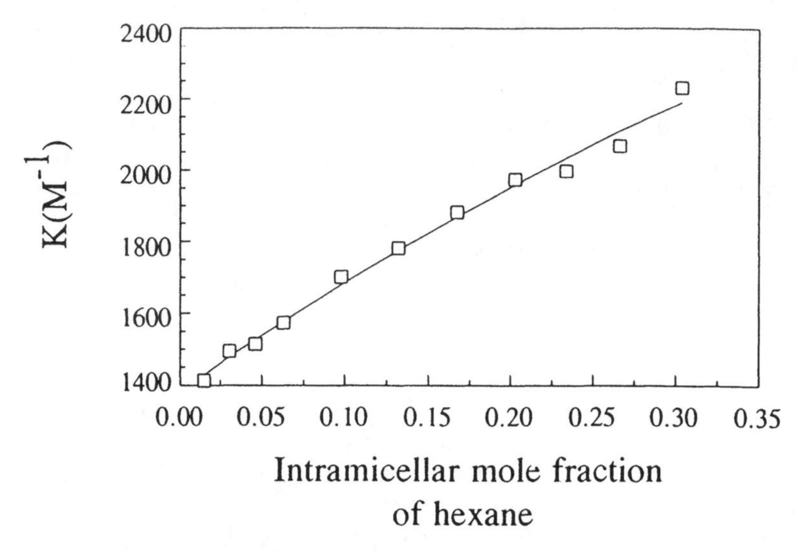

FIG. 1 Dependence of the solubilization equilibrium constant for hexane in *N*-hexadecylpyridinium chloride [CPC] (0.1 M) on the intramicellar mole fraction of hexane at 25°C.

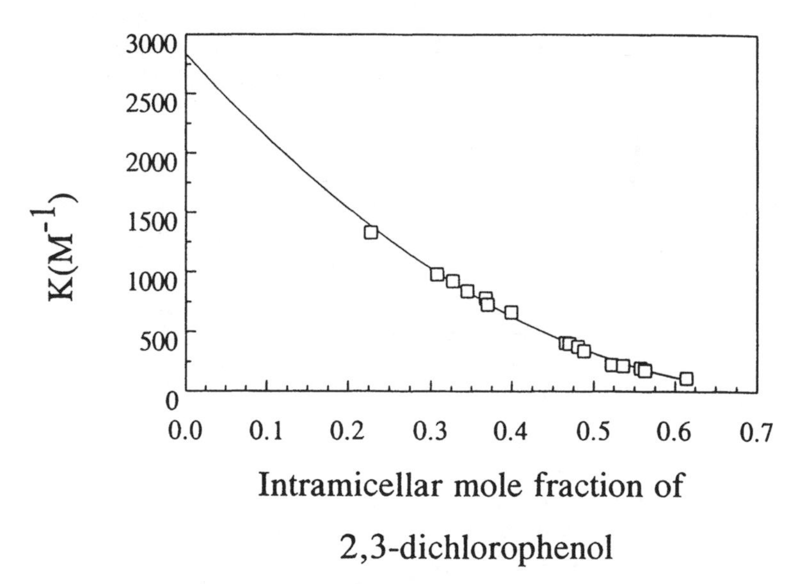

FIG. 2 Dependence of the solubilization equilibrium constant for 2,3-dichlorophenol in *N*-hexadecylpyridinium chloride [CPC] on the intramicellar mole fraction of 2,3-dichlorophenol at 25°C. The solid curve represents $K = K°(1 - BK)^2$ using a value of B equal to 1.26.

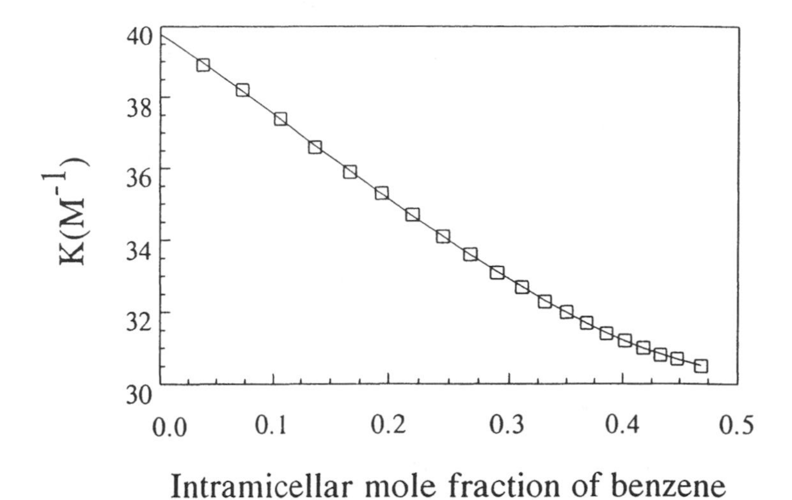

FIG. 3 Dependence of the solubilization equilibrium constant for benzene in *N*-hexadecylpyridinium chloride [CPC] (0.1 M) on the intramicellar mole fraction of benzene at 25°C.

ane inside the micelle will be expected to expand the micelle and provide an environment that is increasingly favorable for the inclusion of still more hexane molecules. The fact that an increase in micelle size may be expected to cause a decrease in the Laplace pressure may account for part of the increase in K as X increases (see Chapter 2), although K should in any case tend to increase as more hexane is incorporated in the micellar core [136, 198, 221, 250, 259, 261, 266–269]. The solubilization data indicate that it is not possible for the micelle to contain as much as 1 mole of hexane per mole of CPC, and an extrapolation of the vapor pressure results indicates that the isotherm will terminate at a value of $X \sim 0.325$, which point K reaches a limiting value of about 2200 M^{-1}.

A very different type of solubilization isotherm is shown in Fig. 2 for 2,3-dichlorophenol in CPC micelles. The largest values of the solubilization equilibrium constant are obtained in the limit as $X \rightarrow 0$, and at X values near 0.7, K has decreased to only a few percent of its limiting value of $\sim$2800 M^{-1} at $X = 0$. It seems apparent that solutes like the chlorinated phenols and other amphiphiles compete for adsorption sites that are located in the vicinity of the headgroups of ionic micelles, so that the addition of more and more of the amphiphile leads to monotonically decreasing values of K as X increases. The Langmuir adsorption model and an extended model that accounts for Langmuir-type adsorption in the low-mole-fraction region have been used to characterize such isotherms [123, 226, 227, 241, 252, 262, 270, 271]. The semiempirical equation $K = K°(1 - BX)^2$ quite accurately represents the K vs. X isotherm for this system, as it does for numerous systems of amphiphiles in ionic surfactant micelles [226, 227, 253]. The limiting value of K at $X = 0$ is $K°$, and the value of the adjustable parameter B is equal to half the number of surfactant molecules constituting a site for the adsorption of 2,3-dichlorophenol in the headgroup region. The solid curve in Fig. 2 is drawn to represent the data, using a value of B equal to 1.26, which also fits data for the other dichlorophenols [226, 227]. K values monotonically decrease from the value $K°$ at $X = 0$ to quite small values as X approaches saturation. The shape of the solubilization isotherm depends entirely on the magnitude of the parameter B; the larger the value of B, the steeper the initial slope and the greater the positive curvature in a plot of K vs. X.

Intermediate, but somewhat more complicated behavior is observed with solutes like benzene and toluene in ionic micelles. Initially, the isotherm for benzene in CPC micelles (Fig. 3) exhibits a moderate decrease in K until a value of X in the vicinity of 0.5 is reached, beyond which point K may increase slightly as X approaches the saturation value (estimated to be approximately 0.7). The tendency of K vs. X curves to reach maximum values near $X = 0.5$ has also been observed for other aromatic solutes in ionic micelles [220]. It should be noted that the benzene/CPC system shows neither the large decrease in K with increasing X that occurs with amphiphiles nor the significant increase in K occurring with alkanes. In fact, using other methods (less accurate than the automated vapor pressure method) it would be difficult to determine that the observed variations in K vs. X do in fact occur. The initial decrease in K with increasing X for the benzene/CPC system may be attributed to a slight tendency for benzene to solubilize preferentially in the headgroup region of the micelle (at least as compared with alkanes), although it also seems obvious that solubilization in the hydrocarbon core must occur at all values of X. Given the ample evidence that the micellar interior is a region resembling a hydrocarbon liquid, one would be hard put to explain why benzene would not dissolve in this aliphatic environment. This conclusion is supported by the fact that values of the activity coefficient of benzene (on the pure component standard state basis) in binary solutions of benzene and hexadecane do not exceed 1.1; similar near-ideal behavior is also observed for toluene/alkane binary systems [272–275]. The gradual increase in K that occurs at X values near saturation indicates that the benzene-swollen micelles provide a somewhat more favorable environment for benzene than do the smaller spherical micelles that exist at small values of X. Again, the decrease in Laplace pressure effects as the micelle increases in size may play a role in causing the benzene to solubilize somewhat more favorably at the highest values of X (see Chapter 2) [267, 269].

VI. ACTIVITY COEFFICIENT ISOTHERMS

An alternative way of presenting solubilization results, which can reveal similarities and differences in solubilization behavior for various solutes, is to plot values of the activity coefficient of the organic solute (γ_o) as a function of X. Utilizing the pure component (liquid or supercooled liquid) standard state for organic solutes, we interpret γ_o as the relative volatility of a solute in a micelle, compared to the volatility of the solute above an ideal solution at the same mole fraction [198, 236, 238]. This may provide a particularly apt comparison for many types of compounds, including hydrocarbons, because binary mixtures of aliphatic/aliphatic, aromatic/aromatic, and aliphatic/aromatic hydrocarbon systems ordinarily show very little deviation from ideality (Raoult's law) throughout the entire mole fraction range (see previous section). Thus, values of the activity coefficient for a series of different hydrocarbons, at any chosen value of X, indicate quite directly which of these solutes are most effectively solubilized by the micelle. Obviously other standard states may be used as references, and it should be mentioned that comparisons based on the transfer of organic compounds from the vapor phase into the micelle [76, 177] and from the solvent octanol into water [244, 276–278] have been suggested as alternatives to considering the transfer from the bulk aqueous phase into the micelle. It is clear that the values of K and P that are commonly reported reflect the unusual effects of hydrophobic interactions and other phenomena occurring in dilute aqueous solutions of the solutes that are solubilized in micelles [190, 279, 280].

Figures 4–6 include activity coefficient plots (γ_o vs. X) for the systems represented in Figs. 1–3. The hexane/CPC data (Fig. 4) indicate a gradual decrease in γ_o as X increases from near 0 to 0.3, corresponding to the increase in K with increasing X. The K vs. X and γ_o vs. X plots are reciprocally related, as is required by Eq. (7). The γ_o vs. X curve seems to show the expected tendency of moving in the direction of the point $\gamma_o = 1$, $X = 1$, although the total range of X is quite limited. At the limit point ($X \sim 0.325$), the activity cofficient has decreased to about $1/0.325 = 3.1$.

The 2,3-dichlorophenol data in Fig. 5 differ qualitatively from data for either hexane or benzene in CPC micelles. Activity coefficients for dichlorophenol are very small compared to unity, indicating that at low values of X the solute in the CPC micelle is in a quite favorable environment compared to its own pure liquid. The very rapid rise of γ_o toward unity

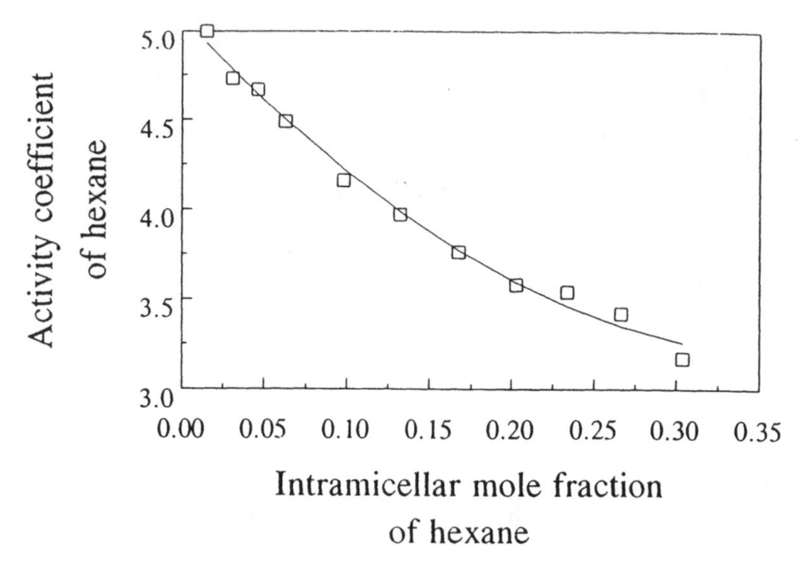

FIG. 4 Dependence of the activity coefficient of hexane in *N*-hexadecylpyridinium chloride [CPC] (0.1 M) on the intramicellar mole fraction of hexane at 25°C.

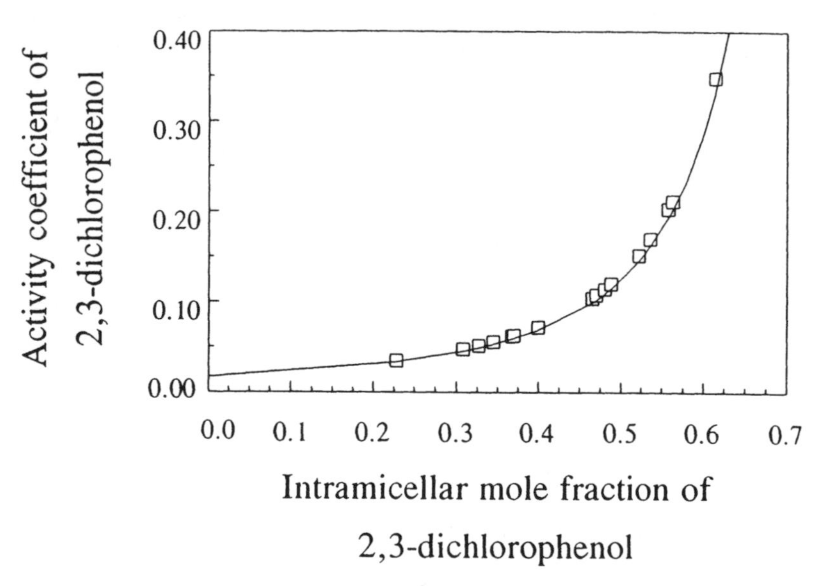

FIG. 5 Dependence of the activity coefficient of 2,3-dichlorophenol in *N*-hexadecylpyridinium chloride [CPC] on the intramicellar mole fraction of 2,3-dichlorophenol at 25°C. The solid curve is consistent with that drawn in Fig. 2, using the relationship $K = K°(1 - BX)^2 = 1/\gamma_o c_o^\circ$.

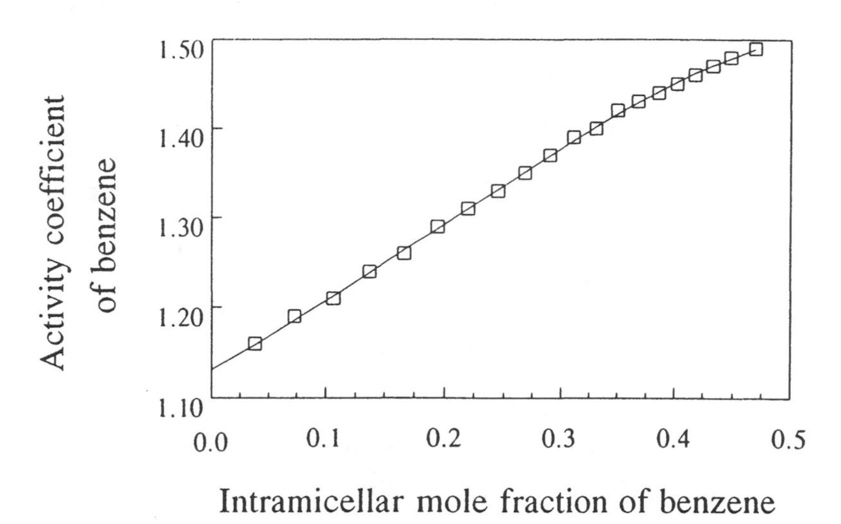

FIG. 6 Dependence of the activity coefficient of benzene in *N*-hexadecylpyridinium chloride [CPC] (0.1 M) on the intramicellar mole fraction of benzene at 25°C.

as X reaches values greater than about 0.6 is a salient feature of this isotherm, as it is of analogous plots for many phenol derivatives and other polar solutes for which detailed solubilization isotherms are available. The solid curve is consistent with that drawn in Fig. 2, corresponding to the equation $K = K^{\circ}(1 - BX)^2 = 1/\gamma_o c_o^{\circ}$.

Activity coefficients for benzene in CPC micelles are plotted in Fig. 6 as a function of X. The initial increase in γ_o with increasing X indicates that the environment becomes somewhat less favorable for benzene, although the changes in γ_o are relatively small. At sufficiently large values of X ($\sim$ 0.5) a maximum is reached in γ_o (corresponding to the minimum in K), and the small decline in γ_o as X approaches the saturation value ($X \sim 0.7$), where γ_o is $\sim$ 1.4, shows the expected gradual decrease toward unity at large values of X. What is probably most remarkable about the activity coefficient results is the fact that γ_o is so close to unity for benzene throughout wide ranges of X. Even increasing the total CPC concentration from 0.1 to 0.6 M does not cause the γ_o vs. X results to deviate much from the data shown in Fig. 6 [221]. Therefore, the environment for benzene in CPC micelles throughout wide variations in both the intramicellar mole fraction (X) and the total concentration of surfactant does not differ much from that expected for an ideal solution of benzene in the CPC micelles. Gibbs free energies, entropies, and enthalpies of transfer of benzene into surfactant micelles, and comparisons with the transfer of amphiphiles and aliphatic hydrocarbons are discussed in Chapter 13 and in Ref. 198.

VII. SOLUBILIZATION IN OTHER TYPES OF SURFACTANT AGGREGATES

Given the vast literature on the solubilization of various solutes by surfactant aggregates, in aqueous as well as nonaqueous systems, no adequate summary can be given here of the variety of physical properties that have been reported. Articles in this monograph include data for solubilization in surfactant bilayers or admicelles adsorbed at the solid/liquid interface, vesicles in aqueous solution, and block copolymers of ethylene oxide and propylene oxide. In addition to solubilization in block copolymers [73, 78, 281], studies have been reported of the application of polymer/surfactant complexes [265, 282–289] and polysoaps [290–303] to solubilize organic compounds. The extent of solubilization of *p-tert*-butylphenol (TBP) in complexes of CPC and sodium poly(styrenesulfonate) (PSS) were determined by using an equilibrium dialysis technique and ultrafiltration separations [265]. Polysoaps, first studied by Strauss and Jackson [290], are water-soluble polymers which have hydrophobic side chains. The polysoap molecules can form intramolecular microdomains capable of solubilizing water-insoluble organic molecules. Various polysoaps have been studied including polyvinylpyridine derivatives and maleic anhydride-alkyl vinyl ether copolymers.

There have been numerous reports of the solubilization of organic compounds by inverse or reverse micelles, and summaries of research in this area have been published [304–309]. The solubilization of organic solutes by nonionic micelles [224, 270, 310–316] in aqueous solution and solubilization in mixed micellar systems have also been extensively reported [317–319]. Many of the general features of the ionic surfactant systems described above are similar to those of nonionic surfactant systems, although it is not common to have simple spherical structures for nonionic micelles.

VIII. OUTLINE OF BOOK

This book discusses solubilization in three types of surfactant aggregates (micelles, vesicles, admicelles). Chapters 2–7 summarize the current knowledge concerning micellar solubilization with Chapters 2–6 covering equilibrium phenomena or thermodynamics of solubilization and Chapter 7 covering kinetics or rate of solubilization. Chapters 2–4 concentrate on single-surfactant systems using widely studied surfactant systems. Chapter 5 summarizes the literature on solubilization in micelles composed of surfactant mixtures. Chapter 6 discusses solubilization in polymeric surfactants, a relatively new area.

Chapters 8–10 discuss solubilization into nonmicellar aggregates; Chapter 8 covers admicelles (surfactant aggregates adsorbed at a liquid/solid interface) and Chapters 9–10 cover vesicles. Chapters 11–14 are concerned with experimental methods of measuring solubilization, including discussions of such techniques as vapor pressure, NMR, and cmc changes caused by solubilizate incorporation into micelles.

Two applications involving solubilization are outlined in Chapters 15 and 16; Chapter 15 discusses the importance of solubilization in detergency, and Chapter 16 outlines the utilization of solubilization in separation processes.

REFERENCES

1. J. W. McBain, *Adv. Colloid Sci. 1*: 99 (1942).
2. H. B. Klevens, *Chem. Rev. 47*: 1 (1950).
3. M. E. L. McBain and E. Hutchinson, *Solubilization and Related Phenomena,* Academic Press, New York, 1955.
4. P. H. Elworthy, A. T. Florence, and C. B. Macfarlane, *Solubilization by Surface-Active Agents and Its Application in Chemistry and the Biological Sciences,* Chapman and Hall, London, 1968.
5. J. W. McBain and T. R. Bolam, *J. Chem. Soc. 113*: 825 (1918).
6. J. W. McBain and M. E. L. McBain, *J. Am. Chem. Soc. 58*: 2610 (1936).
7. J. W. McBain and T.-M. Woo, *J. Am. Chem. Soc. 60*: 223 (1938).
8. J. W. McBain and J. J. O'Connor, *J. Am. Chem. Soc. 62*: 2855 (1940).
9. J. W. McBain, R. C. Merrill, Jr., and J. R. Vinograd, *J. Am. Chem. Soc. 63*: 670 (1941).
10. R. C. Merrill, Jr. and J. W. McBain, *J. Phys. Chem. 46*: 10 (1942).
11. S. A. Johnston and J. W. McBain, *Proc. Roy. Soc. London. A 181*: 119 (1942).
12. J. W. McBain and K. E. Johnson, *J. Am. Chem. Soc. 66*: 9 (1944).
13. J. W. McBain and A. A. Green, *J. Am. Chem. Soc. 68*: 1731 (1946).
14. P. H. Richards and J. W. McBain, *J. Am. Chem. Soc. 70*: 1338 (1948).
15. J. W. McBain and H. McHan, *J. Am. Chem. Soc. 70*: 3838 (1948).
16. J. W. McBain and O. A. Hoffman, *J. Phys. Colloid Chem. 53*: 39 (1949).
17. J. W. McBain, *Colloid Science,* D.C. Heath, Boston, 1950.
18. R. C. Murray and G. S. Hartley, *Trans. Faraday Soc. 31*: 183 (1935).
19. G. S. Hartley, in *Wetting and Detergency,* A. Harvey, New York, 1937.
20. G. S. Hartley, *Trans. Faraday Soc. 34*: 1283 (1938).
21. G. S. Hartley and D. F. Runnicles, *Proc. Roy. Soc. London. A 168*: 420 (1938).
22. G. S. Hartley, *J. Chem. Soc. 141*: 1968 (1938).
23. I. Langmuir, *J. Am. Chem. Soc. 39*: 1848 (1917).
24. I. Langmuir, *J. Chem. Phys. 1*: 756 (1933).
25. I. Langmuir and V. J. Schaefer, *J. Am. Chem. Soc. 59*: 2400 (1937).

26. I. Langmuir, *Proc. Roy. Soc. London. A 170*: 1 (1939).
27. P. Debye, *J. Phys. Colloid Chem. 53*: 1 (1949).
28. K. Shinoda and S. Friberg, *Emulsions and Solubilization,* Wiley, New York, 1986.
29. M. Bourrel and R. S. Schechter, *Microemulsions and Related Systems,* Marcel Dekker, New York, 1988.
30. S.-H. Chen and R. Rajagopalan (Eds.), *Micellar Solutions and Microemulsions,* Springer-Verlag, New York, 1990.
31. L. L. Schramm, *Emulsions: Fundamentals and Applications in the Petroleum Industry,* American Chemical Society, Washington, DC, Advances in Chemistry Series, No. 231, 1992.
32. N. K. Adam, *J. Phys. Chem. 29*: 87 (1925).
33. W. D. Harkins and E. Boyd, *J. Phys. Chem. 45*: 20 (1941).
34. G. L. Gaines, Jr., *Insoluble Monolayers at Liquid-Gas Interfaces,* Interscience, New York, 1966.
35. R. F. Gould (Ed.), *Monolayers,* American Chemical Society, Washington, DC, Advances in Chemistry Series, No. 144, 1975.
36. G. Roberts (Ed.), *Langmuir-Blodgett Films,* Plenum Press, New York, 1990.
37. H. Kuhn and D. Möbius, in *Physical Methods of Chemistry* (B. W. Rossiter and R. C. Baetzold, eds.), Vol. IX B, 1993.
38. K. Hess and J. Gundermann, *Ber. 70*: 1800 (1937).
39. V. K. Hess, W. Philippoff, and H. Kiessig, *Kolloid Z. 88*: 40 (1939).
40. V. J. Stauff, *Kolloid Z. 89*: 224 (1939).
41. V. J. Stauff, *Kolloid Z. 96*: 244 (1941).
42. V. H. Kiessig, *Kolloid Z. 96*: 252 (1941).
43. V. H. Kiessig, *Kolloid Z. 98*: 213 (1942).
44. E. W. Hughes, W. M. Sawyer, and J. R. Vinograd, *J. Chem. Phys. 13*: 131 (1945).
45. S. Ross and J. W. McBain, *J. Am. Chem. Soc. 68*: 296 (1946).
46. W. D. Harkins, R. W. Mattoon, and M. L. Corrin, *J. Colloid Sci. 1*: 105 (1946).
47. J. H. Schulman and D. P. Riley, *J. Colloid Sci. 3*: 383 (1948).
48. R. W. Mattoon, R. S. Stearns, and W. D. Harkins, *J. Chem. Phys. 16*: 644 (1948).
49. W. D. Harkins and R. Mittelmann, *J. Colloid Sci. 4*: 367 (1949).
50. W. Philippoff, *J. Colloid Sci. 5*: 169 (1950).
51. P. Debye and E. W. Anacker, *J. Phys. Colloid Chem. 55*: 644 (1951).
52. E. W. Anacker, *J. Colloid Sci. 8*: 402 (1953).
53. W. P. J. Ford, R. H. Ottewill, and H. C. Parreira, *J. Colloid Interface Sci. 21*: 525 (1966).
54. E. W. Anacker and H. M. Ghose, *J. Am. Chem. Soc. 90*: 3161 (1968).
55. E. A. Hauser, *Colloidal Phenomena,* McGraw-Hill, New York, 1939.
56. W. D. Harkins and G. Jura, *J. Am. Chem. Soc. 66*: 1366 (1944).
57. S. Brunauer, *The Adsorption of Gases and Vapors,* Princeton University, Princeton, 1945.
58. G. Jura and W. D. Harkins, *J. Am. Chem. Soc. 68*: 1941 (1946).

59. T. L. Hill, *J. Chem. Phys. 17*: 520 (1949).
60. T. L. Hill, P. H. Emmett, and L. G. Joyner, *J. Am. Chem. Soc. 73*: 5102 (1951).
61. B. G. Linsen (Ed.), *Physical and Chemical Aspects of Adsorbents and Catalysts,* Academic Press, New York, 1970.
62. W. A. Steele, *The Interaction of Gases with Solids,* Pergamon Press, New York, 1974.
63. W. D. Harkins, F. E. Brown, and E. C. H. Davies, *J. Am. Chem. Soc. 39*: 354 (1917).
64. W. D. Harkins, E. C. H. Davies, and G. L. Clark, *J. Am. Chem. Soc. 39*: 541 (1917).
65. W. D. Harkins, *The Physical Chemistry of Surface Films,* Reinhold, New York, 1952.
66. T. L. Hill, *J. Chem. Phys. 23*: 617 (1955).
67. T. L. Hill, *Statistical Mechanics,* McGraw-Hill, New York, 1956.
68. T. L. Hill, *Statistical Mechanics Principles and Selected Applications,* McGraw-Hill, New York, 1956.
69. J. F. Danielli, K. G. A. Pankhurst, and A. C. Riddiford, *Surface Phenomena in Chemistry and Biology,* Pergamon Press, London, 1958.
70. J. T. Davies and E. K. Rideal, *Interfacial Phenomena,* Academic Press, New York, 1961.
71. J. J. Bikerman, *Physical Surfaces,* Academic Press, New York, 1970.
72. A. W. Adamson, *Physical Chemistry of Surfaces,* Wiley, New York, 1990.
73. R. Nagarajan, M. Barry, and E. Ruckenstein, *Langmuir 2*: 210 (1986).
74. R. Nagarajan and E. Ruckenstein, *Langmuir 7*: 2934 (1991).
75. C. Hirose and L. Sepúlveda, *J. Phys. Chem. 85*: 3689 (1981).
76. G. A. Smith, S. D. Christian, E. E. Tucker, and J. F. Scamehorn, *Langmuir 3*: 598 (1987).
77. R. Nagarajan and K. Ganesh, *Macromolecules 22*: 4312 (1989).
78. P. N. Hurter, J. M. H. M. Scheutjens, and T. A. Hatton, *Macromolecules 26*: 5592 (1993).
79. P. N. Hurter, J. M. H. M. Scheutjens, and T. A. Hatton, *Macromolecules 26*: 5030 (1993).
80. R. Nagarajan and E. Ruckenstein, *J. Colloid Interface Sci. 71*: 580 (1979).
81. G. Gunnarsson, B. Jönsson, and H. Wennerström, *J. Phys. Chem. 84*: 3114 (1980).
82. R. Nagarajan and E. Ruckenstein, in *Surfactants in Solution* (K. L. Mittal and B. Lindman, eds.), Plenum Press, New York, 1984, Vol. 2.
83. M. Aamodt, M. Landgren, and B. Jönsson, *J. Phys. Chem. 96*: 945 (1992).
84. R. Mallikarjun and D. B. Dadyburjor, *J. Colloid Interface Sci. 84*: 73 (1981).
85. R. DeLisi and V. T. Liveri, *Gazzetta Chim. Italiana 113*: 371 (1983).
86. J. W. McBain, *Kolloid Z. 12*: 256 (1913).
87. G. S. Hartley, *Aqueous Solutions of Paraffin-Chain Salts,* Hermann, Paris, 1936.
88. G. S. Hartley, *Nature 163*: 767 (1949).
89. G. S. Hartley, in *Micellization, Solubilization, and Microemulsions* (K. L. Mittal, ed.), Plenum Press, New York, 1977, Vol. 1.

90. W. D. Harkins, *J. Chem. Phys. 16*: 156 (1948).
91. W. D. Harkins, R. Mittelmann, and M. L. Corrin, *J. Phys. Colloid Chem. 53*: 1350 (1949).
92. E. Hutchinson, *J. Colloid Sci. 9*: 2 (1954).
93. K. J. Mysels, *J. Colloid Sci. 10*: 507 (1955).
94. E. W. Anacker, in *Cationic Surfactants* (E. Jungermann, ed.), Marcel Dekker, New York, 1970.
95. D. Stigter, R. J. Williams, and K. J. Mysels, *J. Phys. Chem. 59*: 330 (1955).
96. K. G. Götz and K. Heckmann, *J. Colloid Sci. 13*: 266 (1958).
97. F. Reiss-Husson and V. Luzzati, *J. Phys. Chem. 68*: 3504 (1964).
98. F. Reiss-Husson and V. Luzzati, *J. Colloid Interface Sci. 21*: 534 (1966).
99. J. B. Hayter and J. Penfold, *J. Chem. Soc. Faraday Trans. 1 77*: 1851 (1981).
100. J. B. Hayter and J. Penfold, *Colloid Polym. Sci. 261*: 1022 (1983).
101. R. Zana, C. Picot, and R. Duplessix, *J. Colloid Interface Sci. 93*: 43 (1983).
102. J. B. Hayter, M. Hayoun, and T. Zemb, *Colloid Polym. Sci. 262*: 798 (1984).
103. S.-H. Chen, in *Physics of Amphiphiles: Micelles, Vesicles, and Microemulsions* (V. Degiorgio, ed.), North-Holland, Amsterdam, 1985.
104. J. B. Hayter, in *Physics of Amphiphiles: Micelles, Vesicles, and Microemulsions* (V. Degiorgio, ed.), North-Holland, Amsterdam, 1985.
105. B. Cabane, in *Surfactant Solutions New Methods of Investigation* (R. Zana, ed.), Marcel Dekker, New York, 1987.
106. T.-L. Lin, S.-H. Chen, N. E. Gabriel, and M. F. Roberts, *J. Phys. Chem. 94*: 855 (1990).
107. N. J. Turro, M. W. Geiger, R. R. Hautala, and N. E. Schore, in *Micellization, Solubilization, and Microemulsions* (K. L. Mittal, ed.), Plenum Press, New York, 1977, Vol. 1.
108. N. J. Turro and A. Yekta, *J. Am. Chem. Soc. 100*: 5951 (1978).
109. S. S. Atik, M. Nam, and L. A. Singer, *Chem. Phys. Lett. 67*: 65 (1979).
110. P. Lianos and R. Zana, *J. Phys. Chem. 84*: 3339 (1980).
111. P. Lianos and R. Zana, *Chem. Phys. Lett. 76*: 62 (1980).
112. P. Lianos and R. Zana, *J. Colloid Interface Sci. 84*: 100 (1981).
113. P. Lianos, M.-L. Viriot, and R. Zana, *J. Phys. Chem. 88*: 1098 (1984).
114. A. Malliaris, *Adv. Colloid Interface Sci. 27*: 153 (1987).
115. F. Grieser and C. J. Drummond, *J. Phys. Chem. 92*: 5580 (1988).
116. A. Malliaris, *Int. Rev. Phys. Chem. 7*: 95 (1988).
117. A. Malliaris, J. Lang, and R. Zana, in *Surfactants in Solution* (K. L. Mittal, ed.), Plenum Press, New York, 1989, Vol. 7.
118. M. Van der Auweraer, E. Roelants, A. Verbeeck, and F. C. DeSchryver, in *Surfactants in Solution* (K. L. Mittal, ed.), Plenum Press, New York, 1989, Vol. 7.
119. O. Söderman, M. Jonströmer, and J. van Stam, *J. Chem. Soc. Faraday Trans. 89*: 1759 (1993).
120. E. Ruckenstein and R. Krishnan, *J. Colloid Interface Sci. 71*: 321 (1979).
121. L. Sjöblom, in *Solvent Properties of Surfactant Solutions* (K. Shinoda, ed.), Marcel Dekker, New York, 1967.

122. B. J. Carroll, B. G. C. O'Rourke, and A. J. I. Ward, *J. Pharm. Pharmacol. 34*: 287 (1982).
123. D. Attwood and A. T. Florence, *Surfactant Systems Their Chemistry, Pharmacy, and Biology,* Chapman and Hall, London, 1983.
124. Y. Moroi, H. Noma, and R. Matuura, *J. Phys. Chem. 87*: 872 (1983).
125. T. F. Tadros, in *Surfactants* (T. F. Tadros, ed.), Academic Press, New York, 1984.
126. R. DeLisi, V. T. Liveri, M. Castagnolo, and A. Inglese, *J. Soln. Chem. 15*: 23 (1986).
127. C. M. Nguyen, S. D. Christian, and J. F. Scamehorn, *Tenside Surfactants Deterg. 25*: 328 (1988).
128. A. K. Krishna and D. R. Flanagan, *J. Pharm. Sci. 78*: 574 (1989).
129. C. J. Mertz and C. T. Lin, *Photochem. Photobiol. 53*: 307 (1991).
130. P. Mukerjee and J.-S. Ko, *J. Phys. Chem. 96*: 6090 (1992).
131. B. K. Jha, A. S. Chhatre, B. D. Kulkarni, R. A. Joshi, R. R. Joshi, and U. R. Kalkote, *J. Colloid Interface Sci. 163*: 1 (1994).
132. B. Lindman and H. Wennerström, in *Topics in Current Chemistry,* No. 87, Springer-Verlag, New York, 1980.
133. H. Lange, in *Solvent Properties of Surfactant Solutions,* Marcel Dekker, New York, 1967.
134. E. Kissa, *Textile Res. J. 45*: 736 (1975).
135. T. A. B. M. Bolsman, F. T. G. Veltmaat, and N. M. van Os, *J. Am. Oil Chem. Soc. 65*: 280 (1988).
136. M. J. Rosen, *Surfactants and Interfacial Phenomena,* 2nd ed., Wiley, New York, 1989.
137. J. G. Weers, *J. Am. Oil Chem. Soc. 67*: 340 (1990).
138. C. A. Miller and K. H. Raney, *Colloids Surf. 74*: 169 (1993).
139. M. M. Rieger (ed.), *Surfactants in Cosmetics,* Marcel Dekker, New York, 1985.
140. P. L. Dubin, J. M. Principi, B. A. Smith, and M. A. Fallon, *J. Colloid Interface Sci. 127*: 558 (1989).
141. Y. Tokuoka, H. Uchiyama, M. Abe, and K. Ogino, *J. Colloid Interface Sci. 152*: 402 (1992).
142. M. Abe, K. Mizuguchi, K. Ogino, H. Uchiyama, J. F. Scamehorn, E. E. Tucker, and S. D. Christian, *J. Colloid Interface Sci. 160*: 16 (1993).
143. A. Datyner, *Surfactants in Textile Processing,* Marcel Dekker, New York, 1983.
144. W. V. Valkenburg, in *Solvent Properties of Surfactant Solutions* (K. Shinoda, ed.), Marcel Dekker, New York, 1967.
145. D. Seaman, in *Solution Behavior of Surfactants* (K. L. Mittal and E. J. Fendler, eds.), Plenum Press, New York, 1982, Vol. 2.
146. R. Nagarajan and E. Ruckenstein, *Sep. Sci. Tech. 16*: 1429 (1981).
147. R. O. Dunn, Jr., J. F. Scamehorn, and S. D. Christian, *Sep. Sci. Tech. 20*: 257 (1985).
148. R. O. Dunn, Jr., J. F. Scamehorn, and S. D. Christian, *Sep. Sci. Tech. 22*: 763 (1987).

149. J. F. Scamehorn and J. H. Harwell, in *Surfactants in Emerging Technologies* (M. J. Rosen, ed.), Marcel Dekker, New York, 1987.
150. J. F. Scamehorn and J. H. Harwell (Eds.), *Surfactant-Based Separation Processes*, Marcel Dekker, New York, 1989.
151. D. W. Armstrong, *Sep. Pur. Methods 14*: 213 (1985).
152. S. Terabe, K. Otsuka, and T. Ando, *Anal. Chem. 57*: 834 (1985).
153. D. B. Wetlaufer, in *Surfactants in Emerging Technologies* (M. J. Rosen, ed.), Marcel Dekker, New York, 1987.
154. F. G. P. Mullins, in *Ordered Media in Chemical Separations* (W. L. Hinze and D. W. Armstrong, eds.), American Chemical Society, Washington, DC, 1987.
155. A. Berthod, I. Girard, and C. Gonnet, in *Ordered Media in Chemical Separations* (W. L. Hinze and D. W. Armstrong, eds.), American Chemical Society, Washington, DC, 1987.
156. M. J. Sepaniak, D. E. Burton, and M. P. Maskarinec, in *Ordered Media in Chemical Separations* (W. L. Hinze and D. W. Armstrong, eds.), American Chemical Society, Washington, DC, 1987.
157. A. S. Kord, J. K. Strasters, and M. G. Khaledi, *Anal. Chim. Acta. 246*: 131 (1991).
158. W. L. Hinze, H. N. Singh, Y. Baba, and N. G. Harvey, *Trends Anal. Chem. 3*: 193 (1984).
159. L. J. C. Love, J. G. Dorsey, and J. G. Habarta, *Anal. Chem. 56*: 1132A (1984).
160. G. L. McIntire, *Am. Lab. 18*: 173 (1986).
161. W. D. Harkins, *J. Am. Chem. Soc. 69*: 1428 (1947).
162. B. M. E. van der Hoff, in *Solvent Properties of Surfactant Solutions* (K. Shinoda, ed.), Marcel Dekker, New York, 1967.
163. Z. Gao, J. C. T. Kwak, R. Labonté, D. G. Marangoni, and R. E. Wasylishen, *Colloids Surf. 45*: 269 (1990).
164. E. H. Cordes and R. B. Dunlap, *Accts. Chem. Res. 2*: 329 (1969).
165. E. J. Fendler and J. H. Fendler, *Adv. Phys. Org. Chem. 8*: 271 (1970).
166. C. A. Bunton, *Prog. Solid State Chem. 8*: 239 (1973).
167. E. Cordes (Ed.), *Reaction Kinetics in Micelles*, Plenum Press, New York, 1973.
168. I. V. Berezin, K. Martinek, and A. K. Yatsimirskii, *Russ. Chem. Rev. 42*: 787 (1973).
169. J. H. Fendler and E. J. Fendler, *Catalysis in Micellar and Macromolecular Systems*, Academic Press, New York, 1975.
170. K. L. Mittal and P. Mukerjee, in *Micellization, Solubilization, and Microemulsions* (K. L. Mittal, ed.), Plenum Press, New York, 1977, Vol. 1.
171. Y. Moroi, *J. Phys. Chem. 84*: 2186 (1980).
172. N. Nishikido, M. Kishi, and M. Tanaka, *J. Colloid Interface Sci. 94*: 348 (1983).
173. M. Seno, Y. Shiraishi, S. Takeuchi, and J. Otsuki, *J. Phys. Chem. 94*: 3776 (1990).
174. M. L. Moyá, C. Izquierdo, and J. Casado, *J. Phys. Chem. 95*: 6001 (1991).

175. G. A. Infante, J. Caraballo, R. Irizarry, and M. Rodriguez, in *Solution Behavior of Surfactants* (K. L. Mittal and E. J. Fendler, eds.), Plenum Press, New York, 1982, Vol. 2.
176. L. L. Turai, in *Solution Behavior of Surfactants* (K. L. Mittal and E. J. Fendler, eds.), Plenum Press, New York, 1982, Vol. 2.
177. K. T. Valsaraj, A. Gupta, L. J. Thibodeaux, and D. P. Harrison, *Water Res. 22*: 1173 (1988).
178. D. A. Edwards, R. G. Luthy, and Z. Liu, *Environ. Sci. Technol. 25*: 126 (1991).
179. C. T. Jafvert and J. K. Heath, *Environ. Sci. Technol. 25*: 1031 (1991).
180. C. T. Jafvert, *Environ. Sci. Technol. 25*: 1039 (1991).
181. A. N. Clarke, P. D. Plumb, T. K. Subramanyan, and D. J. Wilson, *Sep. Sci. Tech. 26*: 301 (1991).
182. J. H. Harwell, in *Transport and Remediation of Subsurface Contaminants* (D. A. Sabatini and R. C. Knox, eds.), American Chemical Society, Washington, DC, 1992.
183. D. A. Edwards, S. Laha, Z. Liu, and R. G. Luthy, in *Transport and Remediation of Subsurface Contaminants* (D. A. Sabatini and R. C. Knox, eds.), American Chemical Society, Washington, DC, 1992.
184. H. Schott, *J. Pharm. Sci. 60*: 1594 (1971).
185. L. R. Fisher and D. G. Oakenfull, *Chem. Soc. Rev. 6*: 25 (1977).
186. K. Kalyanasundaram and J. K. Thomas, *J. Phys. Chem. 80*: 1462 (1976).
187. H. Wennerström and B. Lindman, *Phys. Rep. 52*: 1 (1979).
188. R. Zana, in *Surfactant Solutions New Methods of Investigation* (R. Zana, ed.), Marcel Dekker, New York, 1987.
189. C. Y. Young, P. J. Missel, N. A. Mazer, G. Benedek, and M. C. Carey, *J. Phys. Chem. 82*: 1375 (1978).
190. C. Tanford, *The Hydrophobic Effect: Formation of Micelles and Biological Membranes,* Wiley, New York, 1980.
191. S. Hayashi and S. Ikeda, *J. Phys. Chem. 84*: 744 (1980).
192. S. Ikeda, S. Ozeki, and M.-A. Tsunoda, *J. Colloid Interface Sci. 73*: 27 (1980).
193. S. Ozeki and S. Ikeda, *J. Colloid Interface Sci. 77*: 219 (1980).
194. S. Ozeki and S. Ikeda, *J. Colloid Interface Sci. 87*: 424 (1982).
195. G. Porte, Y. Poggi, J. Appell, and G. Maret, *J. Phys. Chem. 88*: 5713 (1984).
196. T. Imae, R. Kamiya, and S. Ikeda, *J. Colloid Interface Sci. 108*: 215 (1985)
197. R. Zieliński, S. Ideda, H. Nomura, and S. Kato, *J. Colloid Interface Sci. 125*: 497 (1988).
198. G. A. Smith, S. D. Christian, E. E. Tucker, and J. F. Scamehorn, *J. Colloid Interface Sci. 130*: 254 (1989).
199. F. Z. Mahmoud, W. S. Higazy, S. D. Christian, E. E. Tucker, and A. A. Taha, *J. Colloid Interface Sci. 131*: 96 (1989).
200. M. E. Cates and S. J. Candau, *J. Phys.: Condens. Matter 2*: 6869 (1990).
201. A. Makayssi, D. Lemordant, and C. Treiner, *Langmuir 9*: 2808 (1993).
202. W. Schorr and H. Hoffmann, in *Physics of Amphiphiles: Micelles, Vesicles, and Microemulsions* (V. Degiorgio, ed.), North-Holland, Amsterdam, 1985.

203. C. Treiner and A. Makayssi, *Langmuir 8*: 794 (1992).
204. D. Nguyen and G. L. Bertrand, *J. Phys. Chem. 96*: 1994 (1992).
205. D. Nguyen and G. L. Bertrand, *J. Colloid Interface Sci. 150*: 143 (1992).
206. P. M. Lindemuth and G. L. Bertrand, *J. Phys. Chem. 97*: 7769 (1993).
207. W. L. Hinze, in *Ordered Media in Chemical Separations* (W. L. Hinze and D. W. Armstrong, eds.), American Chemical Society, Washington, DC, 1987.
208. H. V. Tartar, *J. Phys. Chem. 59*: 1195 (1955).
209. H. V. Tartar, *J. Colloid Sci. 14*: 115 (1959).
210. C. Tanford, *J. Phys. Chem. 76*: 3020 (1972).
211. C. Tanford, *J. Phys. Chem. 78*: 2469 (1974).
212. C. Tanford, *Proc. Natl. Acad. Sci. 71*: 1811 (1974).
213. H. Schott, *J. Pharm. Sci. 62*: 162 (1973).
214. J. N. Israelachvili, D. J. Mitchell, and B. W. Ninham, *Chem. Soc. Faraday Trans. 2 72*: 1525 (1976).
215. S. Karaborni, N. M. van Os, K. Esselink, and P. A. J. Hilbers, *Langmuir 9*: 1175 (1993).
216. R. Friman and J. B. Rosenholm, *Coloid Polym. Sci. 260*: 545 (1982).
217. M. Almgren and S. Swarup, *J. Phys. Chem. 86*: 4212 (1983).
218. A. Malliaris, *J. Phys. Chem. 91*: 6511 (1987).
219. M. Abe, Y. Tokuoka, H. Uchiyama, K. Ogino, J. F. Scamehorn, and S. D. Christian, *Colloids Surf. 67*: 37 (1992).
220. G. A. Smith, Ph.D. dissertation, University of Oklahoma, Norman, Oklahoma, 1986.
221. S. D. Christian, E. E. Tucker, G. A. Smith, and D. S. Bushong, *J. Colloid Interface Sci. 113*: 439 (1986).
222. W. D. Harkins and H. Oppenheimer, *J. Am. Chem. Soc. 71*: 808 (1949).
223. P. Mukerjee, in *Encyclopaedic Dictionary of Physics* (J. Thewlis, ed.), MacMillan, New York, 1962, Vol. 6, p. 553.
224. K. Shinoda, *Principles of Solution and Solubility,* Marcel Dekker, New York, 1978.
225. E. Szajdzinski-Pietek, R. Maldonado, L. Kevan, and R. R. M. Jones, *J. Colloid Interface Sci. 110*: 514 (1986).
226. B.-H. Lee, S. D. Christian, E. E. Tucker, and J. F. Scamehorn, *Langmuir 6*: 230 (1990).
227. B.-H. Lee, S. D. Christian, E. E. Tucker, and J. F. Scamehorn, *J. Phys. Chem. 95*: 360 (1991).
228. A. M. Schwartz, J. W. Berry, and J. Berch, *Surface Active Agents and Detergents,* Interscience, New York, 1958, Vol. 2.
229. E. B. Abuin, E. Valenzuela, and E. A. Lissi, *J. Colloid Interface Sci. 101*: 401 (1984).
230. S. J. Rehfeld, *J. Phys. Chem. 74*: 117 (1966).
231. J. C. Eriksson and G. Gillberg, *Acta. Chem. Scand. 20*: 2019 (1966).
232. J. H. Fendler and L. K. Patterson, *J. Phys. Chem. 75*: 3907 (1971).
233. P. Mukerjee and J. R. Cardinal, *J. Phys. Chem. 82*: 1620 (1978).

234. S. A. Simon, R. V. McDaniel, and T. J. McIntosh, *J. Phys. Chem. 86*: 1449 (1982).
235. D. J. Jobe, V. C. Reinsborough, and P. J. White, *Can. J. Chem. 60*: 279 (1982).
236. E. E. Tucker and S. D. Christian, *Faraday Symp. Chem. Soc. 17*: 11 (1982).
237. R. Nagarajan, M. A. Chaiko, and E. Ruckenstein, *J. Phys. Chem. 88*: 2916 (1984).
238. G. A. Smith, S. D. Christian, E. E. Tucker, and J. F. Scamehorn, in *Ordered Media in Chemical Separations* (W. L. Hinze and D. W. Armstrong, eds.), American Chemical Society, Washington, DC, 1987.
239. A. Heindl, J. Strnad, and H.-H. Kohler, *J. Phys. Chem. 97*: 742 (1993).
240. G. A. Smith, S. D. Christian, E. E. Tucker, and J. F. Scamehorn, *J. Soln, Chem. 15*: 519 (1986).
241. H. Uchiyama, E. E. Tucker, S. D. Christian, and J. F. Scamehorn, *J. Phys. Chem. 98*: 1714 (1994).
242. A. Goto and F. Endo, *J. Colloid Interface Sci. 66*: 26 (1978).
243. H. Høiland, E. Ljosland, and S. Backlund, *J. Colloid Interface Sci. 101*: 467 (1984).
244. C. Treiner and A. K. Chattopadhyay, *J. Colloid Interface Sci. 109*: 101 (1986).
245. R. Pons, R. Bury, P. Erra, and C. Treiner, *Colloid Polym. Sci. 269*:62 (1991).
246. R. DeLisi, C. Genova, and V. T. Liveri, *J. Colloid Interface Sci. 95*: 428 (1983).
247. L. B. Shih and R. W. Williams, *J. Phys. Chem. 90*: 1615 (1986).
248. Z. Gao, R. E. Wasylishen, and J. C. T. Kwak, *J. Phys. Chem. 93*: 2190 (1989).
249. L. Sepúlveda, E. Lissi, and F. Quina, *Adv. Colloid Interface Sci. 25*: 1 (1986).
250. S. D. Christian, E. E. Tucker, and E. H. Lane, *J. Colloid Interface Sci. 84*: 423 (1981).
251. C. Gamboa and A. F. Olea, *Langmuir 9*: 2066 (1993).
252. S. N. Bhat, G. A. Smith, E. E. Tucker, S. D. Christian, J. F. Scamehorn, and W. Smith, *I & EC Research 26*: 1217 (1987).
253. H. Uchiyama, S. D. Christian, J. F. Scamehorn, M. Abe, and K. Ogino, *Langmuir 7*: 95 (1991).
254. S. D. Christian, E. E. Tucker, J. F. Scamehorn, and H. Uchiyama, *Colloid Polym. Sci. 271*: 745 (1994).
255. M. E. Morgan, H. Uchiyama, S. D. Christian, E. E. Tucker, and J. F. Scamehorn, *Langmuir 10*: 2170 (1994).
256. E. Valenzuela, E. Abuin, and E. A. Lissi, *J. Colloid Interface Sci. 102*: 46 (1984).
257. W. S. Higazy, F. Z. Mahmoud, A. A. Taha, and S. D. Christian, *J. Soln. Chem. 17*: 191 (1988).
258. Y. Moroi, K. Sato, and R. Matuura, *J. Phys. Chem. 86*: 2463 (1982).

259. N. Nishikido, K. Abiru, and N. Yoshimura, *J. Colloid Interface Sci. 113*: 356 (1986).
260. Y. Moroi and R. Matuura, *J. Colloid Interface Sci. 125*: 456 (1988).
261. Y. Moroi and R. Matuura, *J. Colloid Interface Sci. 125*: 463 (1988).
262. S. J. Dougherty and J. C. Berg, *J. Colloid Interface Sci. 48*: 110 (1974).
263. S. D. Christian, L. S. Smith, D. S. Bushong, and E. E. Tucker, *J. Colloid Interface Sci. 89*: 514 (1982).
264. E. E. Tucker and S. D. Christian, *J. Colloid Interface Sci. 104*: 562 (1985).
265. H. Uchiyama, S. D. Christian, E. E. Tucker, and J. F. Scamehorn, *J. Colloid Interface Sci. 163*: 493 (1994).
266. P. Mukerjee, ''*Kolloid Z.Z. Polym.* 76 (1970).
267. I. B. C. Matheson and A. D. King, Jr., *J. Colloid Interface Sci. 66*: 464 (1978).
268. P. Mukerjee, *Pure Appl. Chem. 52*: 1317 (1980).
269. W. Prapaitrakul and A. D. King, Jr., *J. Colloid Interface Sci. 106*: 186 (1985).
270. A. Goto, M. Nihel, and F. Endo, *J. Phys. Chem. 84*: 2268 (1980).
271. K. Kandori, R. J. McGreevy, and R. S. Schechfer, *J. Colloid Interface Sci. 132*: 395 (1989).
272. J. H. Hildebrand, H. M. Prausnitz, and R. L. Scott, *Regular and Related Solutions,* Van Nostrand Reinhold, New York, 1970.
273. J. H. Park, A. Hussam, P. Couasnon, D. Fritz, and P. W. Carr, *Anal. Chem. 59*: 1970 (1987).
274. J. H. Park and P. W. Carr, *Anal. Chem. 59*: 2596 (1987).
275. J. Li and P. W. Carr, *Anal. Chem. 65*: 1443 (1993).
276. C. Treiner, *J. Colloid Interface Sci. 93*: 33 (1983).
277. C. Treiner and M. H. Mannebach, *J. Colloid Interface Sci. 118*: 243 (1987).
278. K. T. Valsaraj and L. J. Thibodeaux, *Water Res. 23*: 183 (1989).
279. S. D. Christian and E. E. Tucker, *J. Soln. Chem. 11*: 749 (1982).
280. K. L. Stellner, E. E. Tucker, and S. D. Christian, *J. Soln. Chem. 12*: 307 (1983).
281. D. Kiserow, K. Prochazka, C. Ramireddy, Z. Tuzar, P. Munk, and S. E. Webber, *Macromolecules 25*: 461 (1992).
282. S. Saito, *J. Colloid Interface Sci. 24*: 227 (1967).
283. P. S. Leung, E. D. Goddard, C. Han, and C. J. Glinka, *Colloids Surf. 13*: 47 (1985).
284. K. P. Ananthapadmanabhan, P. S. Leung, and E. D. Goddard, *Colloids Surf. 13*: 63 (1985).
285. E. D. Goddard, *Colloids Surf. 19*: 301 (1986).
286. K. Hayakawa, J. Ohta, T. Maeda, and I. Satake, *Langmuir 3*: 377 (1987).
287. E. A. Sudbeck, P. L. Dubin, M. E. Curran, and J. Skelton, *J. Colloid Interface Sci. 142*: 512 (1991).
288. B.-H. Lee, S. D. Christian, E. E. Tucker, and J. F. Scamehorn, *Langmuir 7*: 1332 (1991).
289. E. D. Goddard, in *Interactions of Surfactants with Polymers and Proteins*

(E. D. Goddard and K. P. Ananthapadmanabhan, eds.), CRC Press, Boca Raton, 1993.

290. U. P. Strauss and E. G. Jackson, *J. Polym. Sci. 6*: 649 (1951).
291. E. G. Jackson and U. P. Strauss, *J. Polym. Sci. 7*: 473 (1951).
292. L. H. Layton, E. G. Jackson, and U. P. Strauss, *J. Polym. Sci. 9*: 295 (1952).
293. U. P. Strauss, S. J. Assony, E. G. Jackson, and L. H. Layton, *J. Polym. Sci. 9*: 509 (1952).
294. U. P. Strauss and L. H. Layton, *J. Phys. Chem. 57*: 352 (1953).
295. L. H. Layton and U. P. Strauss, *J. Colloid Sci. 9*: 149 (1954).
296. U. P. Strauss and N. L. Gershfeld, *J. Phys. Chem. 58*: 747 (1954).
297. U. P. Strauss, N. L. Gershfeld, and E. H. Crook, *J. Phys. Chem. 60*: 577 (1956).
298. U. P. Strauss and S. S. Slowata, *J. Phys. Chem. 61*: 411 (1957).
299. K. Ito, H. Ono, and Y. Yamashita, *J. Colloid Sci. 19*: 28 (1964).
300. U. P. Strauss, in *Polymers in Aqueous Media* (J. E. Glass, ed.), American Chemical Society, Washington DC, 1989.
301. V. S. Zdanowicz and U. P. Strauss, *Macromolecules 26*: 4770 (1993).
302. U. P. Strauss, in *Interactions of Surfactants with Polymers and Proteins* (E. D. Goddard and K. P. Ananthapadmanabhan, eds.), CRC Press, Boca Raton, 1993.
303. O. Anthony and R. Zana, *Macromolecules 27*: 3885 (1994).
304. H.-F. Eicke, in *Topics in Current Chemistry,* No. 87, Springer-Verlag, New York, 1980.
305. A. Kitahara, *Adv. Colloid Interface Sci. 12*: 109 (1980).
306. P. L. Luisi and L. J. Magid, *CRC Critical Rev. Biochem. 20*: 409 (1986).
307. T. A. Hatton, in *Ordered Media in Chemical Separations* (W. L. Hinze and D. W. Armstrong, eds.), American Chemical Society, Washington, DC, 1987.
308. E. B. Leodidis and T. A. Hatton, *J. Phys. Chem. 94*: 6400 (1990).
309. E. B. Leodidis and T. A. Hatton, *J. Phys. Chem. 94*: 6411 (1990).
310. M. J. Schick, *Nonionic Surfactants,* Marcel Dekker, New York, 1966.
311. P. Murkerjee, *J. Pharm. Sci. 60*: 1528 (1971).
312. S. Friberg and I. Lapczynka, *Progr. Colloid Polym. Sci. 56*: 16 (1975).
313. E. C. C. Melo and S. M. B. Costa, *J. Phys. Chem. 91*: 5635 (1987).
314. T. Sobisch and R. Wüstneck, *Colloids Surf. 62*: 187 (1992).
315. Y. Saito, M. Abe, and T. Sato, *Colloid Polym. Sci. 271*: 774 (1993).
316. Y. Saito, M. Abe, and T. Sato, *J. Am. Oil Chem. Soc. 70*: 717 (1993).
317. J. F. Scamehorn (Ed.), *Phenomena in Mixed Surfactant Systems,* ACS Symposium Series, No. 311, Washington, DC, 1986.
318. P. M. Holland and D. N. Rubingh (Eds.), *Mixed Surfactant Systems,* ACS Symposium Series, No. 501, Washington, DC, 1992.
319. K. Ogino and M. Abe (Eds.), *Mixed Surfactant Systems,* Marcel Dekker, New York, 1992.

II
Solubilization in Micelles

2

Solubilization of Gases

ALLEN D. KING, JR. Department of Chemistry, University of Georgia, Athens, Georgia

SYNOPSIS

Gas solubilization is a term used here to describe the enhanced solubility of gases and low molecular weight vapors associated with the presence of micellar aggregates in aqueous solutions. This article (i) reviews the literature associated with this phenomenon, spanning a period of time from 1940 to the present, and (ii) shows that a simple model of a micelle which incorporates a Laplace pressure acting across the micelle-water interface is able to quantitatively account for the variations observed in the sorptive capacity of micelles derived from ionic surfactants. The limited data available for gases and vapors solubilized in aqueous solutions of nonionic surfactants suggest that Laplace pressure effects are much less important in the case of nonionic-type micelles.

I. INTRODUCTION

The term *solubilization* is generally used to describe the ability of surfactants to enhance the aqueous solubility of sparingly soluble organic substances. Solubilization of organic solids and liquids plays an important role in innumerable biological and industrial processes and, as a consequence, a large body of literature devoted to this topic has developed over the years. Common nonpolar gases and low molecular weight vapors are likewise sparingly soluble in water, and it is found that surfactants commonly used to solubilize organic solids and liquids are also capable of enhancing the solubility of gases and vapors in aqueous solutions. This latter property is referred to as *gas solubilization* in the following discussion.

Gas solubilization occupies a very small niche in the voluminous literature devoted to solubilization and Refs. 1–39 constitute a reasonably complete bibliography on this subject. This is no doubt a reflection of the fact that gas solubilization has not played a major role in very many industrial processes and, with the exception of anesthesia and blood substitute chemistry, has also been relatively unimportant to the discipline of biological chemistry. The main reason for considering gas solubilization as a topic for review is that gas solubilization studies provide thermodynamic information about the colloidal species responsible for solubilization in much the same manner that classical gas solubility measurements have enhanced our understanding of ordinary solutions and the liquid state of matter [40, 41].

This review basically consists of three sections. The first part, which encompasses Refs. 1–5 is primarily historical in nature and reviews the earliest work in this area. The second section focuses on solubilization phenomena which involve ordinary permanent gases and low molecular weight saturated hydrocarbon gases ranging in size from methane, CH_4, to *n*-butane, C_4H_{10}. The last section will briefly summarize some pertinent results derived from recent research involving the solubilization of high molecular weight vapors.

II. EARLY STUDIES

The earliest paper known to this author which involves gas solubilization is entitled "A Simple Proof of the Thermodynamic Stability of Materials Taken Up by Solutions Containing Solubilizers Such as Soap," by J. W. McBain and J. J. O'Conner [1], published in 1940. It reports the results of a series of experiments using a simple glass apparatus which employed a differential manometer to monitor the vapor pressure over aqueous solutions of potassium oleate mixed with hexane relative to the vapor pressure of the pure hydrocarbon. With this simple apparatus the authors were able to establish that, when added to water, potassium oleate was capable of absorbing large amounts of the normally insoluble hydrocarbon to produce thermodynamically stable solutions, i.e., solutions for which the partial pressure of the hydrocarbon species was significantly less that that of the pure hydrocarbon at the same temperature. In one modification of this apparatus, the authors let a graduated capillary serve as a reservoir for the hydrocarbon liquid and were able to determine the amounts of hexane and methylcyclopentane taken up at 25°C by a 0.194 *M* solution of potassium oleate as a function of vapor pressure over the range of zero solubilization to saturation. Subsequent studies from this same laboratory established that potassium oleate was capable of solubilizing hydrocarbons of much lower molecular weight, such as isobutane and propene [2], and that a wide variety of surfactants of cationic, nonionic, and anionic types, including sodium deoxycholate, were capable of solubilizing propene at 25°C [3].

These pioneering studies were followed by a paper by Ross and Hudson [4] some 15 years later in which Henry's law constants were determined for butadiene dissolved in aqueous solutions containing various concentrations of a cationic surfactant, Hyamine 1622 (98.8% *p*-diisobutylphenoxyethoxyethyldimethylbenzylammonium chloride monohydrate). This work showed that the Henry's law constant governing the solubility of butadiene in these solutions increased with decreasing surfactant concentration as expected. It was further shown that when extrapolated, the Henry's law constant for butadiene reached that for pure water at a surfactant concentration equal to the critical micelle concentration (CMC) of Hyamine 1622, thus establishing the role played by micelles in the solubilization process. Similar studies by Brady and Huff [5] further established the importance of micelles in determining the extent of solubilization of benzene by sodium dodecylsulfate (SDS) at various temperatures: 24, 35, and 50°C. This work established an important point, namely that at high intramicellar concentrations of benzene, the enthalpy change accompanying the transfer of benzene from a SDS micelle to the vapor phase nearly equals the heat of vaporization of liquid benzene, leading one to conclude that from an energetic standpoint, the micellar environment does not differ greatly from that of pure liquid benzene.

The significance of these early studies cannot be minimized since they clearly established that surfactant micelles present at surfactant concentrations greater than the CMC provide an oil-like environment in which vapor molecules can dissolve, which is the basis of the pseudophase model for solubilization.

III. SOLUBILIZATION OF GASES AND LOW MOLECULAR WEIGHT VAPORS

The next paper to report solubility measurements of low molecular weight gases in micellar solutions seems to have been published by Wishnia [6] in 1963. Solubilities were determined for ethane, propane, *n*-butane, and *n*-pentane in water and a 0.100 *M* NaCl solution containing 1.8% SDS at known pressures using a glass vessel fitted with a manometer. Data were taken at temperatures ranging from 15 to 35°C. Values for the thermodynamic transfer functions, ΔG_{TR}, ΔH_{TR}, and ΔS_{TR} (water → micelle) calculated for each gas at 25°C were found to closely parallel the values obtained for the transfer of the same gas from water to a bulk liquid hydrocarbon, leading the author to a conclusion similar to that of Brady and Huff [5], namely that the interior of an SDS micelle closely resembles a liquid hydrocarbon. It is interesting to note that the values of ΔG_{TR} (water → micelle recorded by Wishnia for each gas are uniformly more positive than the corresponding values for ΔG_{TR} (water → hydrocarbon) by 2.3–3.8 kJ/mol indicating that the standard chemical potential of a gas molecule solubilized in a SDS micelle is considerably greater than that for the same gas in a bulk hydrocarbon (to be defined later as $\Delta G^{*(m,b)}$ in Eq. (7b)).

The gas/vapor solubilities referred to in this review have for the most part been determined at low gas pressures, $P \leq 1$ atm, using all-glass systems. This places a severe limitation on accuracy when low molecular weight gases are involved since the uptake of these gases is quite small due to their low solubilities. Therefore, the experimental method employed to measure the gas solubilities recorded in Refs. 11–13 deserve special note. Here a method is employed whereby the gas to be studied is allowed to equilibrate with the solution of interest at an elevated pressure ($P = 20$ atm or less) in a thermostatted high-pressure vessel. The pressure is then dropped and the volume of gas that effervesces from solution is recorded at ambient pressure using a gas buret. Any disadvantages in precision of temperature control are more than compensated by the increased accuracy afforded by the large volumes of gas released from solution from which the solubilities are calculated. The pressures involved are not high enough to significantly affect the chemical potentials of the dissolved species so that the gas solubilities derived from these data (mol gas/atm) can be taken to represent 1 atm solubilities without incurring a significant error. This method has been used for the micellar solubilities of He, O_2, Ar, CO_2, N_2O, CH_4, C_2H_6, and C_3H_8, which are recorded in Refs. 11–13 as well as Refs. 17, 20, 24, 28–32, 34, 35, 38, 39.

Table 1 lists all the sources of gas solubility data taken with micellar solutions of ionic surfactants and includes all values for 1 atm micellar

TABLE 1 Micellar Gas Solubilities for Common Gases and Low Molecular Weight Saturated Hydrocarbons in Ionic Surfactants

Surfactant	Gas	$X_g^m \times 10^4$ [a]	Ref.
Potassium Oleate	C_3H_6	520	2
Sodium dodecylsulfate/0.1 *M* NaCl	C_2H_6	120	6
Sodium dodecylsulfate/0.1 *M* NaCl	C_3H_8	350	6
Sodium dodecylsulfate/0.1 *M* NaCl	n-C_4H_{10}	1080	6
Sodium dodecylsulfate	He	0	11
Sodium dodecylsulfate	O_2	10	11
Sodium dodecylsulfate	Ar	11	11
Sodium dodecylsulfate	CH_4	19	11
Sodium dodecylsulfate	C_2H_6	111	11
Hexadecyltrimethylammonium bromide	O_2	14[c]	11
Hexadecyltrimethylammonium bromide	Ar	16[c]	11
Sodium dodecylsulfate	C_2H_6	110	17
Sodium dodecylsulfate	C_3H_8	320	17
Sodium decylsulfate	O_2	8	17
Sodium decylsulfate	C_2H_6	94	17
Sodium octylsulfate	O_2	7	17
Sodium octylsulfate	C_2H_6	68	17
Sodium hexylsulfate	CH_4	7	17
Sodium hexylsulfate	C_2H_6	36	17
Sodium hexylsulfate	C_3H_8	77	17
Sodium octanoate	CH_4	—	18
Sodium octanoate	C_2H_6	41	18
Sodium octanoate	C_3H_8	99	18
Sodium octanoate	n-C_4H_{10}	176	18
Sodium octanoate	Ar	—	19
Sodium dodecylsulfate	Ar	11[b]	19
Sodium dodecylsulfate	C_3H_8	330	20
Sodium octylsulfate	C_3H_8	144	20
Hexadecyltrimethylammonium bromide	O_2	10[c]	24
Hexadecyltrimethylammonium bromide	Ar	13[c]	24
Hexadecyltrimethylammonium bromide	CH_4	29[c]	24
Hexadecyltrimethylammonium bromide	C_2H_6	155[c]	24
Hexadecyltrimethylammonium bromide	C_3H_8	480[c]	24
Hexadecyltrimethylammonium bromide	CF_4	9[c]	24
Decyltrimethylammonium bromide	O_2	7	24
Decyltrimethylammonium bromide	Ar	7	24
Decyltrimethylammonium bromide	CH_4	17	24
Decyltrimethylammonium bromide	C_2H_6	80	24
Decyltrimethylammonium bromide	C_3H_8	210	24
Decyltrimethylammonium bromide	CF_4	4	24
Sodium dodecylsulfate	CF_4	5	24

(continued)

TABLE 1 *(continued)*

Surfactant	Gas	$X_g^m \times 10^4$ [a]	Ref.
Sodium dodecylsulfate	CH_4	16[b]	25
Sodium dodecylsulfate	C_2H_6	140[b]	25
Sodium dodecylsulfate	C_3H_8	380[b]	25
Sodium 1-heptanesulfonate	O_2	4	29
Sodium 1-heptanesulfonate	Ar	4	29
Sodium 1-heptanesulfonate	CH_4	10	29
Sodium 1-heptanesulfonate	C_2H_6	45	29
Sodium 1-heptanesulfonate	C_3H_8	102	29
Sodium perfluorooctanoate	O_2	18	29
Sodium perfluorooctanoate	Ar	17	29
Sodium perfluorooctanoate	CH_4	19	29
Sodium perfluorooctanoate	C_2H_6	61	29
Sodium perfluorooctanoate	C_3H_8	148	29
Sodium perfluorooctanoate	CF_4	38	29
Sodium octanoate (pH = 12)	O_2	4	30
Sodium octanoate (pH = 12)	CH_4	10	30
Sodium octanoate (pH = 12)	C_2H_6	49	30
Sodium octanoate (pH = 12)	C_3H_8	100	30
Lithium perfluorooctanoate	O_2	17	31
Lithium perfluorooctanoate	C_2H_6	63	31
Lithium perfluorooctanoate	C_3H_8	135	31
Lithium perfluorodecanoate	O_2	26	31
Lithium perfluorodecanoate	Ar	21	31
Lithium perfluorodecanoate	CH_4	23	31
Lithium perfluorodecanoate	C_2H_6	81	31
Lithium perfluorodecanoate	C_3H_8	210	31
Lithium perfluorodecanoate	CF_4	42	31
Sodium dodecylsulfate/0.3 *N* NaCl	C_3H_8	379	32
Sodium dodecylsulfate/0.6 *N* NaCl	C_3H_8	394	32
Sodium octylsulfate/0.3 *N* NaCl	C_3H_8	142	32
Sodium hexylsulfate/0.6 *N* NaCl	C_3H_8	79	32
Sodium dodecylsulfate	CO_2	93	35
Sodium 1-heptanesulfonate	CO_2	42	35
Hexadecyltrimethylammonium bromide	CO_2	147[c]	35
Sodium perfluorooctanoate	CO_2	139	35
Sodium perfluorooctanoate	N_2O	120	35

[a] T = 25°C unless otherwise noted.
[b] T = 24°C.
[c] T = 26°C.

gas solubilities at 25°C that have been calculated from the experimental data contained in these references. The 1 atm micellar solubilities are given in mole fraction units:

$$X_g^m = \frac{n_g}{n_g + n_s} \tag{1}$$

where n_g denotes the moles of gas (g) dissolved in the micelles at 1 atm and n_s represents the moles of surfactant (s) present in micellar form in the solution of interest. Gas solubility data have been reported in a variety of ways, particularly in the older literature, however, a cursory inspection shows that good agreement exists between X_g^m values taken from gas solubility data reported by different groups: for example, the 25°C, 1 atm micellar solubility of propane in SDS has been determined to be 0.0320, 0.0330, and 0.0380 (24°C) in three separate experiments listed in Table 1.

A variety of other features become evident when one examines the solubility data assembled in Table 1. First, the micellar gas solubilities listed for a given surfactant increase in a regular fashion with the boiling point of the gas (for example see data taken from Ref. 11). This is the behavior expected for situations where dispersion forces represent the dominant mode of interaction between the dissolved gas molecules and the surrounding solvent, the situation normally encountered with solutions of gases in nonpolar solvents. Second, the micellar solubility of a gas increases as the size of the alkyl group of the surfactant increases; e.g., the micellar solubility of ethane increases in the order 0.0036, 0.0068, 0.0094, and 0.0110 as the size of the alkyl portion of a group of sodium alkylsulfates increases in the order hexyl-, octyl-, decyl-, and dodecyl- (see data taken from Ref. 17). A third point of interest is the fact that the specific nature of the headgroup is relatively unimportant in determining the sorptive capacity of a micelle. For example, a comparison of data taken from Refs. 17 and 24 shows that the micellar solubilities of O_2 and ethane are nearly the same in sodium decylsulfate and decyltrimethylammonium bromide. Likewise, the micellar gas solubilities obtained with sodium 1-heptanesulfonate (taken from Ref. 29) agree closely with those determined for micellar sodium octanoate (taken from Refs. 18 and 30). Fourth, as might be expected, the nature of the counterion plays no role in determining gas solubilization, at least for perfluorinated surfactants, since the micellar solubilities of oxygen, ethane, and propane are virtually the same in sodium- and lithium perfluorooctanoate micelles (compare data taken from Ref. 29 to those from Ref. 31). Another feature of interest is that, just as is found with bulk perfluorinated solvents, O_2 is considerably more soluble in perfluorinated micelles (X_g^m is 0.0018 and 0.0017 in sodium- and lithium perfluorooctanoate, respectively) than in analogous micelles having ordinary hydrocarbon tail groups, e.g., sodium octanoate and sodium 1-heptanesulfonate for which the oxygen solubility is measured to be $X_g^m = 0.0004$.

Finally, the micellar solubility of propane in sodium dodecylsulfate increases with increasing salt concentrations, i.e., $X_g^m = 0.0330$ (0 NaCl), 0.0350 (0.1 *N* NaCl), 0.0379 (0.3 *N* NaCl), and 0.0394 (0.6 *N* NaCl) (data from Refs. 6, 20, and 32). These results can be compared with those of Hoskins and King [13] which show that added NaCl does not affect the micellar solubility of ethane in sodium dodecylsulfate to any appreciable extent, and those of Valenzuela et al. [23] and Smith et al. [36] which show that added NaCl enhances the solubilization of *n*-pentane and *n*-hexane in micellar sodium dodecylsulfate. These results follow a trend noted previously by Valenzuela et al. [23] that the degree to which salt enhances the solubilization of simple nonpolar substances appears to depend upon the molecular size of that substance. Interestingly, the solubility of propane in micelles of the smaller sodium alkylsulfates does not seem to be affected by added salt; e.g., the micellar solubility of propane in sodium octylsulfate is $X_g^m = 0.0144$ (0 NaCl, Ref. 20) and 0.0142 (0.3 *N* NaCl, Ref. 32) while that for propane in sodium hexylsulfate is $X_g^m =$ 0.0077 (0 NaCl, Ref. 17) and 0.0079 (0.6 *N* NaCl, Ref. 32). These results can be compared with those of Tucker and Christian [26] who find that added salt has little effect upon the tendency of sodium octylsulfate micelles to take up benzene.

The general picture that emerges from the data of Table 1 is one in which solubilized gas molecules reside in a nonpolar, oil-like, environment which takes on solvent properties associated with the tail groups (hydrocarbon/fluorocarbon) and is only mildly affected by conditions outside the micelle, e.g., salinity. However, in spite of the similarities observed between the sorptive properties of micelles and bulk solvents, the fact remains that the micellar solubility of any given gas is invariably less than that for the same gas in a bulk hydrocarbon (or bulk perfluorocarbon in the case of fluorinated surfactant micelles). This is clearly illustrated in Fig. 1, which shows micellar and bulk solubilities for a variety of gases at 25°C plotted logarithmically as a function of the energy of vaporization of the individual gases, ΔE_b^V, a parameter commonly used in correlating gas solubilities in a common solvent [41].

In Fig. 1a, one sees that the micellar gas solubilities in sodium dodecylsulfate, sodium octanoate (SOc), and sodium 1-heptanesulfonate (SHSo) are all displaced toward lower values relative to the corresponding solubilities in *n*-heptane (n-C_7H_{16}). However, with the exception of CF_4, the micellar solubilities closely parallel the corresponding solubilities in *n*-heptane. This suggests that the intermolecular forces between the solubilized gas molecules and their environment, presumably the alkyl chains

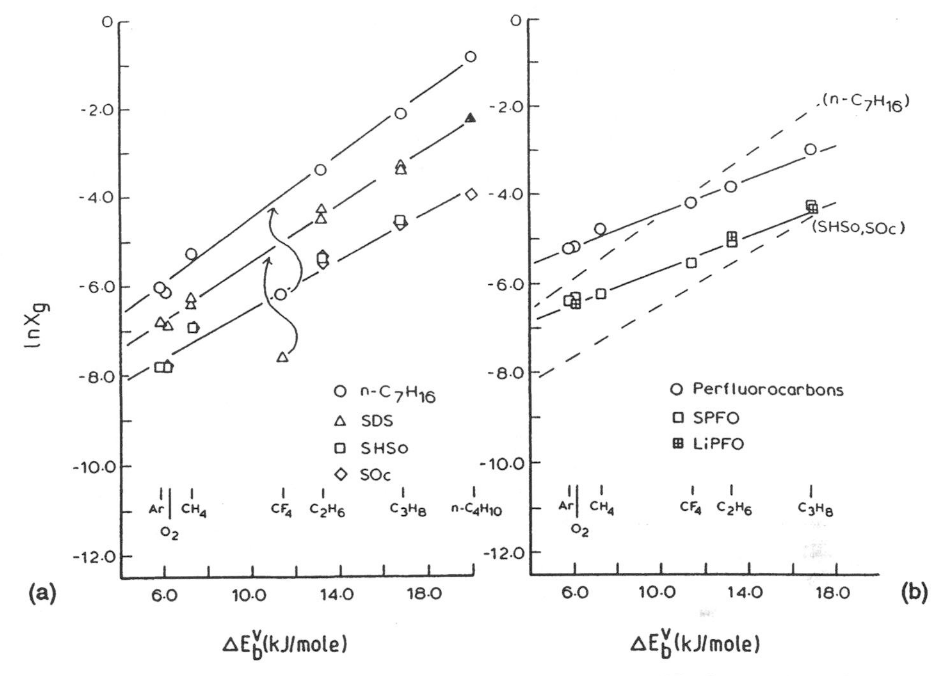

FIG. 1 One-atmosphere gas solubilities measured at 25°C shown plotted as a function of the energy of vaporization at the boiling point, ΔE_b^V, of the respective gases. (a) Micellar gas solubilities, X_g^m, in sodium dodecylsulfate (SDS), sodium octanoate (SOc), and sodium 1-heptanesulfonate (SHSo) compared with gas solubilities in bulk *n*-heptane, X_g^b. Shaded triangle: SDS/0.1 *M* NaCl (6). (b) Micellar gas solubilities in sodium perfluorooctanoate (SPFO) and lithium perfluorooctanoate (LiPFO) compared with gas solubilities in bulk perfluorocarbons: perfluoroheptane (Ar, O_2, CH_4, C_2H_6, C_3H_8) and perfluorotributylamine (CF_4). Gas solubilities in *n*-heptane and bulk perfluorocarbons are taken from sources cited in Ref. 29. The solubility of *n*-C_4H_{10} in *n*-heptane is taken from Ref. 51.

of the micellized surfactant ions, are quite similar to those within solutions of the same gases in *n*-heptane. The nearly equal displacements toward lower solubility of CF_4 in SDS and *n*-heptane are not inconsistent with this picture since it is well known that solubilities of fluorocarbon gases in hydrocarbons are invariably less than predicted by correlation schemes based on ordinary gases [41, 42]. Note that while heptane is a logical choice as a bulk hydrocarbon solvent to use when comparing micellar solubilities in SOc and SHSo, mole fraction solubilities of gases dissolved in *n*-alkanes are generally quite insensitive to the size of the solvent molecules for alkanes ranging from hexane to dodecane [43] so that the upper line in Fig. 1a serves equally well as a point of reference to compare micellar gas solubilities in SDS. Much the same situation exists in Fig. 1b where one sees that the micellar gas solubilities measured for sodium perfluorooctanoate (SPFO) are displaced to lower values but again parallel the corresponding solubilities in bulk perfluorocarbon solvents, suggesting that the force fields experienced by gas molecules in perfluorinated micelles are much the same as those experienced in bulk perfluorocarbon solvents. Thus, the correlations shown in Fig. 1a, b can be taken to indicate that the molecular environment surrounding gas molecules solubilized in micelles of SHSo and SDS is hydrocarbon-like while that surrounding gas molecules within SPFO micelles closely resembles a liquid perfluorocarbon.

The general similarity between the solvent properties of micelles and bulk solvents suggests that gas solubilization can be treated adequately within the context of solution thermodynamics. Accordingly, one can write an expression for the chemical potential, $\mu_i^{(m)}$, of gas (i) solubilized in micelles (m) as

$$\mu_i^{(m)} = \mu_i^{*(m)} + RT \ln a_i^{(m)} \tag{2a}$$

$$a_i^{(m)} = \frac{P_i}{H_i^{(m)}} = \gamma_i^{*(m)} X_i^{(m)} \tag{2b}$$

Here $\mu_i^{*(m)}$ represents the chemical potential of gas i solubilized in a highly dilute Henry's law standard state (a singly occupied micelle). The activity of the solubilized gas, $a_i^{(m)}$, is alternately expressed as the ratio of partial pressure (or fugacity) P_i to the Henry's law constant $H_i^{(m)}$ or as the product of an activity coefficient $\gamma_i^{*(m)}$ multiplied by the mole fraction concentration of solubilized gas, $X_i^{(m)}$. Under this convention, $\gamma_i^{*(m)}$ accounts for deviations from Henry's law that arise from changes in composition of the binary (gas-micelle) solution and approaches unity in the limit of low $X_i^{(m)}$.

One can write an analogous expression for the chemical potential of gas (i) dissolved in a bulk solvent (b) whose properties resemble the micelle interior:

$$\mu_i^{(b)} = \mu_i^{*(b)} + RT \ln a_i^{(b)} \tag{3a}$$

$$a_i^{(b)} = \frac{P_i}{H_i^{(b)}} = \gamma_i^{*(b)} X_i^{(b)} \tag{3b}$$

Here, $a_i^{(b)}$ represents the activity of gas (i) dissolved in the bulk solvent, while $H_i^{(b)}$ and $\gamma_i^{*(b)}$ represent the Henry's law constant and activity coefficient, respectively, for dissolved (i) in the bulk solvent. The chemical potential of i in the gas phase (g) is

$$\mu_i^{(g)} = \mu_i^\circ + RT \ln P_i \tag{4}$$

where μ_i° represents the chemical potential of gas (i) in the usual gas standard state and P_i denotes the partial pressure (fugacity) of gas (i).

At 1 atm, micellar and bulk gas solubilities are low so that the activity coefficients $\gamma_i^{*(m)}$ and $\gamma_i^{*(b)}$ may be set equal to unity without introducing any appreciable error, thus simplifying the expressions relating 1-atm gas solubilities, X_g, to the various standard state chemical potentials:

$$X_g^{(m)} = \exp\left(-\frac{\Delta G^{*(m)}}{RT}\right) \tag{5a}$$

$$\Delta G^{*(m)} = \mu_i^{*(m)} - \mu_i^\circ \tag{5b}$$

$$X_g^{(b)} = \exp\left(-\frac{\Delta G^{*(b)}}{RT}\right) \tag{6a}$$

$$\Delta G^{*(b)} = \mu_i^{*(b)} - \mu_i^\circ \tag{6b}$$

$$\frac{X_g^{(m)}}{X_g^{(b)}} = \exp\left(\frac{-\Delta G^{*(m,b)}}{RT}\right) \tag{7a}$$

$$\Delta G^{*(m,b)} = \mu_i^{*(m)} - \mu_i^{*(b)} \tag{7b}$$

The last of these free-energy expressions, Eq. (7b), represents the difference in free energy of a molecule (i) in a singly occupied micelle and its free energy in a similarily dilute condition within a bulk solvent composed of molecules which are similar chemically to the tail groups of the surfactant ions that make up the micelle.

The vertical spacing between the straight lines shown in Figs. 1a, b can be equated to $\Delta G^{*(m,b)}/RT$, i.e.,

$$\ln X_g^{(b)} - \ln X_g^{(m)} = -\ln\left(\frac{X_g^{(m)}}{X_g^{(b)}}\right) = \frac{\Delta G^{*(m,b)}}{RT} \tag{8}$$

Thus, using the n-heptane and perfluorocarbon data of Figs. 1a and 1b for reference, one sees that the spacing $\ln X_i^{(b)} - \ln X_i^{(m)}$ is approximately 1.1 and 2.0 for a typical gas, e.g., ethane, in SDS and SHSo, respectively, and is approximately 1.3 in the case of SPFO. These correspond to values of $\Delta G^{*(m,b)} = 2.7$, 5.0, and 3.2 kJ/mole for SDS, SHSo, and SPFO, respectively. Thus solution thermodynamics leads one to conclude that the free energy of a typical gas molecule solubilized within a micelle can be expected to be several kJ/mol greater than its free energy in a bulk solvent having similar chemical characteristics.

Several models have been proposed for micelles that would lead to an energetically unfavorable environment for solubilization. One of these models, proposed by Menger [44, 45] envisions a micelle as being a loosely organized assembly of surfactant ions into which water penetrates deeply. A second model, proposed by Dill and Flory [46], considers the micelle interior to be dry but hardly oil-like since crowding of the tail groups is thought to create a highly ordered region in the core which might be expected to resist the intrusion of a foreign molecule.

A third model, proposed by Mukerjee [47, 48], considers a micelle to be a spherical assembly of surfactant ions having a dry oil-like core which is compressed by a Laplace pressure, ΔP_L, which arises from surface tension forces acting across the curved micelle-water interface:

$$\Delta P_L = \frac{2\gamma}{r} \tag{9}$$

where γ denotes the interfacial tension at the micelle-water interface and r is the radius of the micelle. The effect of this compression is to raise the chemical potential of a solubilized molecule i by an amount

$$\Delta G^{*(m,b)} = \Delta P_L \overline{V}_i = \frac{2\gamma \overline{V}_i}{r} \tag{10}$$

where $\overline{V}_i$ is the partial molar volume of the solubilized gas molecule.

The partial molar volume of ethane has been determined to be 68 cm^3/mol in n-hexane [49] and 83 cm^3/mol in perfluoroheptane [50]. Thus, taking the value of $\Delta G^{*(m,b)} = 2.7$ kJ/mole for ethane solubilized in SDS, one calculates that $\Delta P_L = 390$ atm for a SDS micelle which, for a core radius of 1.5 nm, corresponds to an interfacial tension of $\gamma = 29$ mN/m. A similar analysis using an assumed core radius of $(\frac{7}{12}) \times 1.5$ nm $= 0.88$ nm, yields values of $\gamma = 30$ mN/m for SHSo, SOc, and 17 mN/m for SPFO micelles, respectively. The value of γ obtained here for an SDS micelle is somewhat larger than the value $\gamma = 20$ mN/m proposed by Mukerjee in his original paper [47] but agrees well with more recent estimates of $\gamma = 30$ mN/m by Christian and co-workers [14] and $\gamma = 31.5$ mN/m by Mukerjee [48].

It is reasonable to assume that the radius term in Eq. (9) is proportional to the carbon number, n_C, of the alkyl chain of the surfactant of interest: $r = \alpha n_C$, with the proportionality constant α having a value that approximates the carbon-carbon distance in a fully extended alkyl chain of a saturated hydrocarbon; i.e., $\alpha \cong 0.125$ nm. Accordingly, the free-energy increment, $\Delta G^{*(m,b)}$ for a monodisperse system of micelles becomes

$$\Delta G^{*(m,b)} = \frac{2\gamma \overline{V}_i}{\alpha n_C} \tag{11}$$

This in turn allows Eq. (8) to be written as

$$\ln X_g^m = \ln X_g^b - \frac{2\gamma \overline{V}_i}{\alpha RT} \cdot n_C^{-1} \tag{12}$$

Figure 2 shows the graph that results when the 1-atm micellar gas solubilities listed in Table 1 for pure hydrocarbon-type surfactants are plotted logarithmically as a function of n_C^{-1}. It is seen that, when plotted in this manner, the micellar solubilities of the various gases describe a series of straight lines, each of which extrapolates smoothly into the corresponding gas solubility in bulk hydrocarbon, *n*-heptane, at $n_C^{-1} = 0$, in accordance with Eq. (12). Further inspection shows that micellar gas solubilities obtained with surfactants having trimethylammonium-, carboxylate-, sulfate-, and sulfonate head groups all correlate equally well with the straight lines describing $\ln X_g^m$ as a function of n_C^{-1} for the various gases in Fig. 2. This implies, within the framework of the Laplace pressure model, that the interfacial tension, γ, responsible for elevating the chemical potential of solubilized gas molecules is insensitive to the specific nature of the charged headgroups and associated counterions.

The slopes of the straight lines fitted to the solubility data contained in Fig. 2 are shown plotted as a function of gas partial molar volume, $\overline{V}_i$, in the inset to Fig. 2. These data all fall on a straight line passing through the origin, which, in the context of Eq. (12), implies that the distance parameter α and interfacial tension γ are insensitive to the presence of solubilized gas. The slope of the line shown in the inset is -0.18 mol cm^{-3}, which corresponds to a value of $\gamma/\alpha = 2.2 \times 10^8$ J m^{-3} at 25°C. Thus, taking the value $\alpha = 0.125$ nm mentioned earlier, one finds that the slope shown in the inset corresponds to a value of $\gamma = 28$ mN/m, which presumably approximates the interfacial tension present at the micelle-water interface of any micelle composed of ordinary alkyl-type surfactant ions.

A similar analysis of the gas solubility data taken with perfluorocarbon surfactants using $\alpha = 0.125$ nm yields a value of $\gamma = 16$ mN/m for the interfacial tension operative at the surface of fluorocarbon-type micelles [29]. Thus, according to the Laplace pressure model, the somewhat larger sorptive capacities exhibited toward gases by perfluorocarbon-type micelles as compared to hydrocarbon-type micelles results from the interplay of two factors: (1) a greater intrinsic capacity of perfluorocarbon solvents

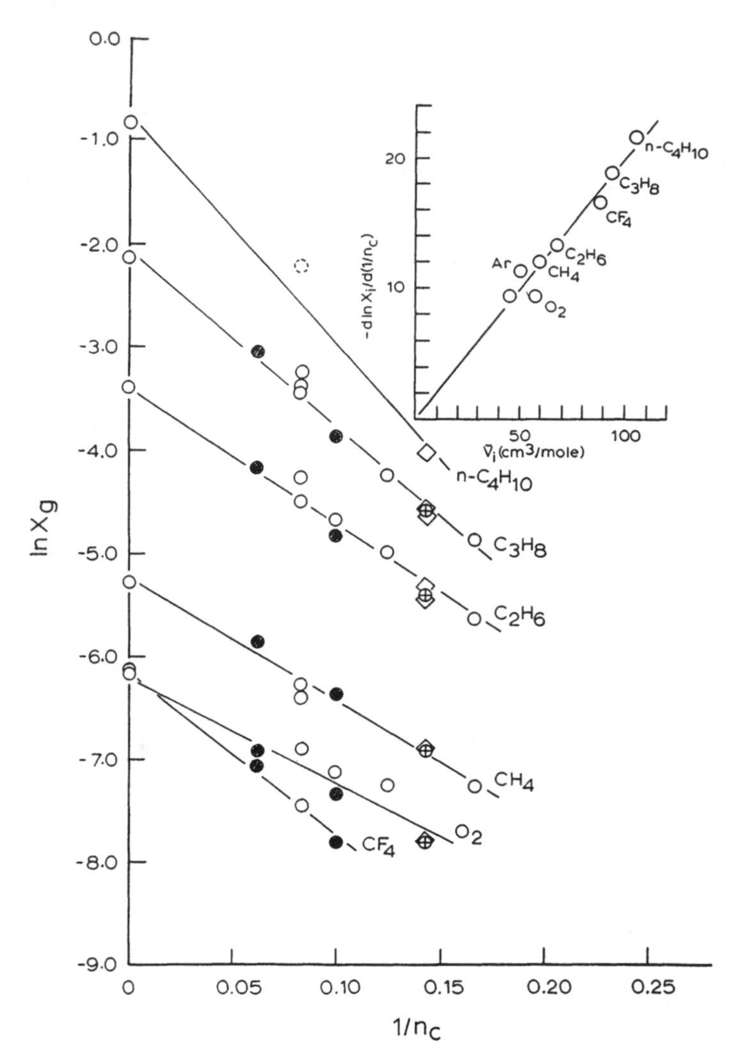

FIG. 2 Logarithm of gas solubilities taken from Fig. 1 shown as a function of reciprocal carbon number. Symbols: ○, sodium alkyl sulfates (and *n*-heptane at $n_C^{-1} = 0$); ● alkyltrimethylammonium bromides; ◇ sodium octanoate; ⊕ sodium 1-heptanesulfonate; ◌ SDS/0.1 *M* NaCl. Argon data omitted for clarity. Inset: Slopes of straight lines shown in Fig. 2 plotted against partial molar volumes of the individual gases. Partial molar volumes taken from sources cited in Ref. 29. In the case of *n*-butane, the molar volume of pure liquid $n\text{-}C_4H_{10}$ at 25°C, taken from Ref. 52 has been used.

to absorb low molecular weight gases as shown in the upper curve of Fig. 1b, and (2) a more favorable energetic situation, i.e., smaller $\Delta G^{*(m,b)}$, arising from the low value of γ found with perfluorocarbon-type micelles.

The Laplace pressure model is simple and can only be applied to systems composed of spherical micelles. However, unlike the more complicated models of micelles mentioned earlier [44–46], it is easily incorporated into the framework of solution thermodynamics and thus is capable of addressing the most salient features of gas solubilization, namely the similarities observed between the solvent behavior of micelles and bulk solvents, the increase in gas solubilization, X_g^m, with micelle size and the fact that large gases (large $\overline{V}_i$) are solubilized to a lesser extent than small molecules in a given type of micelle.

Other than a brief comment regarding the sorptive capacity of Triton NE toward propylene in Ref. 3, no reference has been made in the previous discussion concerning gas solubilization in micellar solutions of nonionic surfactants. In fact, to this author's knowledge, Refs. 38 and 39 are the only two papers published which discuss gas solubilization in simple micellar solutions of nonionic surfactants. Both papers deal with commercially available polyoxyethylene-type surfactants, i.e., polyethoxylated nonyl phenols [38] and polyethoxylated lauryl alcohols [39].

Surfactants of this type are polydisperse in the sense that the polymerized ethylene oxide (EO) groups of such surfactants are not uniform but vary in length about some mean value according to some fixed distribution, usually considered to be a Poisson distribution [53]. As a consequence, no unique molecular weight can be assigned to these surfactants which precludes any analysis of gas solubilization based on mole fraction solubilities. As a result, the 1-atm micellar solubilities taken from Refs. 38 and 39 are reported in Table 2 using molality units (i.e., mol gas dissolved at 1 atm per kilogram of micellized surfactant).

The micellar gas solubilities listed in Table 2 are seen to increase with the boiling point of the gas suggesting that, as in the case of ionic surfactant micelles, dispersion forces are a major factor in determining the solubilities of gases in micelles composed of nonionic surfactants. Furthermore, the micellar solubilities for each gas listed in Table 2A for the IGEPAL surfactants increase as the ethylene oxide content of the surfactant decreases, i.e., the molal gas solubilities increase in the order IGEPAL CO-880 ($NPEO_{30}$) $<$ IGEPAL CO-850 ($NPEO_{20}$) $<$ IGEPAL CO-660 ($NPEO_{10}$). Here the symbol NP is used to denote the lipophilic nonylphenoxy group common to these surfactants while the abbreviation EO_m (m = 10, 20, 30) is used to designate a hydrophilic polyoxyethylene chain attached to the nonylphenoxy group, which, on the average contains 10, 20, or 30 polymerized EO groups.

TABLE 2 Micellar Gas Solubilities in Nonionic Surfactants at 1 atm, 25°C

Gas	IGEPAL CO-880 (mol kg^{-1}) × 10^3 14% NP	IGEPAL CO-850 (mol kg^{-1}) × 10^3 20% NP	IGEPAL CO-660 (mol kg^{-1}) × 10^3 33% NP
	A. Polyethoxylated nonyl phenols (data from Ref. 38)		
O_2	0.8	1.1	2.1
Ar	1.0	1.5	2.3
CH_4	2.0	2.8	4.5
C_2H_6	12	19	26
C_3H_8	32	50	80
	B. Polyethoxylated lauryl alcohols (data from Ref. 39)		
Gas	100% BRIJ 35 (mol kg^{-1}) × 10^3 15.5% C_{12}	12.5% BRIJ 30, 87.5% BRIJ 35 (mol kg^{-1}) × 10^3 20% C_{12}	25.0% BRIJ 30, 75.0% BRIJ 35 (mol kg^{-1}) × 10^3 25% C_{12}
O_2	1.4	—	—
Ar	1.8	1.7	2.6
CH_4	3.0	3.4	4.5
C_2H_6	15	20	25
C_3H_8	43	58	80

The same trend is observed among the data in part B of Table 2 where it is seen that the addition of a polyethoxylated lauryl (C_{12}) alcohol having a low ethylene oxide content, BRIJ 30 ($C_{12}EO_4$), to one having a higher EO content, BRIJ 35 ($C_{12}EO_{23}$), leads to increased micellar gas solubility.

These trends are easily explained by assuming that the polymerized EO groups of the micellized surfactant molecule do not provide sites for gas absorption in the micelle and therefore do not contribute to the sorptive capacity of the micelle, leaving the lipophilic portion of each surfactant (or blend of surfactants) solely responsible for gas absorption in the micelle. As a consequence, micellar solubilities per unit mass of surfactant for any given gas are expected to be proportional to the fraction of lipophilic material in a sample of surfactant or blend of surfactants. Thus, taking the nonylphenoxy group as the lipophile, one predicts that the micellar solubility of a given gas expressed as molality should increase in the order IGEPAL CO-880 ($NPEO_{30}$, 14% NP by weight) < IGEPAL CO-850 ($NPEO_{20}$, 20% NP by weight) < IGEPAL CO-660 ($NPEO_{10}$, 33% NP by weight) as is observed in Table 2A. The assumption that the nonylphenoxy group alone is responsible for gas absorption is reinforced by the fact that the micellar solubilities of each gas listed in Table 2A are found to be proportional to the wt.% NP of each surfactant; i.e., the micellar solubilities of a given gas, when plotted as a function of wt.% NP, fall on a straight line which passes through the origin at 0% NP and extrapolate smoothly into the bulk solubility of that gas at 100% NP (see Ref. 38).

Micelles of BRIJ 35 and its mixtures with BRIJ 30 should be amenable to the same treatment, with dodecanol being the lipophilic moiety. Fig. 3 (taken from Ref. 39) shows the micellar solubilities of argon, methane, ethane, and propane, taken from Table 2B, plotted as a function of wt.% dodecanol for pure BRIJ 35 and its blends. As in the case of the IGEPAL surfactants [38], one sees that the micellar gas solubilities in BRIJ 35 and its blends describe a series of straight lines which pass through the origin at 0% dodecanol and extrapolate smoothly into the bulk solubilities of each gas in pure dodecanol at 100% dodecanol.

Thus, while the micellar gas solubility data of Table 2 hardly constitute a large data base, the information derived from these data suggests that micelles of polyethoxylated nonionic surfactants, mixed and pure, are adequately described by a very simple model in which the extent of gas solubilization is quantitatively accounted for by the bulk sorptive capacity of the lipophile from which the surfactant is derived.

From a chemical standpoint, a surfactant of the sodium alkylsulfate class can be considered to be a sodium salt of an acidic ester derived from sulfuric acid and a long-chain alcohol: *n*-dodecanol in the case of SDS, *n*-octanol in the case of sodium octylsulfate (SOS), and *n*-hexanol in the case of sodium hexylsulfate (SHS). Since gas solubilities in *n*-alkanols are not only lower than those in *n*-alkanes but also decrease markedly with decreasing chain length of the alcohol, it is reasonable to ask whether the model employed to correlate micellar gas solubilities in nonionic surfactants can, in fact, explain the rather low micellar gas solubilities found for the alkylsulfate-type surfactants. Since molecular weights are well defined for these monodisperse species, the model employed for nonionic surfactants simply predicts that the mole fraction solubility of a gas in alkylsulfate micelles of a given type should equal the mole fraction solubility of that gas in the appropriate long-chain alcohol, e.g., dodecanol in the case of SDS. This is not found to be the case however. For example, the 1-atm solubility of methane in dodecanol at 25°C has been measured to be $X_g^b = 38.9 \times 10^{-4}$ [54], which is much larger than the value $X_g^m = 19 \times 10^{-4}$ listed in Table 1 for CH_4 in SDS (Ref. 11). Likewise, the 1-atm solubility of methane in *n*-hexanol at 25°C, $X_g^b = 23.4 \times 10^{-4}$ [55], is much larger than the value $X_g^m = 7 \times 10^{-4}$ listed in Table 1 for CH_4 solubilized in SHS [17]. Similar comparisons can be made with ethane. The 1-atm solubilities of C_2H_6 in *n*-dodecanol, *n*-octanol, and *n*-hexanol at 25°C are $X_g^b = 216 \times 10^{-4}$ [56], 173×10^{-4} [55], and 143×10^{-4} [55],

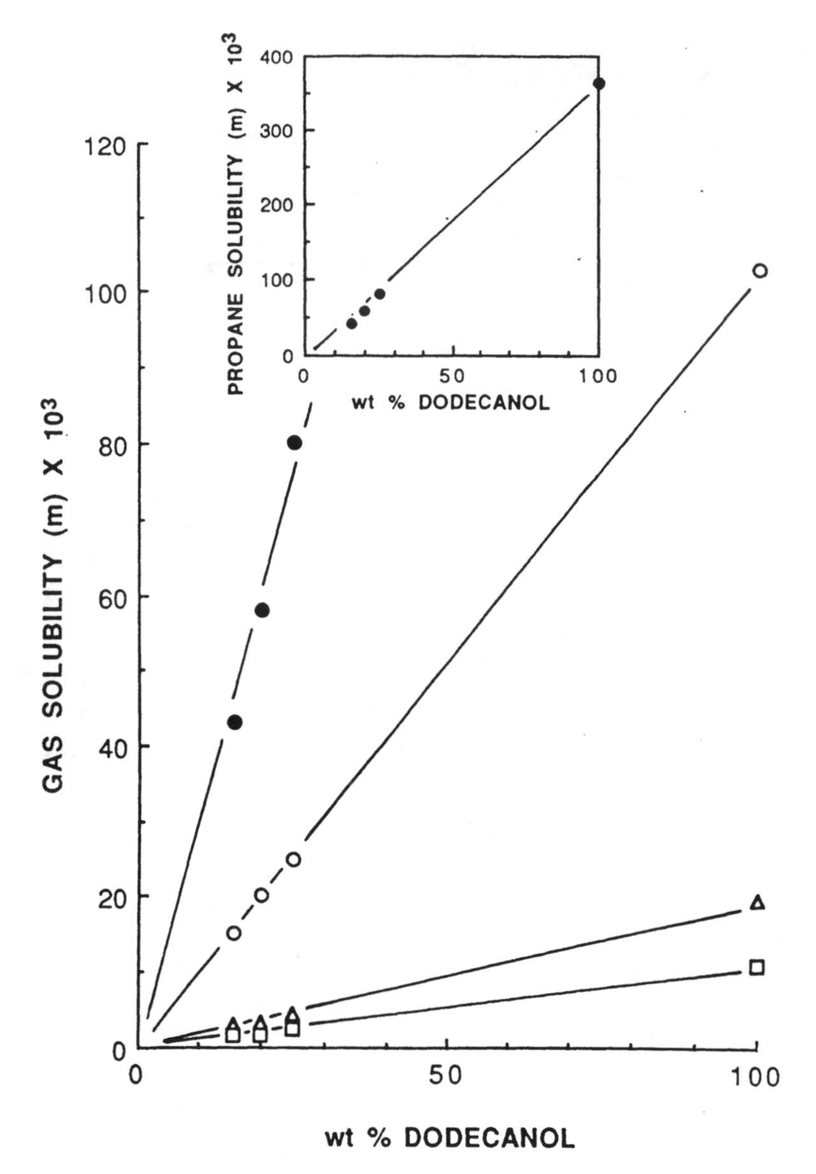

FIG. 3 One-atmosphere micellar gas solubilities and corresponding gas solubilities in pure dodecanol, expressed in molal (m) units, plotted as a function of weight percent dodecanol. $T = 25°C$. Gases: ● C_3H_8; ○ C_2H_6; Δ CH_4; □ Ar. (From Ref. 39.)

respectively. Again one sees that, as was found with methane, the solubilities of ethane in alcohols are much larger than the corresponding values listed in Table 1 for ethane in SDS, SOS, and SHS: for SDS, $X_g^m = 111 \times 10^{-4}$ [11]; for SOS, $X_g^m = 68 \times 10^{-4}$ [17]; and for SHS, $X_g^m = 36 \times 10^{-4}$ [17]. These comparisons make it clear that the simple model used to correlate the gas solubilization data with nonionics fails to account for the extremely low gas solubilities found for alkylsulfate-type micelles. At best, the substitution of the appropriate long-chain alcohol for *n*-heptane as the bulk phase reference state ((b) in Eq. (6a)) to model the micelle interior of alkylsulfate-type micelles reduces $\Delta G^{*(m,b)}$, and by analogy the Laplace pressure ΔP_L, by approximately 38%. These comparisons suggest strongly that the free energy barrier acting to reduce the degree of solubilization of gases in alkylsulfate-type micelles, and presumably ionic-type micelles in general, is much greater than any equivalent barrier that might function to suppress gas solubilization in nonionic micelles, assuming of course that EO groups do not contribute to solubilization.

IV. SOLUBILIZATION OF HIGH MOLECULAR WEIGHT VAPORS

Finally, mention should be made of recent studies by Christian and coworkers involving highly accurate vapor pressure–solubility measurements for liquid organic compounds dissolved in micellar solutions [14, 15, 26, 27, 33, 36, 37]. The results obtained from these studies, when combined with earlier measurements by McBain [1], Brady [5], Abuin [21, 23], and Treiner [22], provide a large body of information concerning the thermodynamics that govern the solubilization of liquids. The results obtained in these experiments complement those derived from gas solubilization studies in that they provide information regarding the energetics of the solubilization process involving large solubilizate molecules in otherwise similar micellar systems.

Since the solubilizates (component *i*) used in these studies are all liquids, the vapor pressures, P_i, measured over the micellar solutions are commonly referenced against the vapor pressure of pure liquid solubilizate, $P_i^\bullet$, allowing the mole fraction micellar solubilities, X_i^m, obtained from these experiments to be reported as a function of solubilizate activity, $a_i^\bullet$, defined according to the Raoult's law convention by equations analogous to Eqs. (2a), (2b):

$$\mu_i^{(m)} = \mu_i^\bullet + RT \ln a_i^\bullet \tag{13a}$$

$$a_i^\bullet = \frac{P_i}{P_i^\bullet} = \gamma_i^{\bullet(m)} X_i^{(m)} \tag{13b}$$

Here, $\mu_i^{(m)}$ denotes the chemical potential of solubilized component (i), while $\mu_i^{\bullet}$ represents the chemical of pure liquid i under ambient conditions. According to this convention, the activity coefficient, $\gamma_i^{\bullet(m)}$, accounts for deviations from ideal solution behavior that arise from dilutional effects within the micelle as well as any free-energy terms required to account for physical differences between the intramicellar environment and pure liquid i, e.g., Laplace pressure effects.

It is convenient to compare the expressions for chemical potential and activity defined by Eqs. (13a, b) with analogous expressions that would apply if the Raoult's law reference state were taken to be pure liquid (i) subject to the same physical conditions as the hydrocarbon-like interior of a micelle:

$$\mu_i^{(m)} = \mu_i^{\ominus} + RT \ln a_i^{\ominus} \tag{14a}$$

$$a_i^{\ominus} = \frac{P_i}{P_i^{\ominus}} = \gamma_i^{\ominus(m)} X_i^{(m)} \tag{14b}$$

In these equations, $\mu_i^{\ominus}$ and $P_i^{\ominus}$ represent a hypothetical chemical potential and vapor pressure that pure liquid (i) would have if it were subject to the above-mentioned physical conditions. Under this latter convention, the activity coefficient $\gamma_i^{\ominus(m)}$ accounts for deviation from ideal solution behavior arising solely from changes in intramicellar composition. Thus, one can equate Eq. (13) with Eq. (14) to obtain an expression for the free energy difference, ΔG^R, between the two Raoult's law standard states:

$$\Delta G^R = (\mu_i^{\ominus} - \mu_i^{\bullet}) = RT \ln\left(\frac{\gamma_i^{\bullet(m)}}{\gamma_i^{\ominus(m)}}\right) \tag{15}$$

It is clear from earlier discussions that the sites for gas solubilization within micelles exhibit many similarities to bulk solvents having similar chemical compositions. Thus, it is reasonable to expect that when nonpolar liquids are involved the solubilized liquids will exhibit ideal solution behavior when dissolved in micelles, so that $\gamma_i^{\ominus(m)} \cong 1$. Therefore, for the systems of most interest here, which involve the solubilization of liquid alkanes, Eq. (15) can be rewritten as

$$\gamma_i^{\bullet(m)} = \frac{P_i}{X_i^{(m)} P_i^{\bullet}} = \exp\left(\frac{\Delta G^R}{RT}\right) \tag{16}$$

The substitution of $X_i^{(m)(\text{ideal})} = P_i/P_i^{\bullet}$ into Eq. (16) establishes the formal analogy between Eq. (16), which is based on the Raoult's law convention, and Eq. (7a), which is based on the Henry's law convention for standard states.

As noted, vapor pressure–solubility measurements have been made for a wide variety of liquid/surfactant systems. The micellar solubility data are commonly reported as values of the activity coefficient, $\gamma_i^{*(m)}$, at the various micellar compositions, X_i^m. While micellar solutions of hydrophilic liquids such as ethers [22] and long-chain alcohols [21, 27] exhibit a variety of unusual features, presumably reflecting very specific micelle-solubilizate interactions, the results obtained with nonpolar liquid hydrocarbons generally exhibit the same trends, irrespective of the specific nature of the surfactant. Namely, it is found that the degree of solubilization increases, i.e., $\gamma_i^{*(m)}$ decreases, as (a) the solubilizate concentration, X_i^m, increases, (b) as the surfactant concentration increases, and (c) as salt is added to the surfactant solutions. These trends suggest that micelle swelling as well as intermicellar interactions, and/or changes in micelle shape play an important role in determining the ability of micelles to absorb large solute molecules. As noted, these trends are for all intents and purposes absent in gas-micelle systems which involve much smaller molecules present at much lower intramicellar concentrations which seldom exceed single occupancy of a given micelle. Nevertheless, one might expect that vapor pressure–solubility data obtained for nonpolar liquids at low solubilizate and surfactant concentrations might exhibit some internal consistency with corresponding data for solubilized gases, and this indeed seems to be the case.

Table 3 summarizes the results of a group vapor pressure–solubility measurements performed at or about 25°C with three liquid hydrocarbons, *n*-pentane, cyclohexane, and *n*-hexane, dissolved in a variety of surfactant solutions. This table lists the values derived from liquid-micelle solubility data for the activity coefficient, $\gamma_i^{*(m)}$, at infinite dilution and the values of ΔG^R calculated for each system according to Eq. (16). Also included in this table are values for the molar volume of each liquid, V_i^*, and the Laplace pressures calculated for each solubilizate/micelle system according to the relationship

$$\Delta G^R = \Delta P_L V_i^* = \frac{2\gamma V_i^*}{r} \tag{17}$$

which results when the Laplace pressure model is applied to systems which are expected to exhibit ideal or near-ideal solution behavior. The last column of Table 3 contains values for the micelle/water interfacial tension, γ, from which the Laplace pressures are thought to arise. These are calculated from Eq. (17) using the same approximation used to analyze the data of Figs. 1a,b, namely that micelle radii are assumed to take on values of $r = (n_C/12)\cdot 1.5$ nm. The similarity between the values of the micelle-water interfacial tension required by the Laplace pressure model

TABLE 3 Activity Coefficients at Infinite Dilution for Nonpolar Liquids Solubilized in Anionic and Cationic Micelles with Associated Parameters

Solubilizate	Surfactant (Conc. (M))	T (°C)	Ref.	$\lim_{X_i \to 0} \gamma_i^{*(m)}$	ΔG^R (kJ mol^{-1})	V_i^* (m^3 mol^{-1}) $\times 10^6$ [a]	ΔP_L (atm)	n_C	γ (mN m^{-1})
Cyclohexane	SOS (0.247)	25	14	17.1	7.0	108	640	8	32
Cyclohexane	KLa (0.1 M)[b]	25	37	7.7(5)	5.1	108	460	11	32
n-Pentane	SDS (0.1 M)	22	23	6.8	4.7	116	400	12	30
Cyclohexane	SDS (0.1 M)	25	36	5.2(5)	4.1	108	380	12	28
n-Hexane	SDS (0.1 M)	25	36	7.5	5.0	131	380	12	28
Cyclohexane	CPC (0.1 M)[c]	25	36	3.5	3.1	108	280	16	28
n-hexane	CPC (0.1 M)	25	27,36	5.2	4.1	131	310	16	31
								Avg	30 ± 2

[a] Data from Refs. 52, 57.
[b] KLa = potassium laurate.
[c] CPC = cetylpyridinium chloride.

to account for the large $\gamma_i^{*(m)}$ found for liquid alkanes dissolved in hydrocarbon-type micelles and the values (29, 30 mN/m) obtained earlier from Fig. 1a for ethane in SDS and SOS using the same approximations for the micelle radius is striking. Although one must exercise great caution using such a simple model as a framework with which to draw conclusions, the close agreement between the γ values mentioned above and those derived from the more general analysis of the data in Fig. 2 suggests that the factors which determine the sorptive capacity of ionic type micelles toward gases are the same as those which determine the degree to which nonpolar liquids are solubilized.

In conclusion, note that this review has focused heavily, perhaps overly so, on an attempt to provide a unified basis for understanding the factors that control the extent to which gases, and by extension nonpolar liquids, are solubilized in dilute micellar solutions. Because of this, a number of very interesting research reports listed in Refs. 1–39 have been slighted or omitted altogether. These include papers by Somasundaran and Moudgil [8] and Zimmels and Metzer [9] which provide evidence for weak association between dissolved nonpolar gases and monomeric surfactant ions and/or premicellar aggregates present at concentrations well below the CMC. Likewise, the solubilities reported for gases in lipid bilayers by Miller and co-workers [10] constitute an important body of information which deserves more attention than it has received in this brief review. Mention should also be made of the spectroscopic studies by DellaGuardia and King [16] which use charge transfer spectra to demonstrate the mutual solubilization of a liquid and a gas within the same micelle. Finally, the work reported in Refs. 28 and 34 is worthy of note since they report gas solubilities measured for aqueous systems containing highly swollen micelles, i.e., water external microemulsions.

REFERENCES

1. J. W. McBain and J. J. O'Connor, *J. Am. Chem. Soc. 62*: 2855 (1940).
2. J. W. McBain and J. J. O'Conner, *J. Am. Chem. Soc. 63*: 875 (1941).
3. J. W. McBain and A. M. Soldate, *J. Am. Chem. Soc. 64*: 1556 (1942).
4. S. Ross and J. B. Hudson, *J. Colloid Sci. 12*: 523 (1957).
5. A. P. Brady and H. Huff, *J. Phys. Chem 62*: 644 (1958).
6. A. Wishnia. *J. Phys. Chem. 67*: 2079 (1963).
7. L. J. Winters and E. Grunwald, *J. Am. Chem. Soc. 87*: 4608 (1965).
8. P. Somasundaran and B. M. Moudgil, *J. Colloid Interface Sci. 47*: 290 (1974).
9. Y. Zimmels and J. Metzer, *J. Colloid Interface Sci. 57*: 75 (1976).
10. K. W. Miller, L. Hammond, and E. G. Porter, *Chem. Phys. Lipids 20*: 229 (1977).
11. I. B. C. Matheson and A. D. King, Jr., *J. Colloid Interface Sci. 66*: 464 (1978).
12. J. C. Hoskins and A. D. King, Jr., *J. Colloid Interface Sci. 82*: 260 (1981).
13. J. C. Hoskins and A. D. King, Jr., *J. Colloid Interface Sci. 82*: 264 (1981).
14. S. D. Christian, E. E. Tucker, and E. H. Lane, *J. Colloid Interface Sci. 84*: 423 (1981).
15. E. E. Tucker and S. D. Christian, *Faraday Symp. Chem. Soc. 17*: 11 (1982).
16. L. DellaGuardia and A. D. King, Jr., *J. Colloid Interface Sci. 88*: 8 (1982).
17. P. L. Bolden, J. C. Hoskins, and A. D. King, Jr., *J. Colloid Interface Sci. 91*: 454 (1983).
18. A. Ben-Naim and J. Wilf., *J. Solution Chem. 12*: 671 (1983).
19. A. Ben-Naim and J. Wilf, *J. Solution Chem. 12*: 861 (1983).
20. D. W. Ownby and A. D. King, Jr., *J. Colloid Interface Sci. 101*: 271 (1984).
21. E. B. Abuin, E. Valenzuela, and E. A. Lissi, *J. Colloid Interface Sci. 101*: 401 (1984).
22. M. Fromon, A. K. Chattopadhyay, and C. Treiner, *J. Colloid Interface Sci. 102*: 14 (1984).
23. E. Valenzuela, E. B. Abuin, and E. A. Lissi, *J. Colloid Interface Sci. 102*: 46 (1984).
24. W. Prapaitrakul and A. D. King, Jr., *J. Colloid Interface Sci. 106*: 186 (1985).
25. A. Ben-Naim and R. Battino, *J. Solution Chem. 14*: 245 (1985).
26. E. E. Tucker and S. D. Christian, *J. Colloid Interface Sci. 104*: 562 (1985).
27. S. D. Christian, E. A. Tucker, G. A. Smith, and D. S. Bushong, *J. Colloid Interface Sci. 113*: 439 (1986).
28. H. L. Flanagan, P. L. Bolden, and A. D. King, Jr., *J. Colloid Interface Sci. 109*: 243 (1986).
29. W. Prapaitrakul and A. D. King, Jr., *J. Colloid Interface Sci. 112*: 387 (1986).
30. W. Prapaitrakul, A. Shwikhat, and A. D. King, Jr., *J. Colloid Interface Sci. 115*: 443 (1987).

31. W. Prapaitrakul and A. D. King, Jr., *J. Colloid Interface Sci. 118*: 224 (1987).
32. N. Nugara, W. Prapaitrakul, and A. D. King, Jr., *J. Colloid Interface Sci. 120*: 118 (1987).
33. G. A. Smith, S. D. Christian, E. A. Tucker, and J. F. Scamehorn, in *Ordered Media in Separations* (W. L. Hinze and D. W. Armstrong, eds.), ACS Symp. Ser. **342**, Amer. Chem. Soc., Washington, DC, 1987, p. 184.
34. A. R. Nanda, N. Nugara, H. L. Flanagan, and A. D. King, Jr., *J. Colloid Interface Sci. 123*: 437 (1988).
35. D. W. Ownby, W. Prapaitrakul, and A. D. King, Jr., *J. Colloid Interface Sci. 125*: 526 (1988).
36. G. A. Smith, S. D. Christian, E. E. Tucker, and J. F. Scamehorn. *J. Colloid Interface Sci. 130*: 254 (1989).
37. F. Z. Mahmoud, W. S. Higazy, S. D. Christian, E. E. Tucker, and A. A. Taha., *J. Colloid Interface Sci. 131*: 96 (1989).
38. A. D. King, Jr., *J. Colloid Interface Sci. 137*: 577 (1990).
39. A. D. King, Jr., *J. Colloid Interface Sci. 148*: 142 (1992).
40. J. H. Hildebrand and R. L. Scott, *The Solubility of Nonelectrolytes*, Van Nostrand Reinhold, New York, 1950; reprinted by Dover, New York, 1964.
41. J. H. Hildebrand, J. M. Prausnitz, and R. L. Scott, *Regular and Related Solutions* Van Nostrand Reinhold, New York, 1970.
42. J. H. Hildebrand and R. H. Lamoreaux, *Ind. Eng. Chem. Fundam. 13*: 110 (1974).
43. R. G. Linford and D. G. T. Thornhill, *J. Appl. Chem. Biotechnol. 27*: 479 (1977).
44. F. M. Menger, *Acc. Chem. Res. 12*: 111 (1979).
45. F. M. Menger, *J. Am. Chem. Soc. 106*: 1109 (1984).
46. K. A. Dill and P. J. Flory, *Proc. Nat. Acad. Sci. USA 78*: 676 (1981).
47. P. Mukerjee. *Kolloid Z. Z. Polym. 236*: 76 (1970).
48. P. Mukerjee, in *Solution Chemistry of Surfactants* (K. L. Mittal, ed.), Plenum, New York, Vol. 1, p. 153.
49. W. Y. Ng and J. Walkley, *J. Phys. Chem. 73*: 2274 (1969).
50. J. Chr. Gjaldbaek and J. H. Hildebrand. *J. Am. Chem. Soc. 72*: 1077 (1950).
51. W. Hayduk and R. Castaneda, *Can. J. Chem. Eng. 51*: 353 (1973).
52. R. R. Dreisbach, *Physical Properties of Chemical Compounds. II*, Advances in Chemistry Series, No. 22, Amer. Chem. Soc., Washington, DC, 1959.
53. N. Shachat and H. L. Greenwald, in *Nonionic Surfactants* (M. J. Schick, ed.), Marcel Dekker, New York, 1967, Vol. 1, Chap. 2.
54. J. Makranczy, L. Rusz, and K. Balog-Megyery, *Hung. J. Ind. Chem. 7*: 41 (1979).
55. F. L. Boyer and L. J. Bircher, *J. Phys. Chem. 64*: 1330 (1960).
56. Unpublished data, measured by the author.
57. R. R. Dreisbach, *Physical Properties of Chemical Compounds*, Advances in Chemistry Series, No. 15, Amer. Chem. Soc., Washington, DC, 1955.

3

Thermodynamics of Solubilization of Polar Additives in Micellar Solutions

ROSARIO DE LISI and STEFANIA MILIOTO Department of Physical Chemistry, University of Palermo, Palermo, Italy

SYNOPSIS

Thermodynamic properties are very useful tools for the understanding the interactions governing mixed micelles formation. The main topic discussed in the present chapter concerns with the mathematical modeling of thermodynamic properties for water-surfactant-additive systems. The solubilizing effect of micelles toward polar additives are easily obtained by applying simple models to a given thermodynamic property of the additive as a function of the surfactant concentration at fixed and low additive concentration. From the measured property, the distribution constant of the additive between the aqueous and the micellar phases and the corresponding property for the additive in the micellar phase are simultaneously obtained. Information on micelle-additive interactions, which can be drawn from these properties, is briefly summarized as it depends on the nature of both the additive and surfactant.

I. INTRODUCTION

From a practical point of view, surfactants are important mainly because of their ability to solubilize organic substances in water. As a consequence, the solubilization of surfactants is relevant in many fields, such as enhanced oil recovery [1, 2], cosmetics [3], micellar catalysis [4], and so on.

In the last 10 years or so, there has been an enormous increase in the number of structural and thermodynamic studies published in this field. Structural investigations give information on the effect of additives on the aggregation number, size, shape, and degree of ionization (in the case of ionic surfactants) of the micelles. Thermodynamics can give information on additive-surfactant interactions governing mixed micelle formation. Our state of knowledge on water-surfactant-additive ternary systems is summarized in many fine and sound monographs and review papers

[5–11]. Because of the great number of such papers, we are inevitably neglecting some and we apologize for this.

Much of the progress achieved on the understanding of the thermodynamics of solubilization of additives in micellar solutions has been derived from the standard free energy of transfer of the additive from water to the micellar phase. Many experimental techniques [12–14], in fact, permit the determination of the fraction of the additive present in either phases. The distribution constant, K, can be also evaluated from the dependence of the cmc [15, 16] and of the counterion concentration [17] on the additive concentration, from kinetic studies [18–20], from solubilities [21, 22], from vapor pressure measurements [23], from Krafft point depression [24], and so on. If the study is carried out as a function of temperature and pressure, the volumes and enthalpies of transfer can be calculated from the first derivative of the Gibbs free energy with respect to pressure and temperature, respectively, while the compressibilities and the heat capacities of transfer from second derivatives. Since the K values are generally evaluated with a relatively large uncertainty and since the change in the properties of transfer is small, this method often yields results which, whenever comparisons are possible, can disagree with each other. In addition, this approach does not distinguish between the interactions in the aqueous phase and those in the micellar phase. Direct measurements are more promising and lead to a more complete understanding of these systems.

Thermodynamic studies of solubilization of additives in micellar solutions can be performed in different ways. For instance, studies are carried out by either increasing the additive or the surfactant concentration at a given concentration of the other, or by fixing the additive/surfactant ratio and by changing the total concentration, and so on. Each of these methods can be considered appropriate depending on the information sought. The simplest and easiest way to approach the thermodynamics of solubilization of the additive in micellar solution is to study the standard thermodynamic property of the additive as a function of the surfactant concentration.

Standard thermodynamic properties, such as volume, heat capacity, expansibility, and compressibility, reflect both the intrinsic property due to the geometry of the solute and the change of the property due to the interactions between the solute particles and their local environment. So, if the above properties of the additive in the micellar phase are known, by comparing them with the corresponding properties in other solvents, conclusions about interactions between micelles and additives can be drawn. For example, the partial molar volume of n-heptane at 298 K [25] is 147.1 and 132.3 $cm^3\ mol^{-1}$ in sodium dodecanoate micellar phase and in water, respectively, while the molar volume of the pure liquid alkane is 147.5 $cm^3\ mol^{-1}$; that for 1-hexanol at 303 K [26] is 124.7, 119.0, and

131.4 cm^3 mol^{-1} in sodium decanoate micellar phase, in water, and in heptane, respectively, while the molar volume is 125.9 cm^3 mol^{-1}. These results can be explained in terms of different sites of solubilization of heptane and hexanol in the micelle: heptane, having a molar volume equal to the partial molar volume in the micellar phase, is located in the micellar core, which possesses the feature of liquid alkanes, while hexanol is located in the palisade layer, since the additive-micelle interactions resemble those in the pure liquid alcohol.

Other thermodynamic properties, such as free energy, enthalpy, and entropy, can be evaluated as excess properties with respect to a given reference state. Therefore, if the thermodynamic properties of solution are considered, the additive in its pure liquid state may be taken as the standard state. Since the energetics dealing with the liquid phase of the additive can play an important role, attention must be paid to interactions in pure liquids when the thermodynamic properties of solutions of different additives are compared. On the other hand, water-surfactant-additive ternary systems are often discussed by comparing the properties of transfer of the additive from water to the micellar phase or, in other words, by taking pure water as reference state. In this case, the solute interactions between unlike solute molecules in the dispersed phase can greatly affect the results especially in the case of the enthalpy whose value is often a few kJ mol^{-1}, free energy and entropy generally being one order of magnitude greater.

Another difficulty arising in the thermodynamic studies of these systems is that the measured properties depend on all the species present in solution and, therefore, the corresponding properties of the additive in the micellar phase can be extracted only through models. In spite of this problem, thermodynamic properties are very useful tools for the understanding these systems, since such properties can be measured with a very good accuracy (at concentrations higher than 0.01 m) and are suitable for mathematical modeling.

In this chapter, neither the treatment of the thermodynamics of solutions, which is adequately treated in many textbooks [27–29], nor experimental techniques, which are reported in the quoted references, will be dealt with. We will concentrate on several approaches of extracting the thermodynamic properties in the micellar phase for polar solutes penetrating micelles from the corresponding bulk properties and on the information which can be drawn from them.

II. THERMODYNAMIC APPROACH

A realistic model describing a given thermodynamic property of an additive in a micellar solution should correctly predict the following contributions: the distribution of the additive between the aqueous and the micellar phases, the shift of the micellization equilibrium due to the additive, the additive-surfactant interactions in the aqueous and the micellar phases, and the effect of the additive on the physicochemical properties of the micelles, i.e., the variation of the aggregation number, the shape, and the degree of ionization.

At the present time, no model accounting for all these effects has been proposed; the present limitations, probably, are not important since an improvement of the models would probably require very involved equations from which, such as is often the case, it would be difficult to extract the quantities of interest. Therefore, in order to minimize the approximations introduced in the model, it is preferable to set up the best experimental approach. It has been mentioned that the most appropriate thermodynamic approach consists of treating the standard (infinite dilution) partial molar properties of the additive so that one can reasonably assume that the additive behaves as a probe not affecting the properties of the micelles [22]. The determination of the standard property of the additive as a function of the surfactant concentration, m_S, is a time-consuming method since it implies that, at each given m_S, measurements are carried out as a function of the additive concentration in order to obtain, by extrapolation, the value for the additive concentration tending to zero; otherwise, one can measure a thermodynamic property at a fixed, sufficiently low, additive concentration so that the derived property can be reasonably assumed to be at infinite dilution.

At a fixed stoichiometric concentration of the additive, by increasing the surfactant concentration, the distribution of the additive between the aqueous and the micellar phases yields a progressive extraction of the additive from the aqueous phase and its solubilization into the micellar phase. For $m_S \rightarrow \infty$, the additive is completely solubilized in the micellar phase where its concentration tends to zero and, therefore, the measured thermodynamic property corresponds to that in the micellar phase. In other words, at high m_S values, no appreciable changes are predicted for the bulk properties which are close to the standard ones for the additive in the micellar phase. For $m_S \rightarrow$ cmc, the thermodynamic properties strongly depend on the surfactant concentration being essentially correlated to the change of the fraction of the additive solubilized in the two phases, i.e., to the distribution constant. Therefore, by studying the partial molar properties in or near the standard state as functions of the surfactant concentration, the corresponding property of the additive in the micellar phase and the distribution constant can be simultaneously evaluated. In addition, as mentioned, under these conditions, the effect of the additive on the physicochemical properties of the micelles can be neglected and the additive distribution, the micellization equilibrium shift due to the additive and the additive-surfactant interactions in both the aqueous and the micellar phases contribute to the bulk property.

The simplest way to extract the thermodynamic property of the additive solubilized in the micellar phase from the bulk one, Y_R, is based on the phase transition model for both the micellization and the additive distribution. By indicating with N_f and N_b the fraction of the additive in the dispersed and in the micellized phases, respectively, and with Y_f and Y_b the property of the additive in the two phases, the following equation can be written:

$$Y_R = N_f Y_f + N_b Y_b \tag{1}$$

By indicating with K_N the ratio N_b/N_f and since $N_f + N_b = 1$, it follows that

$$N_f = \frac{1}{1 + K_N} \tag{2}$$

To a first approximation, the property in the aqueous phase can be assumed equal to that in pure water Y_w so that from Eqs. (1) and (2), one obtains

$$Y_R = Y_b + (Y_f - Y_b)\frac{1}{1 + K_N} \approx Y_b + (Y_w - Y_b)\frac{1}{1 + K_N} \tag{3}$$

Actually, Eq. (3), where the approximation $Y_w \approx Y_f$ has been introduced, does not account for the additive effect on micellization. Moreover, Y_R cannot be simulated as a function of m_S from which K_N depends. In fact, as a normal distribution between two immiscible phases, for a given total additive concentration, the amount of the additive solubilized in the two phases depends on their mass ratio. However, since N_f and/or N_b can be experimentally measured, whenever K_N is known at the surfactant concentration of interest, Y_b can be easily evaluated using Eq. (3).

A more correct thermodynamic approach is obtained by considering equations correlating a given total thermodynamic property of the water-surfactant binary system and of water-surfactant additive ternary system to the partial molar properties of all the present species. At this point, the assistance of models for the micellization and for the additive distribution is not required, while one needs to assume that the thermodynamic property of the micellized surfactant does not depend on the structural properties (aggregation number, shape, size) of the micelle; for the same reason, the symbols $[m_0]$ and $[m]$ are used to indicate the unmicellized surfactant concentration in the absence and in the presence of the additive, respectively. Therefore, by considering, for the sake of simplicity, a thermodynamic property, the first derivative of the Gibbs free energy, according to Euler's theorem, the total property for the water-surfactant, Y_B, and water-surfactant-additive, Y_T, systems can be written as

$$Y_B = 55.5Y_{H_2O} + [m_0]Y_m + (m_S - [m_0])Y_S \tag{4}$$

$$Y_T = 55.5Y'_{H_2O} + [m]Y'_m + (m_S - [m])Y'_S + m_bY_b + m_fY_f \tag{5}$$

where Y_{H_2O} and Y'_{H_2O} are the partial molar properties of water in the absence and in the presence of the additive, respectively; Y_m and Y'_m are the partial molar properties of the surfactant in the unmicellized state in the absence and the presence of additive, respectively; Y_S and Y'_S are the partial molar properties of the surfactant in the micellized state in the absence and the presence of additive, respectively; m_f and m_b are the moles of additive per kilogram of water in the aqueous and in the micellar phases, respectively.

Since the apparent molar property of the additive $Y_{\Phi,R}$ is given by

$$Y_{\Phi,R} = \frac{Y_T - Y_B}{m_R} \tag{6}$$

by assuming $Y_{H_2O} = Y'_{H_2O}$, from Eqs. (4) and (5) one obtains

$$Y_{\Phi,R} = \frac{[m_0]\Delta Y_m - [m]\Delta Y'_m}{m_R} + \frac{m_S(Y'_S - Y_S)}{m_R} + \frac{m_bY_b}{m_R} + \frac{m_fY_f}{m_R} \tag{7}$$

where $\Delta Y_m = Y_S - Y_m$ and $\Delta Y'_m = Y'_S - Y'_m$ correspond to the micellization properties in the absence and the presence of the additive, respectively.

If the additive concentration tends to zero, the first two terms at the right side of Eq. (7) are undetermined. However, as for instance plots in Figs. 1 and 2 show, one can reasonably assess

$$\lim_{m_R \to 0} \Delta Y'_m = \Delta Y_m \qquad \lim_{m_R \to 0} \frac{Y'_S - Y_S}{m_R} = 0 \tag{8}$$

Moreover, by considering that $m_b/m_R = N_b$ and $m_f/m_R = N_f$, Eq. (7) can be written in the form

$$Y_R = \frac{([m_0] - [m])\Delta Y_m}{m_R} + N_bY_b + N_fY_f \tag{9}$$

where, according to the $m_R \to 0$ hypothesis, the standard partial molar property of the additive in the bulk Y_R has been introduced instead of the apparent one. Since $N_b = 1 - N_f$, Eq. (9) becomes

$$Y_R = Y_b + \frac{([m_0] - [m])\Delta Y_m}{m_R} + N_f(Y_f - Y_b) \tag{10}$$

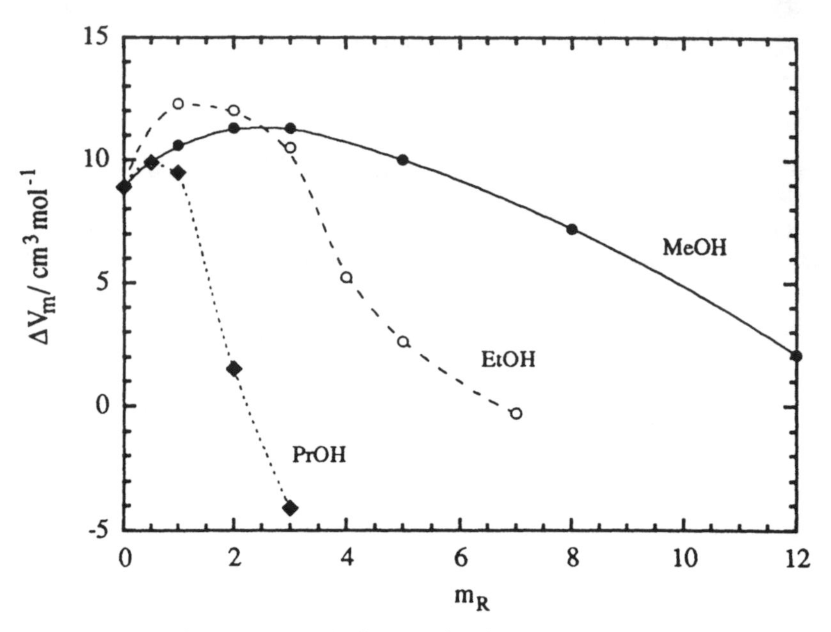

FIG. 1 Plots of the volume of micellization for sodium decanoate in water-alcohol mixtures as a function of the alcohol concentration. (Experimental data from Ref. 63.)

which shows that Y_R includes contributions from the corresponding property in the micellar phase, Y_b, the micellization shift due to the presence of the additive $([m_0] - [m])\Delta Y_m/m_R$ and the additive distribution between the aqueous and the micellar phases $N_f(Y_f - Y_b)$.

Alternatively, by correlating Y_f to the corresponding property in pure water, Y_w, through the McMillan-Mayer approach [30]

$$Y_f = Y_w + 2Y_{RS}\,[m] \tag{11}$$

where only the pair surfactant-additive interaction parameter Y_{RS} has been considered, Eq. (9) can be written in terms of the transfer property of the additive from water to the surfactant solutions $\Delta Y_R(\mathrm{W} \rightarrow \mathrm{W} + \mathrm{S})$:

$$\Delta Y_R(\mathrm{W} \rightarrow \mathrm{W} + \mathrm{S}) = 2Y_{RS}\,[m] + \frac{([m_0] - [m])\Delta Y_m}{m_R} + N_b(Y_b - Y_f) \tag{12}$$

Note that Eqs. (10) and (12) are the same whenever Eq. (11) is valid. Both predict a contribution due to the shift of the micellization equilibrium

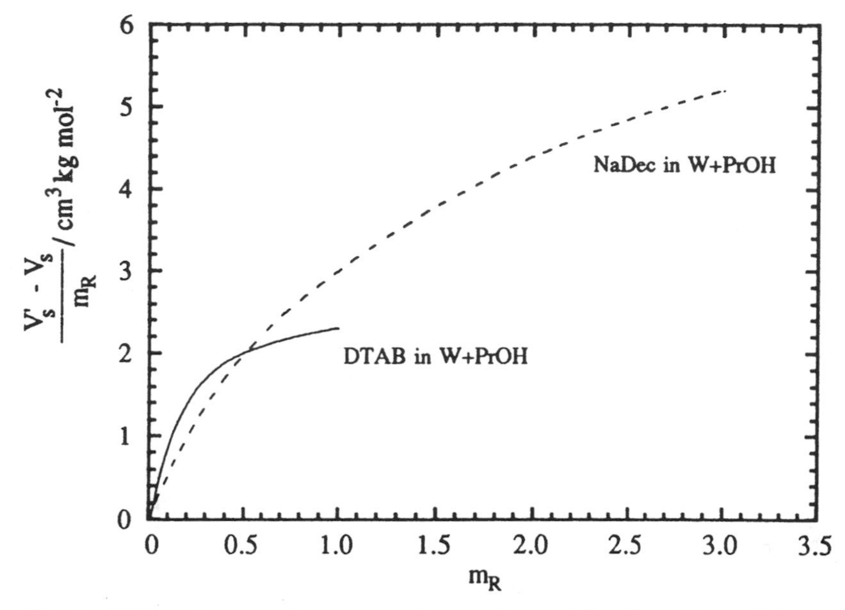

FIG. 2 Effect of propanol concentration on the partial molar volume of the surfactant in the micellized form. (Data for dodecyltrimethylammonium bromide from Ref. 61; data for sodium decanoate from Ref. 63.)

because of the addition of the third component. This is directly proportional to the micellization property and to the change of the unmicellized surfactant concentration and inversely proportional to the additive concentration.

The simulation of experimental data by means of the above equations requires a model for the evaluation of the distribution constant and of the dispersed surfactant concentration. In both cases, mass action and partition models can be used.

A. Micellization

The pseudo phase transition model assumes that for nonionic surfactants the monomer concentration, $[m_0]$, called the critical micellar concentration (cmc), is constant and that the micellized surfactant concentration is m_S − cmc.

According to the two-state equilibrium model for nonionic surfactants [31], the concentrations of the monomeric and the micellized surfactant, whose activity coefficients are taken as unity, depend on the stoichiometry through the micellization constant K_M and the aggregation number n as

$$K_M = \frac{m_S - [m_0]}{n[m_0]^n} \tag{13}$$

In order to interpret Eq. (13), it has been considered that a plot of the fraction of the free monomers as a function of the stoichiometric surfactant concentration has an inflection point, which is an operational definition of the critical micellar concentration. Since the free monomers fraction at this point depends on the aggregation number only, starting from an approximate value of n, for a given stoichiometric concentration, $[m_0]$ can be calculated by successive iterations and used to fit experimental thermodynamic data. The procedure is repeated until the best fit is obtained.

For ionic surfactants the above model is not as successful [32] and the Gunnarson et al.[33] and the Burchfield and Woolley [34] models can be used. However, the former has been applied to the water-sodium dodecylsulfate-pentanol system only [35], while the latter has not been applied to ternary systems.

As far as the thermodynamic properties of micellization ΔY_m are concerned, in the case of the mass action model they are derived from the correlations discussed above, while in the case of the pseudophase transition model they are evaluated as the difference at the cmc between the extrapolated values from the trends in the post- and premicellar regions of the correponding partial molar property as a function of concentration [36]. Note that, although the micellization properties obtained by the latter approach are sometimes difficult to evaluate because of nonregular trends of the partial molar property as a function of m_S, often they yield values which are in a very good agreement with those obtained using mass action models [37].

B. Additive Distribution

It has been reported [38] that for an additive distributing between two immiscible phases, the free energy of transfer has a less questionable physical meaning if the partition constant is expressed in the molarity scale K_C. By assuming that the volume of the aqueous phase is equal to that of water and by indicating with V_M the volume of the micellar phase per liter of water, K_C is correlated to K_N by

$$K_C = \frac{n_b}{n_f V_M} = \frac{N_b}{N_f V_M} = \frac{K_N}{V_M} \tag{14}$$

V_M can be evaluated as the product between the micellized surfactant concentration and its partial molar volume V'_S:

$$V_M = (m_S - [m])V'_S \tag{15}$$

If the additive concentration is low, $[m] = [m_0]$ and $V'_S = V_S$ and, therefore, by combining Eqs. (2), (14), and (15), the fraction of the additive in the aqueous and in the micellar phases is correlated to the surfactant concentration by

$$N_f = \frac{1}{1 + K(m_S - [m_0])} \qquad N_b = \frac{K(m_S - [m_0])}{1 + K(m_S - [m_0])} \tag{16}$$

where K, usually defined as a binding constant, is equal to $K_C V_S$. The $[m_0]$ value depends on the model used for the micellization process. It is a constant quantity, i.e., the cmc, in the case of the pseudo phase transition model, while it depends on the surfactant concentration in the case of the mass action model. However, it is known that the surfactant concentration in the dispersed form does not appreciably change with the stoichiometric concentration even in the case of the mass action model so that the use of the cmc instead of $[m_0]$ is reasonable.

The same equation can be obtained by an equilibrium model for the distribution, assuming that the solubilization of the additive molecules into the micelles is unaffected by other molecules already solubilized, the additive does not affect the physicochemical properties of the micelles, and the equilibrium constant of each step is given by ratio between that of the first step and the number of additive molecules bonded to the micelles.

Alternatively, the distribution constant can be expressed in the mole fraction scale, K_X. In this case, the mole number of the additive and the surfactant in the aqueous phase can be neglected with respect to that of water and, therefore, we can write

$$\begin{aligned} K_X &= 55.51 \frac{m_b}{m_f \{m_b + (m_S - [m_0])\}} \\ &= 55.51 K_N \frac{1}{N_b m_R + (m_S - [m_0])} \end{aligned} \tag{17}$$

Since $K_X/55.51$ is K, from Eq. (2) one obtains

$$N_f = \frac{1}{1 + K\{(1 - N_f)m_R + (m_S - [m_0])\}} \tag{18}$$

which corresponds to Eq. (16) whenever the additive concentration is negligible with respect to that of the surfactant. Actually, since the fixed low additive concentration is finite, there is a difference between the N_f

values calculated through Eqs. (16) and (18). For example, if the additive and the micellized surfactant concentrations are 0.03 m and the K value is 10 (which more or less corresponds to that of pentanol) N_f calculated according to Eq. (16) is 0.77, while that calculated according to Eq. (18) is 0.73. As shown in Fig. 3, the difference tends to slowly vanish with increasing m_S and it increases with K and the additive concentration. In other words, small differences in the K values are predicted if either Eqs. (16) or Eq. (18) are introduced in Eqs. (10) or (12) to fit experimental data.

C. Micellization Shift Contribution

At the present time, there are two methods to rationalize the shift of micellization equilibrium due to the presence of the additive. Desnoyers and co-workers [39, 40] have expanded their equilibrium model for micellization to account for the presence of the additive. The reported equations

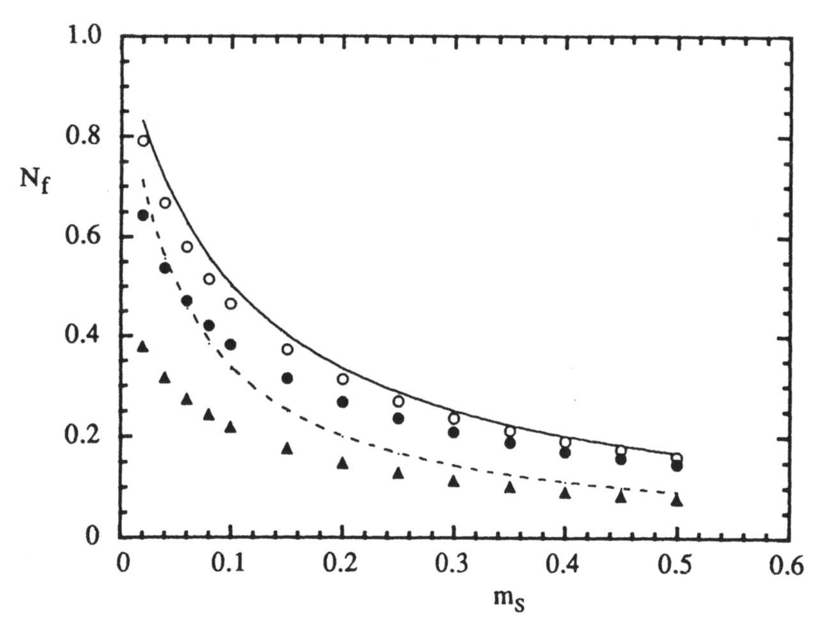

FIG. 3 Dependence of the fraction of the additive in the aqueous phase on the micellized surfactant concentration: (—) from Eq. (16) using $K = 10$ and (--) $K = 20$; (○) from Eq. (18) at $m_R = 0.03$ using $K = 10$; (●) from Eq. (18) at $m_R = 0.1$ using $K = 10$ and (▲) $K = 20$.

for micellization and distribution are

$$-\frac{1}{n}\ln K_M = \ln [m] - \ln \frac{m_S - [m]}{n} + \frac{m_b}{m_b + (m_S - [m])} \tag{19}$$

and

$$-\ln K = -1 + \frac{m_b}{m_b + (m_S - [m])} + \ln \frac{m_f}{m_b}\{m_b + (m_S - [m])\} \tag{20}$$

If K is known, provided that K_M, n, ΔY_m, and $[m_0]$ are evaluated from the fit of the water-surfactant binary system, N_f and $[m]$ for a given pair of m_S and m_R can be derived from Eqs. (19) and (20) by successive approximations. Therefore, the micellization shift contribution can be calculated. Figure 4 shows the dependence of this contribution on the surfactant concentration for pentanol in octyldimethylamine oxide ODAO calculated from literature data [40].

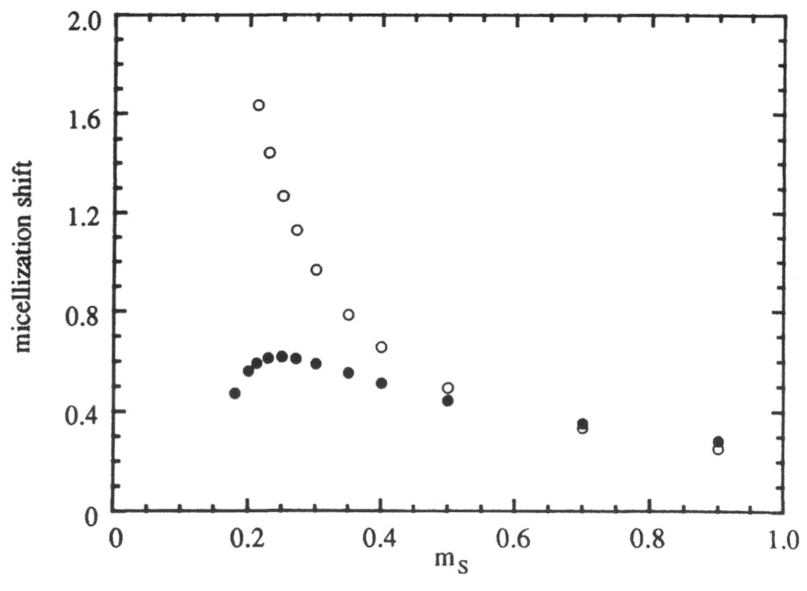

FIG. 4 Effect of additive on the unmicellized surfactant concentration for pentanol in octyldimethylamine oxide. (○) De Lisi et al. model [41]; (●) model of Desnoyers and co-workers [39]. K, K_M, and n from Ref. 40.

The second approach [41] has been developed revising the equation used by Treiner [15] to evaluate the distribution constant of the additive from the dependence of the cmc on the additive concentration. The following equation has been obtained:

$$\ln \frac{[\text{cmc}]}{[m]} = \frac{m_R}{1+\nu}\left\{\frac{2.3K_S}{1+K(m_S-[\text{cmc}])} + \frac{(1+\beta)K}{1+Km_R+K(m_S-[\text{cmc}])}\right\} \tag{21}$$

where K_S is the Setchenov constant, ν and β are, respectively, the degrees of ionization of the surfactant in the unmicellized and micellized states. Calling A the quantity in braces at the right side of Eq. (21), the series expansion of Eq. (21) gives the effect of the additive on the micellization shift

$$\frac{[\text{cmc}] - [m]}{m_R} = A - \frac{A^2 m_R}{2(1+\nu)} + \ldots \tag{22}$$

In Fig. 4, the micellization shift contribution for pentanol in ODAO, calculated by means of Eq. (22), is compared to that evaluated according to the model of Desnoyers and co-workers [39, 40]. As can be seen, both models yield values which agree with each other for micellized surfactant concentrations greater than 2 times the cmc. For lower concentrations, qualitative and quantitative disagrements are observed; in fact, the model of Desnoyers and co-workers predicts a maximum, while the second one does not. The reported system can be considered as one of the worst examples since the high cmc of the surfactant makes the pseudophase transition model inadequate. Accordingly, for surfactants having low cmc, the agreement is extended to concentrations very close to the cmc.

As a general feature, this contribution is not negligible with respect to that of distribution (see Eq. (10), especially when surfactants having high cmc are considered. For instance, for the above system at $m_S = 0.3$ m, the micellization shift contribution to volume is 3 or 5 $cm^3 \, mol^{-1}$, while the distribution contribution is 2 or 3 $cm^3 \, mol^{-1}$ (depending on the model). Obviously, the lower the cmc of the surfactant, the smaller the micellization shift contribution with respect to that of distribution is. In addition, the decreasing trend of the micellization shift contribution as a function of the surfactant concentration can explain the decrease, sometimes observed, of Y_R as m_S increases.

D. Fits of Experimental Data

In the first application of their model, Desnoyers and co-workers [39] assumed K to be equal to the reciprocal of the solubility of the additive in water; this quantity, together with the given thermodynamic property of the additive in its pure state instead of that in the micellar phase, has been introduced in Eq. (12) to simulate the thermodynamic functions of transfer. This approach (defined by the authors "ideal simulation"), which permits the prediction of the behavior of the water-surfactant-additive ternary systems on the basis of the properties of the binary water-surfactant and water-additive binary systems, is quite interesting but is seldom quantitatively valid. More recently, the model has been improved by considering both the thermodynamic property of the additive in the micellar phase and the distribution constant as adjustable parameters [40]. An example of ideal and adjusted simulations for the volume of transfer of hexanol from water to ODAO aqueous solutions is shown in Fig. 5.

As discussed above, the goal of these studies is to obtain reliable values for the distribution constant of the additive between the aqueous and the micellar phases and the property of the additive in the micellar phase. To solve Eq. (10), De Lisi et al. [41] used Eqs. (16) and (22) for N_f and the micellization equilibrium shift, respectively. The resulting equation was

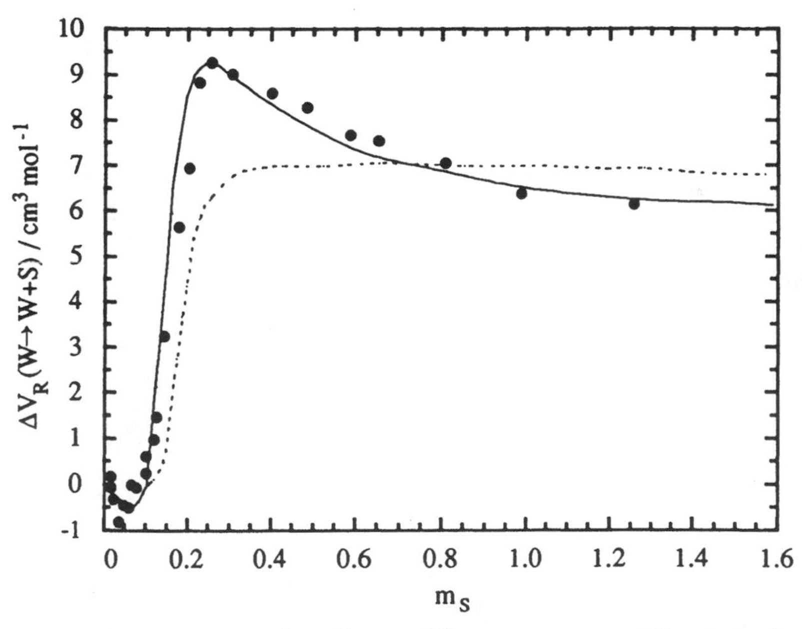

FIG. 5 Volume of transfer of hexanol from water to octyldimethylamine oxide aqueous solutions as a function of the surfactant concentration. (--) ideal simulation; (—) adjusted simulation according to the model of Desnoyers and co-workers [40]. (Data from Ref. 40.)

simplified by neglecting the quadratic term appearing in Eq. (22). For low additive concentration, this approximation involves an uncertainty which in the worst condition (i.e., for short-chain surfactants as ODAO) is about 10% of the value of the micellization shift contribution, but it tends to zero for $m_R \rightarrow 0$. Thus, from Eqs. (16) and (22) one obtains

$$\lim_{m_R \rightarrow 0} \frac{[\text{cmc}] - [m]}{m_R} = A_{\text{cdc}} N_f \tag{23}$$

where

$$A_{\text{cdc}} = \frac{[\text{cmc}]}{1 + \nu} [2.3Ks + (1 + \beta)K] \tag{24}$$

Equation (23) introduced in Eq. (10) gives

$$Y_R = Y_b - [(Y_b - Y_f) - A_{\text{cdc}} \Delta Y_m] N_f \tag{25}$$

which shows that the uncertainty affecting the $A_{\text{cdc}} \Delta Y_m$ term will affect the property of the additive only in the aqueous phase. Equation (25) is a three-parameter equation (K, Y_b, and Y_f) which is solved by linear regression by adjusting the K value in order to minimize the standard deviation of Y_R versus N_f. Figure 6 shows the dependence of the standard deviation on the K values for the enthalpy of solution of pentanol in dodecyltrimethylammonium bromide micellar solutions and the best fit of experimental data [42].

Obviously, all the approximations involved in this method will affect the K values in some way. However, the reliability of this method has been tested by comparing the Gibbs free energy of transfer ΔG_t° values so obtained with those of literature determined using different techniques (see Table 1).

III. EQUATIONS

As said, Eq. (10) is valid for those properties which are first derivatives of the Gibbs free energy. For thermodynamic properties which are second derivatives, in the expressions correlating the property of the additive in micellar solution to the stoichiometric surfactant concentration, the contributions due to the effect of the intensive variable on the micellization and distribution equilibria also appear.

In the following, the equations correlating the thermodynamic properties to the surfactant concentration based on the De Lisi et al. approach [41] will be summarized.

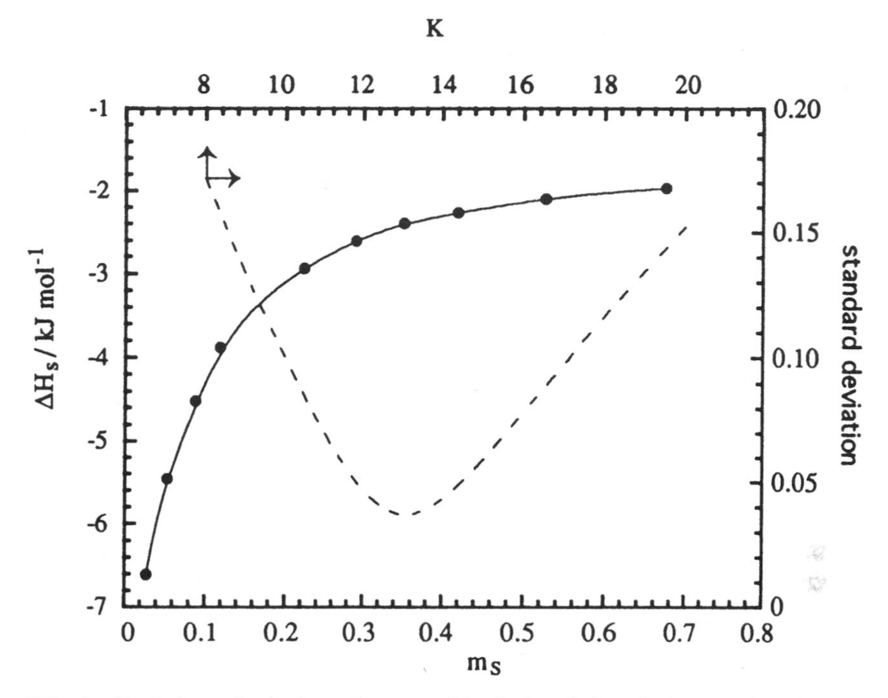

FIG. 6 Enthalpy of solution of pentanol in dodecyltrimethylammonium bromide aqueous solutions as a function of the surfactant concentration. (--) standard deviation as a function of K according to Eq. (27); (—) best fit. (Data from Ref. 42.)

A. Volume

When one considers the standard partial molar volume of the additive in the micellar solution as a function of the surfactant concentration Eq. (25) assumes the following form [41]:

$$V_R = V_b - [V_b - (V_f + A_{cdc}\Delta V_m)]N_f \tag{26}$$

where N_f is given by Eq. (16). From the minimizing procedure, the distribution constant and the partial molar volume of the additive in the micellar phase V_b are obtained; also, V_f can be calculated provided that the $A_{cdc}\Delta V_m$ term is known. Examples of the dependence of V_R on the surfactant concentration for some primary alcohols in sodium dodecylsulfate (NaDS) [43] are shown in Fig. 7, where the lines indicate best fits according to Eq. (26). As can be seen, as the alcohol hydrophobicity increases, the curvature is more pronounced because of the increase of the distribu-

TABLE 1 Standard Free Energy of Transfer (molarity scale) of Primary Alcohols from the Aqueous to the Micellar Phases of Dodecyltrimethylammonium Bromide (DTAB) and Sodium Dodecylsulfate (NaDS) at 298 K

PrOH	BuOH	PentOH	HexOH	HeptOH	Methods
			DTAB		
−1.7	−4.3	−7.1	−9.8		cmc determinations[a]
−4.51	−6.46	−8.67	−11.75	−13.62	Enthalpy of mixing[b]
			NaDS		
−2.9	−5.2	−6.4	−13.0		NMR PSGE self-diffusion[c]
−6.54	−7.99	−11.0	−13.4	−15.8	Enthalpy of mixing[d]
	−7.7	−9.8	−12.6	−15.1	Gas chromatography[e]
	−7.7	−9.7	−12.6	−15.2	Volume[f]
−4.9	−7.9	−10.4	−13.1		Volume[g]
	−8.7	−11.1			Volume[h]
			−13.1		Fluorescence[i]

Units are kJ mol^{-1}.
[a] From Ref. 15.
[b] From Ref. 44.
[c] From Ref. 12.
[d] From Ref. 70.
[e] From Ref. 14.
[f] From Ref. 91.
[g] From Ref. 40.
[h] From Ref. 43.
[i] From Ref. 92.

tion constant of the alcohol between the aqueous and the micellar phases; the V_R values are also larger because of the higher values of the volume of transfer of the alcohol from the aqueous to the micellar phases.

B. Enthalpy

The use of Eq. (25) for the enthalpy of solution gives

$$\Delta H_s = \Delta H_{s,b} - (\Delta H_t^\circ - A_{\text{cdc}} \Delta H_m) N_f \tag{27}$$

from which K and the standard enthalpy of solution of the additive in the micellar phase $\Delta H_{s,b}$ are obtained by least squares analysis; the standard enthalpy of transfer, ΔH_t°, of the additive from the aqueous to the micellar phases is derived from the slope, provided that $A_{\text{cdc}} \Delta H_m$ is calculated.

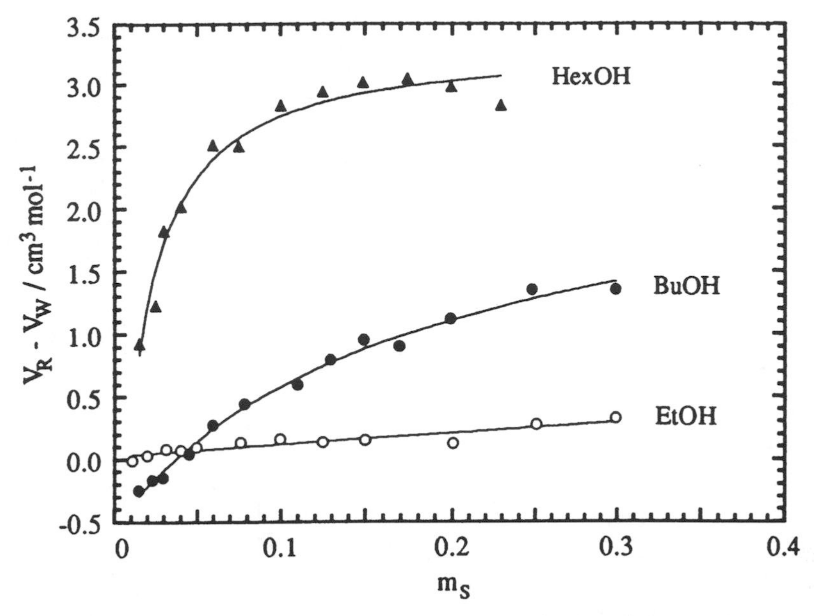

FIG. 7 Excess volume with respect to water of ethanol, butanol, and hexanol in sodium dodecylsulfate aqueous solution as a function of the surfactant concentration. The lines are the best fits of Eq. (26). (Data from Ref. 43.)

Moreover, from ΔG_t° (obtained through K) and ΔH_t° values, the standard entropy of transfer, ΔS_t°, can be calculated.

Figures 6 and 8 show the best fits of Eq. (27) to experimental data for pentanol and butanol, respectively, in dodecyltrimethylammonium bromide (DTAB) [42].

However, there is a simpler method to determine the same quantities. It consists [44] of recording the difference between the thermal effect arising in the mixing process of the surfactant solution with the additive solution and that for the dilution process of the same surfactant solution with water. The experimental quantity, ΔH^{exp}, corrected for the enthalpy of dilution of the additive in water, $\Delta H_{\text{id},R}^{w}$, corresponds to the enthalpy of transfer of the additive from water to the surfactant solution $\Delta H(\text{W} \rightarrow \text{W} + \text{S})$:

$$\Delta H(\text{W} \rightarrow \text{W} + \text{S}) = \Delta H^{\text{exp}} - \Delta H_{\text{id},R}^{w} \tag{28}$$

which is correlated to ΔH_s by means of the equation

$$\Delta H(\text{W} \rightarrow \text{W} + \text{S}) = \Delta H_s - \Delta H_{s,w} \tag{29}$$

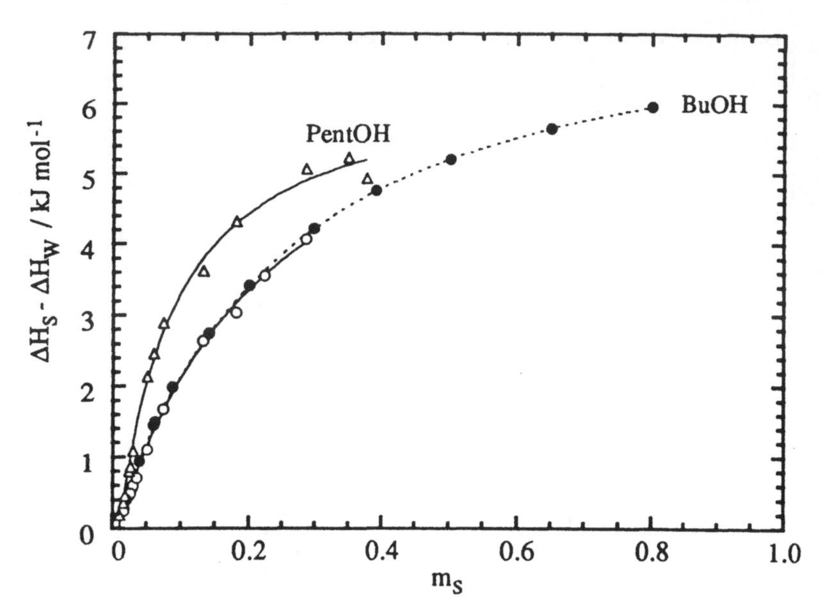

FIG. 8 Excess enthalpy of solution with respect to water of butanol and pentanol in dodecyltrimethylammonium bromide aqueous solution as a function of the surfactant concentration. (●) from enthalpy of solution (data from Ref. 42); (○, △) from enthalpy of mixing (data from Ref. 44); (—) best fit of Eq. (30); (--) best fit of Eq. (27).

By combining Eqs. (28) and (29), the following equation is obtained:

$$\begin{aligned} \Delta H(\mathrm{W} \rightarrow \mathrm{W} + \mathrm{S}) &= \Delta H^{\mathrm{exp}} - \Delta H^{m}_{\mathrm{id},R} \\ &= \Delta H_t - (\Delta H_t^{\circ} - A_{\mathrm{cdc}} \Delta H_m) N_f \end{aligned} \tag{30}$$

where ΔH_t is the standard enthalpy of transfer of the additive from pure water to the micellar phase.

In Fig. 8, $\Delta H(\mathrm{W} \rightarrow \mathrm{W} + \mathrm{S})$ values for butanol and pentanol in DTAB [44] obtained from the enthalpy of mixing (Eq. (28)) are plotted as a function of m_S. The lines are best fits of Eq. (30) to the experimental data. In the same figure, the enthalpy of transfer for butanol in DTAB obtained from the enthalpy of solution (Eq. (29)) and the corresponding best fit are also reported. The good agreement of the two series of data indicates that the two methods are equivalent.

C. Heat Capacity

By differentiating Eq. (25) with respect to temperature, the following equation for the heat capacity Cp_R is obtained:

$$Cp_R = Cp_b - \left[(Cp_b - Cp_f) - \Delta H_m \frac{\partial A_{cdc}}{\partial T} - A_{cdc}\Delta Cp_m\right] N_f + [A_{cdc}\Delta H_m - \Delta H_t^\circ] \frac{\partial N_f}{\partial T} \quad (31)$$

where ΔCp_m is the heat capacity change for the micellization process, while $\partial A_{cdc}/\partial T$ and $\partial N_f/\partial T$ represent, respectively, the effect of temperature on the shift of the micellization equilibrium and on the solute distribution. In order to make explicit these two contributions, the effect of temperature on the cmc, distribution constant, degree of ionization of the micelle, and the Setchenov constant must be taken into account. Accordingly, from Eqs. (16) and (24), one obtains

$$\frac{\partial N_f}{\partial T} = \frac{K(\partial \mathrm{cmc}/\partial T) - (m_S - \mathrm{cmc})(\partial K/\partial T)}{[1 + K(m_S - \mathrm{cmc})]^2} = \left\{K \frac{\partial \mathrm{cmc}}{\partial T} + \frac{\partial \ln K}{\partial T}\right\} N_f^2 - \frac{\partial \ln K}{\partial T} N_f \quad (32)$$

and

$$\frac{\partial A_{cdc}}{\partial T} = \frac{2.3Ks + (1 + \beta)K}{1 + \nu}\left(\frac{\partial \mathrm{cmc}}{\partial T}\right) + \frac{\mathrm{cmc}}{1 + \nu}\left\{(1 + \beta)\frac{\partial K}{\partial T} + K \frac{\partial \beta}{\partial T} + 2.3 \frac{\partial Ks}{\partial T}\right\} \quad (33)$$

In Eqs. (32) and (33), by excluding the $\partial\beta/\partial T$ term, which can only be obtained experimentally, the other derivatives can be related to fundamentals properties of the water-surfactant and water-surfactant-additive systems. In fact, on the basis of the pseudo phase transition model for micellization, the dependence of the cmc on T is given by

$$\frac{\partial \mathrm{cmc}}{\partial T} = -\mathrm{cmc}\left\{\frac{\ln \mathrm{cmc}}{1 + \beta}\frac{\partial \beta}{\partial T} + \frac{\Delta H_m}{(1 + \beta)RT^2}\right\} \quad (34)$$

where ΔH_m is the standard enthalpy of micellization and the cmc is expressed in the mole fraction scale.

The effect of temperature on K is obtained by applying the vant' Hoff equation to the partition constant in the molarity scale K_C, equal to K/V_S:

$$\frac{\partial K}{\partial T} = K\frac{\Delta H_t^\circ}{RT^2} + K\frac{E_S}{V_S} \tag{35}$$

where $E_S = \partial V_S/\partial T$ is the partial molar expansibility of the micellized surfactant.

Finally, since Ks is related to the additive-surfactant pair interaction parameter for the free energy, its dependence on temperature is given by

$$\frac{\partial Ks}{\partial T} = -\frac{2H_{RS}}{RT^2} \tag{36}$$

where H_{RS} is the corresponding parameter for the enthalpy which can be evaluated from studies in the premicellar region.

Since the derivatives in Eqs. (33)–(36) do not depend on the surfactant concentration and Eq. (32) is a quadratic function of N_f, Eq. (31) can be written in the form

$$Cp_R = Cp_b + C_1N_f + C_2N_f^2 \tag{37}$$

where C_1 and C_2 contain the appropriate constants (see above). By solving Eq. (37) with a nonlinear least squares method, the best values of K and Cp_b are obtained. Obviously, more accurate values are obtained if the constant quantities in Eqs. (32)–(36) are known and introduced in C_1 and C_2; in this case, also Cp_f can be calculated.

When Eq. (37) was first applied to butanol and pentanol in DTAB [45], the N_f^2 term was neglected, giving equal Cp_b values for both alcohols. The null methylene group contribution disagrees with that of about 30 J K^{-1} mol^{-1} obtained in several polar solvents [46, 47]. This puzzling contradiction disappears by taking into account the N_f^2 term; in fact, the CH_2 group contribution becomes 35 J K^{-1} mol^{-1}, indicating that the temperature effect on micellization and distribution is not negligible [48]. Fig. 9 shows the plot of Cp_R versus m_S for butanol in DTAB together with the best fits by considering and not considering the N_f^2 term.

D. Compressibility

An equation similar to that above reported for heat capacity (Eq. (37)) can be obtained for the standard partial molar isothermal compressibility, $K_{T,R}$ [49]:

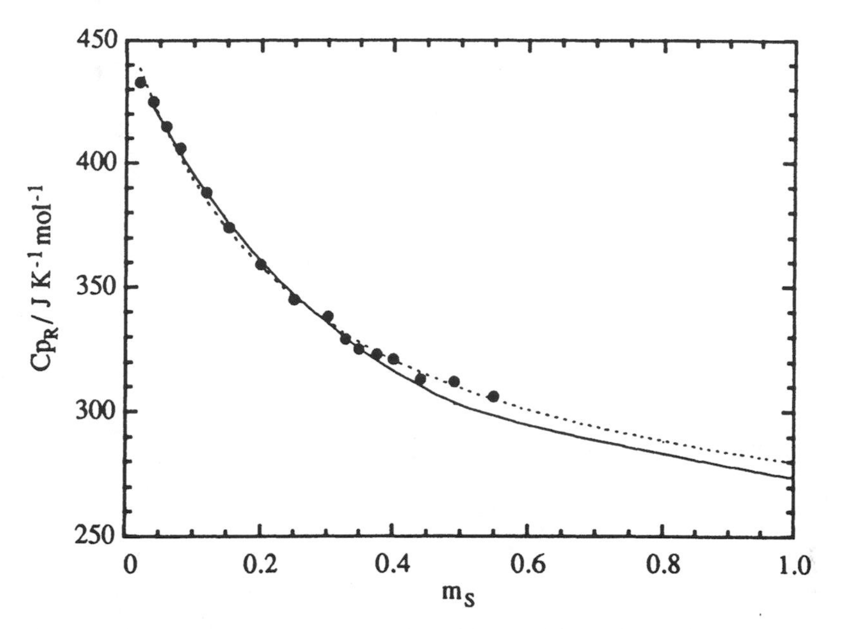

FIG. 9 Apparent molar heat capacity of butanol in dodecyltrimethylammonium bromide aqueous solution as a function of the surfactant concentration. (—) best fit of Eq. (37); (--) best fit of Eq. (37) by neglecting the N_f^2 term. (Data from Ref. 45.)

$$K_{T,R} = K_{T,b} + \left[(K_{T,f} - K_{T,b}) + \Delta V_m \frac{\partial A_{\text{cdc}}}{\partial P} + A_{\text{cdc}} \Delta K_{T,m}\right] N_f + [A_{\text{cdc}} \Delta V_m + V_f - V_b] \frac{\partial N_f}{\partial P} \quad (38)$$

where $K_{T,b}$ and $K_{T,f}$ are the partial molar isothermal compressibilities of the additive in the micellar and the aqueous phases and $\Delta K_{T,m}$ is the isothermal compressibility for the micellization process.

The dependence of the distribution constant and of the micellization equilibrium on the pressure is given by

$$\frac{\partial N_f}{\partial P} = \frac{K(\partial \text{cmc}/\partial P) - (m_S - \text{cmc})(\partial K/\partial P)}{[1 + K(m_S - \text{cmc})]^2} \quad (39)$$

$$= \left(K \frac{\partial \text{cmc}}{\partial P} + \frac{\partial \ln K}{\partial P}\right) N_f^2 - \frac{\partial \ln K}{\partial P} N_f$$

and

$$\frac{\partial A_{\text{cdc}}}{\partial P} = \frac{2.3Ks + (1 + \beta)K}{1 + \nu}\left(\frac{\partial \text{cmc}}{\partial P}\right) + \frac{\text{cmc}}{1 + \nu}\left[(1 + \beta)\frac{\partial K}{\partial P} + K\frac{\partial \beta}{\partial P} + 2.3\frac{\partial Ks}{\partial P}\right] \quad (40)$$

The other derivatives with respect to pressure are

$$\frac{\partial \text{cmc}}{\partial P} = \text{cmc}\,\frac{\Delta V_m - (\partial\beta/\partial P)\,RT \ln \text{cmc}}{(1 + \beta)RT} \quad (41)$$

$$\frac{\partial K}{\partial P} = K\left\{\frac{V_b - V_f}{RT} - \frac{K_{T,S}}{V_S}\right\} \quad (42)$$

and

$$\frac{\partial K_S}{\partial P} = \frac{2V_{RS}}{RT} \quad (43)$$

where $K_{T,S}$ (Eq. (42) is the isothermal partial molar compressibility of the micellized surfactant and V_{RS} (Eq. (43) is the pair additive-surfactant interaction parameter for the volume. As mentioned for the heat capacity, the only quantity which should be known as a function of the intensive variable is β.

Therefore, the equation correlating the partial molar compressibility of the additive in the micellar solutions to N_f is similar to that for the heat capacity (Eq. (37)) and can be solved in the same way:

$$K_{T,R} = K_{T,b} + C_1 N_f + C_2 N_f^2 \quad (44)$$

Fig. 10 shows the plot of $K_{T,R}$ versus m_S together with the best fit to experimental points for pentanol in DTAB [49].

IV. PECULIARITIES IN THE THERMODYNAMIC PROPERTIES

As shown, for example, in Fig. 5, in some cases the experimental trend of a given thermodynamic property of the additive in micellar solution as a function of m_S presents extrema near the cmc [50, 51]. Extrema are also observed in the profiles of the apparent molar properties of surfactants in water-alcohol mixtures as functions of m_S [49, 52].

Some authors have proposed to correlate the presence of these extrema to the relative hydrophobicity of the additive and the surfactant [53];

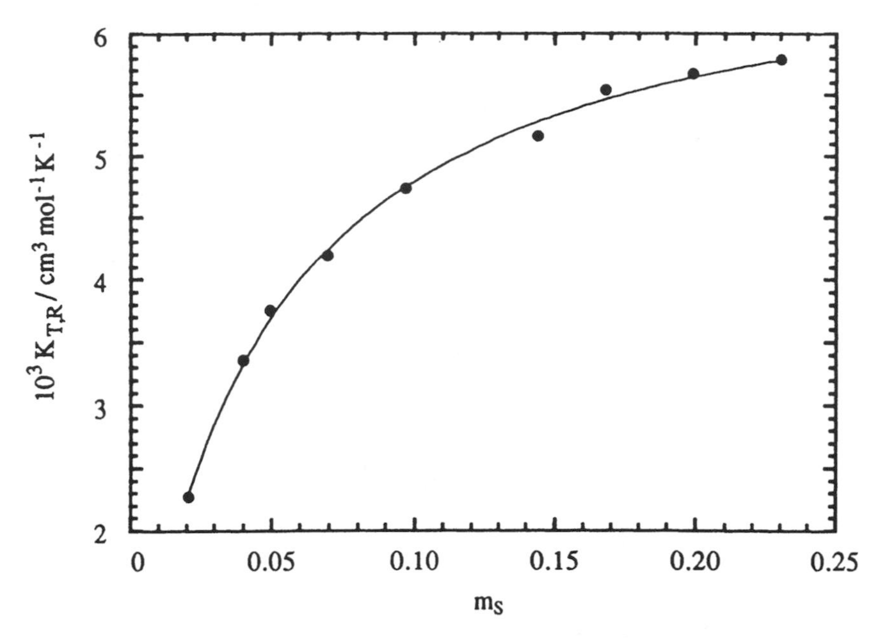

FIG. 10 Apparent molar isothermal compressibility of pentanol in dodecyltrimethylammonium bromide aqueous solutions as a function of the surfactant concentration. (—) best fit of Eq. (44). (Data from Ref. 49.)

namely, it was suggested that the extremum in the plot of Y_R versus m_S appears only when the additive is more hydrophobic than the surfactant. Recently, this idea was confirmed by Hétu and Perron [54] from the study of the transfer volume of butanol, pentanol, and hexanol from water to water-alcohol (2-propanol and butoxyethanol, for instance) mixtures forming microheterogeneities [55–57] which are microstructures not as well defined as micelles. They observed that, for a given solvent mixture, the extrema become more pronounced by increasing the hydrophobicity of the additive. However, this does not seem to be a general rule. In fact, the enthalpy of transfer of ODAO from water to NaDS solutions as a function of NaDS concentration shows an extremum, while that of the higher molecular weight homologue, dodecyldimethylamine oxide (DDAO), does not [58]. Moreover, extrema in the thermodynamic properties of transfer of electrolytes from water to micellar solutions have been also observed [59] even when no mixed micelles are formed. Therefore, it was suggested [53] that the principle of relative hydrophobicity fails when the electrostatic interactions are not negligible with respect to the hydrophobic ones.

Since the extrema are always near the cmc, alternatively, they have been attributed to the formation of mixed microheterogeneities where the additive is the main component. While for the above-mentioned alcohols, there is evidence of microheterogeneities formation, this is not the case for medium-chain-length primary alcohols. According to Tanford [60], this can be due to the hydrogen bonds between the OH groups which lead the associated alcohol molecules to constitute a separate phase rather than to form microstructures. In this respect, in a water-surfactant-alcohol system, the surfactant molecules can promote alcoholic microstructures via a breakdown of the hydrogen bonds which permits the attainment of the force equilibrium between the self-association due to the apolar groups and the repulsion between polar groups [48]. This idea could explain the cases in which surfactants are used also as additives.

This hypothesis seems to be confirmed by data reported in Fig. 11 where the apparent molar heat capacity $C_{\Phi,R}$ of hexanol, at fixed DTAB

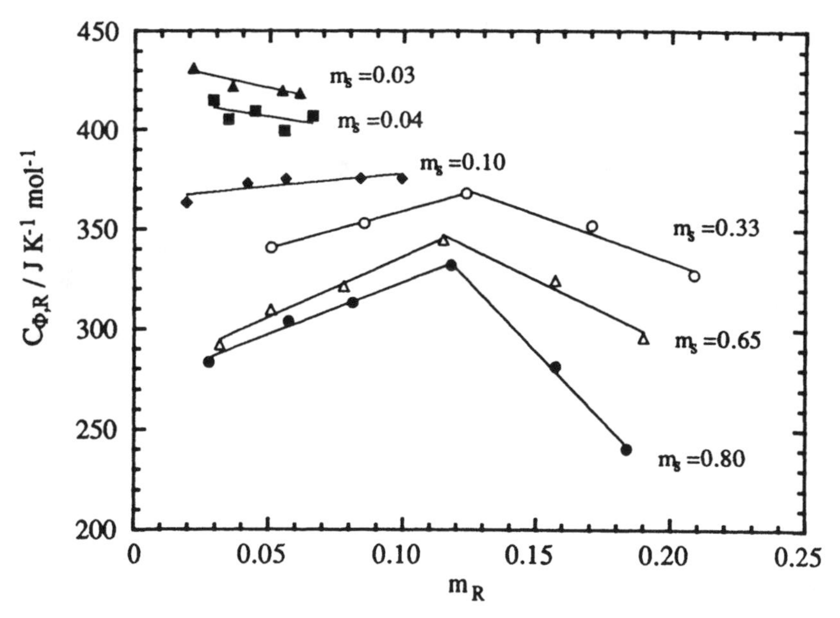

FIG. 11 Apparent molar heat capacity of hexanol in dodecyltrimethylammonium bromide aqueous solutions as a function of the alcohol concentration at fixed surfactant concentrations. (Data from Ref. 48.)

concentrations, is plotted as a function of alcohol concentration up its solubility limit. At low surfactant concentration, $C_{\Phi,R}$ increases linearly with m_R, while at high surfactant concentration a change in the slope is observed at an approximately constant alcohol concentration. These behaviors have been explained [48] by considering that upon addition of a medium-chain-length alcohol to a micellar solution, at low concentration the alcohol distributes between the aqueous and the micellar phases giving rise to mixed micelle formation (positive slope); as the composition becomes richer in alcohol, the micelles lose their identity and the surfactant, assuming the role of an additive, enhances the alcohol solubility and permits the formation of alcoholic microstructures (negative slope). Since the location of this break is slightly dependent on the surfactant concentration but highly dependent on the additive alkyl chain tail (Fig. 12), it was suggested [48] that this peculiar concentration possesses the feature of a critical concentration, called *critical microstructure concentration* cm_sc, at which alcoholic microaggregates are forming. Accordingly, within the

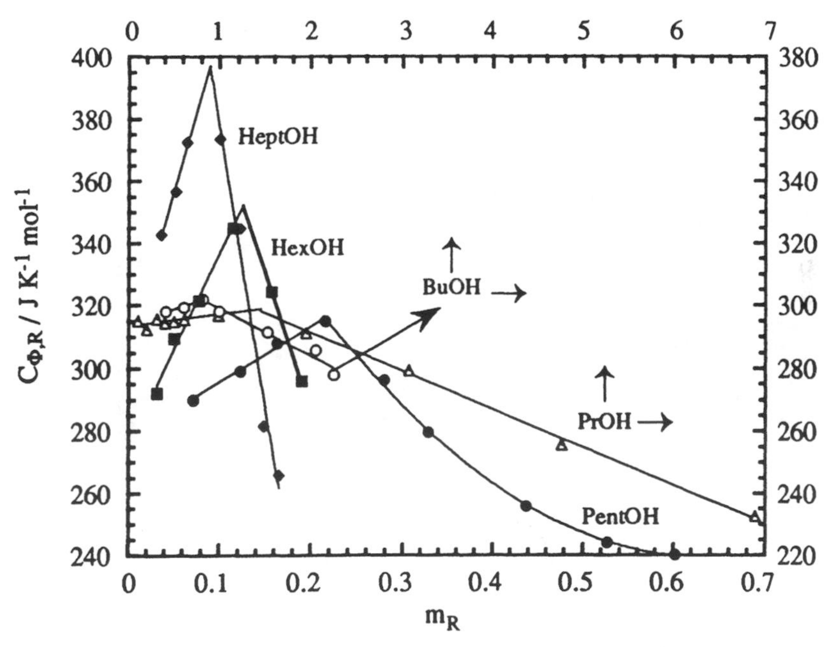

FIG. 12 Apparent molar heat capacity of primary alcohols in dodecyltrimethylammonium bromide aqueous solutions as a function of the alcohol concentration at 0.6 m surfactant concentration. (Data from Ref. 48.)

uncertainties of these cm_sc values, their dependence on the number of carbon atom in the alcohol alkyl chain is similar to that usually observed for the cmc of surfactants, as shown in Fig. 13.

If, on the one hand, the formation of alcoholic microaggregates stabilized by the surfactants seems to be confirmed, on the other it cannot be considered responsible for the extrema since for all alcohols the cm_sc value occurs at concentration very different from that where the extrema are localized.

In addition, microheterogeneities formation does not explain the volume of transfer $\Delta V_{\Phi,S}(W \rightarrow W + R)$ of DTAB from water to water-propanol and water-2-propanol mixtures. In fact, as Fig. 14 shows, $\Delta V_{\Phi,S}(W \rightarrow W + R)$ versus m_S curves at 0.5 m alcohol concentration display maxima localized at about 0.1 mol kg^{-1} for propanol and 2-propanol, although the formation of microheterogeneities in water is reported for 2-propanol only.

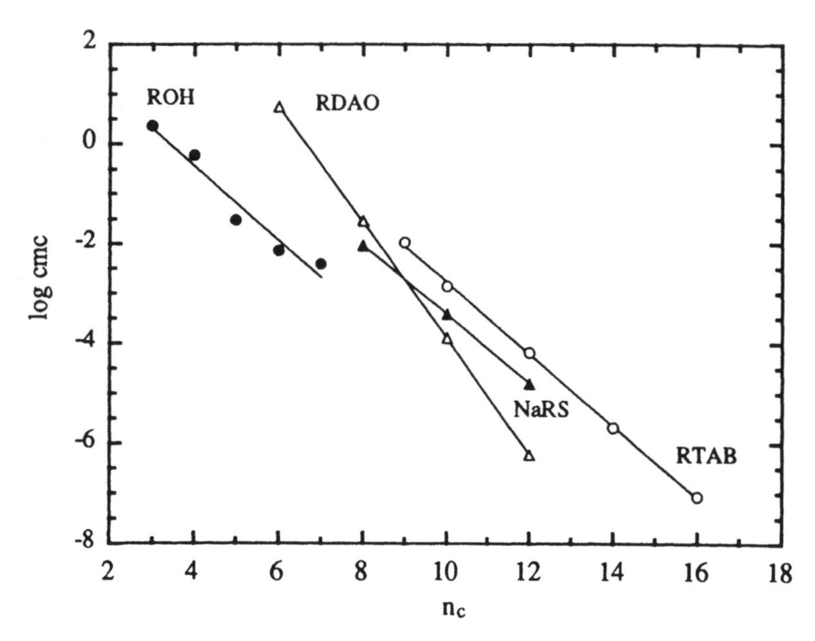

FIG. 13 Critical micellar concentration (molarity scale) against the number of carbon atoms in the alkyl chain. (●) primary alcohols (from Ref. 48); (△) alkyldimethylamine oxides (from Ref. 31); (▲) sodium alkylsulfates (from Ref. 87); (○) alkyltrimethylammonium bromides (from Ref. 88).

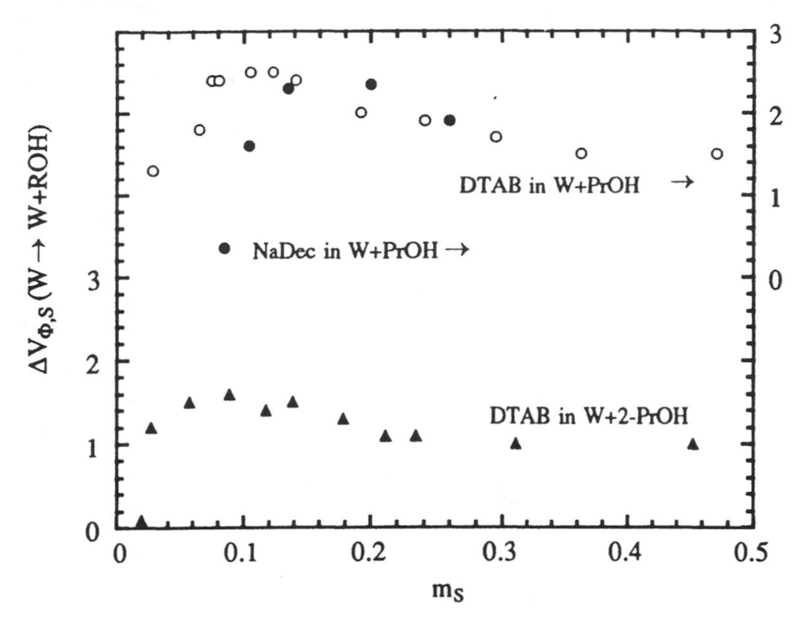

FIG. 14 Volume of transfer ($cm^3\ mol^{-1}$) of sodium decanoate (NaDec) and dodecyltrimethylammonium bromide (DTAB) from water to water-propanol mixtures against the surfactant concentration. (DTAB data from Ref. 61; NaDec data from Ref. 26.)

The above attempts to explain the behavior of additives in water-surfactant and of surfactants in water-additive, although promising, are only qualitative and often ambiguous.

For additives in micellar solutions, it was shown in Sec. II.C. that the presence of extrema can be quantitatively predicted by the models discussed above as due to the predominance of the micellization equilibrium shift on that of distribution. As an example, Fig. 15 shows the dependence of these two contributions (at the cmc) on the number of carbon atoms in the alkyl chain of the additive n_c for the enthalpy of transfer for some primary alcohols in DDAO [51]. As can be seen, the micellization equilibrium shift is larger than that of distribution for $n_c > 5$. Accordingly, as shown in Fig. 16, $\Delta H(\mathrm{W} \rightarrow \mathrm{W} + \mathrm{S})$ monotonically increases with m_S for pentanol, while it presents a maximum for hexanol.

The same models explain the extrema in the curves of the apparent molar properties of surfactants in the water-alcohol mixtures as functions of m_S [41, 49]. In fact, the following equation for the volume of transfer of

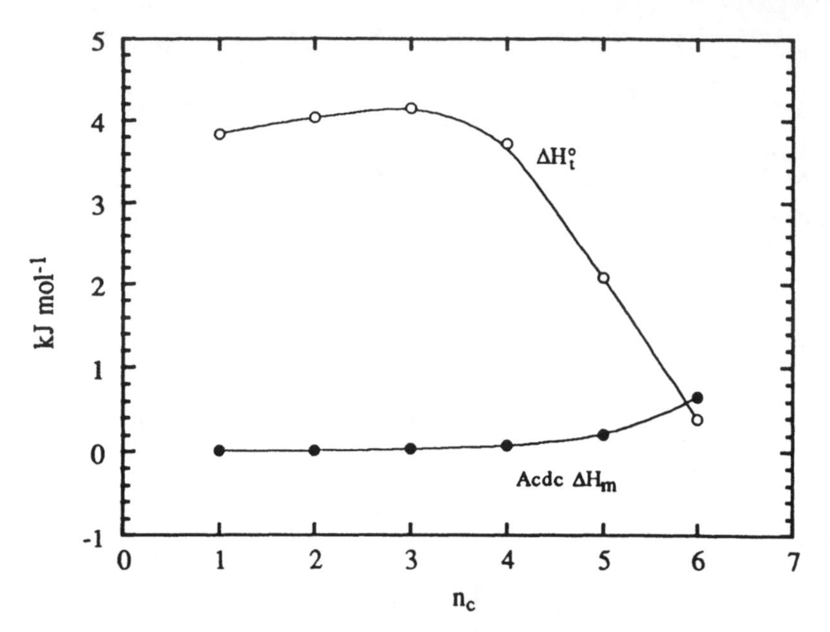

FIG. 15 Enthalpy of transfer from the aqueous to the micellar phases and micellization equilibrium shift contribution for primary alcohols in dodecyldimethylamine oxide as a function of the alcohol alkyl chain length. (Data from Ref. 51.)

the surfactant from water to the water-additive mixtures can be obtained:

$$\Delta V_{\Phi,S}(\mathrm{W} \rightarrow \mathrm{W} + \mathrm{R}) = (1 - N_f)(V_b - V_f)\frac{m_R}{m_S} + \frac{[m]\Delta V'_m - cmc\Delta V_m}{m_S} \tag{45}$$

The second term on the right side of Eq. (45) can be neglected since for $m_R \rightarrow 0$, $\Delta V'_m = \Delta V_m$ and $[m]$ = cmc (according to Eq. (21). Therefore Eq. (45) predicts a maximum occurring at $m_S^{\max} = \mathrm{cmc} + (\mathrm{cmc}/K)^{1/2}$ whose amplitude is given by

$$\Delta V_{\Phi,S}^{\max}(\mathrm{W} \rightarrow \mathrm{W} + \mathrm{R}) = m_R(V_b - V_f)\frac{\sqrt{\mathrm{cmc}\,K}}{(\mathrm{cmc} + \sqrt{\mathrm{cmc}/K})(1 + \sqrt{\mathrm{cmc}\,K})} \tag{46}$$

which shows that, for a given additive, the amplitude of the maximum is

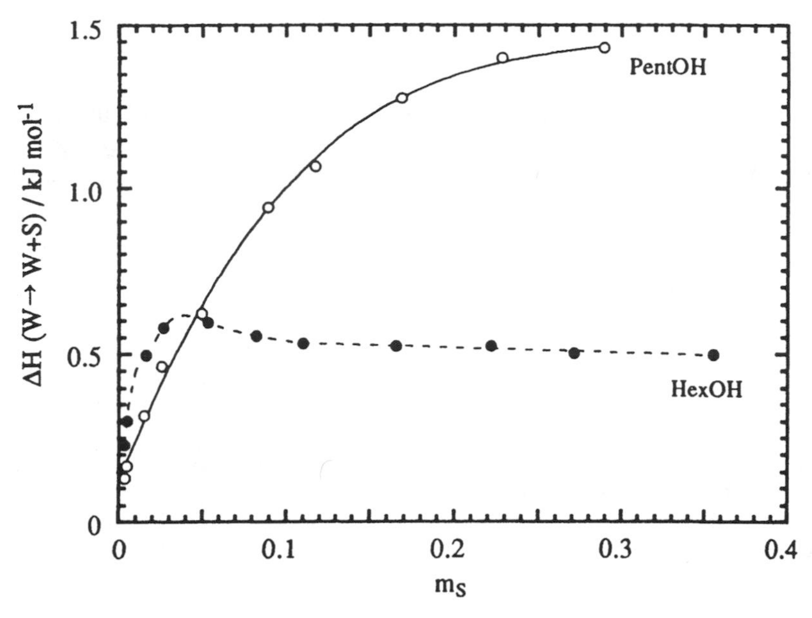

FIG. 16 Enthalpy of transfer from water to micellar solutions against surfactant concentration for pentanol and hexanol in dodecyldimethylamine oxide. (Data from Ref. 51.)

directly proportional to m_R, while for a given m_R it increases with the hydrophobicity of the additive through K, V_b, and V_f (see Fig. 14). By using literature data for propanol and 2-propanol in DTAB [61], the $m_S^{\max}$ values are 0.12 m and 0.10 m to which correspond the values of 2.5 and 1.5 cm^3 mol^{-1} for $\Delta V_{\Phi,S}^{\max}$ at $m_R = 0.5$ m. Since K slightly depends on the alkyl chain length of the surfactant (as it will be seen later), by taking the value for propanol in sodium dodecanoate [62], the $m_S^{\max}$ value for sodium decanoate (NaDec) in water-propanol is 0.21 m. In addition, by reasonably assuming that the volume in the aqueous phase is the same as that in water, from $V_b - V_f$ values for longer alkyl chain length alcohols in NaDec [26], that of propanol has been evaluated. Therefore, the value of 2.6 cm^3 mol^{-1} for $\Delta V_{\Phi,S}^{\max}$ at $m_R = 0.5$ m was calculated. As Fig. 14 shows, the experimental and calculated $m_S^{\max}$ and $\Delta V_{\Phi,S}^{\max}$ agree well. Note that $\Delta V_{\Phi,S}^{\max}$ in water + propanol does not seem to depend on the nature of the surfactant and that it is larger than that in water + 2-propanol despite, as mentioned, microheterogeneity formation is reported for the latter only.

V. VOLUMES, HEAT CAPACITIES, AND COMPRESSIBILITIES IN MICELLAR SOLUTIONS

As seen above, determining the thermodynamic properties of additives in micellar solutions is a useful tool for obtaining knowledge of the interactions governing mixed micelle formation. In particular, the standard partial molar property of the additive in the micellar phase, Y_b, gives information about additive-micelle interactions, while the distribution constant gives information about the solubilizing power of the micelle toward the additive. The latter property together with free energy, enthalpy, and entropy of transfer will be discussed in Sec. VI.

The Y_b values obtained from the minimizing procedure discussed above (Eqs. (10) and/or (12)) usually agree with each other while they are slightly different from those reported at a given, even high, surfactant concentration. However, in spite of this difference, the same qualitative information can generally be derived. There is not an extensive literature on Y_b and, generally, it deals with the volumes of alcohols in dodecylsurfactants. Some of the results are collected in Table 2 together with the molar properties and the standard partial molar properties in water and in octane.

A quite complete thermodynamic picture is reported for the alcohol-water-DTAB ternary systems. For these reasons, in the following we will often refer to these systems, which, compared to other systems, can give insight on the effects of the nature of both the additive and the surfactant on the volume, heat capacity, and compressibility of the additive in the micellar phase.

A. Additive Alkyl Chain Length

Regardless of the nature of the surfactant, from a qualitative point of view, V_b of primary alcohols [26, 40, 41, 43, 63] increases with increasing length of the alkyl chain (see Table 2). For a given alcohol, V_b always ranges between the standard partial molar volume in water and the molar volume, while it is smaller than that in liquid hydrocarbon. As far as the CH_2 group contribution is concerned, it is practically equal (16.7 cm^3 mol^{-1}) to that for liquid pure alcohols (16.8 cm^3 mol^{-1}) and higher than that in octane (16.1 cm^3 mol^{-1}) and in water (15.9 cm^3 mol^{-1}). Plots of V_b in DTAB, standard partial molar volumes in water, V_w, in octane, V_{oct}, and molar volumes, V^*, as functions of the number of carbon atoms in the alkyl chain length are shown in Fig. 17. These results indicate the presence of hydrophilic interactions between the alcohol and the surfactant, suggesting that the solubilization of additives occurs in the palisade layer. However, the different CH_2 group contributions in the micellar phase and in

TABLE 2 Standard Partial Molar Volumes, Heat Capacities, and Isoentropic Compressibilities of Some Primary Alcohols in Water, in Octane, in the Micellar Phase of Different Surfactants and Corresponding Molar Properties

	PrOH	BuOH	PentOH	HexOH	Refs.
V_w	70.8	86.8	102.6	118.7	90
V^*	75.1	91.9	108.7	125.3	89
V_b (NaDS)	72.8	89.2	105.3	122.0	43
V_b (DTAB)	73.7	91.2	107.5	124.0	61
V_b (DDAO)		90.4	107.3		41
V_b (ODAO)		90.8	107.1	124.2	40
V_b (OAB)		90.8	107.0	122.9	40
V_b (NaDec)	72.0	88.8	105.5	122.8	63
V_b (Na_2C_{15})			106	125	26
V_{oct}	83.5	99.8	115.8	131.9	43
Cp_w	353	437	524		93
Cp^*	144	177	208		46
Cp_b (DTAB)	160	197	232	270	48
Cp_{oct}		101	136		45
$10^3 K_{S,w}$	0.05	0.23	0.45	0.58	63
$10^3 K_{S,b}$ (NaDec)	3.91	5.55	7.21	8.92	63
$10^3 K_S^*$	6.46	7.43	8.26	9.06	63

Units are $cm^3\ mol^{-1}$ for the volume; $J\ K^{-1}\ mol^{-1}$ for the heat capacity; $cm^3\ mol^{-1}\ bar^{-1}$ for the compressibility.

octane predict that V_b tends to V_{oct} for very long alkyl chain length alcohols for which hydrophobic contributions largely predominate over the hydrophilic ones. In fact, since the partial molar volumes of alcohols in apolar solvents usually do not depend on the tail of the solvent, the solubilization of the alcohol in the micellar core should involve a partial molar volume in the micellar phase very close to that in octane. As said in Sec. I, this is corroborated by the apparent molar volumes of some alkanes in the micellar phase which are very close to their molar values [25].

Very few data for the standard partial molar heat capacity of additives in the micellar phase, Cp_b, are available; this property has not been extensively studied also because the presence of anomalies often observed in the Cp_R versus m_S curves (see Sec. VII) makes difficult the evaluation of Cp_b. However, some primary alcohols in ODAO and DTAB which, as far we know, are the only surfactants where Cp_b values are reported, seem to confirm the volume results. In fact, Cp_b lies between the standard partial molar heat capacity in water and in organic solvents, being close

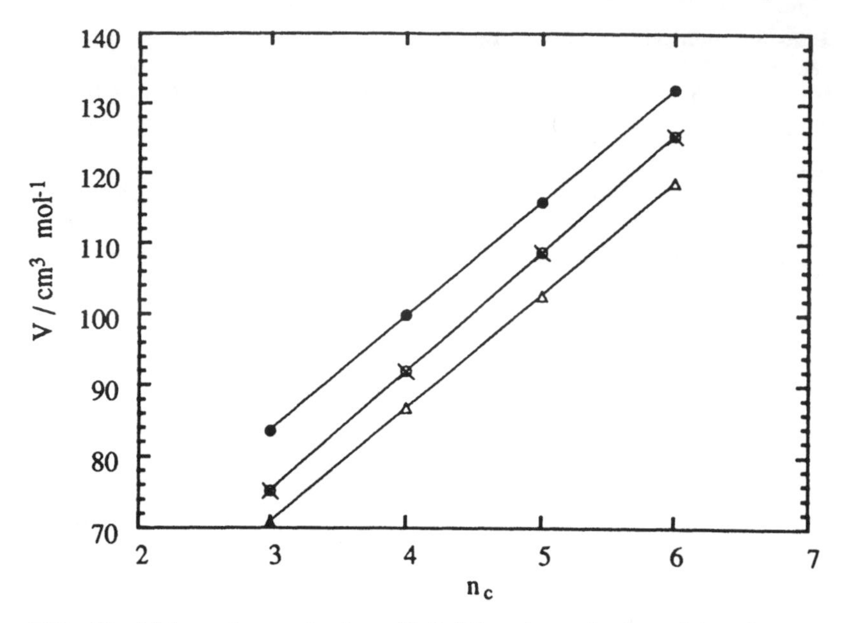

FIG. 17 Molar volumes (x, from Ref. 89) and standard partial molar volume of primary alcohols in dodecyltrimethylammonium bromide micellar phase (●, from Ref. 61), in water (△, from Ref. 90) and in octane (●, from Ref. 43).

to the molar heat capacity of the pure liquid alcohol. From these data, the methylene group contribution is 35–40 J K^{-1} mol^{-1}, which is lower than that in water (89 J K^{-1} mol^{-1}) [64] and close to that in polar and apolar solvents (30–35 J-K^{-1} mol^{-1}) [46, 47].

Isoentropic compressibilities of homologous series of alcohols in some sodium alkylcarboxylates have been investigated [63] at two or three surfactant concentrations. The isoentropic compressibilities of the alcohol in the micellar phase increase with the hydrophobicity of the additive with a methylene group contribution which depends on the alcohol tail. These values confirm, once again, that the environment of the alcohol molecules in the micellar phase is similar to that of the pure liquid alcohol.

B. Surfactant Alkyl Chain Length

For a medium alkyl chain alcohol, the partial molar properties in the micellar phase slightly depend on the alkyl chain length of the surfactant. Accordingly, as Table 3 shows, for the addition of one methylene group to the

TABLE 3 Partial Molar Volume, Heat Capacity, Isoentropic and Isothermal Compressibilities of Pentanol 0.05 m in Micellar Phase of Different Surfactants

	V_b	Cp_b	$10^3K_{S,b}$	$10^3K_{T,b}$
DeTAB[a]	107.4	248		
DTAB	107.5[a]	244[a]	6.72[b]	7.14[b]
TTAB[b]	107.7		7.01	
CTAB[b]	108.0		7.57	
DTAC[c]	107.3	263		
DDAC[c]	106.5	265		
DAC[c]	106.7			

Units are $cm^3\ mol^{-1}$ for V_b; $J\ K^{-1}\ mol^{-1}$ for Cp_b; $cm^3\ mol^{-1}\ bar^{-1}$ for $K_{S,b}$ and $K_{T,b}$.
[a] From Ref. 65.
[b] From Ref. 49.
[c] From Ref. 67.

hydrocarbon chain of the surfactant, the partial molar volume, isoentropic compressibility, and heat capacity of PentOH in alkyltrimethylammonium bromides RTAB micelles slightly change [65]. Also, the absolute values of the above properties are very close to the corresponding values of the pure liquids. The negligible influence of the hydrophobicity of the surfactant on the thermodynamics of solubilization can be understood if it is accepted that the palisade layer is the site of solubilization of the additive.

The effect of the alkyl chain length of some carboxylates RCOONa on the standard partial molar volumes and isoentropic compressibilities of longer alkyl chain alcohols (from heptanol to decanol) [63] is different from that for RTAB. Actually, these systems have been analyzed at two or three high surfactant concentrations and a slight decrease in the standard partial molar property on the surfactant concentration was observed. By assuming that the standard partial molar property in the micellar solution corresponds to that in the micellar phase, whatever the real absolute values are, it is straightforward that these properties decrease with increasing alkyl chain length of the surfactant. This decrease was ascribed to the effect of the alcohol on the micellar structure, which is more important the longer the alkyl chain of the alcohol and the shorter the alkyl chain of the surfactant are. Accordingly, since negligible effects are expected for medium tail alcohols, this explanation fits well both RCOONa and RTAB results.

The standard partial molar expansibility of the additive in micellar solution as a function of m_S could be derived from the dependence of the standard partial molar volume on temperature. This method gives unreliable results and, therefore, does not permit the rationalization by using an equation similar to that above reported for the compressibility. In addition, a straightforward analysis is really difficult since the hydrophobic and hydrophilic contributions of additives are both positive. However, an inspection of V_b of PentOH in the micellar phase as a function of temperature shows that the expansibility of the alcohol in the micellar phase is 0.09 cm^3 mol^{-1} K^{-1} in DTAB and TTAB, and 0.06 cm^3 mol^{-1} K^{-1} in CTAB [49, 66]. The values in DTAB and TTAB are comparable to that of the pure liquid (0.10 cm^3 mol^{-1} K^{-1}) but greater than that in water (0.04 cm^3 mol^{-1} K^{-1}) and smaller than that in octane (0.17 cm^3 mol^{-1} K^{-1}), confirming that the alcohol solubilization in the micelle involves both hydrophilic and hydrophobic interactions.

C. Polar Head of the Surfactant

Since pentanol has been studied in several dodecylsurfactants, it can be considered to be the best additive to obtain information on the effect of the nature of the surfactant polar headgroup on its property in the micellar phase (see Table 2).

The V_b values in DTAB and DDAO are practically equal and larger by about 2 cm^3 mol^{-1} than that in NaDS [41], indicating stronger hydrophilic interactions between alcohols and anionic micelles with respect to the cationic and nonionic ones. This is also true for octylsurfactants. In fact, V_b in ODAO is equal to that in OAB [40], larger than that in NaDS and, obviously, in sodium octylsulfate since, as in Sec. V.B., V_b decreases for decreasing alkyl chain length of the surfactant.

The same V_b values in DTAB and DDAO were tentatively explained in terms of the steric hindrance of the methyl groups at the hydrophilic headgroups of both the surfactants which involves comparable density of charge and void space at the micellar surface [41]. Actually, this idea does not hold in ODAO and OAB [40] since methyl groups are present only in the headgroup of the nonionic surfactant.

The effect of the introduction of methyls in the headgroup of the surfactant has been investigated by analyzing dodecyltrimethylammonium (DTAC), dodecyldimethylammonium (DDAC), and dodecylammonium (DAC) chlorides [67]. The standard partial molar volume of PentOH in DAC micelles is equal to that in DDAC and slightly smaller (0.7 cm^3 mol^{-1}) than that in DTAC. This difference is not really negligible, but it is so small that one can reasonably say that the presence of methyl groups in the polar head of the surfactant has a negligible contribution to V_b. These findings for chloride surfactants can be extended to bromide surfactants since the nature of the counterion does not involve any change in the partial molar volume of PentOH in the micellar phase as data in DTAC and DTAB indicate.

The Cp_b values are not reported for all the systems discussed above. The only information which can be drawn deal with the introduction of methyl groups in the polar head of the surfactant and the effect of the nature of the counterion. According to volume results, a practically null dependence of Cp_b on the polarity of the headgroup was observed. The increase of about 20 J mol^{-1} K^{-1} of Cp_b obtained when bromide ion is replaced by chloride ion can reflect the change in the degree of dissociation of the micelle due to the additive, which tends to affect the heat capacity more than the volume.

The effect of the polar head of the surfactant on the isoentropic compressibilities of additives in the micellar phase can be derived by comparing data in DTAB [49, 66] and in sodium dodecanoate (NaL) [63]. According to the volumetric data for which the V_b value in a cationic micelle is larger than that in an anionic one, the isoentropic compressibility of PentOH in the micellar phase of DTAB (67×10^{-4} cm^3 bar^{-1} mol^{-1}) is larger than that in NaL (61×10^{-4} cm^3 bar^{-1} mol^{-1}), which has been obtained by extrapolating data for longer chain alcohols.

D. Polar Head of the Additive

Volumes and heat capacity data of some penthyl additives in decyltrimethylammonium bromide (DeTAB) show interesting behavior [68]. In fact, as Table 4 shows, the volumes of penthylamine ($PentNH_2$) in the micellar phase and in water are close to those of PentOH and different from those of capronitrile (PentCN) and nitropentane ($PentNO_2$). On the other hand, the standard partial molar volume of $PentNH_2$ in octane and its molar volume are close to those for PentCN and $PentNO_2$ and different from those of PentOH. Therefore, at least from a volumetric point of view, $PentNH_2$ possesses contradictory behavior since it approaches PentOH or PentCN and $PentNO_2$, depending on the nature of the solvent. This situation is not modified if the partial molar volumes are corrected for the intrinsinc contribution, which can be assumed to correspond to the van der Waals volumes.

Heat capacity data are more complicated to explain. In fact, the properties in the micellar phase and the molar heat capacity of pure liquids seem to be independent of the nature of the headgroup of the additive, while values in octane of $PentNH_2$ are close to those of PentCN and $PentNO_2$ and higher by about 50 J K^{-1} mol^{-1} than that of PentOH. The values in water are peculiar since $PentNH_2 < PentOH = PentCN < PentNO_2$.

TABLE 4 Partial Molar Volumes and Heat Capacities of Some Amyl Compounds in Water, in the Decyltrimethylammonium Bromide Micellar Phase, in Octane and Corresponding Molar Properties

	PentOH	PentNH$_2$	PentCN	PentNO$_2$
V_w	101.9	101.6	122.8	127.2
V^*	108.7	116.4	120.4	123.2
V_b	107.4	106.9	118.3	123.0
V_{oct}	115.8	120.1	122.8	127.2
Cp_w	532	502	540	579
Cp^*	208	216	204	213
Cp_b	248	251	234	254
Cp_{oct}	136	186	191	197

Units are cm^3 mol^{-1} for the volume; J K^{-1} mol^{-1} for the heat capacity; cm^3 mol^{-1} bar^{-1} for the compressibility.
Source: Data from Ref. 68.

VI. STANDARD FREE ENERGY, ENTHALPY, AND ENTROPY OF TRANSFER FROM THE AQUEOUS TO THE MICELLAR PHASES

In order to derive complete sets of the standard free energy, ΔG_t°, enthalpy, ΔH_t°, and entropy, ΔS_t°, of transfer of additives from the aqueous to the micellar phases, calorimetric measurements should be combined with other techniques which permit one to determine the distribution constant. This can be made only when both studies are performed under nearly the same experimental conditions. In fact, the distribution constant does not depend on the additive concentration if the mole fraction of the additive in the micellar phase is lower than 0.3, beyond which it drops [69]. Alternatively, they can be simultaneously obtained from the fit of Eqs. (27) or (30) to experimental enthalpimetric data.

As a general feature [44, 51, 70–73], regardless of the nature of the additive and the surfactant, ΔG_t° decreases linearly with the number of carbon atoms in the alkyl chain length of the additive, n_c, while enthalpy and entropy do not.

The ΔG_t° versus n_c linear correlation, which is not specific for surfactant systems,

$$\Delta G_t^\circ = A_g + B_g n_c \tag{47}$$

is a result of hydrophilic (A_g) and hydrophobic (B_g) contributions.

It is to be stressed that the thermodynamic functions of transfer involve the removal of the additive from the aqueous phase and its solubilization in the micelle. Therefore, these properties are straightforward when the energetics of solubilization of a given additive in micelles of different surfactants is considered. When the study is aimed at investigating the nature of the additive, it is preferable to analyze the thermodynamics of solvation since it involves, as the initial reference state, the additive as an ideal gas. In this way, the thermodynamics of solvation gives less questionable information since the additive liquid phase formation, which can play an important role, is taken into account.

From the analysis [43] of the free energies of solution of alkanes in polar and apolar solvents, it was observed that the physical meaning of the intercept of the plot of free energies of solvation as functions of the alkyl chain length of alkanes assumes unquestionable physical meaning if the following process is considered:

$$\text{additive (gas, } P = RT/V) \rightarrow \text{additive (solution, 1 } M) \qquad (48)$$

where V is the volume occupied by one mole of the additive in the final standard state; i.e., the empty space at disposal of the additive molecules in both phases is the same.

According to this scheme, the free energy of solvation in the micellar phase, ΔG^*_M, is correlated to the standard ($P = 1$ atm) free energy of solvation, $\Delta G^\circ_{S,M}$, by means of the equation

$$\Delta G^*_M = \Delta G^\circ_{s,M} + RT \ln \frac{V_b}{RT} \qquad (49)$$

$\Delta G^\circ_{s,M}$ has been correlated to ΔG°_t and to the standard ($P = 1$ atm) free energy of solvation in the aqueous phase, which, in turn, depends on the standard free energy of solvation in water, $\Delta G^\circ_{s,w}$, and on the free energy of transfer from water to the aqueous phase, $\Delta G^\circ_R(\text{W} \rightarrow \text{ap})$. Since $\Delta G^\circ_{s,w}$ depends on the vapor pressure (P_e) of the additive and $\Delta G^\circ_R(\text{W} \rightarrow \text{ap})$ on the Setchenov constant (Ks), the following equation is obtained:

$$\Delta G^*_M = \Delta G^\circ_t + RT \ln \frac{P_e V_b}{RT} + 2.3RTKs \text{ cmc} \qquad (50)$$

Equation (50) shows that to evaluate ΔG^*_M from ΔG°_t, knowledge of P_e, V_b, and Ks is required.

Since the enthalpy of an ideal gas does not depend on pressure, no difference exists between the enthalpy relative to scheme (48) and the standard enthalpy of solvation in micelle, ΔH^*_M, which can be calculated from ΔH°_t, the enthalpy of vaporization, ΔH_v, and the enthalpy of solution in water, $\Delta H^\circ_{s,w}$, and the contribution due to the pair additive-surfactant interactions according to

$$\Delta H_M^* = \Delta H_t^\circ + \Delta H_{s,w}^\circ + \Delta H_v + 2H_{RS}\text{cmc} \tag{51}$$

Because of the lack of data, the properties of solvation can be calculated and discussed only for a few systems, and, therefore, we will focus our attention on the properties of transfer.

A. Primary Alcohols

As Fig. 18 shows, the profiles of the standard thermodynamic properties of transfer as functions of n_c do not depend on the nature of the polar head of the surfactant [44, 51, 70]. In particular, ΔG_t° decreases linearly with n_c, while ΔH_t° and $T\Delta S_t^\circ$ versus n_c are convex curves. In the case of sodium perfluorooctanoate (NaPFO), while ΔG_t° still decreases linearly with n_c, no maxima are present in ΔH_t° and ΔS_t° which could be localized at n_c values higher than those in other surfactants. From a quantitative

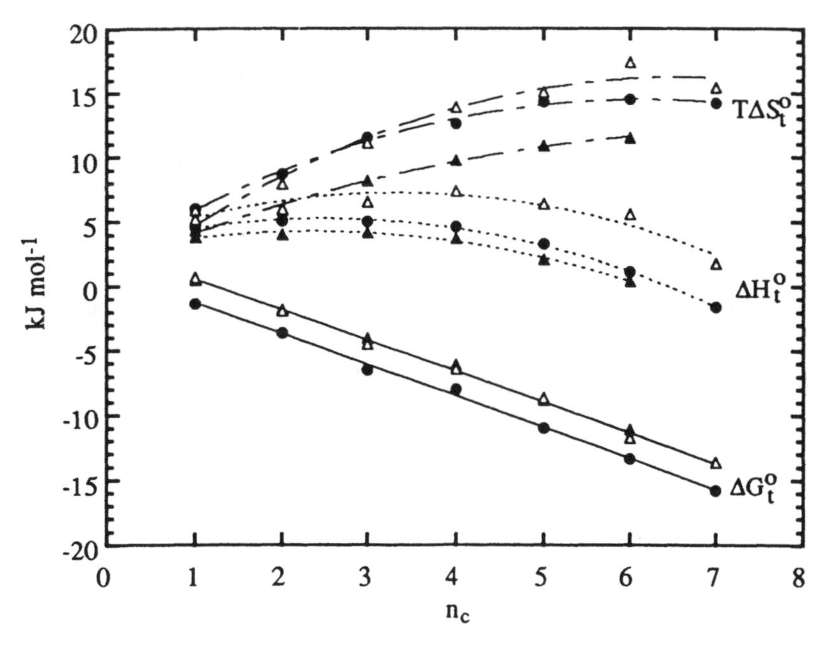

FIG. 18 Standard free energy, enthalpy, and entropy of transfer of primary alcohols from the aqueous to the micellar phases of sodium dodecylsulfate (●, from Ref. 70), dodecyltrimethylammonium bromide (△, from Ref. 44), and dodecyldimethylamine oxide (▲, from Ref. 51).

point of view, the negative slope (-2.4 kJ mol^{-1}) of ΔG_t° versus n_c, which does not depend on the nature of the surfactant, indicates that the transfer of the CH_2 group is energetically favored according to the solubilization of apolar groups in apolar solvents such as the micellar core. The ΔG_t° values in DTAB and DDAO are equal and more positive than those in NaDS, indicating that anionic micelles show a larger affinity toward the OH group than cationic and nonionic ones. Also, the positive values (i.e., A_g in Eq. (47)) for the hydrophilic contribution (2.8 kJ mol^{-1} in DTAB and DDAO and 0.9 kJ mol^{-1} in NaDS) indicate that the solubilization of the OH group in water is favored with respect to that in micelle. The ΔG_t° values in NaPFO are equal to those in NaDS according to the comparable hydrophobicity between the octyl fluorinated and dodecyl hydrogenated chains [74].

As far as the enthalpies and entropies are concerned, in DTAB they are larger than those in DDAO and NaDS, which, in turn, are comparable. Therefore, the same affinity toward alcohols shown by DTAB and DDAO micelles is due to enthalpy-entropy compensative effects.

As a consequence of the nonlinear plots of ΔH_t° and $T\Delta S_t^\circ$ versus n_c, the methylene group contribution depends on the length of the alcohol alkyl chain. In any case, qualitatively one can say that entropy is the driving force of the CH_2 transfer process from the aqueous phase to micelles. The hydrophilic contribution, which can be evaluated with acceptable accuracy, shows that the transferring of the OH group from the aqueous to the micellar phases is driven by the enthalpy.

The positive values of ΔH_t° and $T\Delta S_t^\circ$ seem to support the idea that hydrophobic hydration governs the additive transfer from the aqueous to the micellar phases. However, the ΔG_t° values, different from zero, are not consistent with this idea since there is strong evidence that the removal of a nonpolar group from aqueous solutions has little influence on the free energy [75, 76].

A possible explanation is that the solubilization of additives in the micelles involves a micellar rearrangement to create a cavity suitable to accommodate the alkyl chain. Accordingly, the solubilization of shorter alkyl chain additives in the "void" space at the surface of micelles should not involve rearrangement of the micellar structure, while solubilization of a longer alkyl chain should. This can also explain the peculiar trends of ΔH_t° and $T\Delta S_t^\circ$ as functions of n_c. As far we know, complete sets of ΔG_t°, ΔH_t°, and $T\Delta S_t^\circ$ for other surfactants have not been reported, so that this hypothesis cannot be confirmed.

As noted, the properties of solvation of alcohols in micelles can give direct information on their solubilization in the micellar phase. Unfortunately, they can be evaluated only for primary alcohols in NaDS, DTAB, and NaPFO. As Fig. 19 shows, the free energy of solvation in the micellar phase decreases linearly as a function of n_c with a methylene group contribution of about -1.2 kJ mol^{-1}, independent of the nature of the headgroup and of the hydrophobicity of the surfactant. Note that this contribution is one half that of transfer (-2.4 kJ mol^{-1}), indicating that the removal of the CH_2 group from water involves a free-energy change equal to that due to its solubilization in the micelle. The large negative hydrophilic contributions (-29 kJ mol^{-1} in NaDS and NaPFO and -27 kJ mol^{-1} in DTAB) evidence strong interactions between the OH group and the hydrophilic shell of the micelle. These values are close to that in water (-30 kJ mol^{-1}) and indicate that the OH group contribution to ΔG_t° is a nearly negligible quantity with respect to the free energy of solvation in both phases. As well, as Fig. 19 shows, by increasing the hydrophobicity of the alcohol the free energy of solvation in water increases, while that in the micellar phase decreases such as occurs in octane.

The enthalpies and entropies of solvation in DTAB and NaDS micellar phases, in water and in octane as functions of n_c are shown in Figs. 20

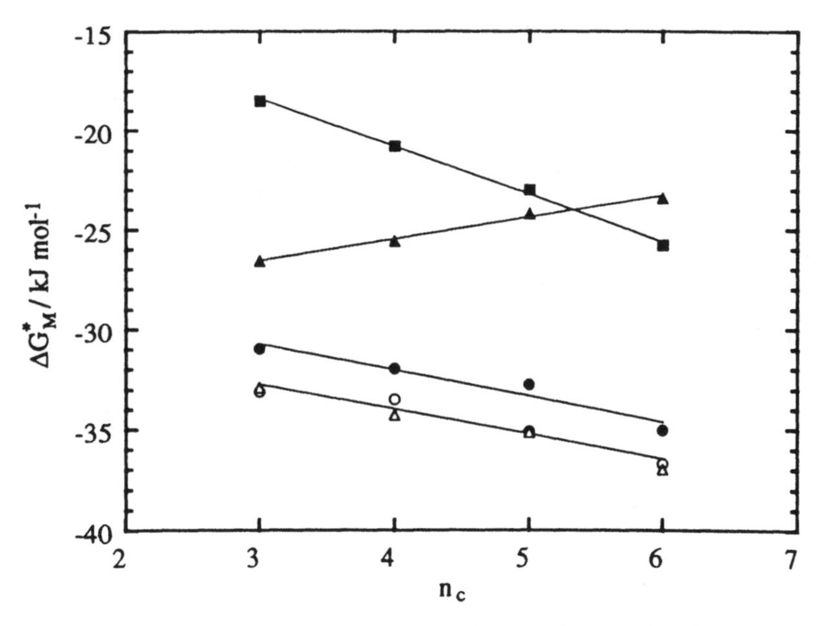

FIG. 19 Free energy of solvation (according to scheme (48)) of primary alcohols in water (▲, from Ref. 70), in octane (■, from Ref. 70), and in micellar phase of sodium dodecylsulfate (○, from Ref. 70), dodecyltrimethylammonium bromide (●, from Ref. 74), and sodium perfluorooctanoate (△, from Ref. 74).

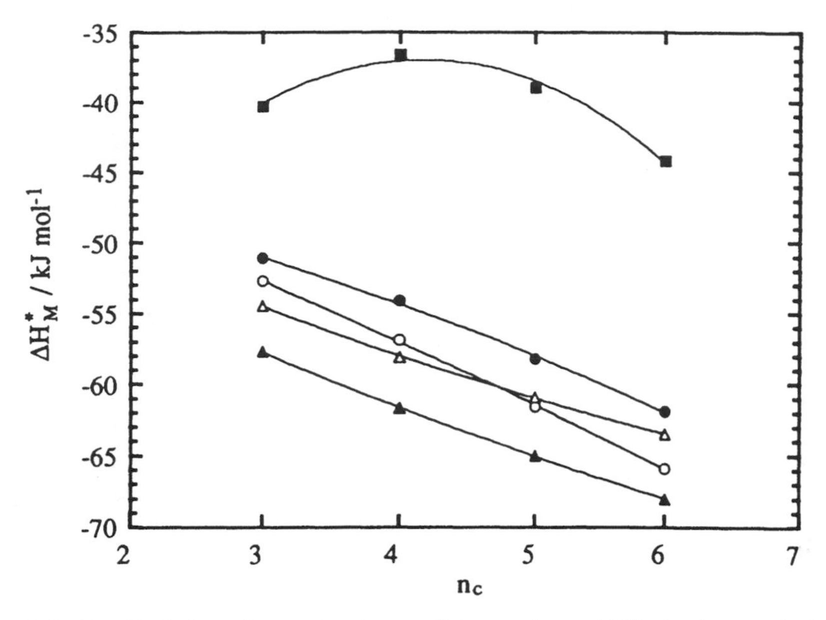

FIG. 20 Enthalpy of solvation (according to scheme (48)) of primary alcohols in water (▲, from Ref. 70), in octane (■, from Ref. 70), and in micellar phase of sodium dodecylsulfate (○, from Ref. 70), dodecyltrimethylammonium bromide (●, from Ref. 74), and sodium perfluorooctanoate (△, from Ref. 74).

and 21. The very small maxima observed in the properties of solvation in the DTAB and NaDS micellar phases justify the maxima observed in the corresponding properties of transfer. They could be ascribed to cavitational effects as supported by data in octane where the maxima are magnified.

B. Secondary Alcohols

The profiles of ΔG_t°, ΔH_t°, and $T\Delta S_t^\circ$ as functions of n_c for secondary alcohols in DTAB [77], NaDS [70], and DDAO [51] are similar to those for primary alcohols. In fact, the free energy of transfer decreases linearly with n_c while the enthalpy and the entropy show maxima which are located at different values of n_c with respect to those of primary alcohols. Similar to primary alcohols, the affinity of these additives toward NaDS micelles is larger than that toward DTAB and DDAO micelles, which, in turn, are comparable. The ΔH_t° and $T\Delta S_t^\circ$ values are comparable in NaDS and DDAO and smaller than those in DTAB.

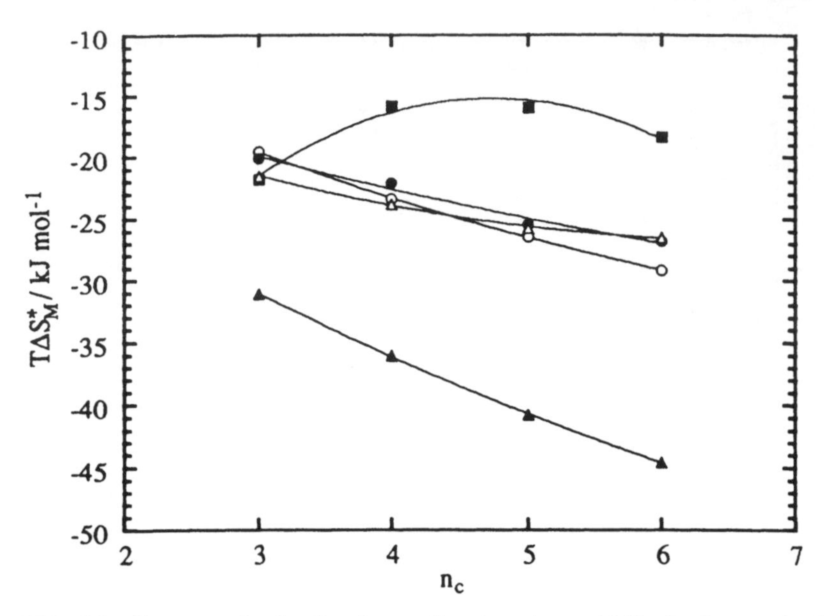

FIG. 21 Entropy of solvation (according to scheme (48)) of primary alcohols in water (▲, from Ref. 70), in octane (■, from Ref. 70), and in micellar phase of sodium dodecylsulfate (○, from Ref. 70), dodecyltrimethylammonium bromide (●, from Ref. 74), and sodium perfluorooctanoate (△, from Ref. 74).

The branching effect of a methyl group in the α position was obtained by subtracting the transfer thermodynamic quantities for primary alcohols from those of secondary alcohols immediately longer. The free energy of the process is negative and slightly increases with n_c. This means that the introduction of the methyl group in the α position in a primary alcohol determines an increase in the solubilizing power of the micelle because of the predominance of the changes in the hydrophilic interactions compared to the hydrophobic ones. In addition, regardless of the nature of the surfactant, the free-energy change of the process is essentially an entropic effect since small and, usually, positive enthalpic contributions are involved. The effect due to the introduction of a second methyl group can be obtained by subtracting the transfer thermodynamic properties for tertiary alcohols from those of secondary alcohols immediately shorter. Because of the lack of data, this is presently possible through *tert*-butanol and 2-propanol [44, 51, 58, 70, 77]. For this process there are free-energy changes less than half those obtained for the introduction of the first methyl group. This derives from large entropic effects which compensate for the enthalpic ones.

We recall that the properties of transfer for primary and secondary alcohols are very nearly the same up to hexanol. This means that the isomerization process, i.e., the removal of a CH_2 group from the linear chain and the introduction of a methyl group in the α position, does not involve important changes in the functions of transfer. However, more information can be drawn by comparing the thermodynamics of transfer of butanol isomers in DDAO, DTAB, and NaDS [44, 51, 58, 69, 77]. Regardless of the nature of the surfactant, the isomerization process is energetically favored when the primary alcohol becomes secondary while the opposite is found when the latter becomes tertiary. Contrarily, the isomerization process from butanol to isobutanol, which is related to the removal of the CH_2 group from the linear chain and the introduction of a CH_3 group in the β carbon atom, does not involve any change in the properties of transfer for DTAB, while for DDAO and NaDS there are small effects and the process is enthalpy driven.

C. Fluorinated Alcohols

The hydrophobicity of the alcohol is greatly affected when the hydrogenated alkyl chain is replaced by the fluorinated one. Sets of ΔG_t°, ΔH_t°, and $T\Delta S_t^\circ$ of a few fluorinated alcohols are available in sodium dodecanoate [62] and NaPFO [74] only, because of their low solubility and the presence of peculiarities in the trends of the experimental enthalpies as functions of m_S (such as those shown in Fig. 16).

Even if they are affected by uncertainties larger than those usually obtained, it is evident that ΔG_t° in both NaL and NaPFO is more negative with respect to those of the corresponding hydrogenated alcohols, while the opposite behavior is displayed by the enthalpy and the entropy. So, the more negative ΔG_t° for the fluorinated alcohols derives from a larger effect of fluorine atoms on $T\Delta S_t^\circ$ than on ΔH_t°. The relative magnitude of the thermodynamic properties of the transfer of fluorinated alcohols with respect to the hydrogenated ones can be ascribed to the larger hydrophobic effect of the fluorinated chain in the aqueous phase than to the larger affinity between hydrogenated chains and/or fluorinated chains in the micellar phase. However, free energy of solvation of hydrogenated alcohols in fluorinated and hydrogenated micelles shows that the affinity of the two kinds of micelles does not depend on the hydrophobic moiety of the micelle.

D. Nitriles and Nitroalkanes

The thermodynamics of transfer of the homologous series of nitriles are known in DTAB and DDAO [72], while that of nitroalkanes is known in DTAB [71] only.

The profiles of ΔG_t°, ΔH_t°, and $T\Delta S_t^\circ$ versus n_c are similar to those of alcohols: free energy decreases linearly, while enthalpy and entropy present maxima. The CH_2 group contribution to ΔG_t° is -1.7 kJ mol^{-1} from nitriles and -1.0 kJ mol^{-1} from nitroalkanes which are smaller than that obtained from alcohols (-2.3 kJ mol^{-1}). A tentative explanation of these results should take into account the swelling of the micelle due to the different hydrophilic interactions between the headgroup of the micelle and of the additive. Accordingly, the contribution to ΔG_t° of the transfer of the headgroup of the additive from the aqueous to DTAB micellar phase decreases in the same order as the CH_2 group contribution increases, namely 2.8, -3.6, and -5.6 kJ mol^{-1} for —OH, —CN, and —NO_2 groups, respectively.

This explanation, however, does not hold when the nitriles transfer in DTAB and DDAO are compared. In fact, as said above, the methylene group contribution is the same, while the hydrophilic one is three times smaller in DDAO (-1.3 kJ mol^{-1}). However, in the first case, additives having different hydrophilic moiety, in a given surfactant are compared, while in the second case additives in surfactants having a different polar moiety are compared.

The enthalpies of transfer of nitroalkanes are very small and about half those of nitriles, which, in turn, are half those of alcohols. Consequently, the transfer processes are entropy driven.

E. Benzene and Phenol

The very high solubility of benzene in hydrocarbons and the presence of π electrons in this molecule, which can interact with the hydrophilic moiety of the surfactant, have generated some discussion [78, 79] about the site of solubilization of benzene in the micellar phase.

Although the values of the thermodynamic properties cannot give insight into the benzene location in the micelle, appropriate comparisons can. The high value of the distribution constant, which is always higher than that of pentanol, can be due to either the strong adsorption on the micellar surface or to strong interactions with the hydrophobic moiety of the micelle. However, comparisons of thermodynamic properties of transfer in different surfactants can be useful in determining the relevant process. Unfortunately, literature data do not seem to give univocal information since the solubilization of benzene in the micellar phase strongly depends on the headgroup of the surfactant in ODAO and OAB [53], whereas it is about the same in DDAO, DTAB, and NaDS [58]. While in the latter case it is suggested that benzene penetrates the micelle, in the former the solubilization in the palisade layer, where benzene can interact with water, has been invoked. Regardless of the solubilization site, since the enthalpy of transfer is very small, the solubilization in the micellar phase is an entropic effect only.

The thermodynamics of transfer of phenol from the aqueous to the micellar phases of DTAB, DDAO, and NaDS compared to that of benzene indicates the importance of the introduction of the —OH group in the aromatic ring. Values of ΔG_t° for phenol in DTAB and DDAO are essentially the same and more negative than that in NaDS. These data agree with the idea that the interactions between the polarizable phenyl groups and headgroups of cationic surfactants are stronger than those involving anionic headgroups. Contrary to what is observed for benzene, the enthalpies of transfer are always largely negative and both enthalpy and entropy contribute to the transfer process for phenol.

VII. POSTMICELLAR TRANSITIONS

Surfactant solutions undergo micelle transitions at different concentrations, depending on the nature of the surfactant, temperature, pressure, and so on. As for micellization, these transitions occur in a broad range of concentrations since they are actually pseudophase transitions. In addition, since they can be detected with several techniques, having different sensitivity to the equilibria in solution, different values of the surfactant concentration at which transitions are detected can be obtained. For example, for CTAB, micellar transitions have been reported in the concentration range between 0.05 and 0.34 mol kg^{-1} [65, 73, 80–82].

Thermodynamics cannot give information on the nature of the transition, but it can detect it, particularly when the thermodynamic properties which are second derivatives of the Gibbs free energy of additives forming mixed micelles are analyzed. The transition is more easily detected at low additive concentrations [83]; the maximum effect is observed at infinite dilution when the additive behaves like a probe that does not affect the physicochemical properties of the micelles.

Volume and enthalpy of primary alcohols in NaDS [43, 70] and DTAB [44, 61] do not show peculiarities, while their second derivatives do. These experimental evidences have been tentatively explained [82] in terms of insensitivity of the volume to the change in counterion binding by invoking structural hydration effects. In fact, according to the Frank and Wen idea, since inorganic ions are surrounded by a coulombic hydration shell and a structure-broken region, the volume is primarily responsive to the first effect, while the second derivative thermodynamic function is more sensitive to the structural hydration. Therefore, the counterions which likely are not so tightly bound to the micelles, do not lose their coulombic hydration while the broken-structure region is altered to some extent. Consequently, an important change in the heat capacity and a rather small variation in the volume is consistent with a loss of structural hydration.

The deviations observed in the trends of the thermodynamic properties of the additive as a function of surfactant concentration due to the transition depend on the nature of both the additive and the surfactant.

If the apparent molar heat capacities of alcohols as a function of NaDS concentration are analyzed, maxima appear in the transition region whose amplitude increases with the alcohol alkyl chain length ranging between 10 J K^{-1} mol^{-1} for PrOH and 250 J K^{-1} mol^{-1} for HeptOH [84]. The amplitude of the maximum also depends on the nature of the polar head of the additive. For example, for $PeNO_2$, $PeNH_2$, and PeCN in DeTAB values of 80, 40, and 20 J K^{-1} mol^{-1}, respectively, are obtained, while no transition has been detected for PentOH [68].

However, the heat capacity of PentOH in longer alkyl chain surfactant indicates the presence of one transition in DTAB and two in TTAB and CTAB [65]. The same peculiarities have been pointed out by the apparent molar isoentropic compressibilities of PentOH in DTAB [49], TTAB [66], and CTAB [66]. In the case in which the two transitions were detected, that at high m_S was ascribed to a transition from a spherical to a rodlike shape, while that at low concentration was tentatively attributed to effects related to the counterion binding and structural hydration since, in the same concentration region, the aggregation numbers of TTAB in the presence of pentanol and hexanol show peculiarities [85].

Micellar transitions can also be promoted by the addition of additives. Recently, enthalpies of solution of some nonelectrolytes in micellar solutions have been carried out in the presence of electrolytes in order to detect micellar transitions [86]. For a given electrolyte, it has been observed that the effectiveness of the additive in inducing micellar transitions depends on the nature of the surfactant. For instance, hexylamine is less effective than PentOH in inducing a transition in TTAB, while the opposite is observed in NaDS. These findings seem to disagree with the above results for which $PentNH_2$ allows detection of DeTAB micellar transition better than PentOH does. Actually, the experimental conditions are quite different, since, in the latter case, standard partial molar properties of the additive in the absence of electrolytes were studied, while in the former one the enthalpies of solution were determined at additive concentration far from that at infinite dilution and in the presence of electrolytes. A further addition of urea and dioxane, which do not solubilize in the micellar phase, increases the amount of PentOH required to cause the transition with urea considerably more effective than dioxane.

Finally, the nature of the electrolytes plays an important role in micellar transitions. In fact, it was reported [86] that the effectiveness in inducing transitions of cationic inorganic ions decreases in the order $K^+ > NH_4^+ > Na^+$, while that of the anionic ones decreases in the order $SCN^- > NO_3^- > Br^- > OH^- > Cl^- > F^-$.

ACKNOWLEDGMENTS

The authors are grateful to the National Research Council of Italy (CNR, Progetto Finalizzato Chimica Fine II) and to the Ministry of University and of Scientific and Technological Research (MURST) for financial support.

SYMBOLS

cmc	critical micellar concentration (pseudo phase transition model)
cm_sc	critical microstructure concentration
$C_{\Phi,R}$	apparent molar heat capacity of the additive
Cp_R	standard partial molar heat capacity of the additive
Cp_b	standard partial molar heat capacity of the additive in the micellar phase
Cp_f	standard partial molar heat capacity of the additive in the aqueous phase
Cp_{oct}	standard partial molar heat capacity of the additive in octane
Cp_w	standard partial molar heat capacity of the additive in water
Cp^*	molar heat capacity of the additive
ΔCp_m	micellization heat capacity
E_S	partial molar expansibility of the micellized surfactant
ΔG_t°	free-energy change for the transfer of the additive from the aqueous to the micellar phases
ΔG_M^*	free energy of solvation in the micellar phase according to scheme (48)
$\Delta G_{s,M}^\circ$	standard free energy of solvation of the additive in the micellar phase

$\Delta G^\circ_{s,w}$	standard free energy of solvation of the additive in water
$\Delta G^\circ_R(\mathrm{W} \rightarrow \mathrm{ap})$	standard free energy of transfer of the additive from water to the aqueous phase
ΔH_s	enthalpy of solution of the additive in micellar solution
$\Delta H_{s,b}$	enthalpy of solution of the additive in the micellar phase
ΔH°_t	standard enthalpy of transfer of the additive from the aqueous to the micellar phases
ΔH_m	micellization enthalpy
ΔH^{exp}	enthalpy of mixing of the additive and surfactant solutions corrected for the enthalpy of dilution of surfactant with water
$\Delta H^w_{\mathrm{id},R}$	enthalpy of dilution of the additive in water
$\Delta H(\mathrm{W} \rightarrow \mathrm{W} + \mathrm{S})$	enthalpy of transfer of the additive from water to surfactant solution
$\Delta H_{s,w}$	enthalpy of solution in water
ΔH_t	enthalpy of transfer of the additive from water to the micellar phase
$\Delta H^\circ_{s,w}$	standard enthalpy of solvation in water
ΔH^*_M	enthalpy of solvation in the micellar phase according to scheme (48)
ΔH_v	enthalpy of evaporation
K	binding constant
K_N	ratio between the fraction of the additive in the micellar phase and that in the aqueous phase
K_M	equilibrium constant for the micellization process
K_C	partition constant of the additive between the aqueous and the micellar phases in the molarity scale
K_X	partition constant of the additive between the aqueous and the micellar phases in the mole fraction scale
K_S	Setchenov constant
$K_{S,w}$	isoentropic partial molar compressibility of the additive in water
$K_{S,b}$	isoentropic partial molar compressibility of the additive in the micellar phase
K^*_S	isoentropic molar compressibility of the additive
$K_{T,R}$	isothermal partial molar compressibility of the additive in micellar solution

$K_{T,b}$	isothermal partial molar compressibility of the additive in the micellar phase
$K_{T,f}$	isothermal partial molar compressibility of the additive in the aqueous phase
$\Delta K_{T,m}$	isothermal compressibility for the micellization process
$K_{T,S}$	isothermal partial molar compressibility of the surfactant in the micellized form at the cmc
$[m_0]$	unmicellized surfactant concentration (mass action model)
$[m]$	unmicellized surfactant concentration in the presence of additive
m_b	moles of the additive solubilized in the micellar phase per kg of water
m_f	moles of the additive solubilized in the aqueous phase per kg of water
m_R	stoichiometric additive concentration
m_S	stoichiometric surfactant concentration
m_S^{max}	concentration where maxima are predicted in the property of transfer from water to water + additive mixtures
n	aggregation number
n_c	number of carbon atoms in the hydrophobic alkyl chain
N_f	fraction of the additive solubilized in the aqueous phase
N_b	fraction of the additive solubilized in the micellar phase
P_e	vapor pressure
ΔS_t°	standard entropy of transfer of the additive from the aqueous to the micellar phase
V_M	micellar volume
V_S	partial molar volume of the surfactant in the micellized form
V_R	partial molar volume of the additive in surfactant solution
V_b	partial molar volume of the additive in the micellar phase
V_f	partial molar volume of the additive in the aqueous phase
V_w	standard partial molar volume of the additive in water

V_{oct}	standard partial molar volume of the additive in octane
V^*	molar volume of the additive
$\Delta V_{\Phi,S}(W \rightarrow W + R)$	volume of transfer of the surfactant from water to water + additive
$\Delta V_{\Phi,S}^{max}$	maximum volume change predicted in the property of transfer of the surfactant from water to water + additive mixture
Y_f	partial molar property of the additive in the aqueous phase
Y_b	partial molar property of the additive in the micellar phase
Y_w	partial molar property of the additive in water
Y_R	partial molar property of the additive in surfactant solution
Y_B	bulk property of binary system
Y_T	bulk property of ternary system
Y_m	partial molar property of the unmicellized surfactant
Y_S	partial molar property of the surfactant in the micellized form
Y_{H_2O}	partial molar property of water in surfactant solution
Y'_{H_2O}	partial molar property of water in water + surfactant + additive solution
Y'_m	partial molar property of the unmicellized surfactant in the presence of additives
Y'_S	partial molar property of the surfactant in the micellized form in the presence of additives
$Y_{\Phi,R}$	apparent molar property of the additive in surfactant solutions
ΔY_m	micellization property at the cmc
$\Delta Y'_m$	micellization property at the cmc in the presence of the additive
$\Delta Y_R(W \rightarrow W + S)$	property of transfer of the additive from water to the surfactant solution
ν	degree of dissociation of the unmicellized surfactant
β	degree of dissociation of micelles
Y_{RS}	additive-surfactant pair interaction parameter
V_{RS}	additive-surfactant pair interaction parameter for volume

H_{RS}	additive-surfactant pair interaction parameter for enthalpy
ODAO	octyldimethylamine oxide
OAB	octylammonium bromide
DDAO	dodecyldimethylamine oxide
DeTAB	decyltrimethylammonium bromide
DTAB	dodecyltrimethylammonium bromide
NaDS	sodium dodecylsulfate
TTAB	tetradecyltrimethylammonium bromide
CTAB	hexadecyltrimethylammonium bromide
DTAC	dodecyltrimethylammonium chloride
DDAC	dodecyldimethylammonium chloride
DAC	dodecylamine hydrochloride
NaDec	sodium decanoate
NaL	sodium dodecanoate
Na_2C_{15}	disodium 2-carboxytetradecanoate
NaRS	sodium alkylsulfate
RTAB	alkyltrimethylammonium bromide
RDAO	alkyldimethylamine oxide

REFERENCES

1. D. O. Shah, Ed., *Surface Phenomena in Enhanced Oil Recovery*, Plenum Press, New York, 1981.
2. C. A. Miller and S. Qutubuddin, in *Interfacial Phenomena in Non-Aqueous Media* (H. E. Eicke and G. D. Parfitt, eds), Marcel Dekker, New York, 1986.
3. T. J. Lin, in *Surfactants in Cosmetics* (M. M. Rieger, ed.), Marcel Dekker, New York, 1985.
4. J. H. Fendler and E. J. Fendler, Eds., *Catalysis in Micellar and Macromolecular Systems*, Academic Press, New York, 1975.
5. R. Zana, Ed., *Surfactant Solutions: New Methods of Investigation* Marcel Dekker, New York, 1987.
6. K. L. Mittal, Ed., *Micellization, Solubilization and Microemulsions* Plenum Press, New York, 1977.
7. K. L. Mittal and B. Lindman, Eds., *Surfactants in Solution* Plenum Press, New York, 1984.
8. K. L. Mittal, Ed., *Surfactants in Solution: Modern Aspects* Plenum Press, New York, 1989.
9. Th. F. Tadros, Ed., *Surfactants* Academic Press, London, 1984.
10. M. J. Schick, Ed., *Nonionic Surfactants. Physical Chemistry* Marcel Dekker, New York, 1987.
11. W. M. Linfield, Ed., *Anionic Surfactants* Marcel Dekker, New York, 1976.
12. P. Stilbs, *J. Colloid Interface Sci. 87*: 385 (1982).

13. Z. Gao, J. C. T. Kwak, R. Labonté, D. G. Marangoni, and R. E. Wasylishen, *Colloid Surf. 45*: 269 (1990).
14. K. Hayase and S. Hayano, *Bull. Chem. Soc. Jpn. 50*: 83 (1977).
15. C. Treiner, *J. Colloid Interface Sci. 93*: 33 (1983).
16. K. Shirahama and T. Kashiwabara, *J. Colloid Interface Sci. 36*: 65 (1971).
17. M. Manabe, H. Kawamura, S. Kondo, M. Kojima, and S. Tokunaga, *Langmuir 6*: 1596 (1990).
18. C. A. Bunton and G. Savelli, *Adv. Phys. Org. Chem. 22*: 213 (1986).
19. J. Burgess and E. Pelizzetti, *Gazz. Chim. Ital. 12*: 803 (1988).
20. G. Calvaruso, F. P. Cavasino, and C. Sbriziolo, *J. Chem. Soc. Faraday Trans. 87*: 3033 (1991).
21. H. Høiland, E. Ljosland, and S. Backlund, *J. Colloid Interface Sci. 101*: 467 (1984).
22. R. Zana, S. Yiv, C. Strazielle, and P. Lianos, *J. Colloid Interface Sci. 80*: 208 (1981).
23. S. D. Christian, E. E. Tucker, and E. H. Lane, *J. Colloid Interface Sci. 84*: 423 (1981).
24. S. Kaneshina, H. Kamaya, and I. Ueda, *J. Colloid Interface Sci. 83*: 589 (1981).
25. E. Vikingstad and H. Høiland, *J. Colloid Interface Sci. 64*: 510 (1978).
26. O. Kvammen, G. Kolle, S. Backlund, and H. Høiland, *Acta Chem. Scand. A 37*: 393 (1983).
27. G. N. Lewis and M. Randall, *Thermodynamics*, McGraw-Hill, New York, 1961.
28. H. S. Harned and B. B. Owens, *The Physical Chemistry of Electrolyte Solutions* Van Nostrand Reinhold, New York, 1958.
29. I. M. Klotz and R. M. Rosenberg, *Chemical Thermodynamics. Basic Theory and Practice* Benjamin, New York, 1972.
30. W. McMillan and J. Mayer, *J. Phys. Chem. 13*: 176 (1945).
31. J. E. Desnoyers, G. Caron, R. De Lisi, D. Roberts, A. Roux, and G. Perron, *J. Phys. Chem. 87*: 1397 (1983).
32. G. Caron, G. Perron, M. Lindheimer, and J. E. Desnoyers, *J. Colloid Interface Sci. 106*: 324 (1985).
33. G. Gunnarson, B. Johnson, and H. Wennerström, *J. Phys. Chem. 84*: 3114 (1980).
34. T. E. Burchfield and E. M. Woolley, *J. Phys. Chem. 88*: 2149 (1984).
35. I. Johnson, G. Olofsson, M. Landgren, and B. Jönsson, *J. Chem. Soc. Faraday Trans. I 85*: 4211 (1989).
36. R. De Lisi, G. Perron, and J. E. Desnoyers, *Can. J. Chem. 58*: 959 (1980).
37. E. M. Woolley and M. T. Bashford, *J. Phys. Chem. 90*: 3038 (1986).
38. A. Ben-Naim, *J. Phys. Chem. 82*, 792 (1978).
39. A. H. Roux, D. Hétu, G. Perron, and J. E. Desnoyers, *J. Solution Chem. 13*: 1 (1984).
40. D. Hétu, A. H. Roux, and J. E. Desnoyers, *J. Solution Chem. 16*: 529 (1987).
41. R. De Lisi, V. Turco Liveri, M. Castagnolo, and A. Inglese, *J. Solution Chem. 15*: 23 (1986).

42. R. De Lisi, S. Milioto, M. Castagnolo, and A. Inglese, *J. Solution Chem. 16*: 373 (1987).
43. R. De Lisi, A. Lizzio, S. Milioto, and V. Turco Liveri, *J. Solution Chem. 15*: 623 (1986).
44. R. De Lisi, S. Milioto, and V. Turco Liveri, *J. Colloid Interface Sci. 117*: 64 (1987).
45. R. De Lisi and S. Milioto, *J. Solution Chem. 16*: 767 (1987).
46. M. Mansson, P. Sellers, G. Stridh, and S. Sunner, *J. Chem. Thermodyn. 8*: 1081 (1976).
47. D. Mirejovsky and E. M. Arnett, *J. Am. Chem. Soc. 105*: 111 (1983).
48. R. De Lisi, S. Milioto, and A. Inglese, *J. Phys. Chem. 95*: 3322 (1991).
49. R. De Lisi, S. Milioto, and R. E. Verrall, *J. Solution Chem. 19*: 97 (1990).
50. J. E. Desnoyers, D. Hétu, and G. Perron, *J. Solution Chem. 12*: 427 (1983).
51. S. Milioto, D. Romancino, and R. De Lisi, *J. Solution Chem. 16*: 943 (1987).
52. E. Vikingstad and O. Kvammen, *J. Colloid Interface Sci. 74*: 16 (1980).
53. D. Hétu, A. H. Roux, and J. E. Desnoyers, *J. Colloid Interface Sci. 122*: 418 (1988).
54. D. Hétu and G. Perron, *J. Solution Chem. 20*: 207 (1991).
55. G. Roux, G. Perron, and J. E. Desnoyers, *J. Solution Chem. 7*: 639 (1978).
56. J. Lara, G. Perron, and J. E. Desnoyers, *J. Phys. Chem. 85*: 1600 (1981).
57. J. F. Alary, M. A. Simard, J. Dumond, and C. Jolicoeur, *J. Solution Chem. 12*: 755 (1983).
58. S. Causi, R. De Lisi, and S. Milioto, *J. Solution Chem. 19*: 995 (1990).
59. D. Hétu and J. E. Desnoyers, *Can. J. Chem. 66*: 767 (1988).
60. C. Tanford, *The Hydrophobic Effect: Formation of Micelles and Biological Membranes* Wiley, New York, 1980.
61. R. De Lisi, S. Milioto, M. Castagnolo, and A. Inglese, *J. Solution Chem. 19*: 767 (1990).
62. S. Milioto, S. Causi, and R. De Lisi, *J. Colloid Interface Sci. 155*: 452 (1993).
63. E. Vikingstad, *J. Colloid Interface Sci. 72*: 75 (1979).
64. P. A. Leduc, J. L. Fortier, and J. E. Desnoyers, *J. Phys. Chem. 78*: 1217 (1974).
65. R. De Lisi, S. Milioto, and R. Triolo, *J. Solution Chem. 17*: 673 (1988).
66. R. De Lisi, S. Milioto, and R. E. Verrall, *J. Solution Chem. 19*: 639 (1990).
67. R. De Lisi, E. Fisicaro, and S. Milioto, *J. Solution Chem. 18*: 403 (1989).
68. R. De Lisi, S. Milioto, and R. Triolo, *J. Solution Chem. 18*: 905 (1989).
69. E. Abuin and E. A. Lissi, *J. Colloid Interface Sci. 95*: 198 (1983).
70. R. De Lisi and S. Milioto, *J. Solution Chem. 17*: 245 (1988).
71. S. Milioto and R. De Lisi, *Thermochim. Acta 137*: 151 (1988).
72. S. Milioto and R. De Lisi, *J. Solution Chem. 17*: 937 (1988).
73. C. Treiner, A. K. Chattopadhyay, and R. Bury, *J. Colloid Interface Sci. 104*: 569 (1985).
74. R. De Lisi and S. Milioto, *Langmuir 10*: 1377 (1994).
75. R. Lumry and S. Rajender, *Biopolymers 9*: 1125 (1970).
76. R. Lumry and H. S. Frank, *Proc. Sixth International Biophysics Congress*, Vol. VII, 1978, pp. 20–54.

77. S. Milioto and R. De Lisi, *J. Colloid Interface Sci. 123*: 92 (1988).
78. P. Mukerjee and J. R. Cardinal, *J. Phys. Chem. 14*: 1620 (1978).
79. C. Hirose and L. Sepulveda, *J. Phys. Chem. 85*: 3689 (1981).
80. P. Ekwall, L. Mandell, and P. Solyom, *J. Colloid Interface Sci. 35*: 519 (1971).
81. G. Lindblom, B. Lindmann, and L. Mandell, *J. Colloid Interface Sci. 42*: 400 (1973).
82. F. Quirion and J. E. Desnoyers, *J. Colloid Interface Sci. 112*: 565 (1986).
83. G. Roux-Desgranges, A. H. Roux, J. P. Grolier, and A. Viallard, *J. Solution Chem. 11*: 357 (1982).
84. G. Roux-Desgranges, A. H. Roux, and A. Viallard, *J. Chim. Phys. 82*: 441 (1985).
85. P. Lianos and R. Zana, *J. Colloid Interface Sci. 101*: 587 (1984).
86. D. Nguyen and G. L. Bertrand, *J. Colloid Interface Sci. 150*: 143 (1992).
87. E. D. Goddard and G. C. Benson, *Can. J. Chem. 35*: 986 (1957).
88. R. Zana, S. Yiv, C. Strazielle, and P. Lianos, *J. Colloid Interface Sci. 80*: 208 (1981).
89. H. Høiland, *J. Solution Chem. 5*: 773 (1976).
90. H. Høiland and E. Vikingstad, *Acta Chim. Scand. A 30*: 182 (1976).
91. M. Manabe, K. Shirahama, and M. Koda, *Bull. Chem. Soc. Jpn. 49*: 2904 (1976).
92. E. B. Abuin, E. Valuenzela, and E. A. Lissi, *J. Colloid Interface Sci. 95*: 198 (1983).
93. C. Jolicoeur and G. Lacroix, *Can. J. Chem. 54*: 624 (1976).

4

Solubilization of Uncharged Molecules in Ionic Micellar Solutions: Toward an Understanding at the Molecular Level

BENGT JÖNSSON and MIKAEL LANDGREN Division of Physical Chemistry 1, Chemical Center, University of Lund, Lund, Sweden

GERD OLOFSSON Division of Thermochemistry, Chemical Center, University of Lund, Lund, Sweden

SYNOPSIS

The theoretical work done at the Chemical Center in Lund on the micellization of ionic surfactants and the solubilization of uncharged molecules in solutions of ionic surfactants is summarized. The aim of this research is to construct a model that may be used to give a quantitative description of the thermodynamic properties of the systems from the knowledge of the molecular properties of the individual components. The experimental observations which give the essential characteristics of ionic amphiphiles are summarized, and the various contributions to the free energy of micellization considered in the model are discussed. Examples are given of properties of ionic amphiphile-water systems for which the model can give a nearly quantitative description. The solubilization of uncharged molecules in ionic micelles has been treated at two different levels of approximation. In the first version of the model all molecules in the micelles were assumed to be anchored at the surface of the micelle. This version was used to evaluate calorimetric measurements on the solubilization of pentanol in sodium dodecylsulfate (SDS) micelles. It was found to give an adequate description of the essential features of micelle formation of SDS in the presence of pentanol. A new model where the interior of the micelle is divided into two regions, an inner core region with no amphiphiles and an outer palisade region, was proposed to describe systems containing nonpolar solutes such as alkanes. The distribution of the solubilizate between the two regions in the micelles can be calculated and as an example the solubilization of octane in decanoate micelles is discussed. Finally, current efforts to improve the model are described.

I. INTRODUCTION

Ionic surfactants have been used for a long time in many technical applications to increase the solubility in aqueous solutions of nonpolar or slightly polar compounds such as different types of oils and fats. Over the years a wide variety of empirical information has been gained about the properties and phase behavior of these systems [1–5]. A considerable amount of theoretical work has also been done to explain and rationalize all this information. We will in this article give a brief summary of the theoretical work done at the Chemical Center in Lund on the micellization of ionic surfactants [6–12] and the solubilisation of uncharged molecules in solutions of ionic surfactants [13–17]. The aim of this research is to construct a model that may be used to give a quantitative description of the thermodynamic properties of this type of systems from the knowledge of the molecular properties of the individual components. Gunnarsson, Jönsson, and Wennerström presented in 1980 a thermodynamic model for the association of ionic surfactants into micelles [6]. The model, in this work called the PBCM model, has after that been improved and extended to be able to describe different properties of systems containing nonpolar or slightly polar solutes in addition to ionic amphiphiles. In the model the free energy of an ionic amphiphile-water system is expressed as a function of the composition, size, and form of the micellar aggregates as well as the concentration of the micelles. This means that the model may be used to describe the influence of the amphiphile concentration on such properties as phase equilibria and component distribution in micellar systems. We will here summarize and discuss the assumptions made in the construction of the PBCM model and mention some of the work that is going on to improve the model.

The solubility of nonpolar or slightly polar molecules in water varies strongly and in a complex way with the addition of surfactants. Such behavior is illustrated in Fig. 1, which shows a section of the phase diagram of the potassium decanoate, water, and octanol system [15] calculated from the PBCM model.

The phase diagram shows three characteristic features of this type of system:

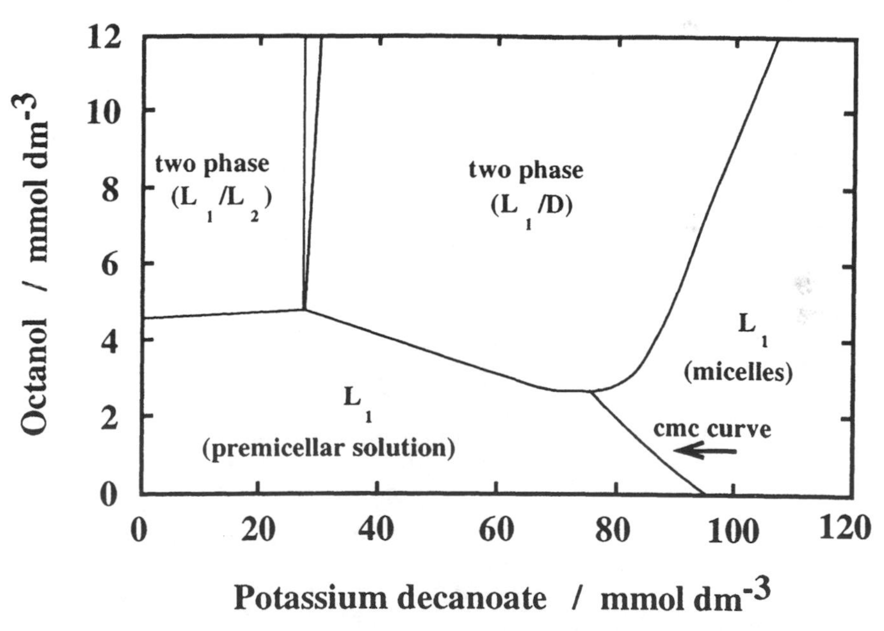

FIG. 1 Part of the calculated phase diagram for the system potassium decanoate-water-octanol. L_1 denotes an isotropic aqueous solution, L_2 a solution of octanol saturated with water, and D a lamellar liquid crystalline phase. (Data from Ref. 15.)

1. The strong increase in the solubility of octanol when micelles start to form.
2. The strong influence of octanol on the cmc (the characteristic concentration when the micelles starts to form).
3. Variation in the nature of the phase that separates when the octanol concentration exceeds the solubility limit. At low amphiphile concentrations an isotropic solution of octanol saturated with water is separated, but at higher amphiphile concentrations a lamellar liquid crystalline phase is formed.

In the first version of the PBCM model [6–8] only two-component systems were treated, but the model was later extended to three-component systems [11]. The solubilization of a third component is in the model regarded as a modification of the aggregation of the pure amphiphile. Therefore, the thermodynamics of the binary amphiphile-water system must first be analyzed before the effect of a third component can be described. However, before we start to discuss the different parts of the PBCM model, we will briefly summarize what is known today about micellar aggregates of ionic amphiphiles.

II. THE IONIC MICELLE

The space-filling model of an octanoate micelle shown in Fig. 2 summarizes the basic features of an ionic micelle. The following experimental observations give the important characteristics of micelles of ionic amphiphiles:

1. Hydrocarbon chains are the dominant constituent in the inner part of a micelle. The concentration of water is low and the concentration of ionic headgroups is negligible in this region [1, 2].
2. The hydrocarbon chains in the core of a micelle are in a fluid state [1, 2, 18].
3. The ionic headgroups, some counterions, water molecules, and parts of the hydrocarbon tails are mixed in a thin layer which surrounds the micelle. The surface layer is some Ångströms thick [19, 20].
4. The geometrical form of the micelle depends on the concentration of amphiphile and salt in the system. The spherical form dominates at concentrations near the cmc and the prolate form at high salt or amphiphile concentrations [1, 2].
5. The size of micelles is not well defined and they are usually polydisperse. The mean size and polydispersity may vary with surfactant and salt concentration [1, 2].

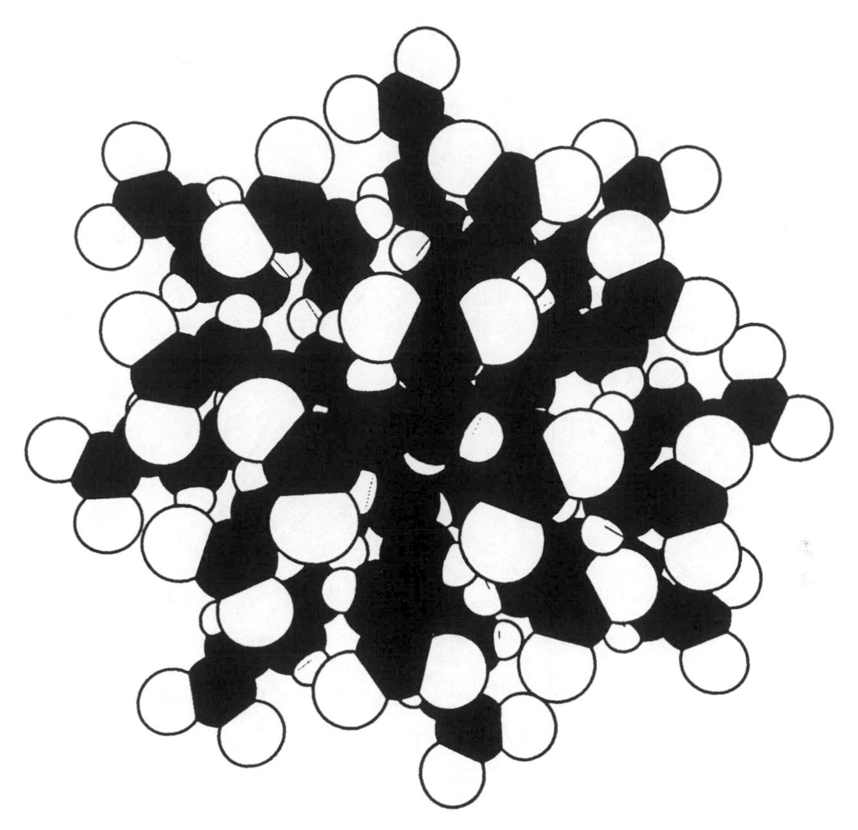

FIG. 2 A model of a spherical octanoate micelle. The model is constructed of space-filling models of octanoate molecules.

6. The micellar phase becomes unstable at high amphiphile or salt content. The phases in equilibrium with the micellar phase are most often the hexagonal and lamellar liquid crystalline phases [3, 4].

Solubilization of a third component may affect the geometrical form of the micelles. If many uncharged molecules are solubilized, the micellar phase may become unstable. As in binary amphiphile-water systems, hexagonal or lamellar liquid crystalline phases are commonly found to be in equilibrium with the micellar phase. Figure 3 shows a typical phase diagram for an ionic amphiphile–long-chain alcohol–water system.

When discussing micelle formation or the solubilization capacity of a micellar system it is not sufficient to treat the thermodynamics of only

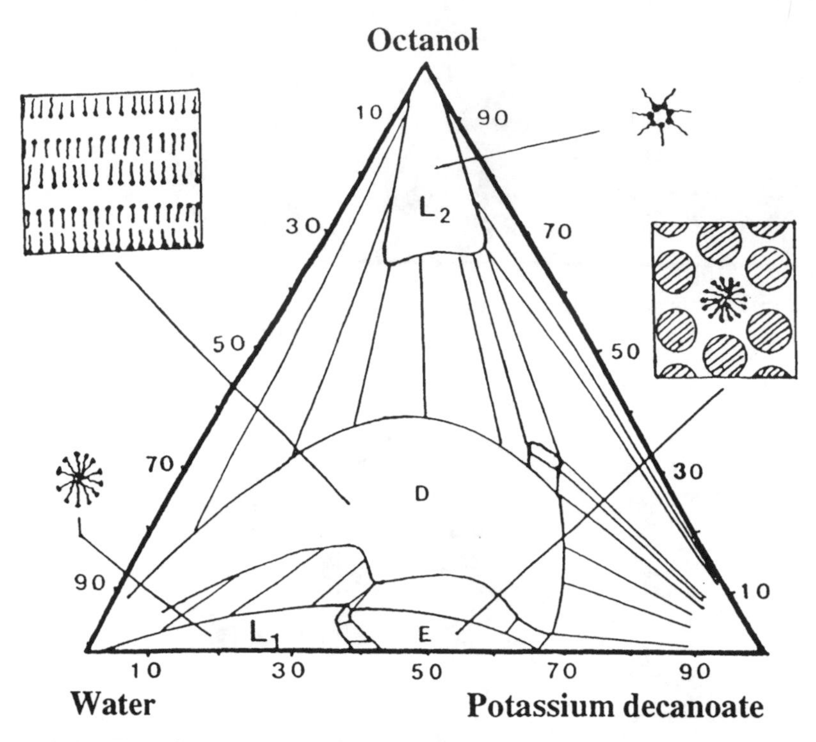

FIG. 3 Experimental phase diagram of the ternary system potassium decanoate-water-octanol at 20°C. L_1 denotes a micellar solution, E a normal hexagonal liquid crystal, D a lamellar liquid crystal, and L_2 a reversed micellar solution. (Adapted from Ref. 21.)

the micellar phase since the stability range will also depend upon the thermodynamic properties of other phases that can be in equilibrium with the micellar phase.

III. THERMODYNAMICS OF MICELLIZATION

A full description of the molecular interactions in an ionic amphiphile water system is at present not feasible. Instead the systems were treated from a semimacroscopic view and the thermodynamic model was developed taking into account the most important free-energy contributions to the micellar aggregates as a whole. It is clear that the electrostatic Coulomb interactions between the ionic groups play a major role in determining the thermodynamic properties of this type of system, and the model emphasizes a consistent treatment of the electrostatic effects. The chemical potentials of the various components of an ionic amphiphile-water system are derived from a model expression for the free energy. The various contributions to the free energy considered in the PBCM model are described below. The ways the estimates of the contributions were derived are outlined, but readers are referred to the original publications for the derivation of the explicit relations and descriptions of calculation procedures. The PBCM model can describe for instance how the monomer activity varies with micelle concentration and the variation of the cmc with alkyl chain length, salt concentration, and counterion valency. Counterion "binding" can be discussed in a conceptually clear way [6]. Stability regions of different phases including two-phase regions can also be derived [8].

A. The Hydrophobic Effect

The main reason for the aggregation of amphiphiles to micelles is the strong interaction between the water molecules in an aqueous solution. This strong water self-association obstructs the mixing of water with nonpolar compounds and hence with the hydrocarbon tails of an amphiphile. The difference between the free energy of different types of nonpolar solutes in water and in hydrocarbon solutions has been carefully measured and much work has been done to gain a deeper theoretical understanding of the so-called hydrophobic effect [22].

By comparing the solubility in water of different compounds containing aliphatic chains the following relation is usually observed for homologous series:

$$\Delta\mu = K_1 + K_2 n(\mathrm{CH_2}) \tag{1}$$

where $\Delta\mu$ indicates the change in the chemical potential when the solute is transferred from hydrocarbon to a water solution, K_1 is a constant that depends on the functional group of the series, K_2 is a constant roughly independent of the functional group, and $n(CH_2)$ is the number of CH_2 groups in the molecule. For charged carboxylic acids the values of K_1 and K_2 are $-4.4kT$ and $1.39kT$ at room temperature [22]; k is the Boltzmann constant and T is the absolute temperature. This means that the contribution to the hydrophobic effect from each CH_2 group is around $1.4kT$ at room temperature.

However, the interior of a micelle differs in some respects from a bulk solution of hydrocarbon. First, a micelle is such a small aggregate that a considerable part of the hydrocarbon tails are in contact with the surrounding water. Second, the amphiphilic molecules are anchored with their headgroups at the micellar surface. The difference between the free energy of amphiphilic compounds in an aqueous solution or in a micellar aggregate may be estimated from their cmc values. The free-energy contribution per CH_2 group in a homologous series of amphiphiles was found in such an investigation [6] to be $1.31kT$ per CH_2 group, a value not very different from the value found from solubility measurements. This means that the interior of a micelle has properties which in many respects are similar to a fluid hydrocarbon solution.

B. Chain Conformation Free Energy

As mentioned, an important difference between the interior of a micelle and the interior of an oil droplet is the anchoring of the ionic headgroup of the amphiphile at the surface of the micelle. The number of conformations that the hydrocarbon chain can occupy is smaller in a micelle than in an oil droplet. In order to compare the stability of different types of aggregates it is important to know how the number of chain conformations changes as a function of form and size of the aggregate. Different models have been proposed to estimate the number of possible chain conformations in micelles and other types of aggregates [9].

The most commonly used theory describing chain conformations in the interior of a micelle is a model put forward by Gruen [23]. In this model the chain conformations are generated by a weighted random walk starting at the surface of the micelle. The different steps in the random walk process are weighted in order to give a constant hydrocarbon concentration in all parts of the micelle. The conformational free energy for some geometries calculated by the Gruen model is shown in Fig. 4. The difference in chain packing entropy between different micellar sizes is rather small for spherical micelles but it is larger for cylindrical and lamellar aggregates as may be seen in Fig. 4. The reason for the small size dependence when the aggregate is spherical is that only a minority of the hydrocarbon chains in a spherical micelle need to be stretched to their maximum length to fill the center of the micelle.

The main drawback of the Gruen model is that the entropy of the amphiphilic headgroup is not considered. All headgroups are either assumed to be in the same layer or in an a priori assumed density profile. The difference in the lateral mobility of the headgroups as a function of the aggregate dimensions is also neglected. An improved theory that does not assume a priori a specific density profile of the headgroups is the self-consistent field lattice theory for adsorption or association of chain molecules (SCFA) [24–26]. This model was originally developed by Scheutjens and

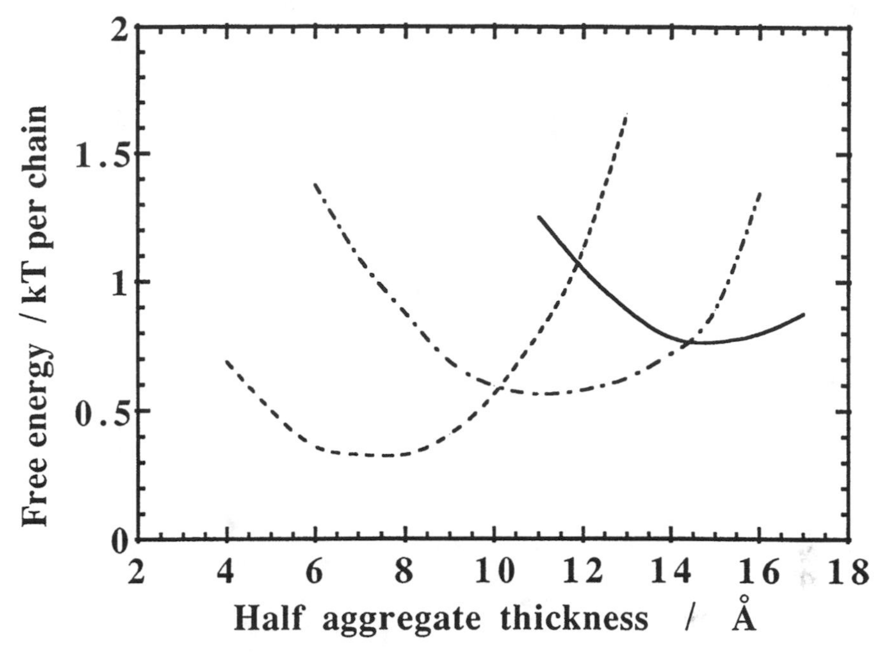

FIG. 4 Free energy cost of packing the hydrocarbon chains of SDS into bilayers (- - -), cylindrical aggregates (-·-·), and spherical micelles (—) as a function of the aggregate dimension. (Data from Ref. 23.)

Fleer to model the adsorption of polymers but was later adapted by Leermakers and Scheutjens [27–29] to describe micelles and bilayers. Electrostatic interactions have now also been incorporated in the model [30, 31]. Böhmer et al. [32] have used this model to calculate the concentration profiles of the different components in a micellar system. The result from such a calculation may be seen in Fig. 5.

An interesting result shown in Fig. 5 is the broad headgroup region. The thickness of this region is estimated to be 5 to 10 Å, which is a considerable part of the micelle. However, the thickness may be overestimated since the strong electrostatic interaction between the ionic headgroups and the surrounding water molecules is suppressed in the model calculations. The concentration of water in the interior of the micelle is also higher than experimentally observed [1, 2].

In the PBCM model no molecular theory has been used to estimate the chain packing restrictions. Instead the free-energy contribution from the chain conformations for a specific type of aggregate, for example micelles,

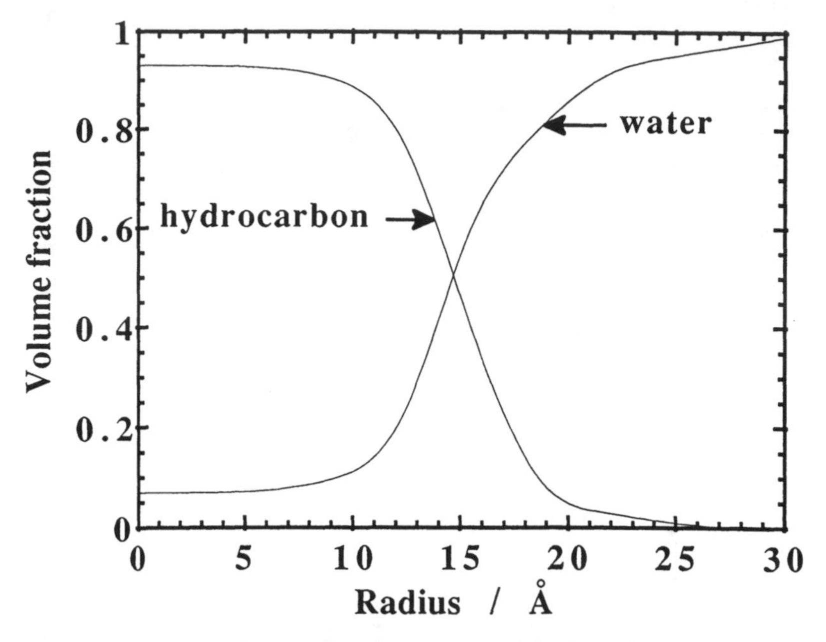

FIG. 5 Concentration profiles for water and hydrocarbon in a spherical SDS micelle as a function of the distance to the center of the micelle. The curves are calculated from the SCFA theory [32].

was assumed to be a function of the interfacial area A per amphiphile in the aggregate. The first two terms in a Taylor series expansion of this energy dependence are then used to describe the free energy of the chain conformations.

$$G_{\text{conf}}(A) = \text{constant} + \left(\frac{\partial G_{\text{conf}}}{\partial A}\right)_{A_0} A \tag{2a}$$

The derivative of the conformational free energy at an area A_0, $(\partial G_{\text{conf}}/\partial A)_{A_0}$, will have the same dimension as a surface tension and is, therefore, denoted by γ_{conf} in the PBCM model.

$$G_{\text{conf}} = \text{constant} + \gamma_{\text{conf}} A \tag{2b}$$

C. Interactions Between Ionic Headgroups and Surrounding Water Molecules

The strong interaction between an ionic headgroup and the surrounding water is of the same importance for the micellization process as the hydrophobic effect. The micelle cannot grow to all types of aggregate geometries because the ionic headgroups must be placed near the surface of the micelle. This means that at least one dimension of the aggregate is restricted by the maximum length of the hydrocarbon tail of the surfactant. This restriction may be circumvented if nonpolar or slightly polar molecules are solubilized in the micelle, as will be discussed later. The free-energy expense of transferring an ionic group from a bulk solution of water to the neighborhood of a hydrocarbon surface is shown in Fig. 6 as a function of the distance to the interface between the two regions.

So far the distribution of the headgroups has been neglected in the PBCM model, and instead they are all placed in a shell that defines the surface of the micelle. This means, for example, that the maximum size of a spherical micelle is restricted by the length of the hydrocarbon tail of the surfactant. However, work is in progress to incorporate into the PBCM model the possibility of having a wider headgroup distribution.

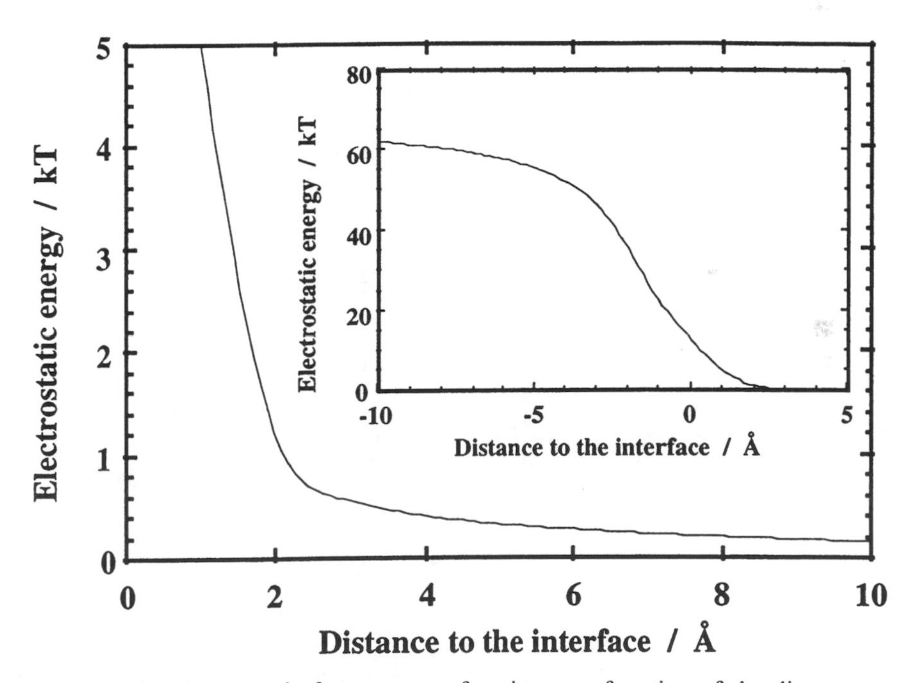

FIG. 6 The electrostatic free energy of an ion as a function of the distance to an interface between a water and a hydrocarbon region. A positive distance means that the ion is in a water solution, a negative that it is in the hydrocarbon solution. The full curve from -10 to $+5$ Å is shown in the reduced figure. (Data from Ref. 33.)

D. The Interfacial Energy

The major difference from a thermodynamic point of view between different types of aggregate geometries is the difference in the interfacial area per amphiphile. For example, the mean area per amphiphile is much larger in a micelle than in a bilayer. The reason for this is the difference in area-to-volume ratio between the different aggregate geometries. Since the volume per amphiphile in an aggregate has been shown from experiments to be nearly independent of the aggregate geometry and since all headgroups are located near the surface of the aggregate, the area-to-volume ratio of the amphiphile will be the same as the area-to-volume ratio of the aggregate.

The large difference between the free energy of a hydrocarbon molecule in an aqueous and a hydrocarbon environment, the hydrophobic effect, was previously expressed as a function of the number of CH_2 groups in the molecule; see Eq. (1). Another possible way to express this dependence is as a function of the increase of the interfacial area between hydrocarbon and water when a molecule moves from a hydrocarbon to an aqueous solution. Because of the hydrocarbon-water contact in the surface layer of a micelle, the increase of the hydrocarbon-water interface when transferring a molecule from a micelle into the surrounding aqueous solution is not as large as if the molecule was transferred from a bulk hydrocarbon solution. Therefore, the difference between the free energy of a surfactant in an aqueous solution or in a micelle is not described by Eq. (1).

Another difference between an oil droplet and a micelle that gives an important contribution to the free energy of the ionic aggregate is the distribution of the headgroups in the interfacial region. The following example may illustrate how this headgroup distribution is related to the interfacial area A and how it may influence the free energy of the ionic aggregate.

Assume that the thickness of the headgroup region, Δ, is independent of the size of the aggregate and that the interaction between the headgroups may be described as the interaction between hard spheres with a volume V_0. The Carnahan-Starling equation [34] may then be used to calculate the chemical potential of amphiphile, μ_a, connected with the headgroup distribution.

$$\mu_a = \text{constant} + kT\left(-\ln(A) + \frac{7\varphi - 3\varphi^2 - \varphi^3}{(1-\varphi)^2}\right) \quad \text{where } \varphi = \frac{V_0}{\Delta A} \qquad (3)$$

It is clear from this equation that the headgroup entropy must be very dependent on the area per amphiphile, A. However, Eq. (3) is too simplified and can only be used as a qualitative example. The interactions between the different components in the surface region of a micelle differ significantly from the hard sphere interaction used to derive Eq. (3).

The easiest way to treat electrostatic interactions between the ionic headgroups is to approximate these ions with a smeared-out surface charge density at the interface between the aggregate and the surrounding water. The electrostatic interactions may then be calculated from the Poisson-Boltzmann equation, as will be discussed later. However, it is important to point out that the entropy connected with the headgroup distribution is neglected in this approximation and this entropy will depend of the interfacial area per amphiphile. More detailed calculations of the headgroup entropy from models where the electrostatic interactions, the hydrophobic effect, and steric interactions are all taken into account can today only be made as SCFA calculations or Monte Carlo simulations [33]. A deeper understanding of the significance of different energy contributions may be obtained by use of the Monte Carlo simulation technique. The distribution of the ionic headgroups in a bilayer of catanionic amphiphiles [35], shown in Fig. 7, is an example of such a calculation [33]. Here the effect on the headgroup distribution of the large difference in the dielectric constant between the interior of a surfactant aggregate and the surrounding aqueous solution was investigated. The considerable narrowing of the headgroup distribution observed here is due to this difference. This effect is also observed for other types of ionic systems and is not specific for catanionic systems.

The effective interaction between the ionic headgroups on the surface of a micellar aggregate is very difficult to model from first principles, and another strategy was therefore used in the construction of the PBCM model. Instead an empirical approach was used, and we assumed that for each type of aggregate (micellar, hexagonal, lamellar, and so on) the total free energy connected with the properties discussed in Secs. III.A.–D. could be approximated by a constant term plus a term that depended on the interfacial area per amphiphile, A, in the aggregate. These two terms may be looked upon as the first two terms in a series evaluation of the real free-energy dependence on the interfacial area.

$$G = G_0 + \gamma A \tag{4}$$

The values of G_0 and γ may, of course, be different for different types of aggregates but may be derived from a comparison between experimental

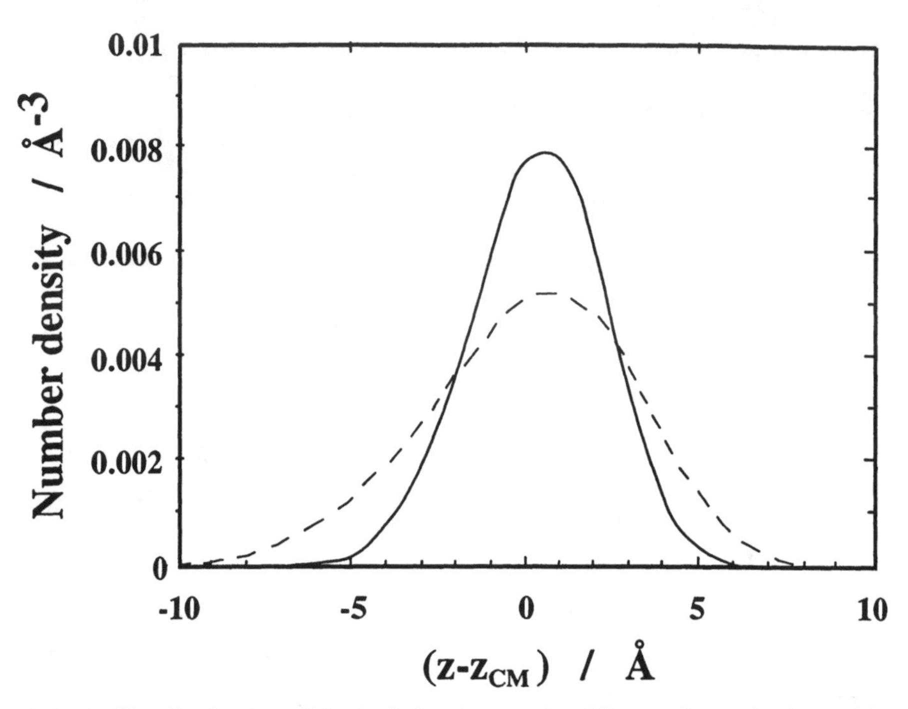

FIG. 7 The distribution of the ionic headgroups in a bilayer of catanionic amphiphiles for two different descriptions of the ion-water interaction. $Z - Z_{CM}$ denotes the distance relative to the center of mass of the headgroup layer. The full- drawn curve was calculated taking into account the dependence of the free energy on the distance to the hydrocarbon-water interface, shown in Fig. 6. This dielectric discontinuity was neglected when calculating the dashed curve. (Data from Ref. 33.)

results and calculated predictions. The critical micelle concentration (cmc) is in the PBCM model strongly dependent of the G_0-value, while the growth of the aggregates as well as the phase equilibria between different phases strongly depends on γ. To our surprise, such calculations have shown that values of both G_0 and γ are nearly independent of the geometry of the aggregate [7]. The value of γ seems also to be rather insensitive to the length of the hydrocarbon chain of the amphiphile [8]. The reason for all this is not understood. The constant γ has the same dimension as surface tension, and in the various calculations using the PBCM model we have used the value 0.018 N/m for pure amphiphile micelles. Note that the constant γ in Eq. (4) cannot be identified with the macroscopic interfacial tension between hydrocarbon and water, which value is about 0.05 N/m at 20°C [36].

E. Electrostatic Interactions

In the PBCM model the mean electrostatic interaction between the ionic headgroups as well as the interaction between the co- and counterions outside a micellar aggregate are described by the Poisson-Boltzmann equation (PB) [6]. The distribution of counterions in a system may easily be calculated from the PB equation if all micellar charges are assumed to be restricted to a single shell at the surface of the micelle. A broad headgroup distribution will complicate the calculations since the number of counterions incorporated into the surface layer is difficult to estimate. One way to obtain such information would be to use the SCFA model.

Electrostatic micelle-micelle interactions give an important contribution to the free energy of concentrated micellar solutions. These interactions and their dependence on amphiphile concentration and salt additions are estimated in the PBCM model by using a cell model [6]. This means that the counterions to a micelle are assumed to be in a cell surrounding the micelle. The size of the cell is chosen as the total volume of the micellar solution divided by the number of micelles. The form of the micelle is usually assumed to be spherical, but spheroidal cells have also been used. The thickness of the counterion layer around the micelles is reduced when the concentration of micelles is increased, and it is this reduction that is modeled by the cell model.

At a first glance the relation between the ion distribution in a cell and the free energy and chemical potentials in the system seems to be a rather difficult problem to solve, but these relations can be made and are presented in Ref. 8. Before the chemical potentials can be calculated, the electrostatic potentials and ion concentrations in the system must be obtained from a solution of the PB equation. For spherical and spheroidal systems, the forms most used to describe micelles, there is no simple analytical solution to the PB equation and therefore numerical solutions must be used. Due to difficulties in solving the PB equation for complex geometries, the form of the aggregates is often assumed to be spherical, cylindrical, or planar. The reason is that the PB equation for these geometries can be written as an equation that only depends on one coordinate, which simplifies the calculations. The PB equation has also been solved for other types of geometries, as for example spheroidal geometries. These calculations are, however, rather involved and the method has only been used occasionally; see Ref. 13.

IV. PROPERTIES OF IONIC AMPHIPHILE-WATER SYSTEMS CALCULATED FROM THE PBCM MODEL

The PBCM model depends as earlier discussed on many approximations, but still it seems to be able to describe many of the properties that characterize ionic amphiphile-water systems. We will indicate some properties of ionic amphiphile-water systems for which the PBCM model gives a nearly quantitative description, but also some properties that are described only in a more qualitative way.

The concentrations of free and micellized amphiphile in an ionic amphiphile-water solution vary with the total amphiphile concentration in a very characteristic way, as can be seen in Fig. 8. The reason for the decrease in the free-amphiphile concentration above the cmc is the increased shielding of the electrostatic interactions at high micellar concentrations. The effect of the micelle-micelle interactions is described in the PBCM model as mentioned earlier by solving the PB equation in a cell model. It seems

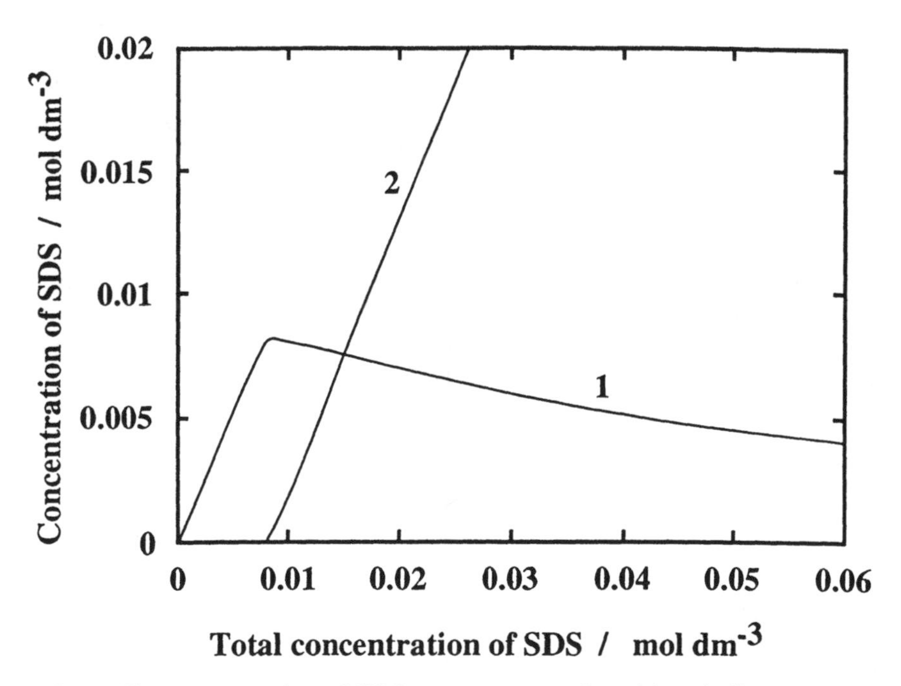

FIG. 8 The concentration of SDS as monomers (1) and in micellar aggregates (2) as a function of the total SDS concentration. (Data from Ref. 15.)

that this procedure gives a rather good estimate of the mean interactions in concentrated micellar solutions.

The cmc value of an ionic surfactant is decreased if molecules or ions that are able to reduce the unfavourable electrostatic interactions in the micellar aggregate are added. This reduction of the electrostatic interactions may be obtained either by the addition of salt to the system or by solubilizing uncharged molecules, such as long-chain alcohols, in the aggregates. A good description of the cmc dependence as a function of the concentration of added salt was already possible in the first version of the PBCM model [6]. The dependence on the addition of a long-chain alcohol will be discussed in the next section.

These are examples of properties that can be derived quantitatively using the PBCM model. A property of ionic amphiphile-water systems that is not so well described by the model is the growth of micelles when the amphiphile concentration is increased. The reason why an ionic micelle has a tendency to grow at high amphiphile concentration is the reduction of the electrostatic interactions. The PBCM model may be used to show that micelles tend to grow, but the present model will only give a qualitative description of the micellar growth and cannot be used to obtain quantitative data since the micelles at high amphiphile concentrations are neither spherical nor monodisperse as assumed in the model. To improve the model rod-shaped and polydisperse aggregates need to be treated.

The main problem when trying to model rod-shaped aggregates is to determine the distribution of the head groups on the micellar surface. The distribution will depend both on the electrostatic interactions, which favor a high charge density at the two end caps of the micelle, and the chain packing energies for which a distribution with a low charge density on the end caps is most favorable.

When modeling micellar systems the polydisperse size distribution of the aggregates is a problem, as it is not clear how to describe the micelle-micelle interactions in such a system. A modified cell model was used in Ref. 9 to model the polydispersity in micellar solutions. The size of the cells surrounding the different micelles was determined from the thermodynamic condition that the chemical potentials of each component in the system must be the same for all cell sizes. However, obtaining the different cell sizes is time consuming and the model has, therefore, not been in much use.

The PBCM model can give a fairly good description of another property, namely the stability regions of the different phases found in ionic amphiphile systems. The single-phase regions as well as the two-phase regions are determined by comparing the chemical potentials of all components in the different types of aggregates that may form. Micelles are often the first type of aggregate formed when an ionic amphiphile is added to an aqueous solution. The reason is that micelles have a lower electrostatic energy than other types of aggregates and the electrostatic interactions are more important in dilute than in concentrated solutions. When the electrostatic interactions are reduced at higher amphiphile concentrations, other types of aggregates are favored, as may be seen in Table 1.

V. SOLUBILIZATION PROPERTIES CALCULATED BY THE PBCM MODEL

The solubilization of uncharged molecules in ionic micelles has been treated at two different levels of simplification. In the first version of the PBCM model all molecules in the micelles, amphiphiles, and solubilizate, were assumed to be anchored at the surface of the micelle. This approximation is of minor importance for systems where the solubilizate is an alcohol or a fatty acid, but is too restrictive for systems where the solubilized molecule is nonpolar. The PBCM model has recently been extended to describe systems that besides ionic amphiphiles and water also contain nonpolar or slightly polar molecules. These molecules are not necessarily anchored at the surface of the micelle but may also be solubilized in the interior. To be able to describe these systems a new model was proposed where the interior of the micelle is divided into two regions [16], an inner

TABLE 1 Calculated and Experimental Phase Boundaries in Potassium Alkylcarboxylate-Water Systems at 86°C

	Phase boundaries (wt.%)			
Soap	Spherical	Cylindrical	Lamellar	
$C_{12}K$	−29	40–62	68−	Calculated[a]
	−36	44–59	66−	Experimental[b]
$C_{14}K$	−28	39–64	70−	Calculated[a]
	−32	40–56	64−	Experimental[b]
$C_{16}K$	−27	38–66	71−	Calculated[a]
	−27	34–55	63−	Experimental[b]
$C_{18}K$	−25	36–66	72−	Calculated[a]
	−22	29–54	62−	Experimental[b]

[a] From Ref. 8.
[b] From Ref. 37.

core region with no amphiphiles and an outer region, here called the palisade region. The proposed model may be used to show how different types of molecules are distributed between the two regions and how this distribution depends on the concentration and molecular properties of the surfactant. The model has also been used to describe the influence of different types of solubilizate on the stability region of the micellar phase.

We will now give an example showing how the PBCM model can be used to reproduce the main features in a previously presented investigation of the solubilization of pentanol in SDS micelles [14]. The model was used primarily to calculate the concentration of SDS and pentanol in the monomer and micellar states at varying total composition of solution, thus allowing an analysis of microcalorimetric titration results.

A. Solubilization of Pentanol in SDS Micelles

Microcalorimetric experiments were made by titrating a concentrated sodium dodecylsulfate, SDS, solution into water solutions containing varying amounts of pentanol. The resulting enthalpic titration curves were interpreted using the PBCM model. The alcohol is distributed between the micelles and the aqueous region, and the composition of the micelles will vary with the composition of the solution. It was assumed that in the micelles the polar OH groups were confined to the micellar surface like the surfactant headgroups. Further, it was assumed that the micelles were spherical and had a constant radius independent of composition equal to the maximum length of the SDS molecule. The thermodynamics of the mixed micelles was modeled by evaluating the energy contributions A–E presented earlier for pure micelles. The surface free-energy contribution, Eq. (4), is modified by the presence of OH groups on the surface of the micelles and there is an additional entropy contribution from the mixing of pentanol and SDS molecules in the micelles. The difference between the free energy of pentanol in water solution and in micelles was derived from the measured value of cmc in an SDS-pentanol solution of known concentration. The ion distribution in the aqueous region and the electrostatic free energy were calculated using the Poisson-Boltzmann approximation by solving the PB equation in spherical symmetry [14]. Electrostatic enthalpy contributions to the enthalpy of formation of the SDS-alcohol micelles were also calculated.

Figure 9 shows the composition of the micellar phase at varying SDS and pentanol concentration as an example of the properties calculated from the PBCM model. The microcalorimetric experiments consisted of measurements of the enthalpy changes (calculated per mole of added SDS) for consecutive additions of a small amount of an SDS solution with a

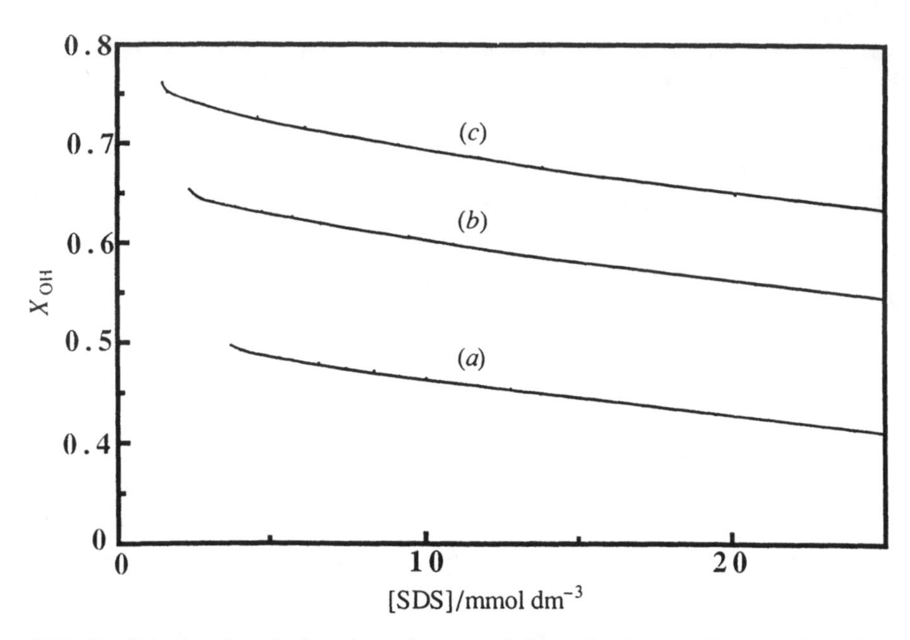

FIG. 9 Calculated mole fraction of pentanol, X_{OH}, in the micelles as a function of total SDS concentration at a total pentanol concentration of (a) 0.059, (b) 0.118, and (c) 0.176 mol dm^{-3}. (From Ref. 14 with permission from the Royal Society of Chemistry.)

concentration well above the cmc to the calorimeter vessel, which initially contained an aqueous pentanol solution. As can be seen from Fig. 10, the resulting enthalpy curves show a complex variation with the composition of the solution. The curves were modeled assuming the enthalpy of formation of the mixed SDS-pentanol micelles to be the sum of the following contributions: (i) the enthalpy of formation of a pure SDS micelle, (ii) enthalpy of formation of a pure (hypothetical) pentanol micelle, ΔH_f^{ref}, (iii) electrostatic and surface enthalpy contributions from the formation of mixed micelles. (There is no hydrophobic enthalpy contribution since the mixed micelles are formed from the pure SDS and pentanol micelles.) The changes in the amount of pentanol and SDS in the micellar phase for each titration step were evaluated from the results of the model calculations. The only adjustable parameter in the expression for the observed enthalpy changes is ΔH_f^{ref}, the transfer enthalpy of pentanol. Values of this parameter at the two experimental temperatures were found by matching the peaks in the calculated curves with the experimental ones.

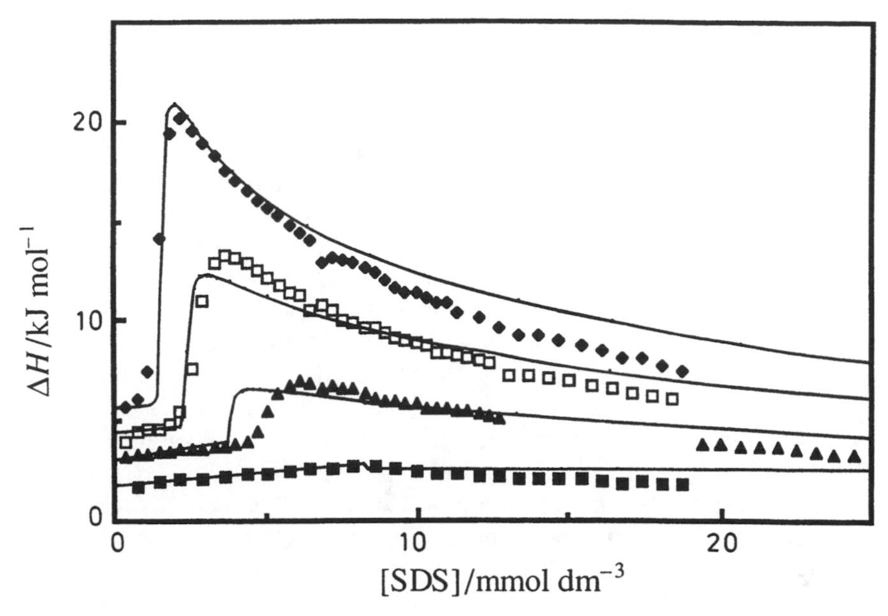

FIG. 10 Differential enthalpies of dilution of a 10 wt.% SDS solution as a function of total SDS concentration. Initial calorimeter solution: ■, pure water and ▲, 0.059; □, 0.118; ◆, 0.176 mol dm^{-3} pentanol solution at 25°C The solid lines represent calculated curves. (From Ref. 14 with permission from the Royal Society of Chemistry.)

The solid lines in Fig. 10 represent the calculated curves at 25°C. As can be seen, the calculated curves agree satisfactorily with experimental results. The major contribution to the endothermic peaks stems from the enthalpy of transfer of pentanol from aqueous solution into the micelles, because the enthalpy of demicellization of pure SDS-micelles happens to be close to zero at 25°C. The shape of the curves reflects the changing amount of pentanol incorporated into the micelles for each injection. The shape of the curves was significantly changed at 35°C due to the large change with temperature of the hydrophobic enthalpy contributions, but the agreement between the experimental and calculated curves was equally satisfactory. We can conclude that the theoretical model adequately describes the essential features of micelle formation of SDS in the presence of pentanol. The assumption of spherical micelles with constant radius and all OH groups anchored at the surface appears to be a sufficient approximation to describe the formation of micelles in dilute SDS-pentanol solutions.

B. The Two-Region Model

In the later version of the PBCM model for three-component systems the description of the micellar phase has been refined by dividing the micelle into two regions, a surface region (the palisade layer) and a core region. With this new description the solubilization of nonpolar molecules like hydrocarbons can also be treated. Depending on the polarity of the solubilizate it is preferably solubilized in either of the two regions. This preference is described by a difference in the standard chemical potential of the solubilizate between the two regions, which in the model is calculated from the interfacial tension between pure solubilizate and water. Thus the distribution of solubilizate between the two regions can be determined.

It is often stated that hydrocarbons always are solubilized in the interior of a micelle, but the model calculations give a slightly different picture, as may be seen in Fig. 11. The fraction of hydrocarbon molecules solubilized in the palisade layer is, especially at low amphiphile concentrations, fairly large after the first hydrocarbon molecule has entered the micelle. The two main reasons why the hydrocarbon molecules also start to solubilize in the palisade layer are the gain in free energy from the mixing with the amphiphiles in the palisade region and the reduction of the electrostatic repulsions between the amphiphilic headgroups.

As mentioned, the electrostatic interactions are most important at low amphiphile concentrations, and that is also the reason why there is a high degree of solubilization in the palisade layer at low amphiphile concentrations, as shown in Fig. 11.

The distribution of the solubilized molecules within a micelle is of course also dependent on the chemical properties of the solubilized molecule. The calculated distribution within a micelle of two different types of molecules, octane and octanol, is presented in Fig. 12.

That polar molecules solubilize in the palisade layer seems in this example to be a fairly good approximation. It is clear from Fig. 12 why the solubility of hydrocarbons in a solution of ionic micelles may be increased by adding a long-chain alcohol as cosurfactant. The hydrocarbon may solubilize in the interior of the micelle, thereby reducing the interfacial area per amphiphile, while the cosurfactant may solubilize in the palisade layer to reduce the electrostatic repulsion between the headgroups of the amphiphiles.

An example of the effect on the cmc of adding a long-chain alcohol to an SDS solution is shown in Fig. 13. There are two reasons for the cmc

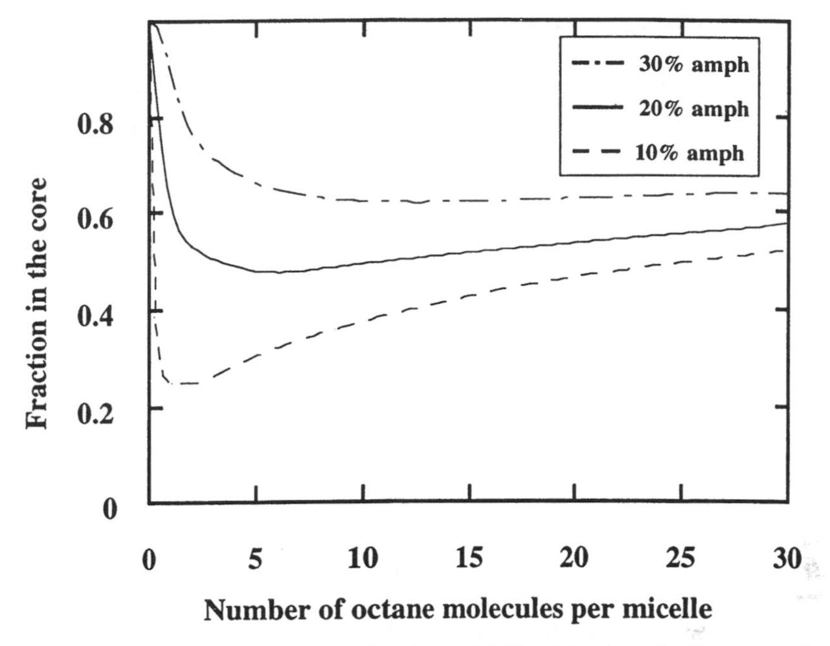

FIG. 11 Fraction of octane molecules solubilized in the micellar core of a decanoate micelle as a function of the number of solubilized molecules per micelle. The amphiphile content before the addition of octane was 10, 20, and 30 wt.%. (Data from Ref. 16.)

depression when octanol is added: (i) an increased micellar entropy due to the mixing of SDS and octanol in the aggregates, and (ii) a decrease of the electrostatic headgroup repulsions on the surface of the micelle due to the reduction of the surface charge density when octanol is added.

The capacity of sodium octanoate micelles to solubilize octane is rather limited at low amphiphile concentrations but increases with increasing amphiphile concentration, as can be seen in Fig. 14. This is one example of a calculated phase diagram using the PBCM model; many other examples may be found in Ref. 17.

VI. PRESENT STUDIES

The PBCM model can be used to give a qualitative description of many typical properties of ionic amphiphile water systems, but the model has

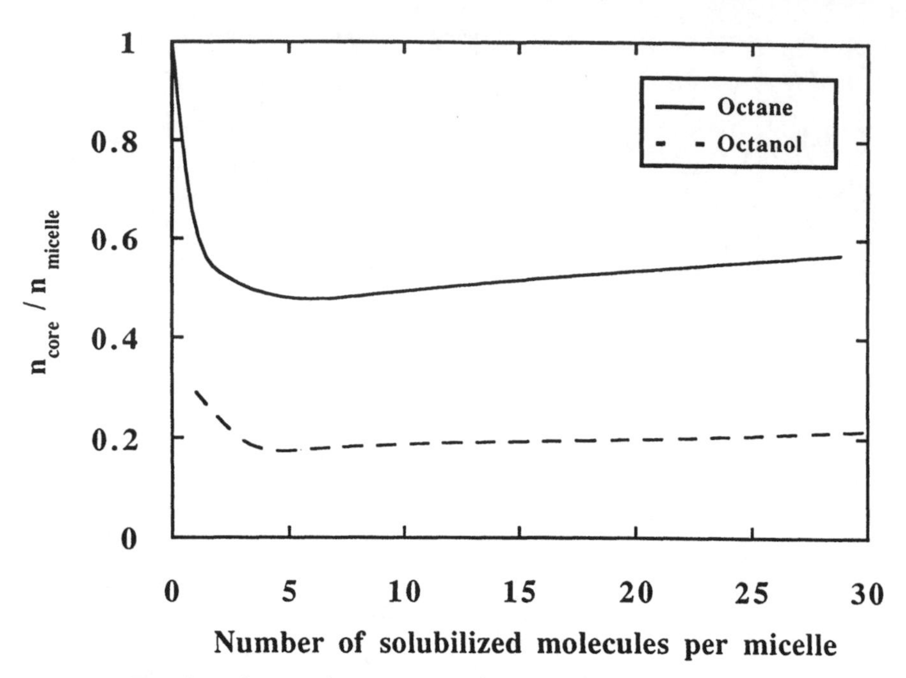

FIG. 12 Fraction of octanol or octane molecules solubilized in the micellar core of a sodium octanoate micelle as a function of the number of solubilized molecules per micelle. (Data from Ref. 16.)

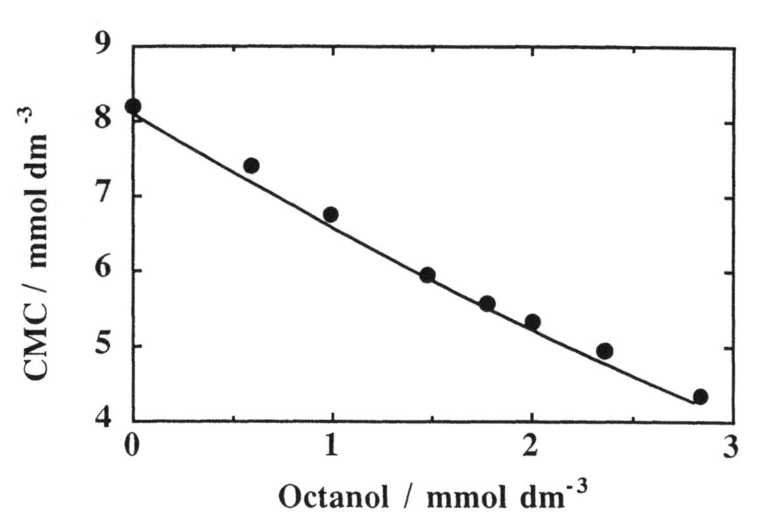

FIG. 13 The effect of octanol on the cmc of SDS: (—) calculated values from the PBCM model; • experimental values from Ref. 38.

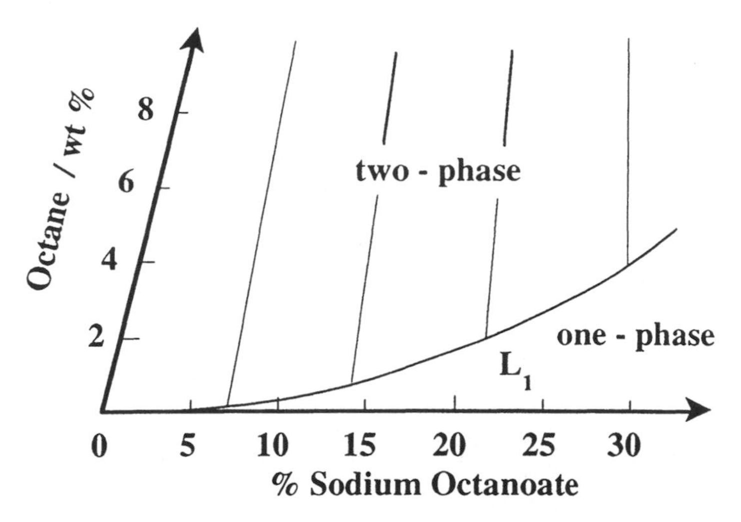

FIG. 14 Calculated phase equilibria between the micellar phase and pure octane for the water corner of the three-component system sodium octanoate-water-octane. L_1 indicates a micellar solution. (Data from Ref. 17.)

to be improved to be able to give a more quantitative description of such properties as micellar growth and aggregate stability. To improve the present model a deeper understanding of the micellar surface layer is needed.

Efforts to improve the PBCM model are proceeding along the following lines:

1. An interface region is introduced where the amphiphilic headgroups, some methylene groups from the tails of the amphiphiles, water, and some counterions are mixed.
2. The spherical end-capped cylinder model is used as a model for rod-shaped micelles. The electrostatic interactions are here calculated from a two-dimensional Poisson-Boltzmann equation and not as the electrostatics from an infinite rod plus the electrostatics from a sphere.
3. The results from the SCFA model is used to improve the description of chain energies.

REFERENCES

1. B. Lindman and H. Wennerström, *Top. Curr. Chem. 87*: 1 (1980).
2. H. Wennerström and B. Lindman, *Phys. Rep. 52*: 1 (1979).
3. P. Ekwall, L. Mandell, and K. Fontell, *Mol. Cryst. Liq. 8*: 157 (1969).
4. P. Ekwall, *Adv. Liq. Cryst. 1*: 1 (1975).

5. G. J. T. Tiddy, *Phys. Rep. 57*: 1 (1980).
6. G. Gunnarsson, B. Jönsson, and H. Wennerström, *J. Phys. Chem. 84*: 3114 (1980).
7. B. Jönsson, Thesis, University of Lund, Lund, Sweden, 1981.
8. B. Jönsson and H. Wennerström, *J. Colloid Interface Sci. 80*: 482 (1981).
9. B. Jönsson, G. Gunnarsson, and H. Wennerström, in *Solution Behaviour of Surfactants: Theoretical and Applied Aspects* (K. L. Mittal and E. J. Fendler, eds.), Plenum, New York, 1982, Vol. 1, p 317.
10. B. Jönsson, P.-G. Nilsson, B. Lindman, L. Guldbrand, and H. Wennerström, in *Surfactants in Solution* (K. L. Mittal and B. Lindman, eds.), Plenum, New York, 1984, Vol. 1, p 3.
11. B. Jönsson and H. Wennerström, *J. Phys. Chem. 91*: 338 (1987).
12. I. Johnson, G. Olofsson, and B. Jönsson, *J. Chem. Soc. Faraday Trans. 1 83*: 3331 (1987).
13. B. Halle, M. Landgren, and B. Jönsson, *J. Phys.* France *49*: 1235 (1988).
14. I. Johnson, G. Olofsson, M. Landgren, and B. Jönsson, *J. Chem. Soc. Faraday Trans. 1 85*: 4211 (1989).
15. M. Landgren, Thesis, University of Lund, Lund, Sweden, 1990.
16. M. Aamodt, M. Landgren, and B. Jönsson, *J. Phys. Chem. 96*: 945 (1992).
17. M. Landgren, M. Aamodt, and B. Jönsson, *J. Phys. Chem. 96*: 950 (1992).
18. H. Wennerström, B. Lindman, O. Söderman, T. Drakenberg, and J. B. Rosenholm, *J. Am. Chem. Soc. 101*: 6860 (1979).
19. J. B. Hayter and J. Penfold, *J. Colloid Polym. Sci. 261*: 1022 (1983).
20. S. S. Berr, M. J. Coleman, R. R. Jones, and J. S. Johnson, *J. Phys. Chem. 90*: 6492 (1986).
21. P. Ekwall, L. Mandell, and K. Fontell, *J. Colloid Interface Sci. 31*: 508 (1969).
22. C. Tanford, *The Hydrophobic Effect. Formation of Micelles and Biological Membranes*, 2nd ed., Wiley, New York, 1980.
23. D. W. R. Gruen, *J. Phys. Chem. 89*: 153 (1985).
24. J. M. H. M. Scheutjens and G. J. Fleer, *J. Phys. Chem. 83*: 1619 (1979).
25. J. M. H. M. Scheutjens and G. J. Fleer, *J. Phys. Chem. 84*: 178 (1980).
26. O. A. Evers, J. M. H. M. Scheutjens, and G. J. Fleer, *Macromolecules 23*: 1619 (1990).
27. F. A. M. Leermakers and J. M. H. M. Scheutjens, *J. Chem. Phys. 89*: 3264 (1988).
28. F. A. M. Leermakers and J. M. H. M. Scheutjens, *J. Phys. Chem. 93*: 7417 (1989).
29. F. A. M. Leermakers and J. M. H. M. Scheutjens. *J. Colloid Interface Sci. 136*: 231 (1990).
30. M. R. Böhmer, O. A. Evers, J. M. H. M. Scheutjens, and G. J. Fleer, *Macromolecules 23*: 2288 (1990).
31. P. A. Barneveld, J. M. H. M. Scheutjens, and J. Lyklema, *Colloids Surf. 52*: 107 (1991).

32. M. R. Böhmer, L. K. Koopal, and J. Lyklema, *J. Phys. Chem. 95*: 9569 (1991).
33. U. Nilsson, Thesis, University of Lund, Lund, Sweden, 1992.
34. N. F. Carnahan and K. E. Starling, *J. Chem. Phys. 51*: 635 (1969).
35. P. Jokela, B. Jönsson, and A. Khan, *J. Phys. Chem. 91*: 3291 (1987).
36. L. A. Girifalco and R. J. Good, *J. Phys. Chem. 61*: 904 (1957).
37. B. Gallot and A. E. Skoulios, *Kolloid Z. Z. Polym. 208*: 37 (1966).
38. M. Manabe, Y. Tanizaki, H. Watanabe, S. Tokunaga, and M. Koda, *Niihama Kogyo Koto Semmon Gakko Kiyo Rikogaku Hen*, *19*: 50 (1983).

5
Solubilization in Mixed Micelles

NAGAMUNE NISHIKIDO Faculty of Science, Fukuoka University, Jonan-ku, Fukuoka, Japan

SYNOPSIS

The articles concerning the solubilization phenomena in mixed micelles were divided into two groups according to the approaches used for the investigation: (i) the microscopic (spectroscopic) method to obtain information on the location and/or the microenvironment of a solubilizate in mixed micelles; (ii) the thermodynamic approach to describe the micelle-water distribution equilibria of a solubilizate in mixed micellar systems.

Most of the reported microscopic investigations are fluorescence studies and have been summarized in Sec. II. The present status of studies concerning the micelle-water distribution equilibria have been outlined in Sec. III.A. In addition, synergistic solubilization effects in mixed micelles have been discussed; thermodynamic equations that describe these effects were proposed and the predictive ability of the equations was examined by applying them to different cases of the mixed-micellar solubilization (Sec. III.B).

I. INTRODUCTION

In this chapter, the term mixed micelle refers to a micelle composed of different (two or more) kinds of surfactant constituents; i.e., each constituent surfactant exhibits ability to form its own micelles in its own dilute solution. The term pure micelle means a micelle composed of the same kind of surfactant constituent (single-component micelle). The solubilization in solutions of pure micelles has been extensively investigated by many authors [1, 2]. A conclusion drawn from these studies is that the following factors influence the solubilization: (1) the surfactant structure, (2) the solubilizate structure, (3) temperature, (4) the addition of electrolytes, and (5) the addition of (organic) nonelectrolytes [1, 2]. It was recently shown by us that pressure, along with factors (1)–(5), also affects the solubilization [3, 4].

The organic nonelectrolytes that have been used in studies of factor (5) are divided into three types: (i) polar monomeric materials such as fatty alcohol, amine, or fatty acid, (ii) hydrophilic polymers (synthetic and proteins), and (iii) surfactants capable of forming their own micelles. The addition of other kinds of surfactants in solutions of a single kind of surfactant results in the formation of mixed micelles, and it frequently occurs that the observed solubilization ability of the mixed micelles is unpredictable from the solubilization ability of the pure micelles of respective constituent surfactants. If the solubilization capacity of mixed micelles is much greater or less than the solubilization capacities of the pure micelles of constituent surfactants, this is important and interesting not only in practical applications such as detergency but also in academic research. At present there are not many articles concerning solubilization in mixed micelles, although many articles have been published about studies of solubilization in pure micelles [5].

Generally speaking, there are two approaches to investigate the solubilization in pure micelles: (i) the microscopic (spectroscopic) method to obtain information on the location and/or the microenvironment of a solubilizate in the micelles; and (ii) the thermodynamic approach to describe the micelle-water distribution equilibria of a solubilizate [1, 2]. As mentioned, there are not many articles concerning the solubilization phenomena in mixed micelles, but the approaches used are also classified into the two categories that are used for the investigation of the solubilization in pure micelles. In this chapter, we divide the relevant articles into two groups according to approaches (i) and (ii), and try to outline the present status of studies of solubilization in mixed micelles.

In this chapter, most of surfactants are represented by rational formulae or abbreviations. Typical abbreviations used for surfactants and groups in them are as follows.

alkylsulfate and alkylsulfonate: C_mSO_4M and C_mSO_3M, respectively (M = Na, Li, $\frac{1}{2}$Mg, etc.)
alkyltrimethylammonium halide: C_mNMe_3X (X = Br, Cl, etc.)
polyoxyethylene alkyl ether: C_mE_n
alkylpolyoxyethylene sulfate: $C_mE_nSO_4M$
 C_m: alkyl group (m is the number of methylene and methyl groups)
 E_n: oxyethylene (OE) groups (n is the number of OE groups)
 Me: methyl group, Et: ethyl group, ϕ: phenyl group
perfluorooctane sulfonate: MFOS (M = Na, Li, etc.)
perfluorooctanoate: MPFO (M = Na, Li, etc.)
bile salts: sodium taurocholate (NaTC), deoxycholate (NaDC), cholate (NaC), and dehydrocholate (NaDHC)
alkylpyridinium halide: C_nPX (X = C for Cl, B for Br)

In Sec. III, sodium dodecylsulfate is in some places represented by SDS, and cetylpyridinium chloride and polyoxyethylene (15) nonylphenyl ether are, respectively, CPC and NPE_{15}.

II. MICROSCOPIC INVESTIGATION FOR THE STATE OF SOLUBILIZATES IN MIXED MICELLES

X-ray diffraction, UV, NMR, ESR spin labeling, and fluorescence spectroscopy have all been used to investigate the microscopic state of solubilizate molecules in pure micelles [1, 2]. Recently, Raman [6], ultrasonic spectroscopy [7], and SANS [8] have also been applied to investigate the state of solubilizates in pure micelles. These techniques provide information on the location and the microenvironment of solubilizate probe molecules in pure micelles. The same techniques can in principle be applied to study the microscopic state of solubilizates in mixed micelles, but such studies have not been carried out extensively [9–36]. Most of the reported spectroscopic investigations are fluorescence studies using probes such as pyrene and 1-anilino-naphthalene-8-sulfonate (ANS) [9–27]. This fluorescence technique has been used to describe qualitatively the microscopic state of the mixed micelles [23–27] and to permit determination of the micelle aggregation number [9–14] and the interior polarity [14–18] of the mixed micelles. Table 1 summarizes the spectroscopic studies, in which the measured properties of solubilizates in mixed micelles are shown.

III. THERMODYNAMIC STUDIES OF THE MICELLE-WATER DISTRIBUTION EQUILIBRIA OF SOLUBILIZATES IN MIXED MICELLAR SYSTEMS

Solubilization in dilute regions of surfactants is macroscopically characterized by the saturation concentration (solubility) or the micelle-water distribution equilibrium of a solubilizate [1, 2]. Originally the term *maximum additive concentration* (MAC) denoted the solubility of a solubilizate for a given surfactant concentration, but in this chapter MAC is defined as the saturation solubilization capacity for a solubilizate expressed as moles of the solubilized solubilizate per mole of micellized surfactant.

The micelle-water distribution equilibrium in pure micellar systems have been analyzed by using the following micelle models: the pseudophase separation model, the mass action model, and the small systems/multiple equilibrium approach [1, 40]. The pseudophase separation model defines the apparent distribution coefficient (mole-fraction basis) of a solubilizate between the bulk and a micellar phase, inferred from the data [1, 41–44]. The mass action model defines the solubilizate-micelle binding constant to be calculated [1, 46–48], and the small systems/multiple equilibrium approach permits calculation of the apparent distribution coefficient between free solution and micelles [1, 40, 45].

Analysis by using either of the above micelle models must determine the amount of the solubilizate incorporated in the (pure) micelles and the amount of the free solubilizate in the intermicellar bulk water. The experimental techniques used to determine the amounts of the solubilizates in the solubilized and the free states are dialysis [1], potentiometric titration [1], molecular sieve [1], ultrafiltration [1], vapor pressure method [41], gel filtration [1], fluorescence quenching [1], calorimetric method [42, 46], pulse radiolysis [43, 47], NMR [49], and the calculation from the variation of CMC with the addition of a solubilizate [41, 50]. The treatment of the solubilization data by means of the micelle models and the experimental techniques used for the solubilization in pure micelles are also applicable to the solubilization in mixed micelles.

TABLE 1 Spectroscopic Studies Exploring the Microscopic State of Solubilizates in Mixed Micelles

Measured	Solubilizate	Mixed micelle	Ref.
Fluorescence (micelle size)	Methylanthracene	$C_{12}SO_4Na$ + C_mNH_3Cl, C_mNMe_3X, $C_{12}E_n$, $C_{14}SO_4Na$	9
Fluorescence (micelle size)	Pyrene	$C_{12}NMe_3Cl$ + $C_{16}NMe_3Cl$, $C_{12}SO_4Na$	10
Fluorescence (micelle size)	Pyrene	$C_8F_{17}SO_4Li$ + $C_{12}SO_4Li$, $C_{12}E_6$, $C_{12}E_8$	11
Fluorescence (micelle size)	Pyrene	C_mNMeBr + $C_{m'}(NMe_3Br)_2$	12
Fluorescence (micelle size)	Pyrene	$C_{12}SO_4Na$ + $C_{10}SO_4Na$	13
Fluorescence (micelle size, interior polarity)	Pyrene	$C_{10}E_8$ + bile salts	14
Fluorescence (interior polarity)	Pyrene, PCA	$C_{12}SO_4Na$ + $C_{12}(NEt_3Br)_2$	15
$E_T(30)$ (interface polarity)	Phenol betaine		
Fluorescence (interior polarity)	Pyrene	Betaine + $C_{12}SO_4Na$	16
λ_{max} (dielectric const)	Azo oil dye		
Fluorescence (Interior polarity)	Pyrene	Triton X-100 + $C_{12}\phi SO_4Na$, $C_{16}NMe_3Br$	17
Photolysis (cage effect)	Dibenzyl ketone		
Fluorescence		$C_{12}SO_4Na$ + NaTC	18
(interior polarity)	Pyrene		
(lifetime)	Perylene		
Fluorescence (surface potential)	ANS	$C_{12}SO_4Na$ + $C_{12}E_7$	19
Fluorescence (Br^- quenching)	Anthracenes, fluorene	$C_{16}NMe_3Br$ + $C_{16}NMe_3Cl$	20
Fluorescence (Br^- quenching)	Biphenyl	$C_{16}NMe_3Br$ + $C_{12}E_{23}$	21
Fluorescence (Br^- quenching)	Naphthalene	$C_{14}NMe_3Br$ + $C_{12}E_{23}$	22
Fluorescence (I_1/I_3, lifetime)	Pyrene	$C_8F_{17}SO_4Li$ + $C_{12}SO_4Li$, $C_{12}E_6$, $C_{12}E_8$ $C_{12}E_6$ + $C_{12}SO_4Li$	11
Fluorescence	ANS	$C_{12}SO_4Na$ + NF[a], $C_{12}E_6$ + $C_8F_{17}SO_3Li$, NF[a]	23, 24

(continued)

TABLE 1 *(continued)*

Measured	Solubilizate	Mixed micelle	Ref.
Fluorescence	Pyrene	$C_8F_{17}COOLi + C_{12}SO_4Na$	25
Fluorescence	1-pyrenecarboxaldehyde	$C_{12}SO_4Na + C_{14}E_7SO_4Na$	26
Fluorescence	ANS, auramine, pyrene	$C_8F_{17}SO_3Li + C_{12}SO_4Na, C_{14}SO_4Na$	27
ESR	Nitroxide	$C_{12}SO_4Na + C_{12}NMe_3Br$	28
ESR, ESEM	TMB^+	$C_{12}SO_4Na + C_{12}NMe_3Cl$,	29
(TMB^+-water inteact.)		$C_{12}E_6 + C_{12}SO_4Na, C_{12}NMe_3Cl$	30
CT	I_3^-	$C_{12}E_5 + C_{12}SO_4Na$,	31
		$C_{12}E_7 + C_{12}SO_4Na$,	32
		$C_{12}E_6 + C_{11}H_{23}COO(CH_2)_4SO_4Na$	33
CT	TCNQ	$C_{12}E_7 + C_{12}SO_4Na$	24, 34
ΔA_{612}	Pinacyanol chloride	$C_{12}SO_4Li + C_8F_{17}SO_3Li$	35
Tautomerism	Benzoylacetoanilide	$C_{12}SO_4Li + C_8F_{17}SO_3Li$	36
Tautomerism	Benzoylacetoanilide	$C_{12}E_6 + C_8F_{17}SO_3Li$	24, 37
Tautomerism	PNDS (acidic dye)	$C_{12}E_8 + C_{12}SO_4Na$	24, 38
Fading, protonation	Azo oil dyes	$C_{12}SO_4Na + C_mE_n$	39

[a] p-$[(CF_3)_2CF]_2C{=}C(CF_3)O(CH_2CH_2O)_7{-}CH_3$.

A. Review of Relevant Studies

As mentioned, there are not many articles concerning thermodynamic studies on the micelle-water distribution equilibria of solubilizates in mixed micellar systems. In what follows, we outline the published articles of relevant studies, which were conveniently divided into seven groups according to kinds of surfactants and their combination in mixed micelles such as ionic/nonionic, anionic/anionic, nonionic/nonionic, anionic/cationic, fluorocarbon/hydrocarbon surfactant mixed micelles, etc.

1. Solubilization in Ionic/Nonionic Mixed Micelles

In 1973, Tokiwa et al. reported the solubilization behavior of $C_m\phi C_nSO_3Na/C_{12}E_9$ ($m = 8, n = 0; m = 4, n = 4; m = 0, n = 8$) and $C_{10}SO_3Na/C_{12}E_9$ mixed micelles toward a water-insoluble dye, yellow OB [51]. The MAC values of yellow OB in the system of $C_8\phi SO_3Na/C_{12}E_9$ mixed micelles were much larger than those of $C_{12}E_9$ pure micelles. They assumed that this "remarkable synergistic effect" is due to increased compactness of polyoxyethylene (POE) shell in the mixed micelles compared to the $C_{12}E_9$ pure micelles because a greater amount of yellow OB is incorporated in the POE shell than in the hydrocarbon core of the micelle.

This increased compactness of the POE shell in the $C_8\phi SO_3Na/C_{12}E_9$ mixed micelles is caused by the strong interaction between the benzene ring of $C_8\phi SO_3Na$ and the POE chain of $C_{12}E_9$.

Nishikido determined the MAC values of yellow OB in aqueous solutions of $C_{12}NH_3Cl/C_{12}E_n$, $M(C_{12}SO_4)_2$ (M = Zn, Mn, Cu, Mg)/$C_{12}E_n$, $C_mE_6/C_{m'}E_6$, and $C_{12}E_n/C_{12}E_{n'}$ mixed micelles (n = 6, 15, 29, 49; m = 8, 10, 12) [52]. Only the MAC values in $M(SO_4Na)_2/C_{12}E_{15,\ 29,\ 49}$ mixed micellar systems were larger than those calculated by using the additivity mixing rule. The additivity rule says that a solubilization capacity X (MAC, solubility, apparent distribution coefficient, etc.) in a mixed micellar system varies linearly with a surfactant mole fraction X_i in the mixed micelle:

$$X = \sum_i \chi_i X_i \tag{1}$$

where X_i is the solubilization capacity in the pure micellar system of surfactant i. We have similarly attributed nonadditivity in such systems to the change in compactness of the POE shell in the mixed micelles compared to in the $C_{12}E_n$ pure micelles; this is caused by the attractive interaction between the slightly positively charged ether oxygen atoms, i.e., oxonium ions, in the POE chains and the anionic headgroups of anionic surfactants [52].

Uchiyama et al. measured the solubilities of oil-soluble azo dyes (azobenzenes) in aqueous solutions of $C_{12}SO_4Na/C_mE_n$ mixed micelles (m = 12, 14, 16, 18; n = 10, 20, 30, 40) [53]. They defined the enhancement ratio (R_i) for the solubility of the solubilizate: $R_i = (C_{exp} - C_{cal})/C_{cal}$, where C_{exp} is the solubility in a mixed micellar solution, and C_{cal} is the solubility calculated by using Eq. (1) from the solubilities in the pure micellar solutions of constituent surfactants. The value of R_i increased with decreasing alkyl chain length or increasing oxyethylene (OE) chain length of C_mE_n, and the extent of increase was larger for the mixed micelles composed of C_mE_n having shorter alkyl chain or longer OE chain. They concluded that the extent of increase was larger in the systems in which two kinds of micelles (one rich in $C_{12}SO_4Na$ and the other rich in C_mE_n) coexist than in systems of mixed micelles of one type.

Muto et al. determined the MAC values of yellow OB and azobenzene in mixed micellar systems of $C_{12}E_n$ (n = 6 or 8) and an anionic surfactant where the anionic surfactant is $C_{12}SO_4Na$, aerosol OT, or LiFOS [54]. They assumed that the ratio of the observed solubility to the solubility calculated from the additivity rule (Eq. (1)) represents the efficiency of solubilization in mixed micelles, and from the ratios they obtained the following results. (1) The efficiency of solubilization decreased when nonionic and anionic surfactants were mixed. The extent of the efficiency depended on the kinds of mixed micelles formed and dyes, and it has a minimum at 0.2–0.4 mole fraction of C_mE_n in the mixed micelles. (2) The efficiency of solubilization decreased with increasing difference in the MAC values for constituent surfactants of the mixed micelles: $C_{12}E_n$/LiFOS < $C_{12}E_n$/aerosol OT < $C_{12}E_n/C_{12}SO_4Na$ for yellow OB; $C_{12}E_n$/LiFOS < $C_{12}E_n$/aerosol OT ~ $C_{12}E_n/C_{12}SO_4Na$ for azobenzene. Analysis of the data of Uchiyama et al. and Muto et al. by using the thermodynamic equations introduced by us leads to some different results; this point is described in Sec. III.B.3.

Treiner et al. determined the values of the apparent partition coefficient (this is defined by them but is essentially identical to the apparent distribution coefficient defined in the pseudophase separation model, see Sec. III.B.1) of barbituric acids or 1-pentanol in aqueous solutions of $C_{12}E_n$/anionic surfactant mixed micelles (anionic surfactant is $C_{12}SO_4Na$ or $C_{14}NMe_3Br$) [55–58]. Then they compared the determined values with the values calculated by the equation based on the O'Connell equation for Henry's constant in a mixed solvent. This equation is described in Sec. III.B.5.

The following solubilization capacities for alkane and alkanol were determined in ionic/nonionic mixed micellar systems, and in Sec. III.B the results are discussed in terms of our thermodynamic equations: the solubilization equilibrium constant (equivalent to the apparent distribution coefficient, see Sec. III.B.1) of 1-hexanol in mixed micellar systems of cetylpyridinium chloride (CPC) and polyoxyethylene (15) nonylphenyl ether (NPE_{15}), and of $C_{12}SO_4Na$ (SDS) and NPE_{15} [59]; the MACs of decane and 1-hexanol in $C_{12}NMe_3Br/C_{12}E_6$ mixed micellar systems [60]; and the MAC of nonane in $C_{12}NMe_3Br/C_{12}E_6$ and $C_{12}SO_4Na/C_{12}E_6$ mixed micellar systems [61].

Other articles published to investigate the solubilization in ionic/nonionic mixed micellar systems are as follows.

Mixed micelle	Solubilizate	Solubilization parameter	Ref.
$C_{12}SO_4Na/C_{12}\phi E_9$	Heptane	Solubility, cloud point	62
$C_{12}E_9SO_4Na/C_{12}E_9$	Yellow OB	MAC	63
$C_mCH(C_{m'})COONa/C_8E_n$	Cyclohexane	Solubility, phase diagram	64
$C_{16}NMe_3Br/C_8\phi E_{9-10}$ $C_{12}SO_4Na/C_8\phi E_{9-10}$	Choresterol	Solubility	65
$C_{12}SO_4Na/C_{16}E_3$	Sudan Red G	Solubility	66
SDS/NPE_{15} CPC/NPE_{15}	Hexane, Cyclohexane	Activity coeff.	67
$C_{16}NMe_3Br/C_{12}E_{23}$	*p*-alkylphenol, *p*-alkylphenoxide	Association const.	68

2. Solubilization in Ionic/Ionic and Nonionic/Nonionic Mixed Micelles

King et al. measured the solubilities of propane gas in aqueous solutions of $C_{12}SO_4Na/C_8SO_4Na$ and $C_{12}SO_4Na/C_6SO_4Na$ mixed micelles [69]. Their data are compared with the calculation by using our thermodynamic equation, as shown in Sec. III.B. 2. King also determined the solubilities of O_2, Ar, CH_3, C_2H_6, and C_3H_8 gases in $C_{12}E_{23}/C_{12}E_4$ mixed micellar systems [70]. Bury et al. determined the values of the apparent partition coefficient of benzyl alcohol in aqueous solutions of $C_{14}NMe_2\phi Cl$/ $C_{14}NMe_3Cl$ mixed micelles [71] and Treiner et al. reported similar results for 1-pentanol in $C_{12}E_{23}/C_{12}E_4$ mixed micellar systems [57]. As mentioned, we reported the MAC of yellow OB in aqueous solutions of $C_mE_n/C_{m'}\ E_{n'}$ mixed micelles [52]. The solubilities of cyclohexane and the phase diagram were determined for aqueous systems of $C_8H_{17}CH(C_6H_{13})COONa/C_8H_{17}CH(C_{10}H_{21})COONa$ mixed micelles [64]. The MAC values of ethylbenzene in $C_{12}E_{6,\ 7,\ 8}$ mixed micellar systems were determined to examine the effect of the POE chain length distribution [72].

3. Solubilization in Anionic/Cationic Mixed Micelles

In aqueous systems of mixed anionic/cationic surfactants, precipitation is frequently observed [57, 58, 60, 73]. This may be due to the partial shielding of micellar charges that confer water solubility upon micelles. The formation of vesicles, instead of precipitation, is also possible in the intermediate composition range of aqueous systems of mixed anionic/cationic surfactants. Weers et al. reported the composition dependence of the MAC of alkane in aqueous systems of anionic/cationic mixed micelles. They observed that in studies of the solubilization of decane in systems of $C_{12}SO_4Na/C_{12}NMe_3Cl$ mixed micelles, both the MAC and the micelle size increase rapidly as the precipitation boundary is approached [60, 73]. A quantitative explanation of this anomalous solubilization behavior by using our thermodynamic equations is shown in Sec. III.B. 4.

Treiner et al. determined the apparent partition coefficients of barbituric acids in $C_{12}SO_4Na/C_{12}NMe_3Cl$ and $C_{10}SO_4Na/C_{10}NMe_3Br$ mixed micellar systems and those of 1-pentanol in $C_{10}SO_4Na/C_{10}NMe_3Br$ systems [57, 58]. They reported that there are minima in the plots of the apparent partition coefficient versus micelle composition. Treiner et al. and Weers and Schuing considered that the Laplace pressure in the hydrocarbon core could contribute to the core solubilization in the mixed micelles [58, 60].

Zhao and Li measured the MAC values of 1-octanol and octane in aqueous solutions of equimolar mixed micelles of $C_8E_1SO_4Na$ and octylpyridinium bromide (C_8PB) [74]. They observed that the MAC value of 1-octanol in the mixed micellar system is less than and that of octane is much larger than the MAC values in the pure micellar systems of the constituent surfactants. In addition, the MAC value of a nonpolar organic compound (octane) is larger than that of a polar one (1-octanol); this is contrary to the results of previous studies [2].

Other studies on the solubilization in anionic/cationic mixed micelles are shown below.

Mixed micelle	Solubilizate	Solubilization parameter	Ref.
SDS/CPC	Cyclohexane	Activity coefficient	67
$C_7SO_3Na/C_{14}PC$[a]	Decane	Micelle dimension	75
$C_3F_7COONa/C_{14}PC$[a]		(light scattering)	
Igepon T/CPC	Orange OT	Solubility	76

[a] Tetradecylpyridinium chloride.

4. Solubilization in Mixed Fluorocarbon/Hydrocarbon Surfactant Micelles

Treiner et al. reported that in $NaPFO/C_{10}SO_4Na$ and $NaPFO/C_{12}SO_4Na$ mixed micellar systems the apparent partition coefficient of 1-pentanol goes through a maximum in the region of the hydrocarbon-rich mixed micelles [77]. Yoda et al. determined the apparent distribution coefficients of 1-hexanol and 2,2,3,3,4,4,4-heptafluorobutanol ($C_4H_2F_7OH$) in three aqueous systems of $LiFOS/C_{12}SO_4Li$, $LiFOS/C_{12}E_6$, and LiFOS/LiPFO mixed micelles [78]. There are maxima in the plots of the apparent distribution coefficient versus micelle composition for both alcohols in $LiFOS/C_{12}SO_4Li$ mixed micellar systems, while the plots exhibit minima for $LiFOS/C_{12}E_6$ mixed micellar systems. The plots for LiFOS/LiPFO mixed micellar systems show no extrema. As mentioned in Sec. III.A.1, the MAC values of yellow OB and azobenzene in $LiFOS/C_{12}E_n$ (n = 6, 8) mixed micellar systems were determined [54]. In addition, the micelle dimension of $C_3F_7COONa/C_{14}PC$ mixed micelles containing solubilized decane was studied by means of light scattering techniques [75].

The results of Treiner et al. and Yoda et al. lead us to speculate that the solubilization in the mixed micelles composed of ionic hydrocarbon and ionic fluorocarbon surfactants is anomalous and remarkably positive synergistic. They determined the apparent distribution (partition) coefficients at dilute alcohol concentrations and compared these values with the values calculated by the equation based on the O'Connell equation (see Secs. III.A.1 and III.B. 5). However, our thermodynamic equations indicate that more information can be obtained from the apparent distribution coefficients at varying alcohol concentrations, i.e., at varying alcohol composition in the mixed micelles. A more definite conclusion could be drawn if the apparent distribution coefficients are obtained at varying alcohol contents in the mixed micelles. This point is referred to in Sec. III.B.

5. Solubilization in Zwitterionic/Anionic Mixed Micelles

To our knowledge, only four articles have been published on solubilization in amphoteric/ionic mixed micelles [79–82]. Amphoteric surfactants are classified as betaine-type, alanine-type, imidazoline-type surfactants, etc. In 1970, Tokiwa et al. synthesized a new-type amphoteric surfactant, sodium *N*-dodecyl *N*,*N*-bisethoxyacetate (NaDEA) ($pK_a \sim 3$) and measured the solubilities of yellow OB in aqueous solutions of NaDEA/anionic mixed micelles (anionics: $C_{12}E_nSO_4Na$, $C_{12}\phi SO_3Na$) [79]. The measured solubilities of yellow OB in both mixed micellar systems were larger than the solubilities calculated from the additivity rule (Eq. (1)). Takai et al. reported the solubilities of Orange OT in mixed micellar solutions (pH 8.5) of the alanine-type amphoteric surfactant ($pK_a = 6.7$) and $C_n\phi SO_3Na$ ($n = 12$–14, 16–18); the solubilities show maxima against the mole ratio in the mixed micelles [80]. To interpret this solubilization and other solution behavior, they assumed that complex formation occurs by electrostatic force between the $—NH^+ =$ cationic group of the amphoteric surfactant and the $—SO_4^-$ anionic group of the anionic surfactant, as well as complex formation between $—COOH^{\delta+}$ and $—SO_4^-$ groups through protonation.

Abe et al. measured the solubilities of oleyl alcohol in aqueous solutions of mixed micelles composed of the lysine-type amphoretic surfactant (DMLL) and $C_{12}SO_4Na$, and they also obtained solubility measurements in mixed micellar systems of DMLL and $C_{16}E_{20}$ [81]. The solubility versus micelle composition plot for the DMLL/$C_{12}SO_4Na$ mixed micellar systems showed a maximum, while the plot for the DMLL/$C_{16}E_{20}$ systems showed a minimum. Iwasaki et al. reported the solubilities of azo oil dye in mixed micellar solutions of the betaine-type amphoteric surfactant and $C_{12}SO_4Na$, and the solubility-composition plot showed a maximum [82]. They assumed that the intermolecular complex is formed in the mixed micelle through the electrostatic interaction of oppositely charged head groups between the betaine and $C_{12}SO_4Na$. A quantitative interpretation of their data by using the derived thermodynamic equation is shown in Sec.III.B.4.

6. Solubilization in Mixed Micelles Containing Bile Salts

Moulik et al. investigated the solubility change of cholesterol in mixed micellar systems of NaDC/$C_{12}SO_4Na$, NaDC/Triton X-100 [83],

NaC/NaDC, NaC/NaDH, NaDC/NaDHC, NaC/NaDC/NaDHC, NaC/$C_{16}NMe_3Br$, and NaDHC/$C_{16}NMe_3Br$ [84]. They observed that the equimolar mixture of NaDHC and $C_{16}NMe_3Br$ can appreciably solubilize cholesterol although individually NaDHC and $C_{16}NMe_3Br$ are poor cholesterol solubilizers. A mixture of NaC and $C_{16}NMe_3Br$ also showed appreciably solubility of cholesterol.

7. Solubilization in Mixed Micelles of Other Types of Surfactant Combinations

Weers et al. determined the composition dependence of the MAC of decane in mixed micellar systems (10 wt.%) of tetradecyldimethylamine oxide ($C_{14}H_{29}N(CH_3)_2O$) and $C_{12}SO_4Na$ [85]. The MAC exhibited a maximum at a composition richer in the amine oxide. The pH studies suggested that the amine oxide is predominantly nonionic, containing at most 1% of the protonated amine oxide. They inferred that ion-dipole interactions between —SO_4^- and the —$N \rightarrow O$ dipole play a dominant role and that the electrostatic interaction ($C_{12}SO_4^-$: $C_{14}NMe_2OH^+$) and hydrogen bonding mediated by water molecules probably also contribute to anomalous solubilization and other properties. They also determined the MAC of decane in aqueous systems of mixed monoalkyl/dialkyl cationic surfactant micelles [86]. The observed MAC-composition relation is similar to that in ionic/nonionic mixed micellar systems; this is a common feature of the core solubilization in mixed micelles other than anionic/cationic mixed micelles (see Sec. III.B. 2). They considered that the Laplace pressure is partly responsible for the observed solubilization behavior. The relation of MAC composition for the core solubilization of alkane in ionic/nonionic mixed micelles is explained quantitatively in Sec. III.B. 2.

Ohno et al. measured the MAC of Orange OT in mixed micellar systems of α, ω-type cationic and anionic surfactants [87]. The MAC-composition relation showed a maximum. This is considered to be a common feature of the solubilization in anionic/cationic mixed micelles showing no precipitation, since there are also maxima in the plots of the solubilizate solubility versus composition for zwitterionic/anionic mixed micellar systems (see Secs. III.A.3, 5). They also determined the solubilities of yellow OB in micellar systems of the nonionic silicone surfactant mixed with $C_{12}SO_4Na$ or LiFOS [88].

B. Thermodynamic Equations Expressing Synergistic Solubilization Effects in Mixed Micelles

As already described, the macroscopic analysis of the solubilization in mixed micelles needs the determination of a solubilization capacity such as the MAC, solubility, or apparent distribution coefficient of solubilizate in the mixed micellar systems. As also noted, the solubilization in mixed micellar systems has frequently been discussed by using the additivity mixing rules, Eq. (1). There are three cases with respect to the magnitude of the observed solubilization capacity in a mixed micellar system: it is larger, smaller, or equal to the magnitude of the capacity calculated by using Eq. (1). From a practical point of view, the first case is called "a positive synergistic solubilization effect" in the mixed micelles (or by the mixed surfactant), whereas the second and the last cases are "a negative and a nonsynergistic solubilization effect," respectively. The (practically defined) positive synergistic solubilization effect is important for purposes of detergent performance. This effect is also interesting from an academic point of view, since some (practically defined) synergistic effect is unpredictable from the solubilization ability of respective constituent surfactants. One of the most important features of the solubilization in mixed micelles is such an unpredictable synergistic solubilization effect. The synergistic effect occurs as a result of the interaction of a solubilizate with each of the constituent surfactants and the interaction between constituent surfactants in mixed micelles. To gain a better understanding of the synergistic solubilization effect, therefore, it is preferable to formulate this effect by means of thermodynamics and/or statistical thermodynamics. It is also important to predict the solubilization ability of certain mixed micelles (mixed surfactants) from that of the pure micelles of constituent surfactants. This could also be accomplished from thermodynamic and/or statistical thermodynamic approaches. It is desirable that the (statistical) thermodynamic formulation should describe the synergistic solubilization effects and also predict the solubilization ability of certain mixed micelles (mixed surfactants) from that of the pure micelles of constituent surfactants. Note that the thermodynamic definition of synergistic solubilization effects may be different from the practical definition mentioned above.

Treiner et al. have presented a thermodynamic equation for analyzing synergistic solubilization effects in mixed micelles [56–58, 77]. This equation is based on the O'Connell equation: the expression of Henry's constant in a mixed solvent as a function of the Henry's constants in the pure solvents [89]. This equation is applicable to only the mixed-micellar solubilization in which the following conditions are satisfied: micelles are assumed to be a solvent phase (the micellar pseudophase model); the solubilities of the solubilizate are low enough in the micellar phases (and in the bulk water); and the nonideal interaction between different constituent surfactants in micelle can be expressed by the regular solution approximation [56–58, 77, 89]. Because of these conditions the equation used by Treiner et al. is considered to be a limiting equation.

The equation based on the conventional pseudophase model does not take into account the Laplace pressure in a micelle core since the assumed micellar pseudophase is a macroscopic phase having no definite curvature. There are reports suggesting that the Laplace pressure would influence appreciably the saturation solubilization of liquid alkanes [4, 45(b), 90] and nonpolar gases [91] in pure micelles. Some authors have attempted to interprete the MAC data of alkanes in mixed micellar systems from the viewpoint of the Laplace pressure effect [57, 58, 60, 61, 86]. However, the role of the Laplace pressure in micellar solubilization is not yet elucidated.

In this section, we propose thermodynamic equations that describe synergistic solubilization effects in mixed micelles, and examine the predictive ability of the equations by applying them to different cases of the mixed-micellar solubilization [92, 93]. By utilizing the equations, i.e., in a way using the simple Laplace equation, we also look at the influence of the Laplace pressure upon the solubilization in mixed micelles (synergistic solubilization effects).

1. A Thermodynamic Definition of the Synergistic Solubilization Effects

We first describe the solubilization equilibria by using the conventional pseudophase model in which the Laplace pressure in a micelle is not allowed. Here we are concerned with solubilizates that are sparingly soluble in water. We can therefore write the following equation for the solubilization equilibrium of the solubilizate (denoted by subscript s) between the bulk water and each micellar phase (the pseudophase composed of surfactant A, of surfactant B, or of their mixture AB):

$$\mu_s^{\ominus \mathrm{b}} + RT \ln \chi_s^{\mathrm{bX}} = \mu_s^{\ominus \mathrm{mX}} + RT \ln \chi_s^{\mathrm{mX}} \gamma_s^{\mathrm{mX}} \qquad (2)$$

where χ and γ are the mole fraction and the activity coefficient, respectively; superscript X represents the kind of surfactant and thus stands for surfactant A, B, or their mixture AB; superscripts b and m refer to the bulk and the micelle phases, respectively. The activity coefficient γ includes effects from the dissociation change of ionic surfactants if the mixed micelles contain them. Note that the standard chemical potential $\mu_s^{\ominus \mathrm{mAB}}(T,P)$ is defined as

$$\mu_s^{\ominus \mathrm{mAB}}(T,P) = \lim_{\chi_{\mathrm{A}}^{\mathrm{mAB}} \to 1} (\mu_s^{\mathrm{mAB}} - RT \ln \chi_s^{\mathrm{mAB}})$$

The other standard chemical potentials $\mu_s^{\ominus \mathrm{mA}}$ and $\mu_s^{\ominus \mathrm{mB}}$ are given by (in this case, superscript or subscript X refers to surfactant A or B)

$$\mu_s^{\ominus \mathrm{mX}}(T,P) = \lim_{\chi_{\mathrm{X}}^{\mathrm{mX}} \to 1} (\mu_s^{\mathrm{mX}} - RT \ln \chi_s^{\mathrm{mX}})$$

By eliminating $\mu_s^{\ominus b}$ from Eq. (2) and using the quantity "the apparent distribution coefficient $K_s^X = \chi_s^{mX}/\chi_s^{bX}$," we obtain the following equation:

$$\ln K_s^{AB} = \chi_A^* \ln K_s^A + \chi_B^* \ln K_s^B + (\chi_A^* \ln \gamma_s^{mA} + \chi_B^* \ln \gamma_s^{mB} - \ln \gamma_s^{mAB}) + \frac{\chi_B^* (\mu_s^{\ominus mB} - \mu_s^{\ominus mA})}{RT} \quad (3)$$

where χ_A^* and χ_B^* are, respectively, the mole fractions of surfactants A and B constituting the mixed micelles that contain the solubilizates ($\chi_A^* + \chi_B^* = 1$ and note that $\chi_A^{mAB} + \chi_B^{mAB} + \chi_s^{mAB} = 1$). Equation (3) is a thermodynamic equation characterizing synergistic solubilization effects in mixed micelles.

Note that the measurable quantities in Eq. (3) are the apparent distribution coefficients K_s^A, K_s^B, and K_s^{AB}, and thus Eq. (3) implies the following fact. When the pure micellar solution of surfactant A of the apparent distribution coefficient K_s^A and that of surfactant B of K_s^B are mixed in such a way that the mole fractions of surfactants A and B are respectively χ_A^* and χ_B^* in the mixed micelles, the apparent distribution coefficient for the resultant mixed micellar solution is K_s^{AB}. In any event, it can be said that Eq. (3) is a thermodynamic equation characterizing synergistic solubilization effects in mixed micelles (or by mixed surfactants) in which the synergistic solubilization effects are expressed by the third and fourth terms of the right-hand side:

$$(\chi_A^* \ln \gamma_s^{mA} + \chi_B^* \ln \gamma_s^{mB} - \ln \gamma_s^{mAB}) + \frac{\chi_B^*(\mu_s^{\ominus mB} - \mu_s^{\ominus mA})}{RT}$$

Under the condition that the contents of a solubilizate in the mixed and the pure micelles are identical ($\chi_s^{mA} = \chi_s^{mB} = \chi_s^{mAB}$), these terms directly represent the direction (positive, negative, or nothing) and the magnitude of the synergistic solubilization effect. This point is described below.

Equation (3) can be written in other forms by using other quantities representing solubilization capacity. In the case of the saturation solubilization of a solubilizate, $\mu_s^{mA}(\text{satn}) = \mu_s^{mB}(\text{satn}) = \mu_s^{mAB}(\text{satn}) = \mu_s^{l}$ (liquid). Thus we write, in this case

$$\ln\left(\frac{\text{MAC}^{AB}}{1 + \text{MAC}^{AB}}\right) = \chi_A^* \ln\left(\frac{\text{MAC}^{A}}{1 + \text{MAC}^{A}}\right) + \chi_B^* \ln\left(\frac{\text{MAC}^{B}}{1 + \text{MAC}^{B}}\right) + (\chi_A^* \ln \gamma_s^{mA} + \chi_B^* \ln \gamma_s^{mB} - \ln \gamma_s^{mAB}) + \frac{\chi_B^*(\mu_s^{\ominus mB} - \mu_s^{\ominus mA})}{RT} \quad (4)$$

By using the partition coefficient P (= $MAC[H_2O]^b/[s]^b$) defined by Treiner et al. [56–58, 76] and the solubilization equilibrium constant K^x (= $K_s^x/([H_2O]^b + [s]^b + \sum [i]^b)$; i = A, B, additive) defined by Nguyen et al. [59], we can write, respectively, under the usual condition that $[H_2O]^b \gg [s]^b$ at constant $\Sigma\ [i]^b$,

$$\ln P^{AB} = \chi_A^* \ln P^A + \chi_B^* \ln P^B + (\chi_A^* \ln \gamma_s^{mA} + \chi_B^* \ln \gamma_s^{mB} - \ln \gamma_s^{mAB}) + \frac{\chi_B^*(\mu_s^{\ominus mA} - \mu_s^{\ominus mA})}{RT} \quad \text{at } MAC^x \ll 1$$

$$\ln K^{AB} = \chi_A^* \ln K^A + \chi_B^* \ln K^B + (\chi_A^* \ln \gamma_s^{mA} + \chi_B^* \ln \gamma_s^{mB} - \ln \gamma_s^{mAB}) + \frac{\chi_B^*(\mu_s^{\ominus mB} - \mu_s^{\ominus mA})}{RT}$$

The experimental determinations of the values of the terms $\mu_s^{\ominus mB} - \mu_s^{\ominus mA}$ and γ_s^{mX} (superscript X refers to A, B, or AB) in Eq. (3) are carried out in the following way. Equation (2) is rewritten as

$$\mu_s^{\ominus mX} - \mu_s^{\ominus b} = -RT \ln(K_s^X \gamma_s^{mX}) \tag{2'}$$

At first, the values of K_s^X (in this case X refers to A or B) are determined for various concentrations of χ_s^{bX}. The determined K_s^X versus χ_s^{bX} plot is then extrapolated to the zero value of χ_s^{bX}, i.e., $\chi_s^{mX} \rightarrow 0$. By utilization of this extrapolated value of K_s^X ($K_s^X(0)$, the value of $\mu_s^{\ominus mX} - \mu_s^{\ominus b}$ can be determined as

$$\mu_s^{\ominus mX} - \mu_s^{\ominus b} = -RT \ln(K_s^X(0))$$

Then the values of γ_s^{mA} and γ_s^{mB} can be determined by using Eq. (2′) from the values of $\mu_s^{\ominus mX} - \mu_s^{\ominus b}$ and K_s^X. The value of γ_s^{mAB} can also be determined in the same way, but the extrapolation of the K_s^{AB} versus χ_s^{bAB} plot must be performed to zero values of χ_s^{bAB} and χ_B^{bAB}; i.e., $\chi_s^{mAB} \rightarrow 0$ and $\chi_B^{mAB} \rightarrow 0$. The value of $\mu_s^{\ominus mB} - \mu_s^{\ominus mA}$ can be easily known from $(\mu_s^{\ominus mB} - \mu_s^{\ominus b}) - (\mu_s^{\ominus mA} - \mu_s^{\ominus b})$. Finally we have the following equations:

$$\gamma_s^{mX} = \frac{K_s^X(0)}{K_s^X} \quad \text{(X is A or B)} \tag{5}$$

$$\gamma_s^{mAB} = \frac{K_s^A(0)}{K_s^{AB}} \tag{6}$$

$$\frac{\mu_s^{\ominus mB} - \mu_s^{\ominus mA}}{RT} = \ln\left[\frac{K_s^A(0)}{K_s^B(0)}\right] \tag{7}$$

By utilizing Eqs. (5)–(7), the stability of solubilizate "s" in the mixed micelles "AB" and that in the pure micelles of surfactant "B," relative to that in the pure micelles of surfactant "A," are respectively expressed as

$$\mu_s^{\mathrm{mAB}}(\chi_s^{\mathrm{mAB}}) - \mu_s^{\mathrm{mA}}(\chi_s^{\mathrm{mA}}) = RT \ln\left(\frac{K_s^{\mathrm{A}}}{K_s^{\mathrm{AB}}}\right) + RT \ln\left(\frac{\chi_s^{\mathrm{mAB}}}{\chi_s^{\mathrm{mA}}}\right) \tag{8}$$

$$\mu_s^{\mathrm{mB}}(\chi_s^{\mathrm{mB}}) - \mu_s^{\mathrm{mA}}(\chi_s^{\mathrm{mA}}) = RT \ln\left(\frac{K_s^{\mathrm{A}}}{K_s^{\mathrm{B}}}\right) + RT \ln\left(\frac{\chi_s^{\mathrm{mB}}}{\chi_s^{\mathrm{mA}}}\right) \tag{9}$$

where $\mu_s^{\mathrm{mX}}(\chi_s^{\mathrm{mX}})$ means μ_s^{mX} at χ_s^{mX}.

Under the condition that the mole fraction of a solubilizate in all types of micelles is identical ($\chi_s^{\mathrm{mA}} = \chi_s^{\mathrm{mB}} = \chi_s^{\mathrm{mAB}} \equiv \chi_s^{\mathrm{m}}$), Eq. (3) becomes, by the use of Eqs. (5)–(9),

$$\begin{aligned} \ln K_s^{\mathrm{AB}} &= \chi_{\mathrm{A}}^* \ln K_s^{\mathrm{A}} + \chi_{\mathrm{B}}^* \ln K_s^{\mathrm{B}} \\ &\quad - \frac{\mu_s^{\mathrm{mAB}}(\chi_s^{\mathrm{m}}) - \mu_s^{\mathrm{mA}}(\chi_s^{\mathrm{m}})}{RT} \\ &\quad + \frac{\chi_{\mathrm{B}}^*(\mu_s^{\mathrm{mB}}(\chi_s^{\mathrm{m}}) - \mu_s^{\mathrm{mA}}(\chi_s^{\mathrm{m}}))}{RT} \end{aligned} \tag{10}$$

Here we consider the situation that the chemical potential of the solubilizate in a hypothetical state of the mixed micelles μ_s^{mAB} (designated by μ_s^{mAB} (nonsyn)) is expressed as, at an identical value of χ_s^{mX} (X is A, B, or AB),

$$\mu_s^{\mathrm{mAB}}(\mathrm{nonsyn}) = \chi_{\mathrm{A}}^* \mu_s^{\mathrm{mA}} + \chi_{\mathrm{B}}^* \mu_s^{\mathrm{mB}} \tag{11}$$

This equation implies that the stability of the solubilizate in the mixed micelles (μ_s^{mAB}) varies linearly with micellar surfactant composition (χ_{A}^* and χ_{B}^*) between the stabilities in the pure micelles of constituent surfactants (μ_s^{mA} and μ_s^{mB}). This "ideal" situation is expected to hold for the solubilization in which the environment of the solubilizate (incorporation sites) is essentially identical among the mixed micelles and the pure micelles of constituent surfactants. Introduction of Eq. (11) into Eq. (10) leads to the equation valid for this ideal micellar solubilization of an identical value of χ_s^{mX} (X is A, B, or AB):

$$\ln K_s^{\mathrm{AB}}(\mathrm{nonsyn}) = \chi_{\mathrm{A}}^* \ln K_s^{\mathrm{A}} + \chi_{\mathrm{B}}^* \ln K_s^{\mathrm{B}} \tag{12}$$

Equation (12) expresses the nonsynergistic solubilization effect in mixed micelles (by mixed surfactants), and Eq. (11) represents the nonsynergistic solubilization state of the solubilizate in the mixed micelles. The positive and negative synergistic solubilization effects can be characterized by, at an identical value of χ_s^{mX} (X is A, B, or AB),

$$\mu_s^{mAB} < \mu_s^{mAB}(\text{nonsyn}) \quad \text{for } K_s^{AB} > K_s^{AB}(\text{nonsyn}) \quad \text{(positive synergism)}$$

$$\mu_s^{mAB} > \mu_s^{mAB}(\text{nonsyn}) \quad \text{for } K_s^{AB} < K_s^{AB}(\text{nonsyn}) \quad \text{(negative synergism)}$$

This implies that the positive (negative) synergistic solubilization effect occurs when the solubilizate in the real state of the mixed micelles is more (less) stable than in the (hypothetical) nonsynergistic solubilization state of the mixed micelles at an identical value of χ_s^{mX} (X is A, B, or AB). Considering Eq. (3), the positive (negative) synergistic effect occurs if the value of the following terms (we call the terms "the synergistic solubilization effect terms" and write it as SYNT):

$$\text{SYNT} \equiv (\chi_A^* \ln \gamma_s^{mA} + \chi_B^* \ln \gamma_s^{mB} - \ln \gamma_s^{mAB}) + \frac{\chi_B^* (\mu_s^{\ominus mB} - \mu_s^{\ominus mA})}{RT}$$

is positive (negative) at an identical value of χ_s^{mX} (χ_s^{m}). The extent of the synergistic effect is given by the absolute value of the SYNT. Equation (3) is symbolically written as, at an identical value of χ_s^{mX} (χ_s^{m}),

$$\ln K_s^{AB} = \chi_A^* \ln K_s^{A} + \chi_B^* \ln K_s^{B} + \text{SYNT}(\chi_s^{m}) \tag{3'}$$

Equations (10)–(12) and the related discussion are effective under the condition that the mole fraction of a solubilizate in the mixed and the pure micelles is identical ($\chi_s^{mA} = \chi_s^{mB} = \chi_s^{mAB} \equiv \chi_s^{m}$). Under this condition, the thermodynamic equation (3) directly describes synergistic solubilization effects in mixed micelles (by mixed surfactants); i.e., the sign and magnitude of the SYNT represent the synergistic effect (Eq. (3′)). Frequently used solubilizates include water-insoluble or slightly water soluble dyes. The solubilization experiments for these dyes usually determine only the MAC (or solubility) rather than the apparent distribution coefficients. The solubilization capacity MAC^X is related to $\chi_s^{mX}(\text{satn})$ (X is A, B, or AB) by

$$\chi_s^{mX}(\text{satn}) = \frac{\text{MAC}^X}{1 + \text{MAC}^X} \quad \text{at the saturation solubilization}$$

If only the MAC values are determined, Eq. (3), i.e., Eq. (4), cannot be applied under the condition of an identical value of χ_s^{mX} (X is A, B, or AB) since the MAC values are generally different among the constituent surfactants and the mixed surfactants. In this case, Eq. (4) does not describe directly the synergistic solubilization effects, since $\ln \gamma_s^{mX}$ is a function of the concentration χ_s^{mX} and the extra terms $[\ln \gamma_s^{mX}(\chi_s^{mX}(\text{satn})) - \ln \gamma_s^{mX}(\chi_s^m)]$ are generally not zero, as shown below.

$$
\begin{aligned}
\ln\left(\frac{\text{MAC}^{\text{AB}}}{1+\text{MAC}^{\text{AB}}}\right)
&= \chi_{\text{A}}^{*} \ln\left(\frac{\text{MAC}^{\text{A}}}{1+\text{MAC}^{\text{A}}}\right) + \chi_{\text{B}}^{*} \ln\left(\frac{\text{MAC}^{\text{B}}}{1+\text{MAC}^{\text{B}}}\right) + \text{SYNT}(\chi_s^m) \\
&\quad + \chi_{\text{A}}^{*}[\ln \gamma_s^{mA}(\chi_s^{mA}(\text{satn})) - \ln \gamma_s^{mA}(\chi_s^m)] \\
&\quad + \chi_{\text{B}}^{*}[\ln \gamma_s^{mB}(\chi_s^{mB}(\text{satn})) - \ln \gamma_s^{mB}(\chi_s^m)] \\
&\quad - [\ln \gamma_s^{mAB}(\chi_s^{mAB}(\text{satn})) - \ln \gamma_s^{mAB}(\chi_s^m)]
\end{aligned}
\tag{4'}
$$

where χ_s^m is equal to the smallest value of $\chi_s^{mX}(\text{satn})$ (X is A, B, or AB), and $\ln \gamma_s^{mX}(\chi_s^m)$ and $\text{SYNT}(\chi_s^m)$ mean the quantities at χ_s^m. If the value of extra terms $\Sigma\chi_{\text{X}}^{*}[\ln \gamma_s^{mX}(\chi_s^{mX}(\text{satn})) - \ln \gamma_s^{mX}(\chi_s^m)]$ are negligibly small compared to the value of other terms, Eq. (4) can express approximately the synergistic solubilization effects:

$$
\begin{aligned}
\ln\left(\frac{\text{MAC}^{\text{AB}}}{1+\text{MAC}^{\text{AB}}}\right) &\simeq \chi_{\text{A}}^{*} \ln\left(\frac{\text{MAC}^{\text{A}}}{1+\text{MAC}^{\text{A}}}\right) \\
&\quad + \chi_{\text{B}}^{*} \ln\left(\frac{\text{MAC}^{\text{B}}}{1+\text{MAC}^{\text{B}}}\right) + \text{SYNT}(\chi_s^m)
\end{aligned}
\tag{4''}
$$

This condition empirically holds if the MAC values for constituent surfactants do not differ greatly [93]. However, in the case where the MAC values for the constituent surfactants differ greatly, there is a case for the empirical reason that Eq. (4″) approximately holds, as shown below.

Equation (11) is expected to hold for the solubilization of alkane in the mixed micelles composed of hydrocarbon surfactants and the solubilization of a solubilizate in the mixed micelles of homologous surfactants of different alkyl chain lengths at an identical value of χ_s^{mX} (X is A, B, or AB), since the incorporation sites of alkane and other solubilizates are essentially identical among the mixed and the pure micelles, as mentioned. In particular, alkanes (notably the alkane of a chain length comparable to the chain lengths of micellized surfactants) dissolve (nearly) ideally in the micellar hydrocarbon core, and therefore we can write [4, 90]

$$
\begin{aligned}
&\mu_s^{\ominus mX}(T, P_0) = \mu_s^{l}(T, P_0), \qquad v_s^{\ominus m}(T, P_0) = v_s^{l}(T, P_0) \\
&\gamma_s^{mA} = 1, \quad \gamma_s^{mB} = 1, \quad \gamma_s^{mAB} = 1
\end{aligned}
\tag{13}
$$

where superscript l refers to the liquid alkane, and P_0 is the bulk pressure. In the case of the solubilization of alkane, substitution of Eqs. (13) in Eq.

(3) leads to the following equation that holds regardless of the value of χ_s^{mX} (X is A, B, or AB):

$$\ln K_s^{AB} = \chi_A^* \ln K_s^A + \chi_B^* \ln K_s^B \tag{14}$$

In the case of the solubilization of a solubilizate in mixed micelles of the homologous surfactants, Eq. (14) is also applicable but at an identical value of χ_s^{mX}. As mentioned, a nonsynergistic solubilization effect (state) is valid for the alkane solubilization in mixed micelles and for the solubilization of a solubilizate in the mixed micelles of homologous surfactants of different alkyl chain lengths. In general, a simple nonsynergistic solubilization state is not valid for other solubilization cases in mixed micelles. However, we can discuss any solubilization in mixed micelles in terms of positive, negative, or zero deviation (positive, negative, or nonsynergistic solubilization effect) from the hypothetical nonsynergistic solubilization effect. Aside from this we can define a reference solubilization state for a mixed micellar solubilization, and we can also discuss the solubilization in terms of positive, negative, or zero deviation from the reference solubilization state; we hereinafter show a reference solubilization state for the solubilization of alkanol in mixed micelles (Sec. III.B. 5).

2. Nonsynergistic Solubilization of Alkane in Mixed Micelles

Here we consider the solubilization of alkane in mixed micelles. Equation (14) applicable to this solubilization was derived with the conventional pseudophase model in which the Laplace pressure in the micelle is not allowed. We also have the following equation derived from the corpuscular pseudophase model taking into account Laplace pressures P_m^X inside micelles [4, 90]:

$$\ln K_s^{AB} = \chi_A^* \ln K_s^A + \chi_B^* \ln K_s^B + v_s^l(T, P_0)\frac{\chi_A^* P_m^A + \chi_B^* P_m^B - P_m^{AB}}{RT} \tag{15}$$

The Laplace pressure P_m is assumed to be expressed, as a first approximation, by the simple Laplace equation

$$P_m - P_0 = \frac{2\sigma}{\gamma'}$$

where σ is the micelle core/water interfacial tension, and γ' is the mean radius of curvature of the micelle core. For the saturation solubilization of alkane, Eqs. (14) and (15) are rewritten, respectively, as

$$\ln\left(\frac{\mathrm{MAC^{AB}}}{1 + \mathrm{MAC^{AB}}}\right) = \chi_{\mathrm{A}}^{*} \ln\left(\frac{\mathrm{MAC^{A}}}{1 + \mathrm{MAC^{A}}}\right) + \chi_{\mathrm{B}}^{*} \ln\left(\frac{\mathrm{MAC^{B}}}{1 + \mathrm{MAC^{B}}}\right) \tag{16}$$

$$\ln\left(\frac{\mathrm{MAC^{AB}}}{1 + \mathrm{MAC^{AB}}}\right) = \chi_{\mathrm{A}}^{*} \ln\left(\frac{\mathrm{MAC^{A}}}{1 + \mathrm{MAC^{A}}}\right) + \chi_{\mathrm{B}}^{*} \ln\left(\mathrm{MAC^{B}}/1 + \mathrm{MAC^{B}}\right) + v_s^{1}(T, P_0)\,\frac{\chi_{\mathrm{A}}^{*} P_{\mathrm{m}}^{\mathrm{A}} + \chi_{\mathrm{B}}^{*} P_{\mathrm{m}}^{\mathrm{B}} - P_{\mathrm{m}}^{\mathrm{AB}}}{RT} \tag{17}$$

In a way using the simple Laplace equation, we hereinafter look at the Laplace pressure effect on the synergism in the saturation solubilization of alkane in mixed micelles. For this purpose, we calculate the MAC values of alkane in certain mixed micellar systems by means of Eqs. (16) and (17) and compare them with the observations.

We calculate the MAC values of decane in $C_{12}NMe_3Br$(DTAB)/$C_{12}E_6$ aqueous solutions in which mixed micelles with maximal decane content are present. Then we compare these MAC values with the observations by Weers and Schuing (21°C, 100 mM) [60]. The Laplace pressure in the micelle core with maximal decane content was calculated, by using the simple Laplace equation, from the micelle core size and the core/water interfacial tension, 22 mN/m [4, 90]. The core size of $C_{12}NMe_3Br/C_{12}E_6$ mixed micelles of varying compositions (containing no decane) were calculated by means of our proposed equations [94], and then the core size of the mixed micelles with maximal decane contents were calculated by means of our other equations [95]. In Table 2 are listed the estimated aggregation numbers and Laplace pressures of the $C_{12}NMe_3Br/C_{12}E_6$ mixed micelles with maximal decane contents. These values were used for calculating the MAC values of decane in $C_{12}NMe_3Br/C_{12}E_6$ solutions by means of Eq. (17). The MAC values calculated by using Eqs. (16) and (17) are tabulated in Table 2, together with the observed values. There is only small difference between the MAC values calculated by Eq. (16) and those by Eq. (17). In addition, fair agreement between the MAC values calculated by Eqs. (16) and (17) and the observed values is obtained within the experimental uncertainty. These indicate that the value of the third term of the right-hand side of Eq. (17) is rather small compared to the value of the first two terms. It can, therefore, be considered that the Laplace pressure effect on the synergistic solubilization in ionic/nonionic mixed micelles is almost canceled out in Eq. (17). In other words, Laplace pressure is hidden in the synergistic solubilization effect even if this pressure exists in micelle. Consequently, the conventional pseudophase model alone is adequate to predict a solubilization capacity in ionic/nonionic mixed micellar systems ($\mathrm{MAC^{AB}}$, K_s^{AB}, etc.) from the same quantities in

TABLE 2 Comparison of the Calculated MAC of Decane with the Observed One in $C_{12}NMe_3Br/C_{12}E_6$ Mixed Micellar Solutions, as a Function of $C_{12}NMe_3Br$(DTAB) Mole Fraction in Surfactants Constituting Mixed Micelles, χ^*_{DTAB}, together with the Estimated Aggregation Number N and Laplace Pressure P_m of the Micelles

	MAC (mol/mol)				
χ^*_{DTAB}	Obsd. [60]	Calcd. (Eq. (16))	Calcd. (Eq. (17))	N	P_m (MPa)
0	0.84			710	11.7
0.1	0.55	0.58	0.56	456	13.5
0.3	0.31	0.32	0.30	240	16.7
0.5	0.18	0.19	0.17	142	19.9
0.7	0.13	0.12	0.11	100	22.4
0.9	0.11	0.072	0.077	76	24.5
1	0.058			59	26.7

Source: Refs. 92 and 93.

the pure micellar systems of constituent surfactants. This conclusion can be considered to be valid also for ionic/ionic (the same charges) and nonionic/nonionic mixed micellar systems [92].

In Fig. 1 is shown the comparison of the MAC values of nonane calculated by means of Eq. (16) (the conventional pseudophase model) with the observed values of Ward and Quigley for $C_{12}SO_4Na/C_{12}E_6$ mixed micellar solutions (25°C, 25 mM) [61]. Figure 2 shows the calculated (Eq. (16)) and observed [61] MAC values of propane gas in aqueous solutions of $C_8SO_4Na/C_{12}SO_4Na$ mixed micelles (25°C, 0.03 M NaCl). It can be seen that the calculations are in fair agreement with the observations within the experimental uncertainty. As mentioned, an alkane solubilization is an ideal micellar solubilization and thus the MAC values of alkane calculated by Eq. (16) (Eq. (17)) exhibit nonsynergistic solubilization effects. It can, therefore, be seen from Table 2 and Fig. 1 that the thermodynamic definition of nonsynergistic solubilization effects is evidently different from the practical definition using the additivity mixing rule of Eq. (1).

3. Nonsynergistic Solubilization in Mixed Micelles of Homologous Surfactants and Synergistic Solubilization

The solubilization of a solubilizate in the mixed micelles of homologous surfactants of different chain lengths obeys Eq. (14) at an identical value of χ_s^{mX} (X is A, B, or AB). At saturation, Eq. (16) (nonsynergistic solubili-

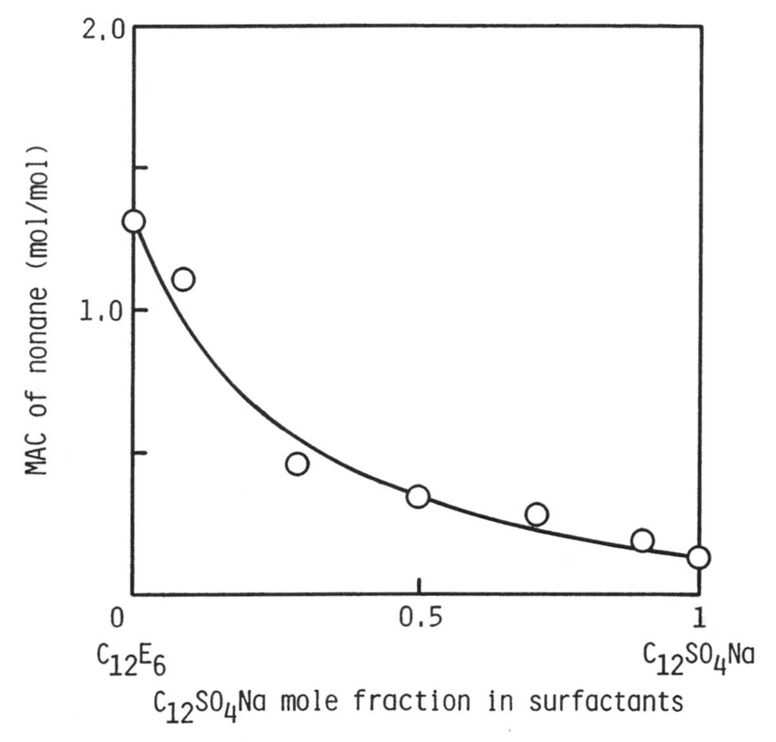

FIG. 1 Comparison of the calculated MAC of nonane with the observed one in $C_{12}SO_4Na/C_{12}E_6$ mixed micellar solutions, as a function of $C_{12}SO_4Na$ mole fraction in surfactants constituting mixed micelles: —, calculated by Eq. (16); ○, observed data of Ref. 61. (From. Ref. 92.)

zation in Eq. (4)) can describe this solubilization, if the MAC values for constituent surfactants do not differ greatly. To confirm this, we compare the MAC values calculated by Eq. (16) with the MAC data of yellow OB in aqueous solutions of $C_8E_6/C_{12}E_6$ and $C_{10}E_6/C_{12}E_6$ mixed micelles in which the MAC values for constituent surfactants do not differ greatly [52].

Figure 3 shows the MAC values of yellow OB calculated by Eq. (16) and the observed data for aqueous systems of $C_8E_6/C_{12}E_6$ and $C_{10}E_6/C_{12}E_6$ mixed micelles as a function of C_8E_6 or $C_{10}E_6$ mole fraction in surfactants constituting the mixed micelles [52]. In Fig. 3 the calculation agrees well with the observation within the experimental uncertainty, and hence it can be said that Eq. (16) predicts the observation well. If the

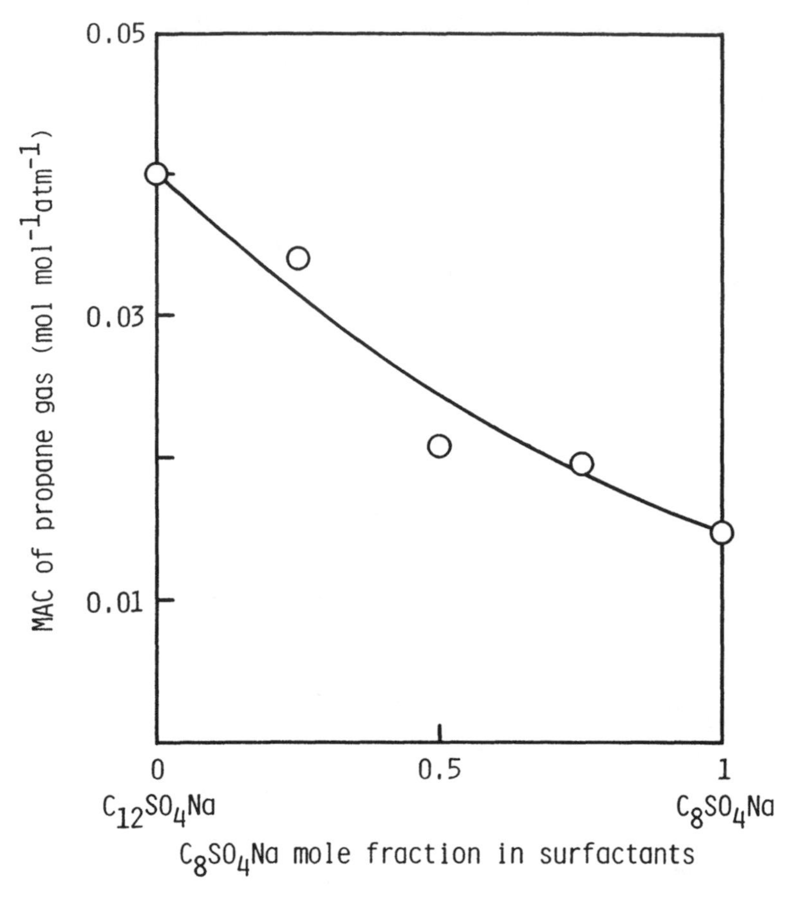

FIG. 2 Comparison of the calculated MAC of propane gas with the observed one in $C_8SO_4Na/C_{12}SO_4Na$ mixed micellar solutions, as a function of C_8SO_4Na mole fraction in surfactants constituting mixed micelles: —, calculated by Eq. (16); ○, observed data of Ref. 69.

incorporation sites of a solubilizate are different between the pure micelles of different kinds of constituent surfactants, it is possible to observe a deviation from the nonsynergistic solubilization capacity calculated by Eq. (16) or (14). Figure 4 shows the comparison of the nonsynergistic MAC values calculated by Eq. (16) with the observed ones for the solubilization of yellow OB in aqueous solutions of mixed micelles of homologous surfactants having different polyoxyethyelene (POE) groups: three mixed surfactants of $C_{12}E_n$, $C_{12}E_6/C_{12}E_{29}$, $C_{12}E_6/C_{12}E_{49}$, and $C_{12}E_{29}/C_{12}E_{49}$ [52].

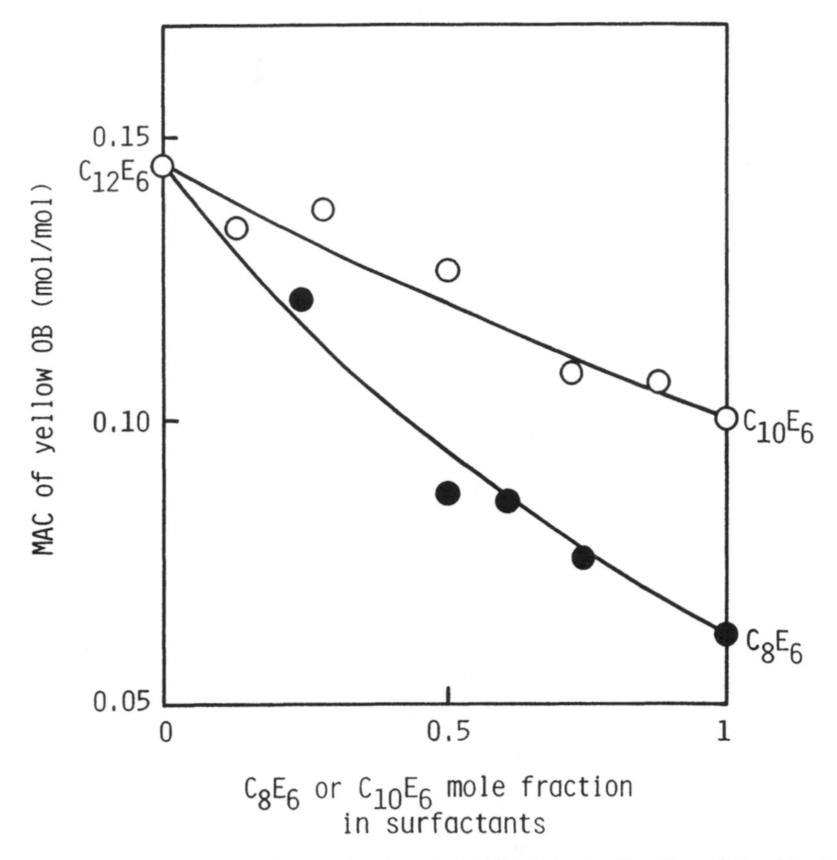

FIG. 3 Comparison of the calculated MAC of dyestuff yellow OB with the observed one in aqueous solutions of $C_8E_6/C_{12}E_6$ and $C_{10}E_6/C_{12}E_6$ mixed micelles, as a function of C_8E_6 or $C_{10}E_6$ mole fraction in surfactants constituting mixed micelles: —, calculated by Eq. (16); ○, ●, observed data of Ref. 52. (From Ref. 93.)

In Fig. 4, the observed MAC values for $C_{12}E_6/C_{12}E_{29}$ and $C_{12}E_6/C_{12}E_{49}$ mixed systems were almost identical within the experimental uncertainty. The observation for the three mixed micellar systems exhibits negative deviation from the calculation, i.e., negative synergistic solubilization effects. The incorporation sites of dyestuff yellow OB in $C_{12}E_n$ micelles involve the POE shell as well as the hydrocarbon core [52]. If POE shells in different kinds of $C_{12}E_n$ micelles provide different incorporation sites for the dye, it is reasonable to observe negative synergistic solubilization

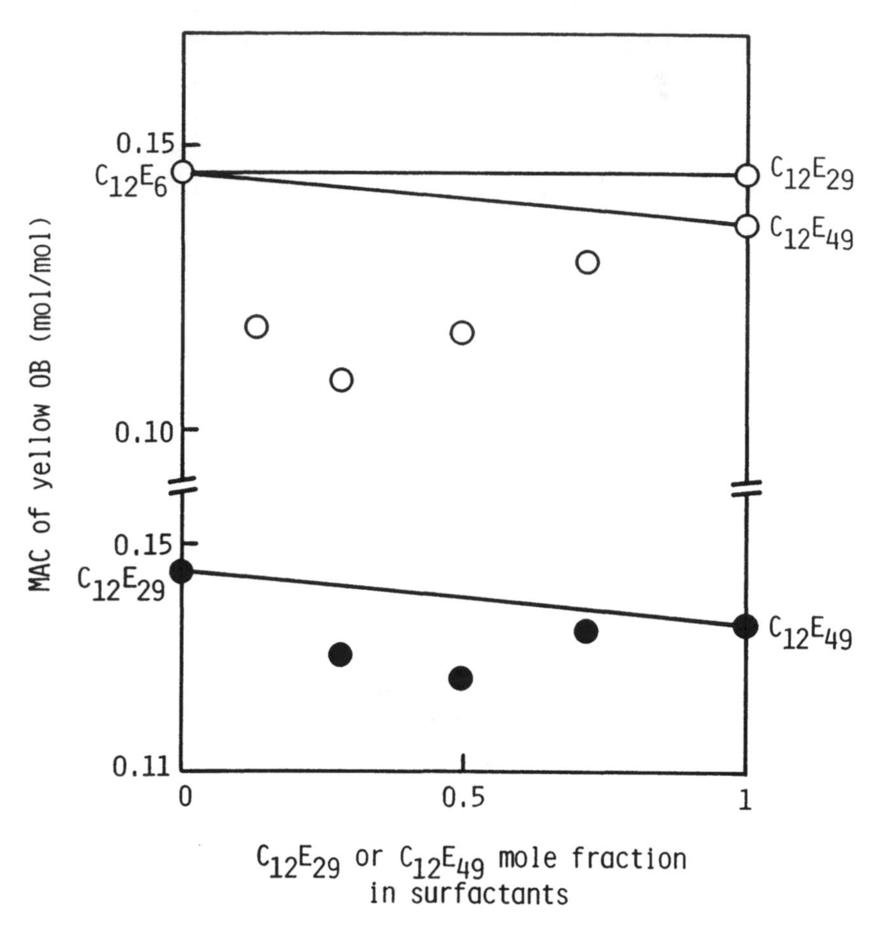

FIG. 4 Comparison of the calculated MAC of dyestuff yellow OB with the observed one in aqueous solutions of $C_{12}E_{29}/C_{12}E_6$ and $C_{10}E_{49}/C_{12}E_6$ mixed micelles, as a function of $C_{12}E_{29}$ or $C_{12}E_{49}$ mole fraction in surfactants constituting mixed micelles: —, calculated by Eq. (16); ○, ●, observed data of Ref. 52. (From Ref. 93.)

effects shown in Fig. 4. Contrary to these negative synergistic solubilization effects, a positive synergistic effect is observed in Fig. 5 for the system of $C_{12}NH_3Cl/C_{12}E_{49}$ mixed micelles. In Fig. 5 the system of $C_{12}NH_3Cl/C_{12}E_6$ mixed micelles exhibits a nonsynergistic solubilization effect. It is likely that these results relate to the change in the compactness of the POE shell in the mixed micelles compared to in the pure $C_{12}E_6$ or $C_{12}E_{49}$ micelles, but it needs further investigation to draw a definite conclusion.

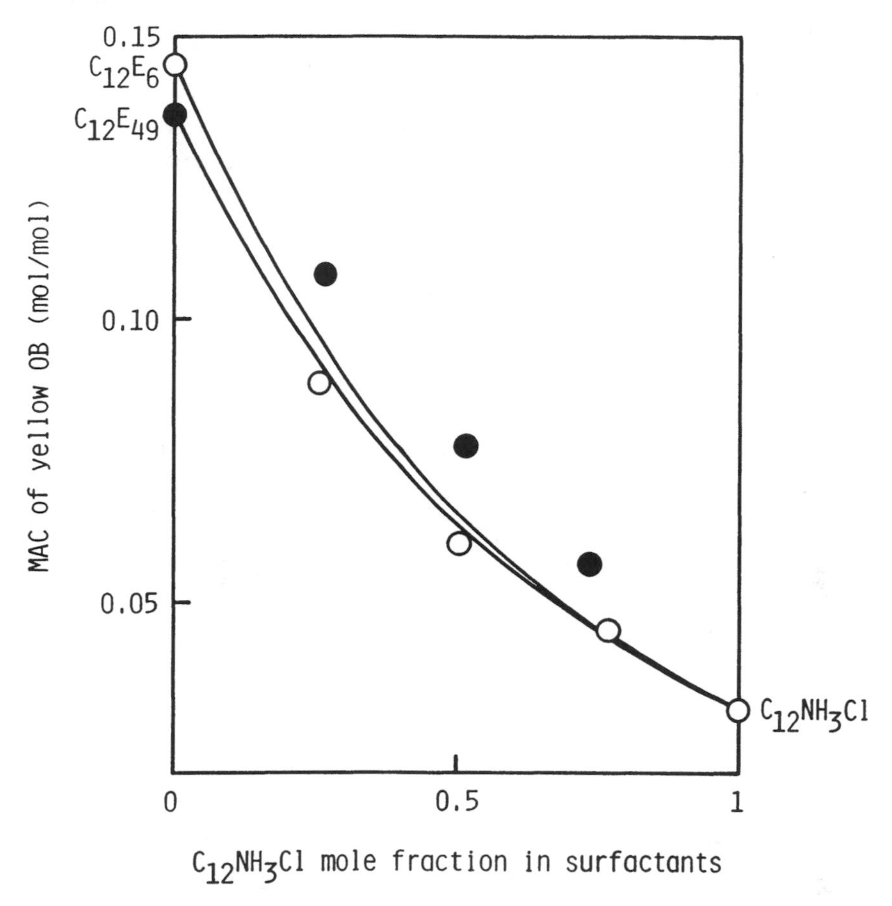

FIG. 5 Comparison of the calculated MAC of dyestuff yellow OB with the observed one in aqueous solutions of $C_{12}NH_3Cl/C_{12}E_6$ and $C_{12}NH_3Cl/C_{12}E_{49}$ mixed micelles, as a function of $C_{12}NH_3Cl$ mole fraction in surfactants constituting mixed micelles: —, calculated by Eq. (16); ○, ●, observed data of Ref. 52. (From Ref. 93.)

Equation (4) or (4″) enables us to consider synergistic solubilization effects in mixed micelles if the MAC values for constituent surfactants do not differ greatly. Even if the MAC values for constituent surfactants differ greatly, there are some MAC data which are relatively close to the calculation by Eq. (16) (nonsynergistic solubilization in Eq. (4)): the solubilization of yellow OB in $C_{12}E_6$/aerosol OT and $C_{12}E_6$/LiFOS mixed micelles and of azobenzene in $C_{12}E_6$/LiFOS mixed micelles investigated by Muto et al. [54]. This suggests the possibility that Eq. (4) could express synergsitic solubilization effects in general mixed micellar systems. Equation (4) or (4″) could, therefore, be a criterion for quantitative evaluation of the MAC values for solubilization in mixed micelles. Complete evaluation requires determination of activity coefficients of the solubilizate in the micelles γ_s^{mX} (X is A, B, or AB) and/or their concentration dependence (see Eqs. (4) and (4′)).

Finally, the following comment is worth noting. The thermodynamic equation (3) or (4) characterizes synergistic solubilization effects in mixed micelles and the extent of the synergistic effect is given by the sign and magnitude of the SYNT (Eqs. (3′) and (4″)). The thermodynamic definition of synergistic effects based on Eq. (3) or (4) is generally different from the practical definition using the additivity mixing rules, Eq. (1). The discussion using the additivity rule might, therefore, lead to a wrong conclusion with regard to synergistic solubilization effects. For example, in the case of the solubilization of azobenzene in $C_{12}SO_4Na/C_mE_{20}$ (m = 10, 18) mixed micelles, the extent of the positive deviation from the additivity rule in the observed solubility increased with decreasing alkyl chain length of C_mE_{20} [53]. However, the positive value of the SYNT in Eq. (4″) decreases with decreasing alkyl chain length; this is an opposite situation.

4. Anomalous Solubilization Behavior of Alkane in Anionic/Cationic Mixed Micelles

As mentioned in Sec. III.A. 3, the MAC of decane and the micelle size in $C_{12}SO_4Na$(SDS)/$C_{12}NMe_3Cl$ mixed micellar systems increase rapidly as the precipitation boundary is approached (21°C, 100 mM) [60, 73]. In attempting to interprete this anomalous solubilization behavior by means of Eqs. (16) and (17) we must estimate the Laplace pressures P_m in $C_{12}SO_4Na/C_{12}NMe_3Cl$ mixed micelles with maximal decane contents. To do this, we calculate the aggregation numbers of such mixed micelles by making the assumption that the mean area per headgroup on the micelle surface is unchanged by alkane solubilization, as is assumed in the case of alkane-solubilized micelles of single ionic surfactants [95]. In Table 3 are listed the aggregation numbers N of $C_{12}SO_4Na/C_{12}NMe_3Cl$ mixed micelles with maximal decane contents calculated by using our equation [95] from the data by Malliaris et al. for the aggregation numbers of $C_{12}SO_4Na/C_{12}NMe_3Cl$ mixed micelles containing no alkane [73]. The estimated Laplace pressures P_m and the MAC values of decane in $C_{12}SO_4Na/C_{12}NMe_3Cl$ solutions calculated by Eqs. (16) and (17) are also listed in Table 3. It can be seen in Table 3 that both of the calculated MAC values disagree with the observation. Direct application of the equations of the conventional and corpuscular pseudophase models, Eqs. (16) and (17), fail to explain the MAC data of decane in $C_{12}SO_4Na/C_{12}NMe_3Cl$ mixed

TABLE 3 Comparison of the Calculated MAC of Decane with the Observed One in $C_{12}SO_4Na/C_{12}NMe_3Cl$ Mixed Micellar Solutions, as a Function of $C_{12}SO_4Na$ (SDS) Mole Fraction in Surfactants Constituting Mixed Micelles, χ^*_{SDS}, together with the Estimated Aggregation Number N and Laplace Pressure P_m of the Micelles

	MAC (mol/mol)				
χ^*_{SDS}	Obsd. [60]	Calcd. (Eq. (16))	Calcd. (Eq. (17))	N	P_m (MPa)
0	0.053			65	25.8
0.05	0.086	0.056	0.073	100	22.4
0.13	0.13	0.061	0.098	154	19.4
0.17	0.21	0.064	0.13	256	16.4
0.19	0.37	0.065	0.17	481	13.3
		Precipitation region			
0.87	0.49	0.14	0.39	1300	9.6
0.88	0.40	0.14	0.31	696	11.8
0.95	0.28	0.15	0.22	281	15.9
1	0.16			157	19.3

Source: Refs. 92 and 93.

micellar systems. Next, we apply the pseudophase model's equations by introducing the following assumption:

A micellar complex of very large size forms in the vicinity of the precipitation boundary, and this complex exhibits a very large solubilization capacity. The complex bears the composition of constant molecular ratio of cationic to anionic surfactant, and this ratio is close to the characteristic molecular ratio of the precipitation boundary. The component species of the micellar complex could be regarded as a kind of molecular compound. If this is the case, we can assume that, in the micellar solution of compositions richer in cationic (anionic) surfactants, the mixed micelle contains both the monomeric cationic (anionic) surfactants and such molecular compounds. In the case of the solubilization in the micellar solution of compositions richer in cationic surfactants, Eq. (16), Eq. (14), and Eq. (17) are, respectively, modified as follows:

$$\ln\left(\frac{\mathrm{MAC}}{1+\mathrm{MAC}}\right) = \chi^*_{cp} \ln\left(\frac{\mathrm{MAC_{cp}}}{1+\mathrm{MAC_{cp}}}\right) + \chi^*_{ca} \ln\left(\frac{\mathrm{MAC_{ca}}}{1+\mathrm{MAC_{ca}}}\right) \tag{18}$$

$$\ln\left(\frac{\mathrm{MAC}}{1+\mathrm{MAC}}\right) = \chi^*_{cp} \ln\left(\frac{\mathrm{MAC_{cp}}}{1+\mathrm{MAC_{cp}}}\right) + \chi^*_{ca} \ln\left(\frac{\mathrm{MAC_{ca}}}{1+\mathrm{MAC}_{ca}}\right) + v^l_s\,(T, P_0)\,\frac{\chi^*_{cp}P^{cp}_m + \chi^*_{ca}P^{ca}_m - P_m}{RT} \tag{19}$$

where subscripts and superscripts cp and ca denote the molecular compound and the monoeric cationic surfactant, respectively. Next we consider the composition and concentration of the assumed molecular compound in the solution of compositions richer with cationic surfactants. The micellar complex is expressed by C_pA_q in which C and A denote cationic and anionic surfactants, respectively. The fraction $\chi/(p + q)$ ($\chi = p$ or q) is closed to the (micelle) composition at the precipitation boundary, and thus the molecular compound could be expressed as $C_{p/q}A$. The mole fraction of the molecular compound, χ^*_{cp}, and that of the monomeric cationic surfactant, χ^*_{ca}, constituting the mixed micelles are expressed, respectively, by

$$\chi^*_{cp} = \frac{\chi^{m^*}_{an}}{1 - (p/q)\chi^{m^*}_{an}} \tag{20}$$

$$\chi^*_{ca} = \frac{1 - (1 + p/q)\chi^{m^*}_{an}}{1 - (p/q)\chi^{m^*}_{an}} \tag{21}$$

where $\chi^{m^*}_{an}$ is the mole fraction of the anionic surfactant in the surfactants of monomeric form that constitute mixed micelles; i.e., it is identical with χ^*_A or χ^*_B in Eqs. (14), (16), or (17). In the case of the solubilization in the solution of compositions richer with anionic surfactants, equations similar to Eqs. (18)–(21) can be easily written.

When Eqs. (18)–(21) are applied to the solubilization of an alkane in anionic/cationic mixed micelles, it can be assumed that

$$MAC_{cp} \gg 1 \text{ (mol/mol)} \qquad P^{cp}_m \simeq P_0$$

Table 4 lists the MAC values calculated by means of Eqs. (18) and (19), and the molecular ratios of cationic/anionic surfactants in the molecular compounds, p/q. The values of p/q were determined in such a way as to fit the calculation of the observation. Table 4 shows that both Eqs. (18) and (19) give the calculated MAC values in good agreement with the observed values. In addition, the values of p/q are also reasonable since the mole fractions of monomeric $C_{12}SO_4Na$ (SDS) (χ^*_{SDS}) at the precipitation boundaries are around 0.22 and 0.86 [60]. We summarize as follows. We explained quantitatively the anomalous solubilization behavior of alkane in anioic/cationic mixed micelles by introduction of the following assumptions into the pseudophase models: the solubilization is nonsynergistic, and the mixed micelles formed consist of monomeric surfactants and the molecular compounds having a composition close to that of the precipitation boundary. The phase models used were the conventional and the corpuscular models; the former does not take into account the Laplace pressure in the micelle, whereas the latter allows for it. Both models ex-

TABLE 4 Calculated MAC (mol/mol) of Decane in $C_{12}SO_4Na$(SDS)/$C_{12}NMe_3Cl$ Mixed Micellar Solutions, as a Function of χ^*_{SDS}, together with the Mole Fraction of the Assumed Molecular Compound, χ^*_{cp}, in the Micelles and Its Molecular Ratio of Cationic/Anionic Surfactants, p/q

	Eq. (18)		Eq. (19)	
χ^*_{SDS}	χ^*_{cp}	MAC	χ^*_{cp}	MAC
0	p/q = 77/23		p/q = 79/21	
	0		0	
0.05	0.06	0.064	0.06	0.075
0.13	0.23	0.11	0.27	0.12
0.17	0.42	0.21	0.57	0.21
0.19	0.55	0.35	0.85	0.36
	Precipitation region			
	p/q = 16/84		p/q = 18/82	
0.87	0.45	0.52	0.31	0.54
0.88	0.30	0.34	0.23	0.40
0.95	0.06	0.19	0.06	0.23
1	0		0	

Source: Ref. 93.

plained the solubilization of alkane in anionic/cationic as well as ionic/nonionic mixed micelles, and hence the existence of the Laplace pressure is not a sufficient condition, at least, for this solubilization. As also mentioned in Sec. III.A.3, the formation of vesicles, other than precipitation, is possible in the intermediate composition range of aqueous systems of anionic/cationic mixed surfactants. We have assumed in the pseudophase model that a micellar complex of very large size forms in the micellar solution close to the precipitation boundary. Even if vesicles form in these systems, we can also assume the formation of such micellar complexes in the micellar solution close to the phase boundary against the vesicular solution (in the place of the precipitation boundary). In this case, it could be viewed that the assumed micellar complexes transform into vesicles, i.e., a phase transition from a micelle phase to a vesicle phase (instead of a phase transition to the surfactant solid).

Anomalous solubilization behavior is observed for aqueous systems of azo oil dye solubilized in *N*-tetradecyl-*N*,*N*-dimethylbetaine/$C_{12}SO_4Na$ mixed micelles (25°C, 0.01 *M* NaCl, total conc. 0.05 *M*); there is a sharp maximum in the MAC versus composition plot [82] (Sec. III.A.5). In the same way as above, we presume that micellar complex is formed at the micelle composition showing the MAC maximum through electrostatic interaction between oppositely charged headgroups of the betaine and $C_{12}SO_4NA$. The MAC values calculated by using Eqs. (18), (20), and (21) are shown in Fig. 6, together with the data by Iwasaki et al. [82]. The fair agreement between the calculation and the observation is also obtained.

5. Solubilization of Alkanol in Mixed Micelles

The thermodynamic equation, Eq. (3), expressed synergistic solubilization effects in mixed micelles, at an identical value of χ_s^{mX} (X is A, B, or

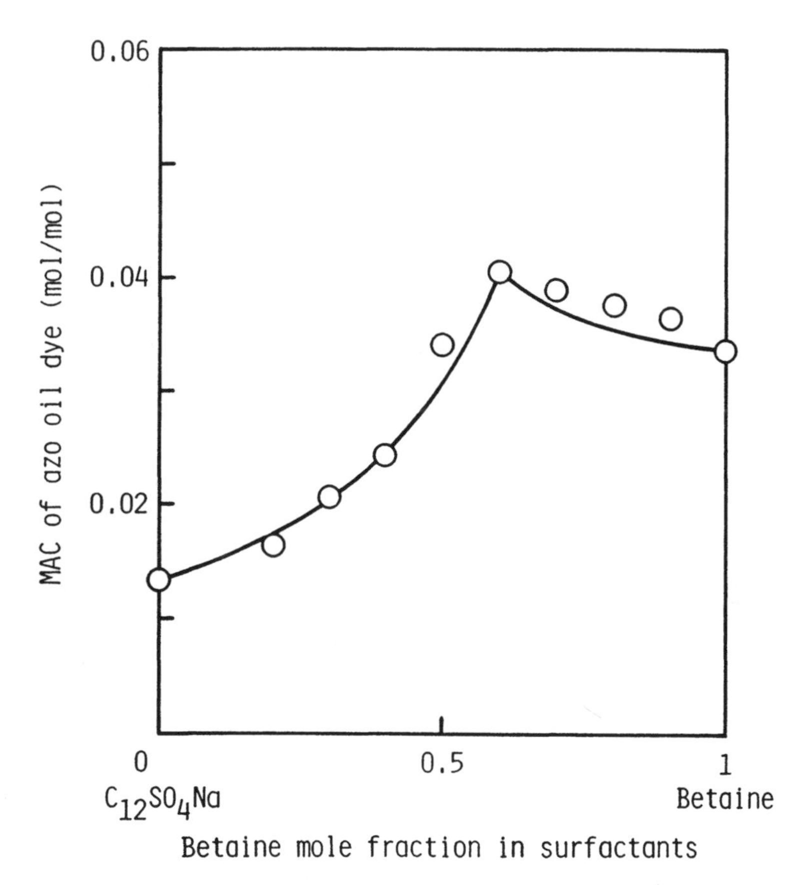

FIG. 6 Comparison of the calculated MAC of azo oil dye with the observed one in aqueous solutions of mixed micelles of *N*-tetradecyl-*N*,*N*-dimethylbetaine and $C_{12}SO_4Na$, as a function of betaine mole fraction in surfactants constituting mixed micelles: —, calculated by Eq. (18); ○, observed data of Ref. 82.

AB). The nonsynergistic solubilization effect and state are, respectively, expressed by Eq. (12) and Eq. (11). As mentioned, a reference solubilization state (effect) can be specified for the solubilization of alkanol in mixed micelles. We consider a reference solubilization state in which the mole fraction of the solubilizate is sufficiently low in the mixed micelles and in the pure micelles of the constituent surfactants. The following equation obtained from Eq. (3) represents this reference solubilization state:

$$\ln K_s^{AB}(\text{ref}) = \chi_A^* \ln K_s^A + \chi_B^* \ln K_s^B - \lim_{\chi_s^{mAB} \to 0} \ln \gamma_s^{mA} + \frac{\chi_B^* (\mu_s^{\ominus mB} - \mu_s^{\ominus mA})}{RT} \tag{22}$$

Here we assume that the nonideal interactions in the micellar phase composed of surfactants A and B and solubilizate are approximately written by the regular solution expression [96]. Thus we have [87]

$$\lim_{\chi_s^{mAB} \to 0} \ln \gamma_s^{mAB} = (\alpha_{Bs} - \alpha_{As})\chi_B^* - \alpha_{AB}\chi_A^*\chi_B^* \tag{23}$$

$$\mu_s^{\ominus mB} - \mu_s^{\ominus mA} = (\alpha_{Bs} - \alpha_{As})RT \tag{24}$$

where α_{XY} is the interaction parameter between molecules X and Y. Introducing Eqs. (23), and (24) into Eq. (22), we have

$$\ln K_s^{AB}(\text{ref}) = \chi_A^* \ln K_s^A + \chi_B^* \ln K_s^B + \alpha_{AB}\chi_A^*\chi_B^* \tag{25}$$

Equation (25) is the equation which Triener et al. have used for analyzing synergistic solubilization effects, as mentioned. The reference solubilization state expressed by Eq. (25) at an identical value of χ_s^{mX} (X is A, B, or AB) is thermodynamically expressed by

$$(\mu_s^{mAB}(\text{ref}) - \mu_s^{mA}) - (\mu_s^{mAB}(\chi_s^{mAB} \to 0) - \mu_s^{\ominus mA}(\chi_s^{mA} \to 0)) = \chi_B^* [(\mu_s^{mB} - \mu_s^{mA}) - (\mu_s^{\ominus mB} - \mu_s^{\ominus mA})] \tag{26}$$

where $\mu_s^{mAB}(\text{ref})$ is the chemical potential of the solubilizate alkanol in the reference solubilization state for the mixed micelles. Positive and negative deviations from the reference solubilization are characterized by

$\mu_s^{mAB} < \mu_s^{mAB}(\text{ref})$ for $K_s^{AB} > K_s^{AB}(\text{ref})$ (positive deviation)

$\mu_{smAB} > \mu_s^{mAB}(\text{ref})$ for $K_s^{AB} < K_s^{AB}(\text{ref})$ (negative deviation

Nguyen et al. determined the values of the solubilization equilibrium constant K^x (= $K_s^x/([H_2O]^b + [s]^b + \Sigma\,[i]^b)$; i = A, B, additive) (X is A, B, or AB) of 1-hexanol in aqueous solutions of mixed micelles of polyoxyethylene (15) nonylpheny ether (NPE_{15}) and cetylpyridinium chloride (CPC) (40°C, 20 mM) [59]. The $K^A{}_B$ values were determined as a function of micellar mole fractions of the 1-hexanol and the surfactants (χ_s^{mX} and χ_A^*). By using the solubilization equilibrium constant, Eqs. (12) and (25) are rewritten, respectively, as

$$\ln K^{AB}(\text{nonsyn}) = \chi_A^* \ln K^A + \chi_B^* \ln K^B \tag{12'}$$

$$\ln K^{AB}(\text{ref}) = \chi_A^* \ln K^A + \chi_B^* + \alpha_{AB}\chi_A^*\chi_B^* \tag{25'}$$

The values of K^{AB} (nonsyn) and K^{AB}(ref) for 1-hexanol in NPE_{15}/CPC mixed micellar solutions were, respectively, calculated by means of Eq. (12′) and (25′), at varying mole fractions of 1-hexanol (χ_s^{mX}) and those of NPE_{15} (χ_{NPE15}^*) in the micelles ($\alpha_{AB} = -1.3$ [76, 96]). In Figs. 7–10 are

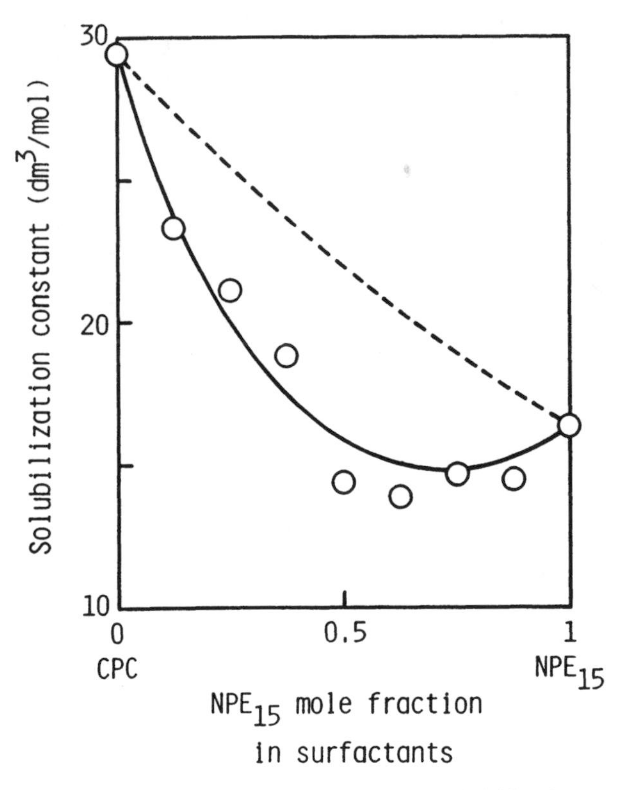

FIG. 7 Comparison of the calculated solubilization equilibrium constant of 1-hexanol with the observed one in NPE_{15}/CPC mixed micellar solutions at $\chi_s^{mX} \to 0$, as a function of NPE_{15} mole fraction in surfactants constituting mixed micelles: —, calculated by Eq. (25′); ---------, calculated by Eq. (12′); ○, observed data of Ref. 59. (χ_s^{mX}, mole fraction of 1-hexanol in the micelles). (From Ref. 92.)

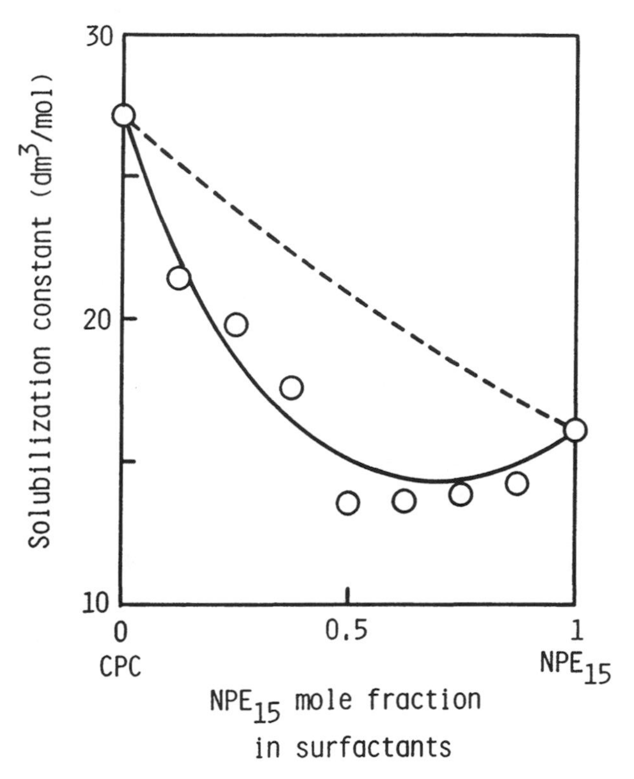

FIG. 8 Comparison of the calculated solubilization equilibrium constant of 1-hexanol with the observed one in NPE_{15}/CPC mixed micellar solutions at $\chi_s^{mX} = 0.1$, as a function of NPE_{15} mole fraction in surfactants constituting mixed micelles: —, calculated by Eq. (25′); ---------, calculated by Eq. (12′); ○, observed data of Ref. 59. (From Ref. 92.)

shown these calculated values and the data of Nguyen et al., as a function of NPE_{15} mole fraction in surfactants (NPE_{15} + CPC) constituting mixed micelles (χ^*_{NPE15}) at constant value of χ_s^{mX} (X is A, B, or AB).

Figure 7 shows the K^{AB} values calculated by Eq. (25′) at $\chi_s^{mX} \rightarrow 0$ agree well with the observations within the experimental uncertainty. When $\chi_s^{mX} \rightarrow 0$, Eq. (25′) holds in principle and so the good agreement is expectable. By comparison of Eq. (3) with Eq. (25′), deviations from the reference solubilization expressed by Eq. (25′) are represented by the terms

$$(\chi^*_A \ln \gamma_s^{mA} + \chi^*_B \ln \gamma_s^{mB} - \ln \gamma_s^{mAB}) + (\alpha_{Bs} - \alpha_{As})\chi^*_B - \alpha_{AB}\chi^*_A\chi^*_B$$

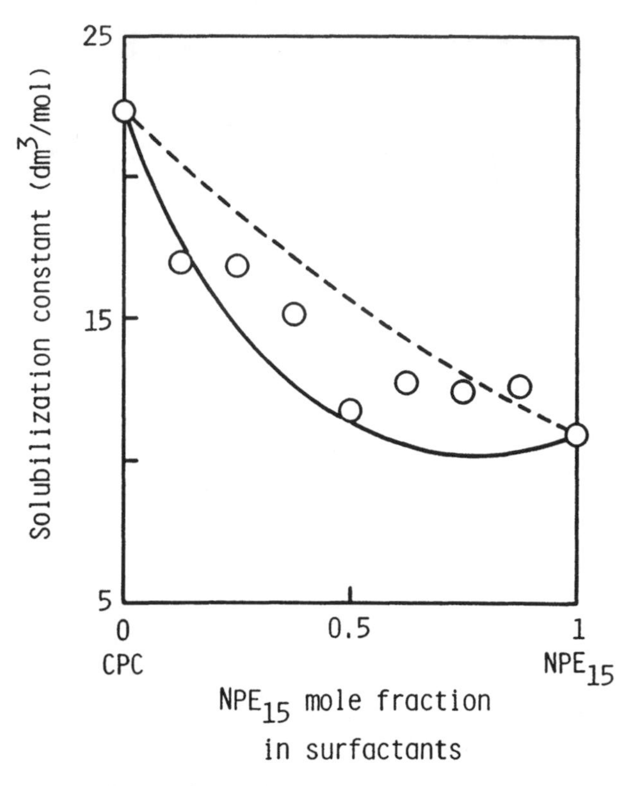

FIG. 9 Comparison of the calculated solubilization equilibrium constant of 1-hexanol with the observed one in NPE_{15}/CPC mixed micellar solutions at $\chi_s^{mX} = 0.3$, as a function of NPE_{15} mole fraction in surfactants constituting mixed micelles: —, calculated by Eq. (25′); ---------, calculated by Eq. (12′); ○, observed data of Ref. 59. (From Ref. 92.)

As the mole fraction of 1-hexanol in the micelles χ_s^{mX} becomes larger, it is probable that the magnitude of the first three terms becomes more significant. Figures 9 and 10 show positive deviations from the reference solubilization and the magnitude of the deviation seems to be larger for the micelles with more 1-hexanol content (of a greater χ_s^{mX} value). Figure 8 essentially shows no deviation from the reference solubilization expressed by Eq. (25′). In this case, it can be considered that Eq. (25′) apparently holds since the value of $(\chi_A^* \ln \gamma_s^{mA} + \gamma_B^* \ln \gamma_s^{mB} - \ln \gamma_s^{mAB})$ is small enough to make the value of the deviation terms nearly zero (canceled out) because of the low content of 1-hexanol in the micelles. It

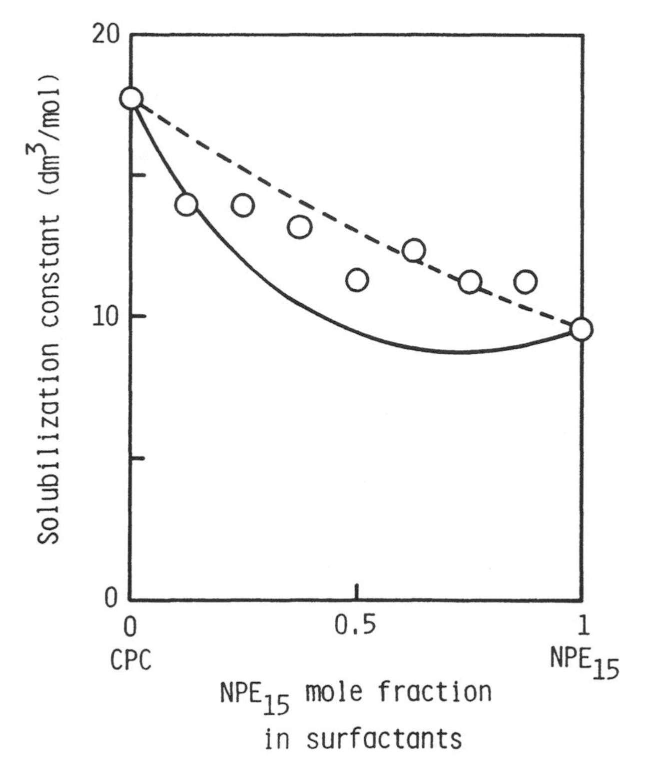

FIG. 10 Comparison of the calculated solubilization equilibrium constant of 1-hexanol with the observed one in NPE_{15}/CPC mixed micellar solutions at $\chi_s^{mX} = 0.5$, as a function of NPE_{15} mole fraction in surfactants constituting mixed micelles: —, calculated by Eq. (25′); ---------, calculated by Eq. (12′); ○, observed data of Ref. 59. (From Ref. 92.)

may, therefore, be possible to predict the solubilization capacity of an alkanol up to ca. 10% of its micellar content, by means of Eq. (25) or its analogous equation such as Eq. (25′).

On the other hand, as can be seen from Figs. 7 to 10, the observed K^{AB} values becomes closer to the K^{AB} values calculated by using Eq. (12′) (nonsynergistic solubilization) as the value of χ_s^{mX} becomes larger. In particular, fair agreement between the calculation by Eq. (12′) and the observation is obtained at micelle compositions richer in the nonionic surfactant NPE_{15} for the NPE_{15}/CPC mixed micelles of $\chi_s^{mX} = 0.5$. It is, therefore, inferred that the nonsynergistic solubilization capacities (calculated by Eq. (12) or (16)) are much closer to the observed ones at the saturation solubilization. In Fig. 11 are shown the MAC values calculated by the nonsynergistic solubilization Eq. (16) and the observed data of Weers for solubilized systems of 1-hexanol in ionic/nonionic ($C_{12}NMe_3Br$/$C_{12}E_6$) mixed micelles (21°C, 100 mM) [60(a)]. It is seen from Fig. 11 that the nonsynergistic MAC values are close to the observation but agree well with the observed ones at micelle compositions richer in the nonionic surfactant $C_{12}E_6$.

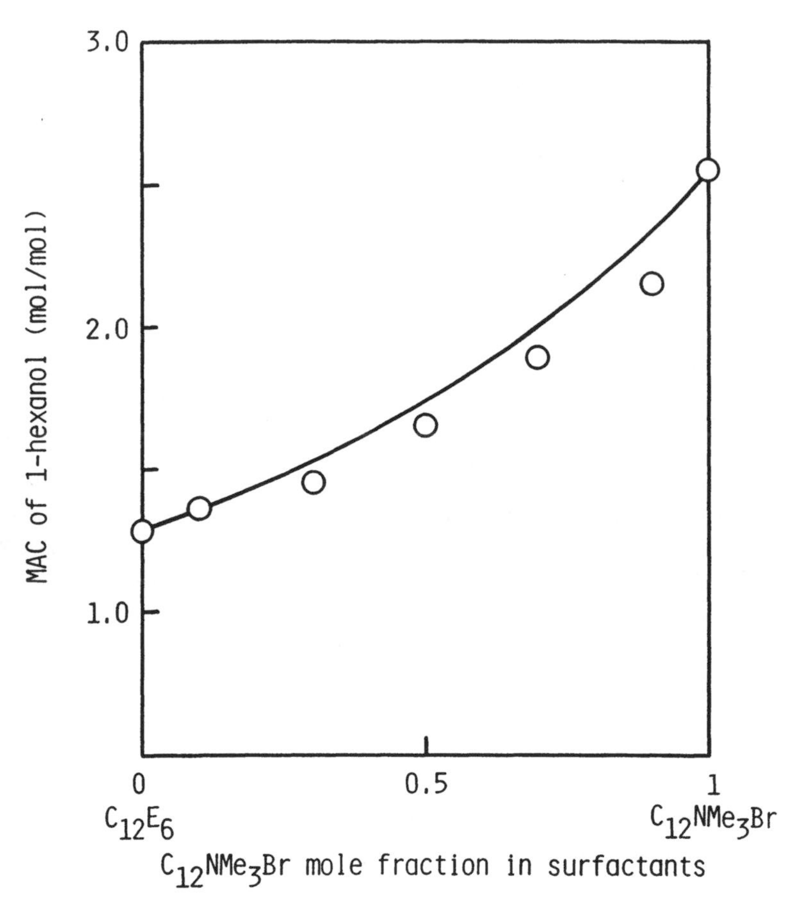

FIG. 11 Comparison of the calculated MAC of 1-hexanol with the observed one in aqueous solutions of $C_{12}NMe_3Br/C_{12}E_6$ mixed micelles, as a function of $C_{12}NMe_3Br$ mole fraction in surfactants constituting mixed micelles: —, calculated by Eq. (16); ○, observed data of Ref. 60(a).

Equations (12) and (25) are predictive equations to estimate the solubilization capacities in mixed micellar systems. As above, the predictive ability of these equations was evaluated by applying them to the solubilized system of 1-hexanol in NPE_{15}/CPC mixed micelles. Next, by application of Eqs. (5), (6), (8), and (9) to the same solubilized system, we determine the micellar activity coefficients (γ_s^{mAB}, γ_s^{mA}, and γ_s^{mB}) and the relative stabilities (($\mu_s^{mAB} - \mu_s^{mA})/RT$ and ($\mu_s^{mB} - \mu_s^{mA})/RT$) of the alkanol in micelles at an identical value of χ_s^{mX} (X is A, B, or AB), as a function of micelle composition. In this case, the surfactants A and B are taken as NPE_{15} and CPC, respectively. The values of γ_s^{mAB}, γ_s^{mA}, and γ_s^{mB} of 1-hexanol are listed in Table 5. The values of ($\mu_s^{mAB} - \mu_s^{mA})/RT$ and ($\mu_s^{mB} - \mu_s^{mA})/RT$ are tabulated in Table 6, as a function of χ_A^* (χ_{NPE15}^*). At the investigated alkanol contents in the micelles, Tables 5 and 6 demonstrate the following results. The alkanol 1-hexanol is more stable in the pure CPC micelles than in the pure NPE_{15} micelles. The alkanol in the mixed and the pure micelles becomes stabilized to a greater extent with increasing content in the micelles. At the alkanol contents below 10% in the micelle ($\chi_s^{mX} \leq 0.1$), the alkanol in the mixed micelles of CPC-rich compositions is more stable than in the NPE_{15} pure micelles while the alkanol in the mixed micelles of NPE_{15}-rich compositions is less stable than in the NPE_{15} pure micelles. At the alkanol compositions >10% in the micelle ($\chi_s^{mX} > 0.1$), the alkanol is more stabilized in the mixed micelles

TABLE 5 Evaluated Values of the Activity Coefficients of 1-Hexanol γ_s^{mX} in NPE_{15}/CPC Mixed Micelles and the Pure CPC and NPE_{15} Micelles, as a Function of Micelle Composition, χ_{NPE15}^* and χ_s^{mX} [a]

	γ_s^{mX} at various χ_s^{mX}					
χ_{NPE15}^*	0	0.1	0.2	0.3	0.4	0.5
0 (CPC)	1.00	1.09	1.20	1.32	1.47	1.66
0.125	0.70	0.76	0.85	0.96	1.09	1.17
0.25	0.77	0.82	0.88	0.97	1.05	1.17
0.375	0.87	0.93	0.99	1.08	1.17	1.24
0.50	1.14	1.21	1.28	1.39	1.44	1.46
0.625	1.18	1.20	1.25	1.28	1.29	1.33
0.75	1.12	1.18	1.24	1.32	1.38	1.46
0.875	1.13	1.15	1.22	1.29	1.37	1.46
1 (NPE_{15})	1.00	1.01	1.41	1.50	1.58	1.71

[a] Evaluated by using the data of Ref. 59.
Source: Ref. 93.

TABLE 6 Evaluated Values of the Stabilities of 1-Hexanol $(\mu_s^{mX} - \mu_s^{mA})/RT$ in NPE_{15}/CPC Mixed Micelles and CPC Pure Micelles Relative to That in NPE_{15} Pure Micelles, as a Function of Micelle Composition, χ^*_{NPE15} and χ_s^{mX} [a]

	$(\mu_s^{mX} - \mu_s^{mA})/RT$ at various χ_s^{mX}					
χ^*_{NPE15}	0	0.1	0.2	0.3	0.4	0.5
0 (CPC)	−0.50	−0.51	−0.75	−0.72	−0.66	−0.62
0.125	−0.36	−0.28	−0.50	−0.44	−0.37	−0.38
0.25	−0.26	−0.21	−0.47	−0.43	−0.41	−0.38
0.375	−0.14	−0.09	−0.35	−0.33	−0.30	−0.32
0.50	0.13	0.18	−0.09	−0.07	−0.09	−0.16
0.625	0.17	0.17	−0.11	−0.15	−0.20	−0.25
0.75	0.11	0.15	−0.13	−0.13	−0.14	−0.16
0.875	0.12	0.13	−0.14	−0.14	−0.15	−0.16

[a] Evaluated by using the data of Ref. 59.
Source: Ref. 93.

than in the NPE_{15} pure micelles and the extent of the stabilization becomes greater with increasing content of CPC in the micelles. At all of the micelle compositions investigated (surfactant and alkanol contents in the micelles), the alkanol in the mixed micelles is not more stable than in the CPC pure micelles. The above results are consistent with the following idea [59]. A strong ion-dipole attractive interaction between the alkanol (1-hexanol) hydroxyl group and the ionic surfactant (CPC) headgroup contributes to the stabilization of the alkanol in the micelles. There is also an ion-dipole interaction between the POE group of the nonionic surfactant (NPE_{15}) and the CPC headgroup in the NPE_{15}/CPC mixed micelles. In the NPE_{15}/CPC mixed micelles that contain the alkanol, the NPE_{15} POE group competes with the alkanol hydroxyl group in interacting attractively with the CPC headgroup, but the former interaction is stronger than the latter one.

6. Summary

1. The synergistic solubilization effects in mixed micelles are characterized by a thermodynamic equation derived with the pseudophase model:

$$\begin{aligned}\ln K_s^{AB} &= \chi_A^* \ln K_s^A + \chi_B^* \ln K_s^B \\ &\quad + (\chi_A^* \ln \gamma_s^{mA} + \chi_B^* \ln \gamma_s^{mB} - \ln \gamma_s^{mAB}) \\ &\quad + \frac{\chi_B^*(\mu_s^{\ominus mB} - \mu_s^{\ominus mA})}{RT}\end{aligned} \tag{3}$$

In particular, under the condition that $\chi_s^{mA} = \chi_s^{mB} = \chi_s^{mAB}$, Eq. (3) directly expresses the synergistic solubilization effects; i.e., the sign and the magnitude of the sum of the following SYNT terms estimate a (positive, negative, or non-) synergistic solubilization effect:

$$\text{SYNT} \equiv (\chi_A^* \ln \gamma_s^{mA} + \chi_B^* \ln \gamma_s^{mB} - \ln \gamma_s^{mAB}) + \frac{\chi_B^*(\mu_s^{\ominus mB} - \mu_s^{\ominus mA})}{RT}$$

2. The following equation for the solubilization capacity MAC describes the synergistic solubilization effects if the MAC values for constituent surfactants do not differ greatly (in some case, the equation holds even if the MAC values differ greatly):

$$\ln\left(\frac{\text{MAC}^{AB}}{1 + \text{MAC}^{AB}}\right) = \chi_A^* \ln\left(\frac{\text{MAC}^{A}}{1 + \text{MAC}^{A}}\right) + \chi_B^* \ln\left(\frac{\text{MAC}^{B}}{1 + \text{MAC}^{B}}\right) + \text{SYNT} \quad (4)$$

3. In general, the thermodynamic definition of the synergistic solubilization effects is different from the practical definition using the additivity mixing rules, Eq. (1).

4. The solubilization capacities K_s^{AB} and MAC^{AB} of alkane in mixed micellar systems of hydrocarbon surfactants and those of a solubilizate in mixed micellar systems of homologous surfactants are predictable from the following equations expressing nonsynergistic solubilization effects:

$$\ln K_s^{AB} = \chi_A^* \ln K_s^A + \chi_B^* \ln K_s^B \quad (14)$$

$$\ln\left(\frac{\text{MAC}^{AB}}{1 + \text{MAC}^{AB}}\right) = \chi_A^* \ln\left(\frac{\text{MAC}^{A}}{1 + \text{MAC}^{A}}\right) + \chi_B^* \ln\left(\frac{\text{MAC}^{B}}{1 + \text{MAC}^{B}}\right) \quad (16)$$

5. At 10% or less content of solubilized alkanol in micelles, the solubilization capacities of alkanol can be predicted by the equation

$$\ln K_s^{AB} = \chi_A^* \ln K_s^A + \chi_B^* \ln K_s^B + \alpha_{AB}\chi_A^*\chi_B^* \quad (25)$$

At the saturation solubilization or at a sufficiently high content of alkanol in ionic/nonionic mixed micelles, the observed solubilization capacities are close to the nonsynergistic solubilization capacities calculated by using Eq. (14) or (16).

6. The activity coefficients of a solubilizate in the pure micelles and mixed micelles are evaluated by means of the equations

$$\gamma_s^{mX} = \frac{K_s^X(0)}{K_s^X} \qquad \text{(X is A or B)} \quad (5)$$

$$\gamma_s^{\mathrm{mAB}} = \frac{K_s^{\mathrm{A}}(0)}{K_s^{\mathrm{AB}}} \tag{6}$$

7. The stability of a solubilizate in the mixed micelles "AB" and that in the pure micelles of surfactant "B," relative to that in the pure micelles of surfactant "A," are calculated from the equations

$$\mu_s^{\mathrm{mAB}}(\chi_s^{\mathrm{mAB}}) - \mu_s^{\mathrm{mA}}(\chi_s^{\mathrm{mA}}) = RT\ln\left(\frac{K_s^{\mathrm{A}}}{K_s^{\mathrm{AB}}}\right) + RT\ln\left(\frac{\chi_s^{\mathrm{mAB}}}{\chi_s^{\mathrm{mA}}}\right) \tag{8}$$

$$\mu_s^{\mathrm{mB}}(\chi_s^{\mathrm{mB}}) - \mu_s^{\mathrm{mA}}(\chi_s^{\mathrm{mA}}) = RT\ln\left(\frac{K_s^{\mathrm{A}}}{K_s^{\mathrm{B}}}\right) + RT\ln\left(\frac{\chi_s^{\mathrm{mB}}}{\chi_s^{\mathrm{mA}}}\right) \tag{9}$$

8. The anomalous solubilization behavior of an alkane in anionic/cationic mixed micelles can be explained quantitatively by introducing the following assumption into Eq. (16): the mixed micelles formed consist of the monomeric surfactant and the molecular compound having a composition close to that at the precipitation boundary.

9. The present treatment based on the pseudophase model and the simple Laplace equation indicates that the influence of a Laplace pressure on the synergistic solubilization effects is nearly canceled, even if the Laplace pressure exists in micelles.

SYMBOLS

A, B (X)	surfactants constituting mixed micelles
AB (X)	mixed surfactants composed of surfactants A and B
s	solubilizate
b	bulk water phase (solution)
m	micellar pseudophase
l	liquid
$\chi_{\mathrm{A}}^{*}, \chi_{\mathrm{B}}^{*}$	mole fractions of surfactants A and B constituting mixed micelles ($\chi_{\mathrm{A}}^{*} + \chi_{\mathrm{B}}^{*} = 1$)
$\chi_{\mathrm{A}}^{\mathrm{X}}, \chi_{\mathrm{B}}^{\mathrm{X}}, \chi_s^{\mathrm{X}}$	mole fractions of surfactants A and B and solubilizate s in micelle ($\chi_{\mathrm{A}}^{\mathrm{X}} + \chi_{\mathrm{B}}^{\mathrm{X}} + \chi_s^{\mathrm{X}} = 1$)
$\gamma_{\mathrm{A}}^{\mathrm{mX}}, \gamma_{\mathrm{B}}^{\mathrm{mX}}, \gamma_s^{\mathrm{mX}}$	activity coefficients of surfactants A and B and solubilizate s in micelle
$\mu_s^{\mathrm{mX}}, \mu_s^{\ominus\mathrm{mX}}$	chemical potential and standard chemical potential of solubilizate s in micelle
$\mu_s^{\ominus\mathrm{b}}$	standard chemical potential of solubilizate s in bulk
K_s^{X}	apparent distribution coefficient
$\mathrm{MAC}^{\mathrm{X}}$	saturation solubilization capacity, moles of solubilized material per mole of micellized surfactant

$\chi_s^{\mathrm{mX}}(\mathrm{satn})$	mole fraction of solubilizate s in micelle at saturation solubilization
$\alpha_{\mathrm{A}s}, \alpha_{\mathrm{B}s}, \alpha_{\mathrm{AB}}$	interaction parameters defined in the regular solution theory
P_m^{X}	Laplace pressure in micelle
P_0	bulk pressure
SYNT	synergistic solubilization effect terms: $(\chi_{\mathrm{A}}^* \ln \gamma_s^{\mathrm{mA}} + \chi_{\mathrm{B}}^* \ln \gamma_s^{\mathrm{mB}} - \ln \gamma_s^{\mathrm{mAB}}) + \dfrac{\chi_{\mathrm{B}}^*(\mu_s^{\ominus\mathrm{mB}} - \mu_s^{\ominus\mathrm{mA}})}{RT}$
nonsyn	nonsynergistic solubilization effect or state
ref	reference solubilization state

REFERENCES

1. D. Attwood and A. T. Florence, *Surfactant Systems*, Chapman and Hall, London, 1983, Chap. 5.
2. M. J. Rosen, *Surfactant and Interfacial Phenomena*, 2nd ed., Wiley, New York, 1989, Chap. 4.
3. N. Nishikido, M. Kishi, and M. Tanako, *J. Colloid Interface Sci. 94*: 348 (1983).
4. N. Nishikido, K. Abiru, and N. Yoshimura, *J. Colloid Interface Sci. 113*: 356 (1986).
5. N. Nishikido and G. Sugihara, in *Mixed Surfactant Systems* (K. Ogino and M. Abe, eds.), Marcel Dekker, New York, 1993, Chap. 10.
6. T. Takenaka, K. Harada, and T. Nakanaga, *Bull. Inst. Chem. Res. Kyoto Univ. 53*: 173 (1975).
7. D. J. Jobe, V. C. Reinsborough, and P. J. White, *Can. J. Chem. 60*: 279 (1982).
8. R. Zana, C. Picot, and R. Duplessix, *J. Colloid Interface Sci. 93*: 43 (1983); B. Hayter, M. Hayoun, and T. Zemb., *Colloid Polym. Sci. 262*: 798 (1984).
9. M. Almgren and S. Swarup, *J. Phys. Chem. 87*: 881 (1983); M. Almgren and S. Swarup, in *Surfactants in Solution* (K. L. Mittal and B. Lindman, eds.), Plenum, New York, 1984, Vol. 1, pp. 613–626.
10. A. Malliaris, W. Binana-Limbele, and R. Zana, *J. Colloid Interface Sci. 110*: 114 (1986).
11. Y. Muto, K. Esumi, K. Meguro, and R. Zana, *J. Colloid Interface Sci. 120*: 162 (1987), *Langmuir 5*: 885 (1989).
12. R. Zana, Y. Muto, K. Esumi, and K. Meguro, *J Colloid Interface Sci. 123*: 502 (1988).
13. M. M. Velázquez and S. M. B. Costa, *J. Chem. Soc. Faraday Trans. 86*: 4043 (1990).
14. M. Ueno, Y. Kimoto, Y. Ikeda, H. Momose, and R. Zana, *J. Colloid Interface Sci. 117*: 179 (1987); H. Asano, K. Aki, and M. Ueno. *Colloid Polym. Sci. 267*: 935 (1989); H. Asano, A. Murohashi, and M. Ueno. *J. Am. Oil*

Chem. Soc. 67: 1002 (1990); H. Asano, M. Yamazaki, A. Fujima, and M. Ueno. *J. Jpn. Oil Chem. Soc. 40*: 293 (1991).
15. M. Ishikawa, K. Matsumura, K. Esumi, and K. Meguro, *J. Colloid Interface Sci. 141*: 10
16. T. Iwasaki, M. Ogawa, K. Esumi, and K. Meguro, *Langmuir 7*: 30 (1991).
17. X. G. Lei, X. D. Tang, and Y. C. Lin, *Langmuir 7*: 2872 (1991).
18. L. B. McGown and K. Nithipatikom, *J. Res. Nat. Bur. Stand. (U.S.) 93*: 443 (1988).
19. M. Nakagaki, S. Yokoyama, and I. Yamamoto, *Nippon Kagaku Kaishi* 1865 (1982).
20. T. Wolff and G. von Bünau, *Ber. Bunsenges. Phys. Chem. 86*: 225 (1982).
21. M. Meyer and L. Sepúlveda, *J. Colloid Interface Sci. 99*: 536 (1984).
22. E. Abuin and E. Lissi, *J. Colloid Interface Sci. 143*: 97 (1991).
23. K. Meguro, Y. Muto, F. Sakurai, and K. Esumi, in *Phenomena in Mixed Surfactant Systems* (J. F. Scamehorn, ed.), ACS, Washington, DC, 1986, pp. 61–67.
24. K. Meguro and K. Esumi, in *Surfactants in Solution* (K. L. Mittal, ed.), Plenum, New York, 1989, Vol. 7, pp. 385–396.
25. K. Kalyanasundaram, *Langmuir 4*: 942 (1988).
26. R. L. Hill and L. D. Rhein, *J. Dispersion Sci. Technol. 9*: 269 (1988).
27. T. Asakawa, M. Mouri, S. Miyagishi, and M. Nishida, *Langmuir 5*: 343 (1989).
28. P. Baglioni, in *Surfactants in Solution* (K. L. Mittal and P. Bothorel, eds.) Plenum, New York, 1986, Vol. 4, pp. 393–404.
29. E. Rivara-Minten, P. Baglioni, and L. Kevan, *J. Phys. Chem. 92*: 2613 (1988).
30. P. Baglioni, E. Rivara-Minten, C. Stenland, and L. Kevan, *J. Phys. Chem. 95*: 10169 (1991).
31. C. S. Chao, S. Muto, and K. Meguro, *Nippon Kagaku Kaishi* 1572 (1972).
32. C. S. Chao, S. Muto, and K. Meguro, *Nippon Kagaku Kaishi* 2013 (1972).
33. J. H. Chang, M. Ohno, K. Esumi, and K. Meguro, *J. Jpn. Oil Chem. Soc. 37*: 1122 (1988).
34. H. Akasu, A. Nishii, M. Ueno, and K. Meguro, *J. Colloid Interface Sci. 54*: 278 (1976); K. Meguro, H. Akasu, and M. Ueno, *J. Am. Oil Chem. Soc. 53*: 145 (1976).
35. M. Ueno, K. Shioya, T. Nakamura, and K. Meguro, in *Colloid and Interface Science* (M. Kerker, ed.), Academic Press, New York, 1976, Vol. 2, pp. 411–420.
36. K. Meguro, M. Ueno, and T. Suzuki, *J. Jpn. Oil Chem. Soc. 31*: 909 (1982).
37. T. Suzuki, K. Esumi, and K. Meguro, *J. Colloid Interface Sci. 93*: 205 (1983).
38. K. Meguro, Y. Tabata, N. Fujimoto, and K. Esumi, *Bull. Chem. Soc. Jpn. 56*: 627 (1983); N. Kawashima, N. Fujimoto, and K. Meguro, *J. Colloid Interface Sci. 103*: 459 (1985).
39. K. Ogino and M. Abe, in *Phenomena in Mixed Surfactant Systems* (J. F. Scamehorn, ed.), ACS, Washington, DC, 1986, pp. 68–78; M. Abe, M. Ohsato, H. Uchiyama, N. Tsubaki, and K. Ogino, *J. Jpn. Oil Chem. Soc. 35*:

522 (1986); K. Ogino, H. Uchiyama, M. Ohsato, and M. Abe, *J. Colloid Interface Sci. 116*: 81 (1987); K. Ogino, H. Uchiyama, and M. Abe, *Colloid Polym. Sci. 265*: 52, 838 (1987); K. Ogino, H. Uchiyama, T. Kakihara, and M. Abe, in *Surfactants in Solution* (K. L. Mittal, ed.), Plenum, New York, 1989, Vol. 7, pp. 413–429; H. Uchiyama, S. Akao, M. Abe, and K. Ogino, *Langmuir 6*: 1763 (1990).

40. N. Nishikido, in *Mixed Surfactant Systems* (K. Ogino and M. Abe, eds.), Marcel Dekker, New York, 1993, Chap. 2.
41. K. Hayase and S. Hayano, *Bull. Chem. Soc. Jpn. 50*: 83 (1977); K. Hayase, S. Hayano, and H. Tsubota, *J. Colloid Interface Sci. 101*: 336 (1984).
42. R. De Lisi, C. Genova, and V. Turco Liveri, *J. Colloid Interface Sci. 95*: 428 (1983).
43. E. L. Evans, G. G. Jayson, and I. D. Roff, *J. Chem. Soc. Faraday 1 76*: 528 (1980).
44. S. J. Dougherty and J. C. Berg, *J. Colloid Interface Sci. 48*: 110 (1974); A. Goto, M. Nihei, and F. Endo, *J. Phys. Chem. 84*: 2268 (1980); E. A. Abuin and E. A. Lissi, *J. Colloid Interface Sci. 95*: 198 (1983); E. A. Abuin, E. Valenzuela, and E. A. Lissi, *J. Colloid Interface Sci. 101*: 401 (1984); R. Zana, S. Yiv, C. Strazielle, and P. Lianos, *J. Colloid Interface Sci. 80*: 208 (1981); S. A. Simon, R. V. McDaniel, and T. J. McIntosh, *J. Phys. Chem. 86*: 1449 (1982); C. Treiner and A. K. Chattopadhyay, *J. Colloid Interface Sci. 98*: 447 (1984), *109*: 101 (1986); H. Høiland, E. Ljosland, and S. Backlund, *J. Colloid Interface Sci. 101*: 467 (1984); C. Gamboa, A. Olea, H. Rios, and Henriquez, *Langmuir 8*: 23 (1992); and references cited therein.
45. (a) P. Mukerjee, *J. Pharm. Sci. 60*: 1531 (1971); (b) P. Mukerjee, in *Solution Chemistry of Surfactants* (K. L. Mittal, ed.), Plenum, New York, Vol. 1, pp. 153–174.
46. R. De Lisi, C. Genova, R. Testa, and V. Turco Liveri, *J. Solution Chem. 113*: 121 (1984); R. De Lisi, V. Turco Liveri, M. Castagnolo, and A. Inglese, *J. Solution Chem. 15*: 23 (1986); R. De Lisi, S. Milioto, and V. Turco Liveri, *J. Colloid Interface Sci. 117*: 64 (1987).
47. M. Almgren, F. Grieser, and J. K. Thomas, *J. Chem. Soc. Faraday 1 75*: 1674 (1979).
48. M. Almgren and R. Rydholm, *J. Phys. Chem. 83*: 360 (1979); M. Almgren, F. Grieser, and J. K. Thomas, *J. Am. Chem. Soc. 101*: 279 (1979); C. A. Bunton and L. Sepulveda, *J. Phys. Chem. 83*: 680 (1979); C. Hirose and L. Sepulveda, *J. Phys. Chem. 85*: 3689 (1981); E. Pramauro and E. Pelizzetti, *Ann. Chim. 72*: 117 (1982).
49. P. Stilbs, *J. Colloid Interface Sci. 87*: 385 (1982); J. Carlfors and P. Stilbs, *J. Colloid Interface Sci. 103*: 332 (1985); D. G. Marangoni and J. C. T. Kwak, *Langmuir 7*: 2083 (1991).
50. C. Treiner, *J. Colloid Interface Sci. 93*: 33 (1984); M. Abu-Hamdiyyah and I. A. Rathman, *J. Phys. Chem. 94*: 2518 (1990).
51. F. Tokiwa and K. Tsujii, *Bull. Chem. Soc. Jpn. 46*: 1338 (1973).
52. N. Nishikido, *J. Colloid Interface Sci. 60*: 242 (1977).

53. H. Uchiyama, M. Abe, and K. Ogino, *J. Jpn. Oil Chem. Soc. 35*: 1031 (1986); H. Uchiyama, Y. Tokuoka, M. Abe, and K. Ogino, *J. Colloid Interface Sci. 132*: 88 (1989).
54. Y. Muto, M. Asada, A. Takasawa, K. Esumi, and K. Meguro, *J. Colloid Interface Sci. 124*: 632 (1988).
55. C. Treiner, C. Vaution, E. Miralles, and F. Puisieux, *Colloids Surf. 14*: 285 (1985).
56. C. Treiner, M. Nortz, C. Vaution, and F. Puisieux, *J. Colloid Interface Sci. 125*: 261 (1988).
57. C. Treiner, A. Amar Khodja, and M. Fromon, *Langmuir 3*: 729 (1987).
58. C. Treiner, M. Nortz, and C. Vaution, *Langmuir 6*: 1211 (1990).
59. C. M. Nguyen, J. F. Scamehorn, and S. D. Christian, *Colloids Surf. 30*: 335 (1988).
60. (a) J. G. Weers, *J. Am. Oil Chem. Soc. 67*: 340 (1990); (b) D. R. Schuing and J. G. Weers, *Langmuir 6*: 665 (1990).
61. A. J. I. Ward and K. Quigley, *J. Dispersion Sci. Technol. 11*: 143 (1990).
62. H. Saito and K. Shinoda, *J. Colloid Interface Sci. 24*: 10 (1967).
63. F. Tokiwa, *J. Colloid Interface Sci. 28*: 145 (1968).
64. H. Sagitani, T. Suzuki, M. Nagai, and K. Shinoda, *J. Colloid Interface Sci. 87*: 11 (1982).
65. S. Pal and S. P. Moulik, *J. Lip. Res. 24*: 1281 (1983).
66. F. Jost, H. Leiter, and M. J. Schwuger, *Colloid Polym. Sci. 266*: 554 (1988).
67. G. Smith, S. D. Christian, E. E. Tucker, and J. F. Scamehorn, *J. Colloid Interface Sci. 130*: 254 (1989).
68. J. W. Cabrera and L. Sepúlveda, *Langmuir 6*: 240 (1990).
69. N. Nugara, W. Prapaitrakul, and A. D. King, Jr., *J. Colloid Interface Sci. 120*: 118 (1987).
70. A. D. King, Jr., *J. Colloid Interface Sci. 148*: 142 (1992).
71. R. Bury, E. Souhalia, and C. Treiner., *J. Phys. Chem. 95*: 3824 (1991); C. Treiner and R. Bury, Prog. *Colloid Polym. Sci. 84*: 108 (1991).
72. J. Xia and Z. Hu, in *Surfactants in Solution* (K. L. Mittal and P. Bothorel, eds.), Plenum, New York, 1986, Vol. 5, pp. 1055–1066.
73. A. Malliaris, W. Binana-Limbele, and R. Zana, *J. Colloid Interface Sci. 110*: 114 (1986).
74. G. X. Zhao and X. G. Li, *J. Colloid Interface Sci. 144*: 185 (1991).
75. H. Hoffmann and W. Ulbricht, *J. Colloid Interface Sci. 129*: 388 (1989).
76. J. M. Lambert and W. F. Busse, *J. Am. Oil Chem. Soc. 26*: 289 (1949).
77. C. Treiner, and J. F. Bocquet, and C. Pommier, *J. Phys. Chem. 90*: 3052 (1986).
78. K. Yoda, K. Tamori, K. Esumi, and K. Meguro, *Colloids Surf. 58*: 87 (1991).
79. F. Tokiwa and K. Ohki, *J. Jpn. Oil Chem. Soc. 9*: 901 (1970).
80. M. Takai, H. Hidaka, S. Ishikawa, M. Takada, and M. Moriya, *J. Am. Oil Chem. Soc. 57*: 382 (1980).
81. M. Abe, T. Kubota, H. Uchiyama, and K. Ogino, *Colloid Polym. Sci. 267*: 365 (1989).

82. T. Iwasaki, M. Ogawa, K. Esumi, and K. Meguro, *Langmuir 7*: 30 (1991).
83. S. Pal and S. P. Moulik, *Indian J. Biochem. Biophys. 21*: 17 (1984).
84. A. Bandyopadhyay and S. P. Moulik, *J. Phys. Chem. 95*: 4529 (1991).
85. J. G. Weers, J. F. Rathman, and D. R. Scheuing, *Colloid Polym. Sci. 268*: 832 (1990).
86. J. G. Weers and D. R. Scheuing, *J. Colloid Interface Sci. 145*: 563 (1991).
87. M. Ishikawa, K. Matsumura, K. Esumi, and K. Meguro, *J. Colloid Interface Sci. 141*: 10 (1991).
88. M. Ohno, K. Esumi, and K. Meguro, *J. Am. Oil Chem. Soc. 69*: 80 (1992).
89. J. P. O'Connell and J. M. Praunitz, *Ind. Eng. Chem. Fundam. 3*: 347 (1964); J. P. O'Connell. *AIChE J. 17*: 658 (1971).
90. N. Nishikido, *J. Colloid Interface Sci. 127*: 310 (1989).
91. P. L. Bolden, J. C. Hoskins, and A. D. King, Jr., *J. Colloid Interface Sci. 91*: 454 (1983); W. Prapaitrakul and A. D. King, Jr., *J. Colloid Interface Sci. 106*: 186 (1985), *112*: 387 (1986).
92. N. Nishikido, *Langmuir 7*: 2076 (1991).
93. N. Nishikido, *Langmuir 8*: 1718 (1992).
94. N. Nishikido, *J. Colloid Interface Sci. 120*: 495 (1987).
95. N. Nishikido, *J. Colloid Interface Sci. 131*: 340 (1989).
96. D. N. Rubingh, in *Solution Chemistry of Surfactants* (K. L. Mittal, ed.), Plenum, New York, 1979, Vol. 1, pp. 337–354; J. F. Scamehorn, R. S. Schechter, and W. H. Wade, *J. Dispersion Sci. Technol. 3*: 261 (1982); P. M. Holland, *Adv. Colloid Interface Sci. 26*: 111 (1986); A. Graciaa, M. Ben Ghoulam, G. Marion, and J. Lachaise, *J. Phys. Chem. 93*: 4167 (1989).

6

Solubilization in Amphiphilic Copolymer Solutions

PATRICIA N. HURTER, PASCHALIS ALEXANDRIDIS, and T. ALAN HATTON Department of Chemical Engineering, Massachusetts Institute of Technology, Cambridge, Massachusetts

SYNOPSIS

The solubilization of hydrophobic compounds in aqueous solutions of amphiphilic copolymers is important in many practical applications. We provide here an overview of micelle formation by copolymers, and then discuss the effect of polymer type and structure and solute hydrophobicity on micellar solubilization. The influence of solution conditions, such as temperature, pH, and ionic strength, on the micellar properties and, consequently, on the solubilization behavior are also examined. Finally, phenomenological and mean-field lattice theories of micelle formation and solubilization in block copolymer solutions are described, and the predictions of these theories are compared to the experimental observations. The solubilization in amphiphilic copolymer micelles is affected by the detailed microstructure of the aggregates, which in turn depends on both polymer and solute structure, and solution conditions. This allows a great deal of flexibility in designing novel solvents which can be used as extractants and solubilizers in a variety of pharmaceutical, biological, and industrial applications.

I. INTRODUCTION

Copolymers are synthesized by the simultaneous polymerization of more than one type of monomer. The result of such a synthesis is called a block copolymer if the monomers occur as blocks of various lengths in the copolymer molecule. The different types of blocks within the copolymer are usually incompatible with one another and, as a consequence, block copolymers self-assemble in melts and in solutions. In the case of amphiphilic copolymers in aqueous solutions, the copolymers can assemble in microstructures that resemble micelles formed by low molecular weight surfactants. An interesting property of aqueous micellar systems is their ability to enhance the solubility in water of otherwise water-insoluble hydrophobic compounds. This occurs because the core of the micelle provides a hydrophobic microenvironment, suitable for solubilizing such molecules. The phenomenon of solubilization forms the basis for many practical applications of amphiphiles.

The enhancement in the solubility of lyophobic solutes in solvents afforded by amphiphilic copolymer micelles has shown promise in many industrial and biomedical applications. The solubility and stability of hydrophobic drugs can be improved by solubilizing the drug in an aqueous micellar solution [1, 2]. Nonionic surfactants (including amphiphilic block copolymers) are considered suitable for use as drug delivery agents since they are less toxic to biological systems than cationic or anionic surfactants, and their lower critical micelle concentration means that smaller concentrations are required. Photochemical reactions can be carried out in micellar systems: copolymers are used which both solubilize the molecules of interest and sensitize the photochemical reaction [3]. Other promising applications are in separations: hydrophobic pollutants, such as polycyclic aromatic hydrocarbons, can be selectively solubilized in aqueous solutions of block copolymer micelles [4, 5].

In all these applications, it is important to understand how the polymer structure, solution condition, and solute properties affect the solubilization behavior. One of the advantages of using amphiphilic copolymers over traditional surfactants is that, to a certain extent, an optimal polymer can be designed for a given application, since the polymer molecular weight, composition, and microstructure can all be varied. In this chapter, experimental observations on the effect of polymer structure on micellar solubilization are reviewed. The effect of solute hydrophobicity and size are also discussed, as is the effect of solute uptake on the micelle structure. The effect of solution conditions, such as temperature, pH, and ionic strength, on the micellar properties, e.g., size and hydrophobicity of the core, are examined. Finally, theories of micelle formation and solubilization are described, and the predictions of these theories are compared to the experimental observations.

II. MICELLE FORMATION IN AMPHIPHILIC COPOLYMER SOLUTIONS

A. PEO-PPO Block Copolymers

The formation of micelles by triblock poly(ethylene oxide)-poly(propylene oxide)-poly(ethylene oxide) (PEO-PPO-PEO) copolymers has been investigated rather extensively (the PEO-PPO-PEO copolymers are commercially available in different molecular weights and compositions under the name Pluronics; properties of various Pluronic copolymers are listed in Table 1). Early studies to determine the critical micelle concentrations (CMC) of these polymers reported results varying over two orders of magnitude [6–10], casting doubt on whether these polymers form micelles

TABLE 1 Properties of Pluronic and Tetronic PEO-PPO Block Copolymers

A	B	C	D	E	F	G	H	I
Pluronics								
L62	2500	20	−4	450	43	25	32	1–7
L63	2650	30	10	490	43	30	34	7–12
L64	2900	40	16	850	43	40	58	12–18
P65	3400	50	27	180	46	70	82	12–18
F68	8400	80	52	1000	50	35	>100	>24
P84	4200	40	34	280	42	90	74	12–18
P85	4600	50	34	310	42	70	85	12–18
F88	11400	80	54	2300	48	80	>100	>24
F98	13000	80	58	2700	43	40	>100	>24
P103	4950	30	30	285	34	40	86	7–12
P104	5900	40	32	390	33	50	81	12–18
P105	6500	50	35	750	39	40	91	12–18
F108	14600	80	57	2800	41	40	>100	>24
P123	5750	30	31	350	34	45	90	7–12
F127	12600	70	56	3100	41	40	>100	18–23
Tetronics								
T904	6700	40	29	320	35	70	74	12–18
T1107	15000	70	51	1100	43	50	>100	18–23
T1304	10500	40	35	650	36	50	85	12–18
T1307	18000	70	54	2700	44	40	>100	18–23
T1504	12000	40	41	725	37	40	>100	12–18

A: Copolymer
B: Average molecular weight
C: PEO content (wt.%)
D: Melt/pour point (°C)
E: Viscosity (Brookfield) (cps; liquids at 25°C, pastes at 60°C, solids at 77°C)
F: Surface tension at 0.1%, 25°C (dynes/cm)
G: Foam height (mm) (Ross Miles, 0.1% at 50°C)
H: Cloud point in aqueous 1% solution (°C)
I: HLB (hydrophilic-lipophilic balance)
Source: BASF Corp.

at all. There are a number of reasons for the inaccuracy of the different measurements. The polymers used in these studies are commercial polymers of relatively high molecular weights, and one would expect a fairly high degree of polydispersity in a given sample, and that batch variability would be fairly significant. Any trace impurities in the polymer solutions would have a significant effect on surface tension measurements.

More recent light-scattering and fluorescence experiments, however, have shown that PEO-PPO-PEO block copolymers of suitable composition and molecular weight do indeed form polymolecular aggregates in solution. Micellar growth was observed for L64 (a Pluronic polymer with 40 wt.% PEO and a total molecular weight of 2900) with an increase in polymer concentration at 25°C, but the micelle radius remained constant at 35°C [11]. Micelle formation by P85 (a Pluronic with 50 wt.% PEO and a molecular weight of 4500) was investigated by Brown et al. [12] using static and dynamic light scattering. It was found that monomers, micelles, and micellar aggregates coexist, the relative proportions of each species depending strongly on temperature and polymer concentration. Micellar growth was found to occur with increasing temperature. This is shown in Fig. 1 where the scattering intensity of a P85 solution is plotted as a function of temperature; the increase of scattering intensity at ~30°C signifies the formation of micelles. A significant temperature dependence of the micellization behavior of F68 (a Pluronic polymer with 80 wt.% PEO and a total molecular weight of 8350) was also observed [13]. Using static

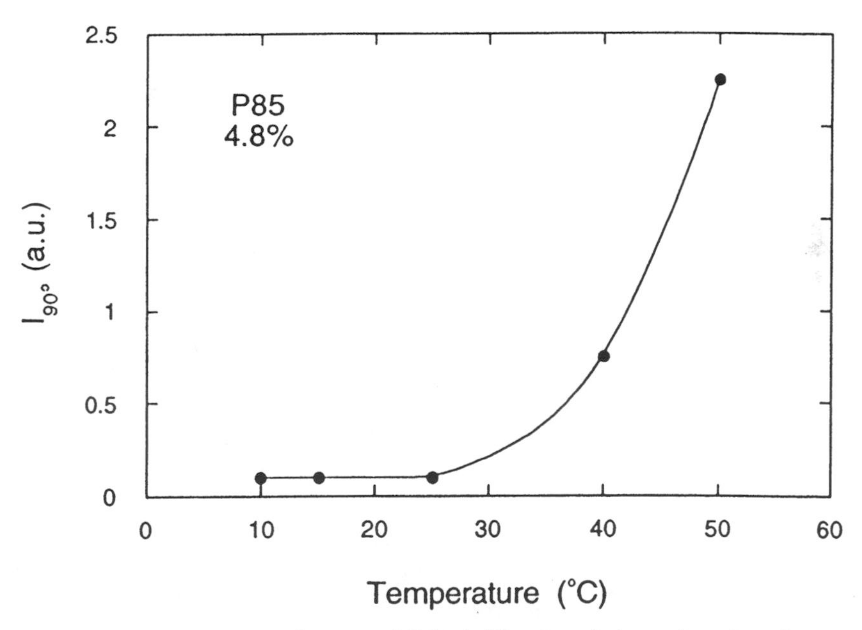

FIG. 1 Plot of intensity of scattered light (arbitrary units) as a function of temperature for 4.8% Pluronic P85 in water; the increase of scattering intensity at ~30°C signifies the formation of micelles. (Data extracted from Ref. 12.)

and dynamic light scattering, Zhou and Chu detected three temperature regions, which they called the unimer, transition, and micelle regions. At room temperatures, "particles" with a hydrodynamic radius of around 2.3 nm and a broad polydispersity were detected. Above 50°C, micelle molecular weights were found to increase linearly with temperature, while the hydrodynamic radius remained constant, at about 8 nm. It was suggested that the constant micelle size accompanying an increase in aggregation number in the micelle region was due to exclusion of water from the micelle interior with an increase in temperature. Wanka et al. [14] confirmed the finding of Zhou and Chu, that aggregation numbers increase with temperature, while the micelle radius remains approximately constant. These results will be discussed in detail later and compared to predictions of micelle structure obtained using a self-consistent field lattice theory.

Photoluminescent probes were employed by Turro and Chung [15] to study the behavior of a poly(ethylene oxide)-poly(propylene oxide) copolymer in water. The polymer they used had a total molecular weight of 3000, with a molar ratio of PEO to PPO of 0.8:1.0. Regions of no aggregation, monomolecular micelles, and polymolecular micelles were identified with increasing polymer concentration. The aggregation number of the micelles formed at high polymer concentration was determined to be 52. With an increase in temperature, a hydrophobic fluorescence probe, which would be expected to reside in the micelle core, experienced a more hydrophilic environment, whereas a hydrophilic probe, expected to reside in the interfacial region, experienced a more hydrophobic environment. The authors suggested that, as temperature is increased, greater thermal agitation of the chains could cause more mixing of the PEO chains in the core of the micelle. A hydrophobic fluorescence probe (DPH) was used by Alexandridis et al. [16] to determine the onset of micellization for a number of Pluronic PEO-PPO-PEO block copolymer aqueous solutions. The fluorescence efficiency of the DPH is zero in a hydrophilic environment and unity in a hydrophobic environment (such as the core of a micelle), thus providing a sensitive indicator of micelle formation with increasing solution temperature and copolymer concentration. Critical micelle concentration and critical micelle temperature data were derived and correlated to the Pluronic molecular weight and PPO/PEO ratio [16]. Fig. 2 shows a plot of intensity of absorbed light (arbitrary units) due to DPH as a function of temperature for aqueous P104 solutions at various polymer concentrations; the critical micelle temperature values were obtained from the first break of the intensity versus temperature sigmoidal curves. Critical micelle concentrations can be obtained in a similar manner from intensity versus log concentration plots.

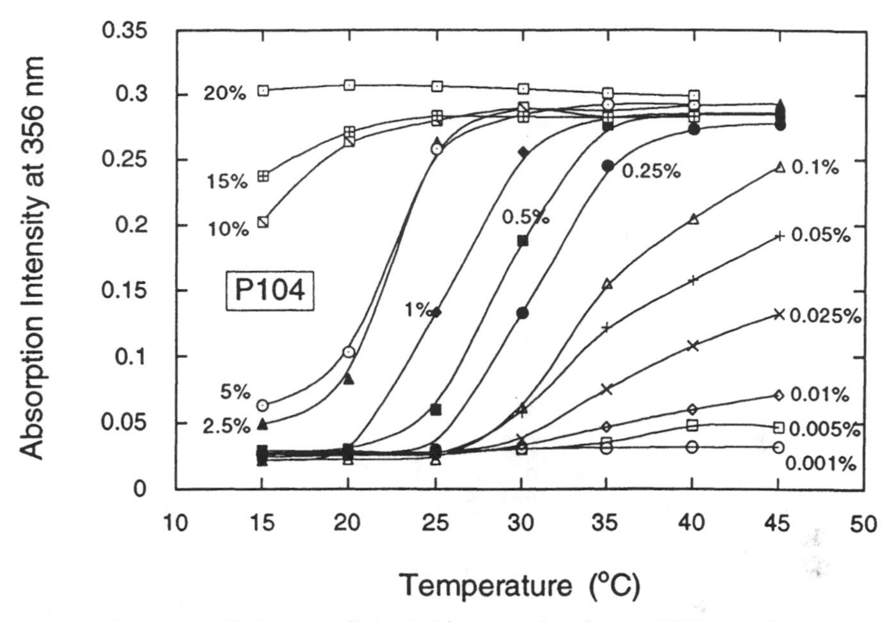

FIG. 2 Intensity of absorbed light (arbitrary units) due to DPH as a function of temperature for solutions of Pluronic P104 in water at various concentrations; the critical micelle temperature values were obtained from the first break of the intensity vs. temperature sigmoidal curves. (Adapted from Ref. 16.)

It is apparent from the experimental studies that PEO-PPO copolymers which are relatively less hydrophobic, either due to a high ethylene oxide content or a low molecular weight, do not form micelles at room temperature, but do start to aggregate at higher temperatures. This can be explained by the fact that water becomes a poorer solvent for both ethylene oxide and propylene oxide at higher temperatures, due to the breaking of hydrogen bonds with increasing temperature. In fact, it has been shown by Alexandridis et al. [16] that the micellization process is strongly driven by entropy, and that the free energy of micellization is mainly a function of the PPO, the hydrophobic part of the copolymer. The increased tendency of Pluronics to aggregate and form micelles as the temperature of the solution increases is expected to influence the partitioning and will be discussed in more detail later, in light of the observed solubilization behavior in these Pluronic polymers.

B. PS-PEO Block Copolymers

There has also been evidence for the formation of micelles in aqueous solutions of poly(styrene)-poly(ethylene oxide) (PS-PEO) block copolymers [17, 18]. The polymers were synthesized with ethylene oxide as the major component in order to make the copolymers water soluble, the polystyrene being extremely hydrophobic. The apparent molecular weight of PS-PEO micelles was found by Riess and Rogez [18] and Xu et al. [19, 20] to increase with increasing copolymer molecular weight and decreasing PEO content. Diblock copolymers formed larger micelles than triblock copolymers. Bahadur and Sastry [17] found the micelle size to decrease with an increase in PS content, which indicates that the micelle size also depends on the insoluble block. Increasing the temperature resulted in the formation of micelles with a smaller hydrodynamic radius. This is probably because water becomes a poorer solvent for ethylene oxide at higher temperatures, so that the corona becomes less swollen with water, resulting in a decrease in the hydrodynamic radius. Critical micelle concentrations of these PS-PEO block copolymers were estimated from the increase of the fluorescence intensity of pyrene with an increase in polymer concentration and found to be of the order of 1–5 mg/L [21]. The micelle/water partition coefficient of pyrene was found to be of the order of 3×10^5, based on the polystyrene volume [21].

Gallot et al. [22] investigated micelle formation in methanol/water mixtures for several series of PEO-PS block copolymers each having a constant PS chain length and varying lengths of the PEO block. Light-scattering and small-angle neutron-scattering measurements carried out in methanol showed that the critical micelle concentrations were primarily dependent on the molecular weight of the PS (hydrophobic) block and that, for a given molecular weight of the PS block, the aggregation number decreased with an increase in the PEO chain length. Despite the decrease in aggregation number, the hydrodynamic radius of the micelles was found to increase with PEO chain length. The SANS results showed the polystyrene core to be compact and small compared to the overall dimensions of the micelles and the micelles to be spherical.

C. Ionic Block Copolymers

In addition to the well-studied formation of micelles in nonionic PEO-PPO and PS-PEO copolymer solutions, a number of investigators probed ionic copolymer solutions. Valint and Bock [23] synthesized poly(*tert*-butylstyrene-*b*-styrene) using anionic polymerization, and then selectively sulfonated the styrene block to make the polymer water soluble with amphiphilic properties. Block copolymers and random copolymers were synthesized, with styrenesulfonate comprising the major component of the block to ensure water solubility. Micelle formation was inferred from viscosity measurements, and the effect of solubilization of toluene on the solution viscosity was also investigated. The results indicated changes in the size, shape, or aggregation numbers of the micelle, but these effects were not examined in detail. Di- and triblock copolymers of poly(methacrylic acid) and polystyrene have also been shown to form micellar aggregates in solution [24–26]. The solubilization of pyrene in these micelles will be discussed below. Ikemi et al. [27, 28] used small-angle x-ray-scattering and fluorimetric measurements to examine the structure of micelles formed by poly(ethylene oxide) and poly(2-hydroxyethyl methacrylate) (PHEMA) triblock copolymers in water. Aggregates were formed at concentrations greater than 0.1% (w/v), and fluorescence measurements indicated the formation of hydrophobic domains (Fig. 3). The x-ray-scattering data showed that there is a large interfacial boundary region between the micellar core and corona; the width of this interface decreased with a decrease in temperature.

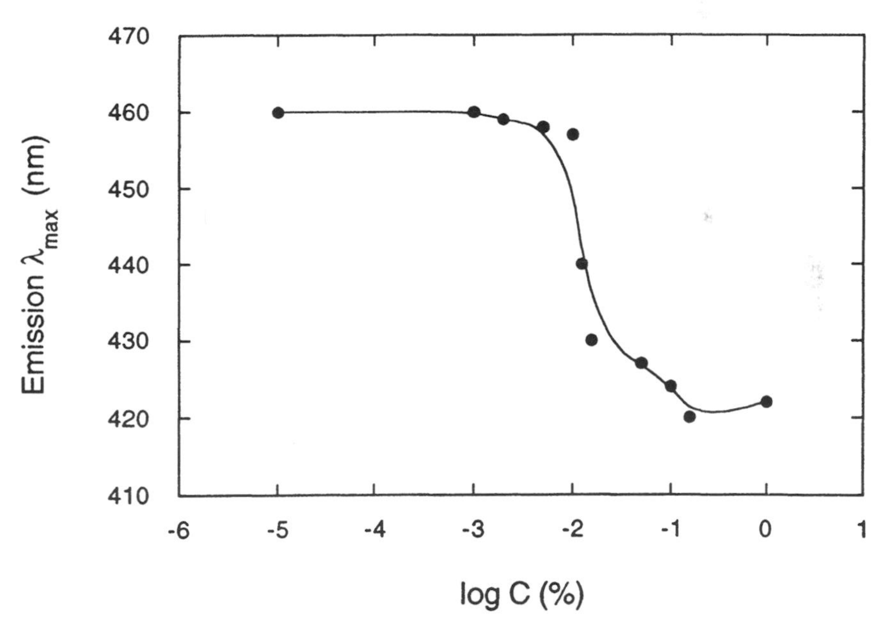

FIG. 3 Polymer concentration dependence of the emission λ_{max} for aqueous solutions of the HEMA-EO block copolymer in the presence of a fluorescence dye at 30°C. [Dye] = 1.0×10^{-5} mol/L. (Data extracted from Ref. 28.)

III. A SURVEY OF SOLUBILIZATION STUDIES IN COPOLYMER SOLUTIONS

We include here a compilation of solubilization studies in aqueous solutions of amphiphilic copolymers. The copolymer used in each study and the compound solubilized are reported in the table. Although the list is by no means comprehensive, it certainly demonstrates the applications potential amphiphilic copolymer solutions have in solubilizing molecules that are otherwise sparingly soluble in water.

Polymer	Solute	Reference
Pluronic F68	Hexane	Al-Saden et al. [11]
Pluronics L62, L63, L64, P65, F68	p-Substituted acetanilides	Collett and Tobin [2]
Pluronics F68, F88, F108	Indomethacin	Lin and Kawashima [1]
Pluronic L64	*o*-Xylene	Tontisakis et al. [29]
Pluronics P103, P104, P105, F108, P123	Naphthalene, phenanthrene, pyrene	Hurter and Hatton [30]
Tetronics T904, T1304, T1504, T1107, T1307	Naphthalene, phenanthrene, pyrene	Hurter and Hatton [30]
Pluronics L64, P104, F108	Pyrene	Hurter et al. [31]
Pluronics P65, F68, P84, P85 F88, P103, P104, P105, F108, P123, F127	Diphenylhexatriene	Alexandridis et al. [32]
Pluronic L64	Diazepam	Pandya et al. [33]
Pluronic P85	Fluorescein isothiocyanate, haloperidol	Kabanov et al. [34]
PEO-PPO-PEO	Cumyl hyperoxide	Topchieva et al. [35]
PEO-PPO (70:30), 12500 M_w poly (*N*-vinylpyrrolidone-styrene)	Benzene, toluene, *o*-xylene, ethylbenzene, *n*-hexane, *n*-heptane, *n*-octane, *n*-decane, cyclohexane	Nagarajan et al. [4]
Poly(*N*-vinylpyrrolidone-styrene)	Toluene, naphthalene, phenanthrene	Haulbrook et al. [36]
Poly(sodium styrenesulfonate-co-2-vinylnaphthalene)	Anthracene, perylene, 9,10-dimethylanthracene, 9,10-diphenylanthracene, 9-methylanthracene,	Nowakowska et al. [37]

Polymer	Solute	Reference
Poly(sodium styrenesulfonate-co-2-vinylnaphthalene)	2-Undecanone	Nowakowska et al. [38]
Poly(sodium styrenesulfonate-co-2-vinylnaphthalene)	Styrene	Nowakowska and Guillet [39]
Poly(sodium styrenesulfonate-co-9-(*p*-vinylphenyl)-anthracene)	Perylene, tetraphenylporphine 1,3-diphenyliso-benzophuran	Sustar et al. [40]
Poly(styrene)-poly(methacrylic acid)	Benzene	Kiserow et al. [41]

IV. EFFECT OF POLYMER TYPE AND STRUCTURE ON SOLUBILIZATION

The solubilization of para-substituted acetanilides in a series of Pluronic PEO-PPO-PEO triblock copolymers depended on both the polymer structure and the nature of the solubilizate [2]. The more hydrophobic halogenated acetanilides showed a decrease in solubility with an increase in the ethylene oxide content, whereas the opposite trend was noted with less hydrophobic drugs such as acetanilide and the hydroxy-, methoxy-, and ethoxy-substituted compounds. The solubility of indomethacin in aqueous solutions of Pluronic was studied by Lin and Kawashima [1]. This hydrophobic anti-inflammatory drug was solubilized in significant amounts only above a certain threshold polymer concentration, most likely related to the CMC. The polymer concentration at the transition point was found to decrease with an increase in temperature. Also, the solubilizing capacity increased with polymer molecular weight and with temperature. Approximately 0.5 mole of indomethacin per mole of surfactant could be solubilized in solutions of the polymer.

An extensive study of the effect of copolymer structure on the solubilization of naphthalene in a range of poly(ethylene oxide)-poly(propylene oxide) block copolymers has been reported by Hurter and Hatton [30]. Enhanced solubility of naphthalene in Pluronic copolymer solutions can be achieved; for example, the naphthalene concentration in a 6% P103 copolymer solution was 140 times that in pure water. Fig. 4 shows the relationship between the micelle/water partition coefficient of naphthalene, K_{mw}, and the PPO content of the polymer ($K_{mw} = C^s_m/C^s_w$, and

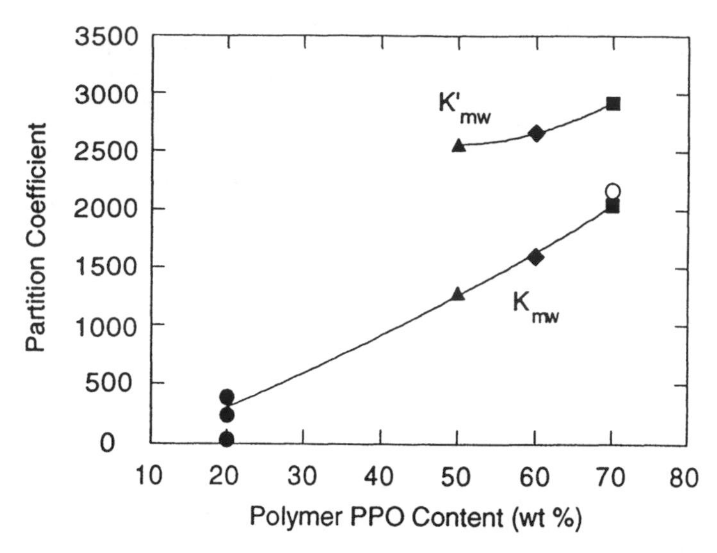

FIG. 4 The variation in the micelle-water partition coefficient for naphthalene in Pluronic polymers as a function of the PPO content of the polymer. (From Ref. 30.)

C_m^s, C_w^s are the concentrations of naphthalene in the micelle, based on total polymer mass, and in the water respectively). As the proportion of PPO in the polymer increases, at a given total polymer concentration, the solubility of naphthalene is enhanced. Since naphthalene is hydrophobic, it is expected to associate with the PPO; increasing the hydrophobicity of the polymer by increasing the proportion of PPO should lead to an increase in the uptake of naphthalene. If K_{mw} is normalized with PPO content (denoted by K_{mw}^1), there is still a small increase in the partition coefficient with increased PPO content of the polymer, which suggests that structural changes within the micelle, as a result of changes in polymer composition, must be influencing the partitioning behavior. For the majority of the copolymers studied in Ref. 30, the partition coefficient was constant over the entire polymer concentration range probed. A linear relationship between the polymer concentration and the amount of naphthalene solubilized is an indication that the micellar structure does not change significantly in this concentration range, and that increasing the polymer concentration leads to an increase in the number of micelles rather than micellar growth. Leibler et al. [42] showed that at concentrations far from the critical micelle concentration, micellar properties such as aggregation number are indeed only weakly dependent on the copolymer concentration.

The partitioning of naphthalene was found to increase with increasing molecular weight for a series of Pluronic polymers with 40% PEO [30]. The observed dependence of the partition coefficient on polymer molecular weight could either be due to a more favorable interaction between the solute and longer PPO chains or to changes in the micelle structure. Using the relationship for the chemical potential of a solute in a polymer solution, according to Flory [43], it can be shown that solutes would partition favorably into a solution where the polymer was of smaller chain length/lower molecular weight [30]. The observed increase in the partition coefficient with polymer molecular weight thus cannot be explained by simply considering the interaction between the polymer segments forming the core and the solutes. Clearly the molecular weight of the polymer affects the micelle structure, which has an effect on the partitioning behavior of the solute. Theoretical studies of block copolymer micelle formation have suggested that the micelle size increases with molecular weight [42, 44, 45]. There are a number of possible reasons why larger micelles would solubilize more naphthalene. The ratio of core volume to interfacial volume increases with micelle size, so that if the naphthalene is effectively excluded from the interfacial region, solubilization will increase with an increase in micelle size. The excess pressure within a copolymer-covered oil-in-water droplet increases with decreasing micelle radius, which would increase the chemical potential within the droplet and decrease the partition coefficient. Finally, changing the molecular weight of the polymer could change the microstructure of the micelle, which would affect the solubilization behavior. The effect of polymer molecular weight on micelle structure has been investigated using the self-consistent mean-field lattice theory and is discussed in more detail in Sec. VIII.

The partition coefficients of naphthalene in the linear Pluronic copolymers with 40% PEO were compared with those for the branched Tetronic copolymers (see Table 1 for properties of the Tetronic copolymers) of the same PPO/PEO composition [30]. For the same PPO content and molecular weight, the Pluronic polymers have a significantly higher capacity for naphthalene than the Tetronic polymers. This indicates that all facets of the polymer architecture play a role in determining the affinity of the solute for the block copolymer; not only are the composition and molecular weight important, but also the microstructure of the polymer. The solubilization of 1,6-diphenyl-1,3,5-hexatriene (DPH), a fluorescence dye, in Pluronic copolymer micelle solutions was investigated by Alexandridis et al. [32]. DPH partitioned favorably in the micellar phase, with the partition coefficient higher for P123 (a Pluronic copolymer with 70% PPO), and decreasing in the order of P123 > P103 > P104 = P105 > F127 > P84 > P85 = F108 > F88 > F68 > P65. It can be concluded from these data that the solubilization was influenced by both the relative (with respect to PEO) and absolute size of the hydrophobic PPO block.

Nagarajan et al. [4] studied the solubilization of aromatic and aliphatic hydrocarbons in solutions of poly(ethylene oxide)-poly(propylene oxide) and poly(*N*-vinylpyrrolidone-styrene) (PVP-PS) copolymers. A 10 wt.% solution of 12500 molecular weight PEO-PPO (70:30) copolymer and a 20% solution of poly(*N*-vinylpyrrolidone-styrene) (40:60) were used. For both aromatic and aliphatic hydrocarbons, more hydrocarbon was solubilized per gram of polymer in the PVP-PS copolymer than in the PEO-PPO copolymer. This is probably due to the fact that the PEO-PPO copolymer contained only 30% of the hydrophobic PPO constituent, whereas the PVP-PS copolymer contained 60% PS, and polystyrene is more hydrophobic than PPO. Nagarajan et al. [4] also noted that aromatic hydrocarbons were solubilized in larger amounts than the aliphatic hydrocarbons (see also Sec. V).

Sustar et al. [40] have synthesized poly(sodium styrenesulfonate-co-9-vinylphenyl anthracene) copolymers (PSSS-VPA) with 10.6–51 wt.% VPA. The polymers containing up to 40.5% VPA were water soluble. This copolymer was believed to adopt a pseudomicellar, "hypercoiled" structure in aqueous solution, in which the hydrophobic vinylnaphthalene groups cluster in the interior of the coil and are shielded from the solvent by a shell of carboxyl groups and counterions. The pendant naphthalene groups on the polymer absorb light in the UV-visible range and transfer energy to solubilized molecules in the hydrophobic microdomains within the polymer coil; these copolymers thus display photocatalytic activity. The solubilizing ability of the polymer was investigated using perylene and tetraphenyl porphine as probes, both of which are extremely hydrophobic, and fluoresce negligibly in water but very efficiently in organic solvents. The hydrophobic probes were found to be solubilized in greater amounts in the polymers with a higher VPA content, since the hydrophobicity of the polymer coil increases with an increase in the VPA fraction (Fig. 5a). The amount solubilized by a given polymer was found to depend linearly on polymer concentration, and the distribution coefficient of perylene was found to increase from 2×10^5 for a polymer with 11% VPA to 2.6×10^6 for a polymer with 51% VPA (Fig. 5b). Nowakowska and Guillet [39] reported the solubilization of styrene in aqueous solutions of poly(sodium styrene-sulfonate-co-2-vinylnaphthalene) (PSSS-VN) and found that the solubility of styrene in polymer solutions depends linearly on the polymer concentration. The distribution coefficient for styrene (which has a fairly high water solubility) between polymer and water was calculated to be 10^3.

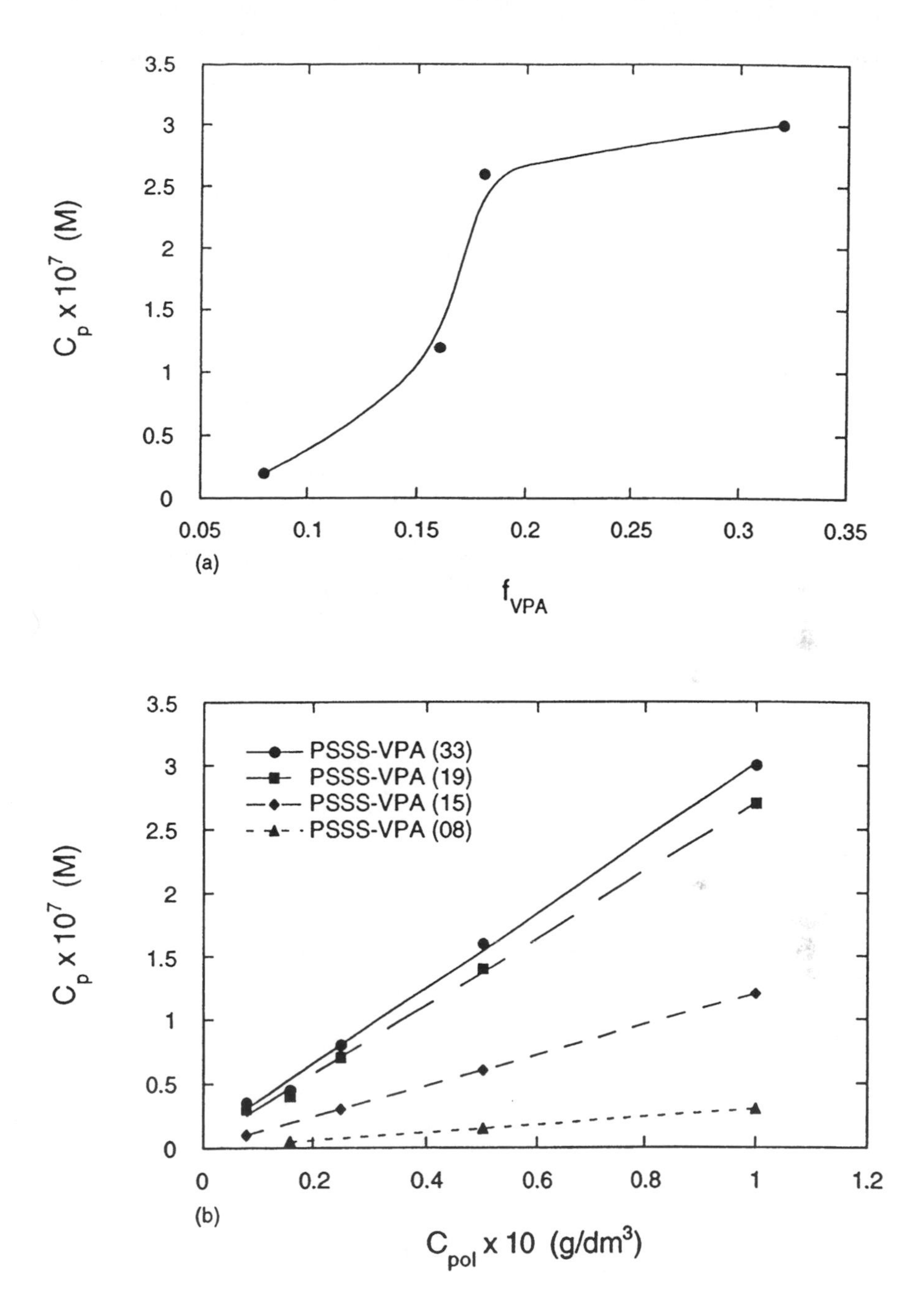

FIG. 5 (a) Dependence of the concentration of perylene solubilized in aqueous solutions of PSSS-VPA (c_{pol} = 0.1 g/L) on the mole fraction of VPA in the copolymer, and (b) effect of the polymer concentration on the amount of perylene solubilized in aqueous solutions of various PSSS-VPA polymers. (Data extracted from Ref. 40.)

V. SOLUTE EFFECTS ON SOLUBILIZATION

The phase behavior of the ternary system *o*-xylene/triblock poly(ethylene oxide)-poly(propylene oxide)-poly(ethylene oxide) copolymer/water was studied by Tontisakis et al. [29]. The polymer was Pluronic L64, with 40 wt.% ethylene oxide and a total molecular weight of 2900. The phase diagram exhibited stable regions of both oil-in-water and water-in-oil microemulsions. A region of the phase diagram consisting of polymer concentrations up to 10% with small amounts of solubilized *o*-xylene was investigated in detail using light-scattering techniques and viscosity measurements. The aggregation number of the micelles was determined as a function of temperature and the ratio of *o*-xylene to polymer. The addition of *o*-xylene resulted in micellar growth, and the aggregation numbers also increased significantly with temperature (see also Sec. VI). The finding that micellar growth occurs with solute uptake is in agreement with the results of Al-Saden et al. [11], who measured the variation in the hydrodynamic radius of Pluronic micelles with hexane uptake, and found that the micelle radius increased with an increase in solubilizate uptake (saturation loading of 0.08 *g* hexane/*g* polymer at 35°C for F68) (Fig. 6). This was

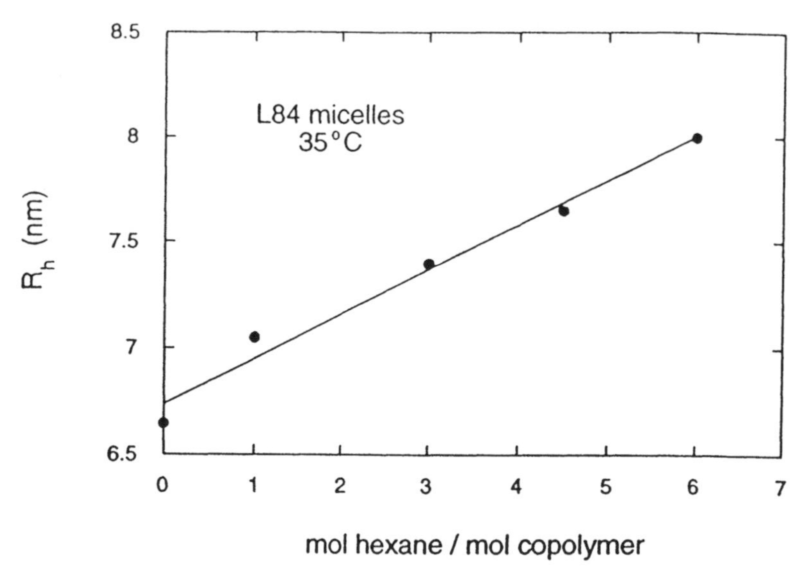

FIG. 6 Variation of hydrodynamic radius of L84 aggregates with hexane uptake at 35°C. (Data extracted from Ref. 11.)

postulated to result not only from the incorporation of the solutes in the micellar core but also from a simultaneous increase in the aggregation number of the micelle.

The solubilization of polycyclic aromatic hydrocarbons (PAHs) in aqueous solutions of poly(sodium styrene-sulfonate-co-2-vinylnaphthalene) copolymers has been studied by Nowakowska et al. [37]. Distribution coefficients between the polymer pseudophase and the aqueous phase were determined for five PAHs: perylene, 9,10-dimethylanthracene, 9,10-diphenylanthracene, 9-methylanthracene, and anthracene. The average number of molecules of probe solubilized by one polymer molecule ranged from 1 (for perylene) to 10 (for 9-methylanthracene) and the distribution coefficients ranged from 10^6 to 10^{10}. The more hydrophobic PAHs had higher distribution coefficients, but the number of molecules solubilized per polymer was lower, i.e., the large increase in the distribution coefficient with an increase in solute hydrophobicity was chiefly due to the decrease in the water solubility of the PAH compounds.

This is in agreement with the results of Hurter and Hatton [30, 31] on the solubilization of three PAHs (naphthalene, phenanthrene, and pyrene) in aqueous solutions of Pluronic block copolymers. A significant increase in the solubility enhancement for the more hydrophobic PAHs was observed, as there is a strong correlation between the octanol/water partition coefficient, K_{ow}, of the solute (a standard measure of hydrophobicity) and the micelle-water partition coefficient. Similar relationships have been found for micelle/water partition coefficients of various hydrophobic solutes in commercial surfactants such as Triton X-100 (an octylphenol polyoxyethylene surfactant) [46] and sodium dodecyl sulfate [47]. Note that, while the partition coefficient increased with the hydrophobicity of the solute, the micellar concentration of solute remained of the same order of magnitude (in fact it decreased slightly with an increase in hydrophobicity) and the dramatic increase in the partition coefficient was chiefly due to the decrease in PAH water solubility.

Nagarajan et al. [4] found that the solubilization of aliphatic hydrocarbons was negligible in comparison to that of aromatics. A Flory-Huggins-type interaction parameter between the solubilizate and the polymer block constituting the core, χ_{sc}, was estimated using Hildebrand-Scatchard solubility parameters. It was found that the aromatic hydrocarbons, which had smaller values of χ_{sc}, were solubilized to a greater extent than the aliphatic molecules, which had larger values of χ_{sc}. However, there was not a direct correlation between χ_{sc} and the amount solubilized, particularly for the aliphatic molecules which had large χ_{sc} values. For binary mixtures of benzene and hexane, it was shown that the micelles selectively solubilized the aromatic benzene.

VI. TEMPERATURE EFFECTS ON SOLUBILIZATION

Experimental results suggested that raising the temperature of the solution could induce micelle formation in polymers that are relatively hydrophilic, either due to their low molecular weight or to a high PEO content [11, 48]. The phase diagrams of both poly(ethylene oxide) and poly(propylene oxide) in water exhibit a lower critical solution temperature [49], indicating that raising the temperature causes water to become a poorer solvent for these polymers. The decreased solubility of PEO could be due to the breaking of hydrogen bonds between the ether oxygen and water or to conformational changes in the PEO chain structure as temperature is increased [50, 51]. Thus, while these hydrophilic polymers might not aggregate at room temperature, an increase in the temperature could induce aggregation as water becomes a poorer solvent. The formation of micelles should lead to enhanced solubilization, resulting from the hydrophobic environment afforded by the micelle core, which is attractive to organic solutes such as naphthalene. These observations indicate that solution temperature could have a significant effect on solubilization.

Tontisakis et al. [29] studied the solubilization of *o*-xylene in an aqueous solution of Pluronic L64. The aggregation number of the micelles was determined as a function of temperature and of the *o*-xylene/copolymer ratio. As illustrated in Fig. 7, the addition of *o*-xylene resulted in micellar growth, and the aggregation numbers also increased significantly with temperature. For the pure polymer in water, the aggregation number increased from 14 to 90 as the temperature was increased from 30 to 42°C. Similar increases in the aggregation number with temperature were observed for the copolymer systems with solubilized *o*-xylene. While the aggregation number increased monotonically with temperature, the hydrodynamic radius of the micelle initially decreased, passed through a minimum, and then increased with further increase in temperature [29]. This was attributed to exclusion of water from the micelle, as water became a poorer solvent for both PEO and PPO at higher temperatures. As temperature was raised even further, a point was reached where effectively all the water was excluded from the micelle; since the micelle aggregation number continued to increase, the micelle radius started increasing.

The results of McDonald and Wong [48] and Al-Saden et al. [11] discussed in Sec. II, suggest that, while Pluronic L64 does not form polymolecular micelles at 25°C, stable micelles are formed at 35°C. Fig. 8 compares the results of Hurter and Hatton [30] for the solubility of naphthalene in L64 as a function of polymer concentration at these two temperatures. At 25°C, not much naphthalene is solubilized, and the curve is nonlinear, whereas the solubility enhancement at 35°C is significant. The formation of micelles thus appears to significantly enhance the solubilizing ability

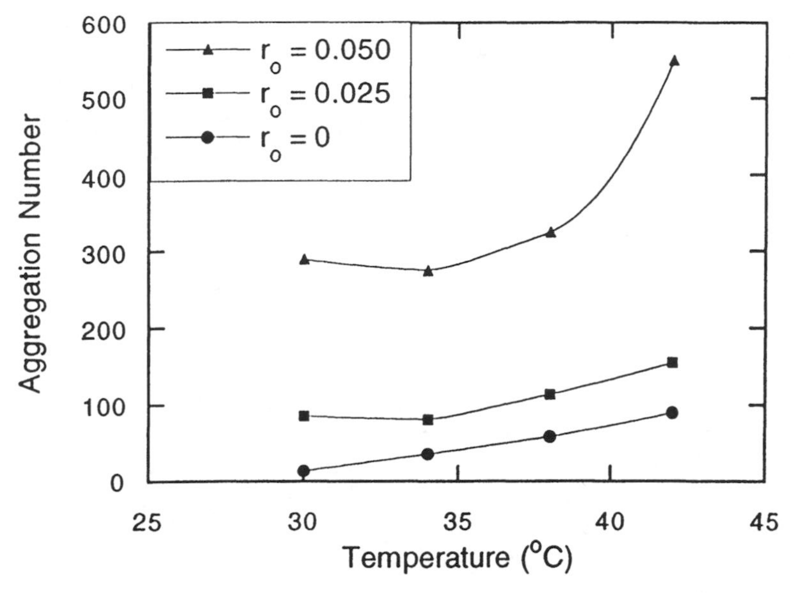

FIG. 7 Effect of temperature and solute uptake on the aggregation number of Pluronic L64 micelles: r_o is the ratio of solute (*o*-xylene) to polymer (on a mass basis). (Data extracted from Ref. 29.)

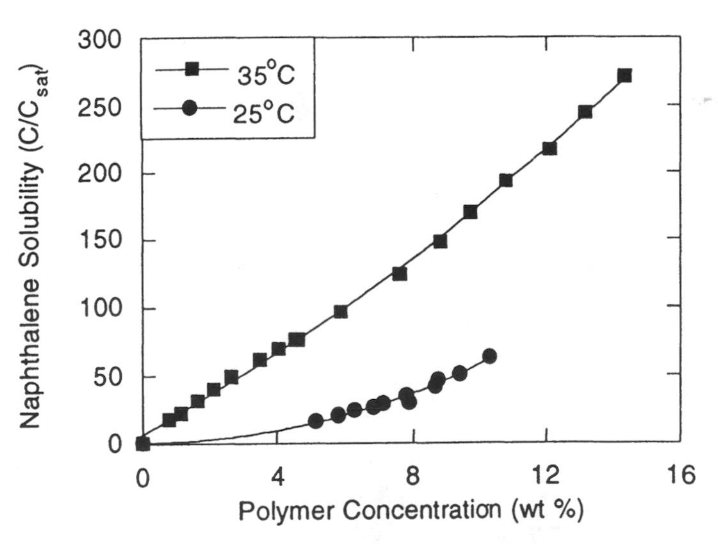

FIG. 8 The effect of temperature on the dependence of naphthalene solubility on polymer concentration for solutions of Pluronic L64. The naphthalene concentration has been normalized by the solubility of naphthalene in water at the appropriate temperature. (From Ref. 5.)

of L64. Al-Saden et al. [11] showed that for L64, micellar growth occurs with an increase in polymer concentration at 25°C, which would explain the nonlinearity at 25°C. At 35°C there is still a somewhat nonlinear relationship between the amount of naphthalene solubilized and the polymer concentration, indicating that more naphthalene is solubilized per polymer molecule at higher polymer concentrations. These results are consistent with the results of Al-Saden et al., which show that the micelle radius is essentially constant at 35°C (Fig. 9), but at high polymer concentrations the hydrodynamic radius of the micelle decreases slightly. This could indicate that water is being excluded from the micelles at high polymer concentrations, which would explain the increase in naphthalene solubilization at high polymer concentrations observed at 35°C.

Similar results were observed for F108, a Pluronic polymer which is relatively hydrophilic, having 80% PEO [30]. The solubility of naphthalene was significantly increased by raising the temperature to 35°C, with a constant partition coefficient being obtained at the higher temperature. The increased temperature makes water a poorer solvent for both PEO and PPO blocks, inducing aggregate formation and enhancing the solubilization of naphthalene. The effects illustrated in Fig. 8 are not due to the increase in water solubility of naphthalene with an increase in tempera-

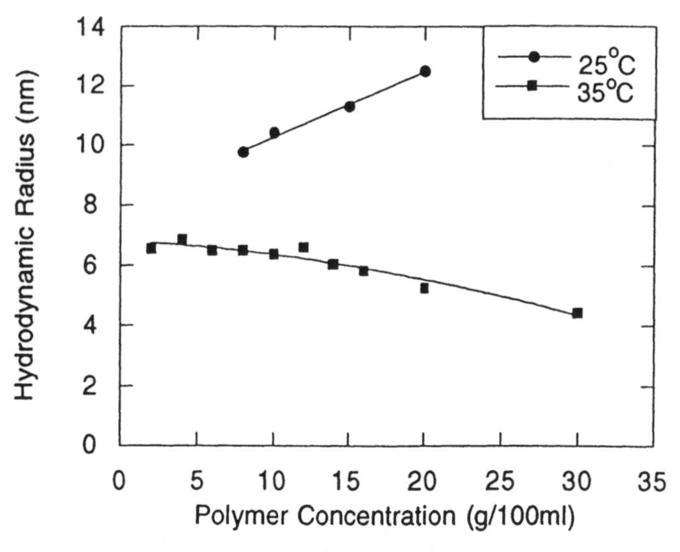

FIG. 9 The hydrodynamic radius of L64 aggregates at 25°C and 35°C obtained from the light-scattering results of Ref. 11.

ture, since the concentration of naphthalene in the micellar solution has been normalized with the saturated concentration of naphthalene in water at the appropriate temperature. The partitioning results observed for F108 are consistent with the light-scattering results for F68 of Zhou and Chu [13] that showed that F68 does not form micelles at 25°C, but at higher temperatures polymolecular micelles with a constant hydrodynamic radius are formed. F68 is similar to F108 in that it is a Pluronic polymer with 80% PEO, but it has a lower molecular weight (8350).

In Sec. IV, it was reported that Pluronic P104 significantly enhanced the solubility of naphthalene in water at 25°C, and the micelle-water partition coefficient was not dependent on polymer concentration. The light-scattering results of Wanka et al. [14] indicate that P104 forms micelles with an aggregation number of approximately 50 at 25°C. In Fig. 10, the solubilization of naphthalene as a function of temperature is compared for 10% solutions of Pluronics P104, L64, and F108 [5]. The solubility of naphthalene in both F108 and P104 appears to increase rapidly at first, and then above a transition temperature of 21°C for P104 and 26°C for F108, the solubility enhancement with temperature becomes less significant. It can be shown that the solubility enhancement at higher temperatures is chiefly

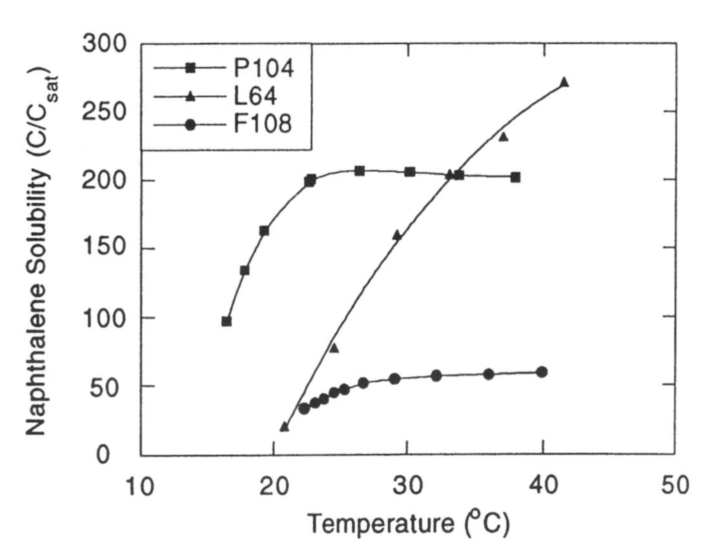

FIG. 10 The temperature effect on the concentration of naphthalene in 10% aqueous solutions of Pluronics P104, L64, and F108. The naphthalene concentration has been normalized by the solubility of naphthalene in water at the appropriate temperature. (From Ref. 5.)

due to the increase in the naphthalene water solubility; the normalized naphthalene concentration becomes independent of temperature above the transition point. The results for L64 show a monotonic increase in solubility with temperature. It is possible that the saturation phenomenon would occur at a higher temperature than that investigated, but the solution of L64 became turbid above 41.5°C. Similar trends were observed in the solubilization of 1,6-diphenyl-1,3,5-hexatriene in various Pluronic PEO-PPO-PEO copolymer solutions [32].

The results in Fig. 10 for P104 and F108 seem to indicate that the micelles become saturated with naphthalene at higher temperatures. It is possible that, for P104 and F108, micellar growth occurs up to the transition temperature, and above this temperature the micelle size no longer increases, resulting in no further increase in the partition coefficient between the micelles and water. McDonald and Wong [48] showed that the aggregation number of L64 aggregates was unity at 24°C, 5.9 at 30°C, and 29.9 at 35°C, i.e., micellar growth occurs with an increase in temperature. An alternative explanation, or contributing factor, could be that the hydrophobicity of the micellar core (or the polymer coil, in the case of unimolecular "micelles") increases with an increase in temperature. Al-Saden et al. [11] measured the hydrodynamic radius of aggregates formed in L64 8% solution as a function of temperature and found that the radius initially decreased and then increased above 40°C. They suggest that as temperature is increased, the polymer chains contract, squeezing out water, as the latter becomes a poorer solvent for both PEO and PPO. As the temperature approaches the cloud point, no more contraction is possible (if effectively all the water has already been excluded) and the increase in aggregation leads to an increase in the hydrodynamic radius. The transition temperature observed for P104 and F108 could be the temperature at which the micelle has become as hydrophobic as possible, owing to the exclusion of virtually all water from the core. Raising the temperature further cannot significantly alter the hydrophobicity of the core, so that the normalized naphthalene concentration no longer increases with temperature. The transition temperature for L64 would be approximately 41°C, which is above the range of these experiments, so that the saturation phenomenon was not fully observed, though Fig. 10 does show that the increase in solubility with temperature becomes less pronounced above 35°C.

VII. EFFECT OF SOLUTION IONIC STRENGTH AND pH ON SOLUBILIZATION

Amphiphilic block copolymers, with a polyelectrolyte comprising the hydrophilic portion, can also be effective at solubilizing organic molecules in aqueous solutions. Since polyelectrolytes have a high affinity for water, highly nonpolar polymers, such as polystyrene, can be used for the hydrophobic block in larger proportions, while still maintaining water solubility. The chain conformation of the polyelectrolyte constituent of the copolymer is obviously strongly affected by factors such as the ionic strength and pH of the solution, which leads to some interesting effects on the micelle structure and on the solubilization behavior.

Kiserow et al. [41] have synthesized diblock copolymers of polystyrene and polymethacrylic acid and studied the solubilization of benzene in aqueous micellar solutions of these polymers. The polymers had molecular weights ranging from 45,000 to 70,000 and molar fraction of polystyrene varying from 46% to 60%. The weight-average molar masses of the micelles were determined using static and quasielastic light scattering; the micelle aggregation numbers ranged from 53 to 290. While most of the benzene was solubilized within the polystyrene core of the micelle, some benzene was solubilized in hydrophobic domains in the shell, particularly at low pH where hydrophobic hypercoiling of the methacrylic acid occurs in the shell. At high pH, the methacrylic acid blocks are ionized, resulting in an extended shell region, which leads to a more stable micelle. In fact, viscosity measurements indicated that the mutual repulsion of these highly charged micelles resulted in some structural organization in the micellar solution. The solubilization of benzene was found to increase with increasing pH, with mass ratios of benzene to polystyrene of up to 5:1 being obtained in these systems. Solubilization limits of benzene in copolymer micelles (expressed as the weight ratio of benzene to polystyrene) are plotted in Fig. 11 as a function of pH.

Nowakowska et al. [52] have synthesized a high molecular weight random copolymer of poly(sodium styrenesulfonate) and poly(2-vinylnaphthalene) (PSSS-VN) and studied the conformation of this polymer in water. The polymer had a molecular weight of 310 000 and contained 60 mol% of the hydrophobic vinylnaphthalene. In aqueous solution these polymers adopt a pseudomicellar, "hypercoiled" structure, like the poly(sodium styrenesulfonate-co-9-vinylphenyl)anthracene polymers discussed in Sec. IV. It has been shown [38] that 2-undecanone is effectively solubilized in aqueous solutions of PSSS-VN. The distribution coefficient was maximized at neutral pH and at high ionic strength, where the coil was determined to be in a more expanded form. This is in contrast to the solubilization behavior observed using more hydrophobic solutes such as PAHs [30], which favor a more compact coil conformation, as will be discussed below.

Photochemical reactions of PAHs solubilized in aqueous solutions of these PSSS-VN copolymers have been studied by Nowakowska et al. [37]. The pendant naphthalene groups on the polymer absorb light in the

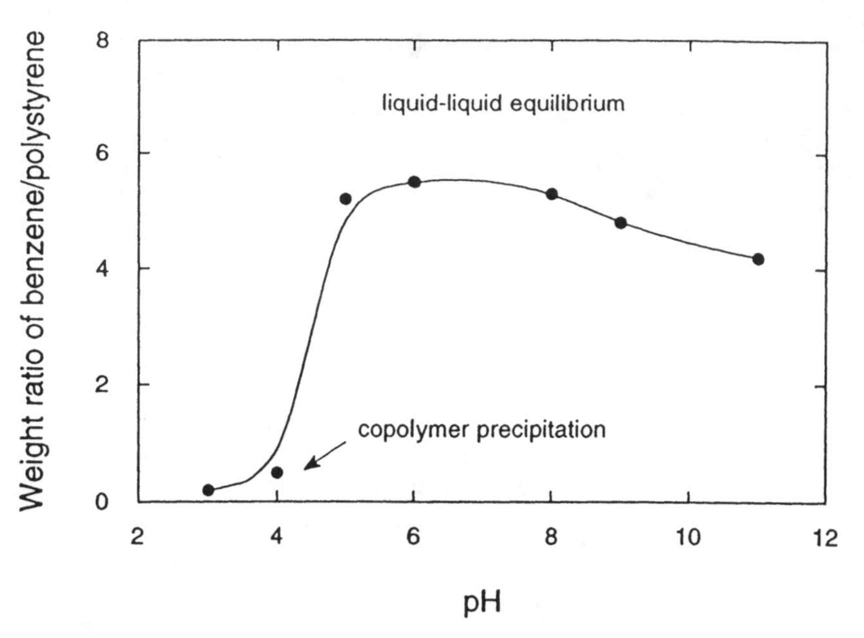

FIG. 11 Solubilization limits of benzene into micelles (expressed as the weight ratio of benzene to polystyrene) as a function of pH. (Data extracted from Ref. 41.)

UV-visible range and transfer energy to the PAHs, which are solubilized in the hydrophobic microdomains within the polymer coil. Perylene was used as a fluorescence probe to determine the location of the solute. Perylene has a very low emission intensity in water, and it was shown that solubilization of perylene resulted in a decrease in naphthalene emission and an increase in perylene emission, which indicates that the excited pendant naphthalene groups in the core of the polymer coil are transferring energy to the solubilized perylene molecules, and that perylene is trapped within the core of the PSSS-VN pseudomicelle. Similar results have been obtained for 9,10-dimethylanthracene, 9,10-diphenylanthracene, 9-methylanthracene, and anthracene. The effect of polymer composition, pH, and ionic strength on the solubilization behavior has also been investigated. Perylene was found to be solubilized more effectively in polymers with a lower number of sulfonate groups. The decrease in repulsive interactions in the coil resulted in a more compact conformation, which probably lead to a more hydrophobic environment, and consequently favored solubilization of PAHs. Similarly, maximum solubilization occurred at acidic and alkaline pH, where the coil is most compact (Fig. 12). With an increase in the ionic strength, solubilization was initially enhanced, however, at higher ionic strengths the amount solubilized decreased (Fig. 13). This was explained by comparing the dependence of the hydrodynamic volume of the polymer coil on ionic strength: raising the ionic strength causes the coil to become more compact, which enhances solubilization; however, at very high ionic strengths the coil becomes too small, and solubilization decreases.

Sustar et al. [40] investigated the solubilization of tetraphenyl porphine in poly(sodium styrenesulfonate-co-9-vinylphenyl)anthracene copolymers (PSSS-VPA) as a function of ionic strength. The results were similar to those obtained for PSSS-VN, with the solubilization increasing up to an ionic strength of 0.5, and then decreasing at higher ionic strengths. It was suggested that the extremely compact conformation formed at high ionic strengths limited the number of molecules that could be solubilized.

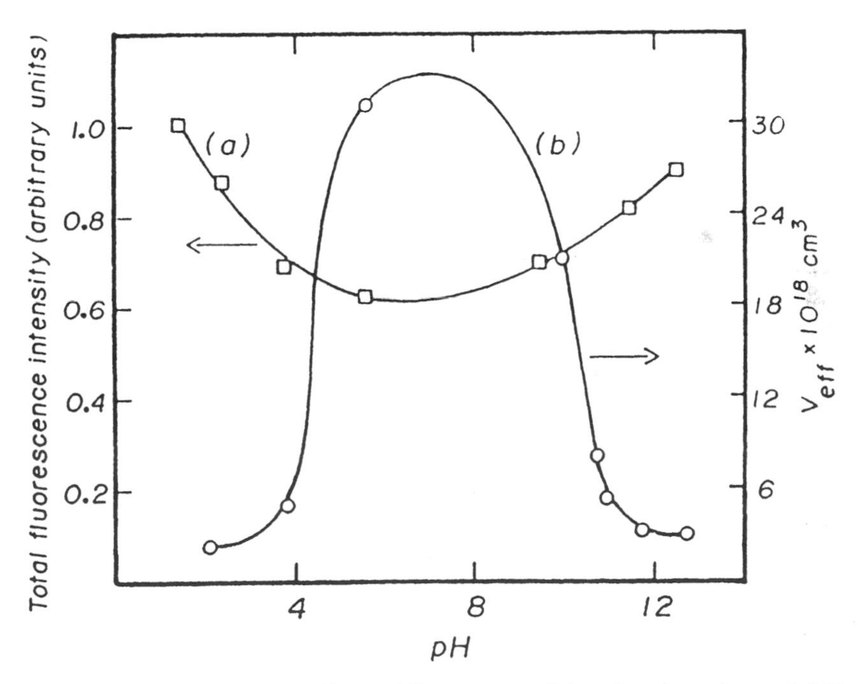

FIG. 12 (a) Dependence of the total fluorescence intensity of perylene solubilized in aqueous solutions of PSSS-VN (λ_{ex} = 415 nm) on pH and (b) dependence of the effective hydrodynamic volume of PSSS-VN on pH. (From Ref. 37.)

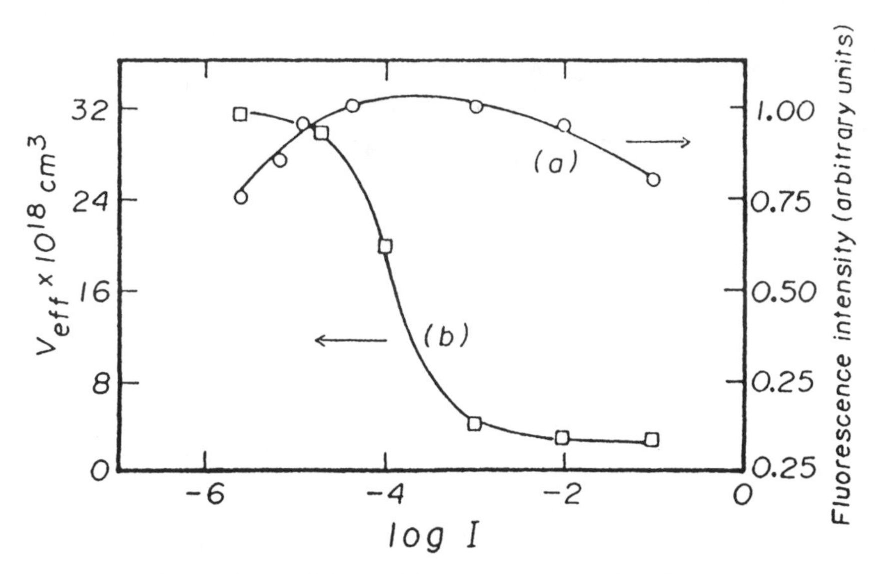

FIG. 13 (a) Dependence of the total fluorescence intensity of perylene solubilized in aqueous solutions of PSSS-VN (λ_{ex} = 415 nm) on ionic strength and (b) dependence of the hydrodynamic volume of PSSS-VN on ionic strength ($\log_{10}$). (From Ref. 37.)

VIII. MODELING OF SOLUBILIZATION IN BLOCK COPOLYMER MICELLES

Theories of micelle formation in solutions of block copolymers have been advanced by a number of researchers over the past decade. Leibler et al. [42], Noolandi and Hong [45], Munch and Gast [53], and Nagarajan and Ganesh [44] computed the free energy of micelle formation assuming uniform copolymer concentrations in the core and corona regions, respectively. Scaling theories for polymeric micelles with an insoluble core and an extended corona were developed by Halperin [55], Marques et al. [56], Semenov [57], and Zhulina and Birshtein [58]. In a third approach, van Lent and Scheutjens [59] used a self-consistent field theory to determine the detailed segment density profiles within a micelle, making no a priori assumptions as to the locations of the micellar components. The micelle structures determined from this model have been confirmed recently by Monte Carlo simulations [60] and show that the interfacial region between the core and corona of the micelle is diffuse and not sharp as is assumed in the other modeling approaches.

A. Phenomenological Theory of Solubilization in Block Copolymer Micelles

Nagarajan and Ganesh [54] extended their theory of micellization [44] to describe solubilization in micelles formed by A-B diblock copolymers. A micelle of spherical shape was assumed, with a core consisting of solvent-incompatible A-blocks and solute, and a corona which contained solvent and the solvent-compatible B-blocks. No allowance was made for solute molecules in the corona of the micelle, which lead trivially to their conclusion that "almost all" of the solute was confined to the core. Constant compositions of all components were assumed in both the core and the corona. The interfacial free energy was related to the interfacial tension between the core components and water, assuming a sharp interface between them and neglecting the presence of the B-blocks in the corona. The total free energy of solubilization was assumed to be due to contributions from the change in the state of dilution and deformation of the A blocks in the core and of B blocks in the corona, the localization of the joints between the blocks at the interface, and the free energy of the core-corona interface.

On the basis of this free-energy model, it was possible to predict the CMCs, the size and composition distribution of the micelles containing solubilizates, the aggregation number of the micelles, the maximum extent of solubilization, and the core radius and the shell thickness of the micelles. Illustrative calculations show that, in general, the micelles are practically monodispersed both in their size and in the extent of solubilization. The solubilization behavior of the micelles and their geometrical characteristics were found to be influenced significantly by the interactions between the solubilizate and the solvent-incompatible A-block of the copolymer as well as by the solubilizate-solvent interfacial tension. The various free-energy contributions were analyzed as a function of the volume fraction of the solubilizate in the micelle core. The (negative) free-energy contribution provided by the change in state of dilution of block A decreases with an increase in the volume fraction of the solubilizate, thus favoring the uptake of the solubilizate within the micelle. The magnitude of this contribution is larger if the solubilizate-core block interaction parameter χ_{AJ} is lower and if the molecular size of the solubilizate is smaller. The free-energy contributions arising from the deformation of block A and the core-solvent interfacial energy restrict the swelling of the micellar core by the solubilizate, and, consequently, the extent of solubilization. The increase in the positive interfacial free energy accompanying the uptake of solubilizates by the micelles is also dependent on the solubilizate-solvent interfacial tension σ_{SJ}. As a result, the micellar capacity will be

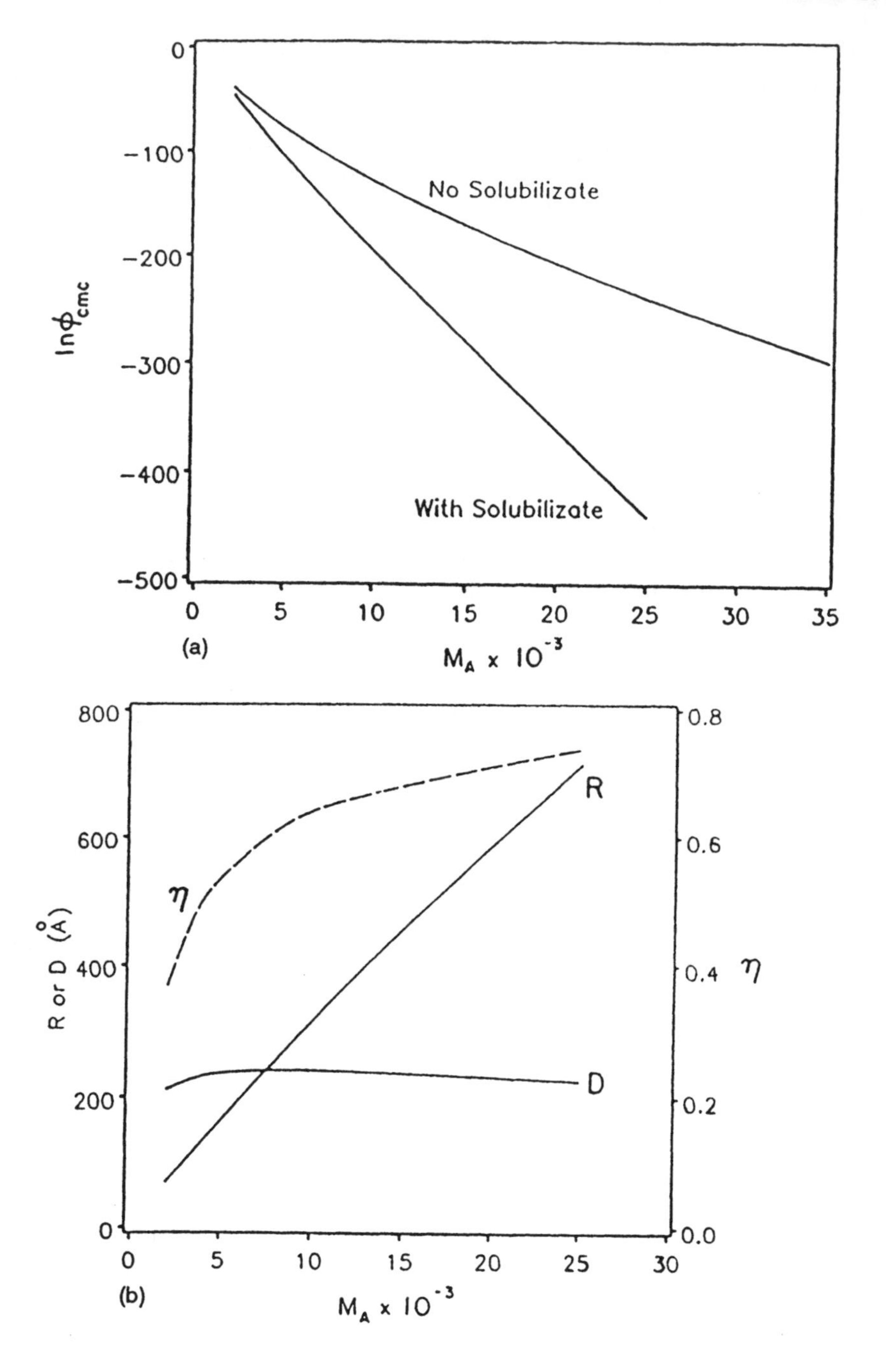

0
−100
−200
−300
−400
−500
$\ln\phi_{cmc}$
No Solubilizate
With Solubilizate
0
5
10
15
20
25
30
35
(a)
$M_A \times 10^{-3}$
800
600
400
200
0
R or D (Å)
η
R
D
0.8
0.6
0.4
0.2
0.0
η
0
5
10
15
20
25
30
(b)
$M_A \times 10^{-3}$

larger for solubilizates with lower σ_{SJ}. The micellar structural parameters are also affected by the interactions between the solvent S and the solvent-compatible B-block of the copolymer though to a lower degree when compared to the corresponding solubilizate-free systems. The solubilization process was found to decrease the critical micellization concentration (Fig. 14a) and increase the micellar core radius substantially (Fig. 14b). The larger the solubilization capacity, the more significant are the changes in the radius and the CMC. The increase in the core radius results not only from the incorporation of the solubilizate but also because of the increasing number of block copolymer molecules that are accommodated within the micelle. This increase in the aggregation number is more dramatic for a solubilizate whose uptake by the micelles is large. The ratio of shell thickness to micelle radius was found to decrease with increasing solubilization capacity of the micelles, while the shell thickness was not very much affected by solubilization.

The predictions of this theory have been compared against the earlier experimental data of Nagarajan and co-workers [4] on the solubilization of hydrocarbons in aqueous solutions of diblock copolymers. These measurements suggested that the extent of solubilization is dependent on the nature of interactions between the solubilizate and the block that constitutes the core of the micelle. More interestingly, the measurements revealed unusual selectivity in the solubilization behavior when mixtures of hydrocarbons were solubilized; for example, aromatic molecules were found to be solubilized preferentially compared to aliphatic hydrocarbons. In general, the agreement between the experimental and theoretically predicted values of solubilization capacities were satisfactory for all solubilizates studied (aromatic hydrocarbons and straight-chain alkanes). Generalized scaling relations have also been developed that explicitly relate the micelle core radius, the shell thickness, and the volume fraction of the solubilizate within the micelle core to the molecular features of the copolymer, the solvent, and the solubilizate. The domain of the existence of spherical micelles (on the basis of the Nagarajan and Ganesh theory [54]) for the system PEO-PPO copolymer (M_w: 12,500, 50% PEO)-water-benzene was determined. The theory predicted micelles forming in benzene up to 20% water content (with respect to benzene-water mixture) and

FIG. 14 (a) Critical micelle concentration of PEO-PPO micelles containing solubilizates and solubilizate-free micelles. The molecular weight of PEO is constant at 8750, while the molecular weight of PPO is varied as shown. (b) Core radius (R), shell thickness (D), and volume fraction (η) of solubilizate benzene (at saturation) within the core of PEO-PPO micelles in water. The system is identical with that of (a). (From Ref. 54.)

micelles forming in water up to 5% benzene (with respect to benzene-water mixture). The effect of temperature on micelle size, structure, and solubilization (very important for copolymers containing PEO as discussed in Sec. VI) was not considered in the theory of Ref. 54.

B. Self-Consistent Mean-Field Lattice Theory of Solubilization in Block Copolymer Micelles

Another approach in modeling block copolymer micelles has been through the Scheutjens-Fleer theory, an extension of the Flory-Huggins analysis of homogeneous polymer solutions [43], in which the polymer chains are allowed to assume different conformations on a lattice. In the Flory-Huggins analysis, a first-order Markov approximation is used, so that the position of a segment depends only on that of the preceding segment in the chain, the result being that the chain conformation follows the path of a random walk. In addition, a mean-field assumption is invoked to describe the interactions between unlike segments. The free energy of the system is minimized in order to calculate the equilibrium thermodynamic properties of the polymer solution. For spatially inhomogeneous systems such as those containing interfaces, Scheutjens and co-workers [59, 61, 62] restricted the mean field approximation to two dimensions, i.e., within parallel or concentric lattice layers, and applied a step-weighted random walk to account for the inhomogeneities normal to the layers. The polymer and solvent molecules are assumed to be distributed over a lattice, such that solvent molecules and polymer segments occupy one lattice site each. Each polymer chain can assume a large number of possible conformations, defined by the layer numbers in which successive segments are found. There can be many different arrangements for each conformation; if the number of polymer chains in each conformation is specified, the configurational entropy contribution to the system free energy can be evaluated, the other contributions to this free energy being due to the interactions between the polymer molecules, solvent molecules, and the surface, which are characterized by Flory-Huggins χ-parameters. If the free energy of the system is minimized with respect to the number of polymer chains in each conformation, it is possible to calculate the equilibrium segment density profiles. The self-consistent mean-field theory can be used to calculate the segment density profiles in a micelle once the aggregation number of the micelle is known [59, 62, 63]. To find this aggregation number, small-system thermodynamics can be used [64–66], in which the change in the free energy due to the change in the number of micelles (at constant temperature, pressure and number of molecules) must be zero at equilibrium. This "excess free energy" is the sum of the energy required to create micelles and the energy due to the translational entropy of the micelles. Since the segment density profiles are required in order to calculate the energy of micelle formation, an iterative process is employed.

The self-consistent mean field lattice theory of Scheutjens and Fleer [61] has been used to study many colloidal systems, including homopolymer and block copolymer adsorption on surfaces [67], interactions between adsorbed polymer layers [68], and the formation of micelles [62], vesicles [63], and membranes [69]. For various aggregation structures, the detailed segment density profiles are obtained, and macroscopic quantities such as critical micelle concentration, aggregation number, and micelle size can be calculated. Cogan et al. [70] applied the Scheutjens-Fleer theory to describe spherical block copolymer micelles in the presence of two solvents, each selective for one of the blocks. In particular, they performed calculations describing their earlier experimental results of PEO-PS block copolymers in cyclopentane with trace amounts of water. The size of the core block (in their case PEO) was found to dominate the micelle solubilization capacity (Fig. 15). However, the coronal block size can be an important factor if the size change is large enough to significantly modify the copolymer symmetry (Fig. 16). The structural changes illustrated by the constant N_E = 34 series in Fig. 16 correspond to a change in copolymer symmetry from 1:3 to 1:15 [70].

Recently, the Scheutjens and Fleer theory has been extended to study the formation of micelles by linear and branched, starlike amphiphilic block copolymers, and the solubilization of naphthalene in these micelles has been analyzed as a function of polymer structure, composition, and molecular weight [71, 72]. The micellar solubilization of PAHs was investigated by modeling the PAHs as chains of benzene molecules. Fig. 17 shows the segment density profiles for naphthalene solubilized within a micelle formed by Pluronic P104. The naphthalene is confined chiefly to the core of the micelle; the amount of naphthalene solubilized falls off sharply after about the tenth layer, which corresponds to the radius of the homogeneous core region in Fig. 17. Significantly less naphthalene is solubilized within the interfacial region. Fig. 18 compares the experimental results of Hurter and Hatton [30] for the micelle/water partition coefficient (K_{mw}) of naphthalene solubilized in aqueous solutions of Pluronic polymers with model predictions. The theory reproduces the experimental finding of a strong correlation between the PPO content of the polymer and K_{mw}. As the polymer becomes more hydrophobic, the micelle-water partition coefficient increases, as expected for hydrophobic naphthalene. The effect of molecular weight on solubilization is reproduced qualitatively by the theory, but the experimental results show a far larger dependence of K_{mw} on polymer size. The calculations show that higher molecu-

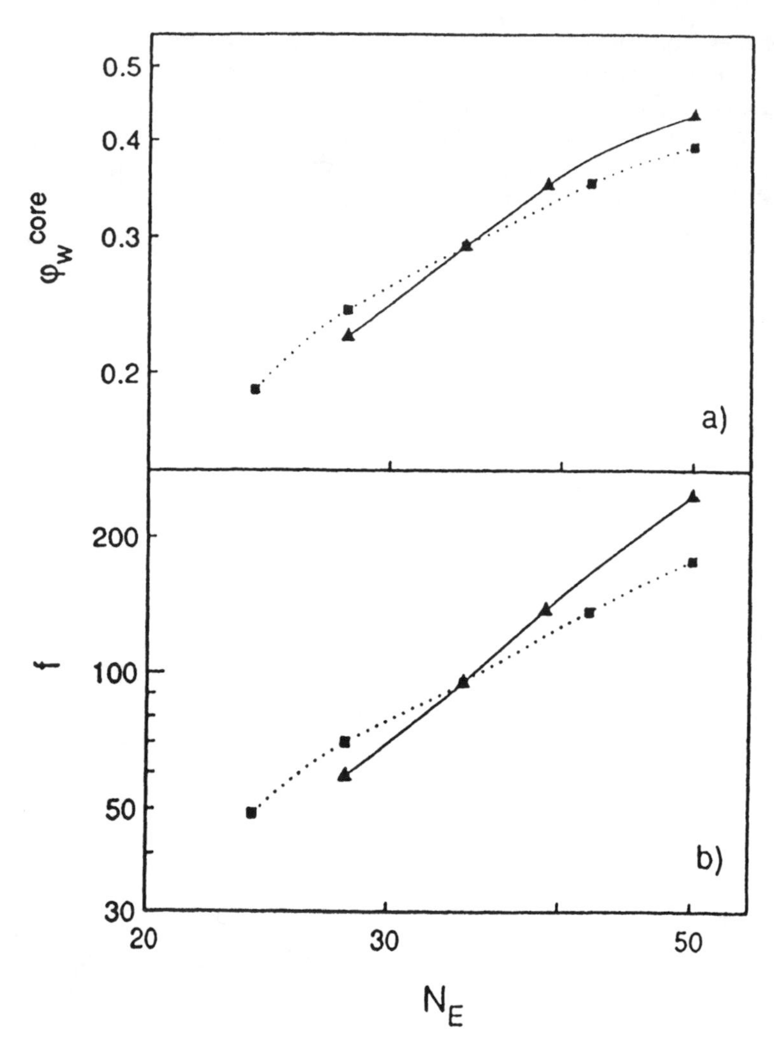

FIG. 15 (a) Volume fractions of water in the micelle cores and (b) aggregation numbers for micelles with varying copolymer composition: ■, constant symmetry series; ▲, varying PEO block, N_E with constant PS block, N_S = 346. (From Ref. 70.)

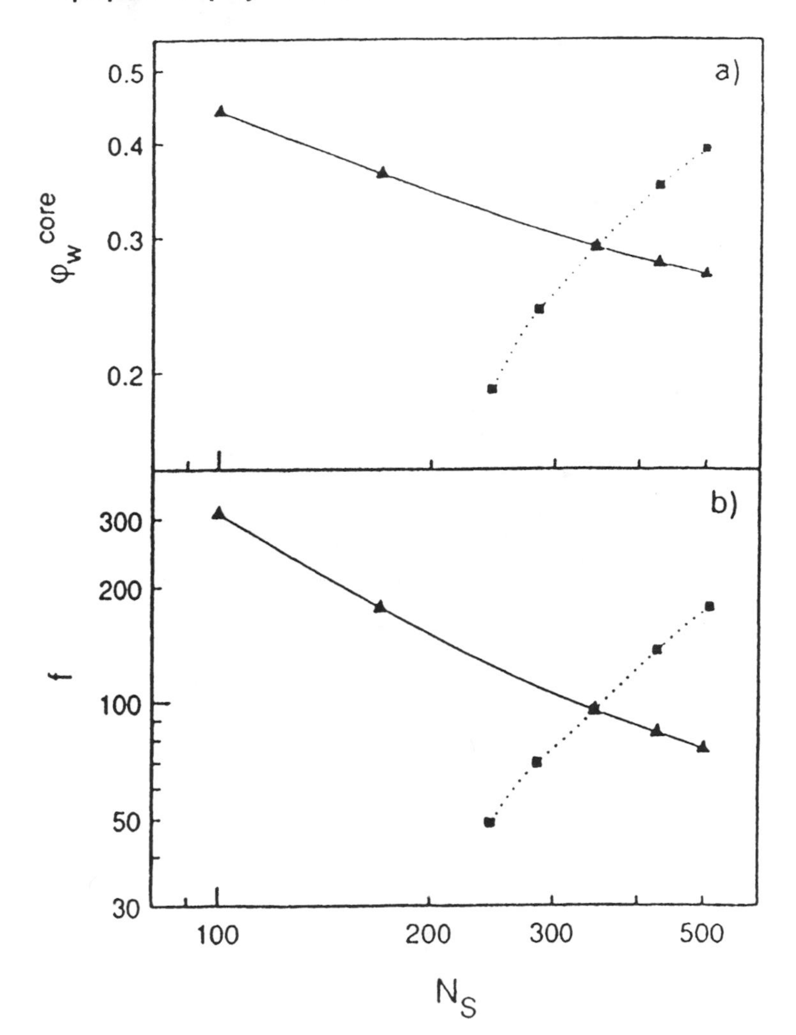

FIG. 16 (a) Volume fractions of water in the micelle cores and (b) aggregation numbers for micelles with varying copolymer composition: ■, constant symmetry series; ▲, varying PS block, N_S with constant PEO block, $N_E = 34$. (From Ref. 70.)

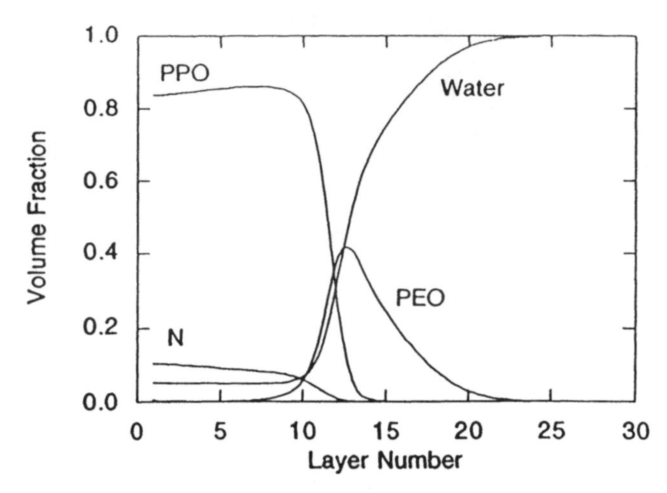

FIG. 17 The segment density profiles for naphthalene solubilized within a micelle formed by Pluronic P104. The polymer concentration is 2%, and the volume fraction of naphthalene in the bulk is 1×10^{-5}. The curve for naphthalene is labeled *N*. (From Ref. 5.)

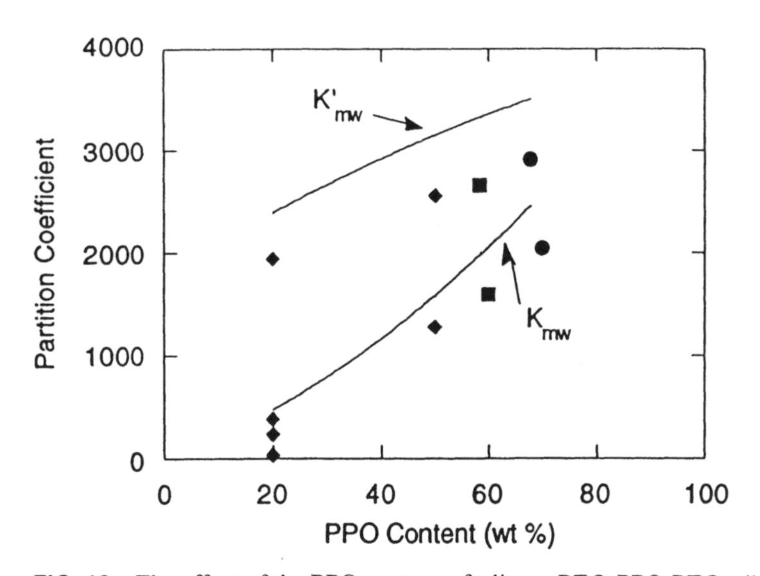

FIG. 18 The effect of the PPO content of a linear PEO-PPO-PEO triblock copolymer on the micelle-water partition coefficient of naphthalene. The data points are experimental results, and the solid lines are theoretical predictions. (From Ref. 5.)

lar weight polymers form larger micelles. The formation of larger micelles will result in a decrease in the interfacial volume compared to the core volume. If naphthalene is effectively excluded from the interfacial region, larger micelles would result in higher micelle-water partition coefficients. The theoretical predictions show that, for both the Pluronic and Tetronic copolymers, the PPO content of the polymer has a strong effect on the micelle-water partition coefficient; this is in agreement with experiments. The theory also confirms the experimental finding that naphthalene partitions more favorably into the linear Pluronic polymers than into the branched Tetronic molecules. The Tetronic molecules aggregate to form micelles with a lower concentration of PPO in the core, resulting in a less hydrophobic core environment, which would inhibit naphthalene solubilization. In addition, the Tetronic molecules form smaller micelles; this is expected to result in a lower micelle-water partition coefficient, as discussed above. The segment density profiles for a micelle with naphthalene solubilized in P104 at 300 K are shown in Fig. 19; the profiles for a solute-free micelle at the same polymer concentration are also shown in Fig. 19.

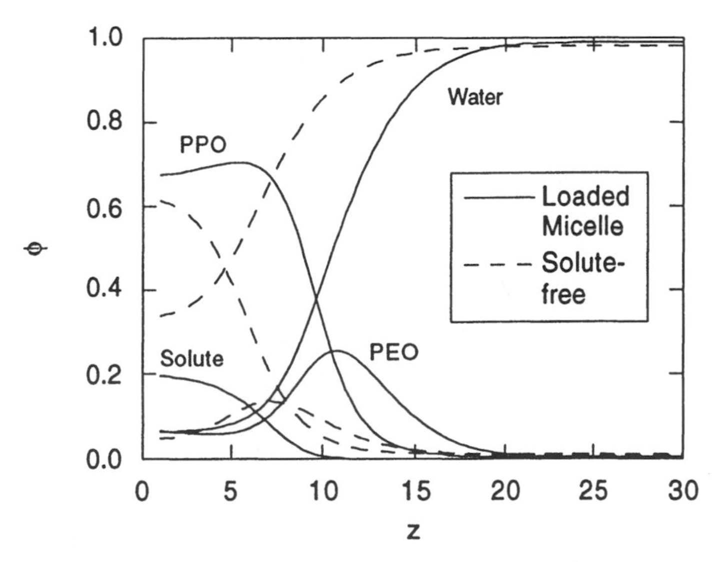

FIG. 19 Segment density profiles for a micelle of $EO_{30}PO_{61}EO_{30}$, with solubilized naphthalene (solid lines), compared to the profiles for a solute-free micelle (dashed lines). The micelle with solubilized naphthalene is close to the saturation point. Average polymer volume fraction in solution is 0.05 in both cases, and the temperature is 300 K. (From Ref. 5.)

The naphthalene is confined chiefly to the core of the micelle. The addition of the solute causes the micelle to become larger, with a lower core concentration of water and a higher concentration of PPO in the core.

C. Lattice Theory for Monomers with Internal Degrees of Freedom

The simple lattice theory for flexible chain molecules cannot capture effects such as the phase behavior of poly(ethylene oxide) and poly(propylene oxide) in water, where a lower critical solution temperature is observed. To predict such behavior, the *gauche* and *trans* bond orientations of the polymer chain must be accounted for. Leermakers [73] used the rotational isomeric state scheme, which accounts for the *gauche-trans* orientations in a chain and eliminates backfolding, combined with the self-consistent field theory to predict the formation of lipid bilayer membranes and lipid vesicles. An approach which is computationally simpler than the rotational isomeric state scheme, but which accounts for the temperature and composition dependence of the interaction parameter χ in a physically acceptable manner, has recently been presented by Karlstrom [50]. This model for PEO recognizes that certain sequences of the *gauche-trans* orientations in a PEO monomer will lead to a polar conformation, while others will be essentially nonpolar. Using this model, the solubility gap in PEO-water and PPO-water phase diagrams can be reproduced.

A model for solubilization in block copolymer micelles has been developed [72] which incorporates Karlstrom's ideas to account for the conformational distribution in PEO and PPO. Using this model, the effect of temperature and the solubilization of naphthalene on the micelle structure have been investigated and the results compared to experimental observations. An increase in temperature caused the model-predicted aggregation number of the micelles to increase and the critical micelle concentration to decrease. The aggregation number increased rapidly with polymer concentration in the dilute regime, but was approximately invariant to polymer concentration at concentrations far from the CMC. Raising the PPO content or the molecular weight of the polymer increased the aggregation number and decreased the CMC (Fig. 20). The conformation of the polymer was affected by both the temperature and the composition of the surrounding solution; both PEO and PPO had a lower fraction of polar segments in the core of the micelles, and the polar fraction decreased with an increase in temperature. The micelle aggregation number was found to increase with an increase in the solute bulk concentration. This is in agreement with the experimental results of Al-Saden et al. [11], who found that the increase in the hydrodynamic radius of Pluronic L64 aggregates

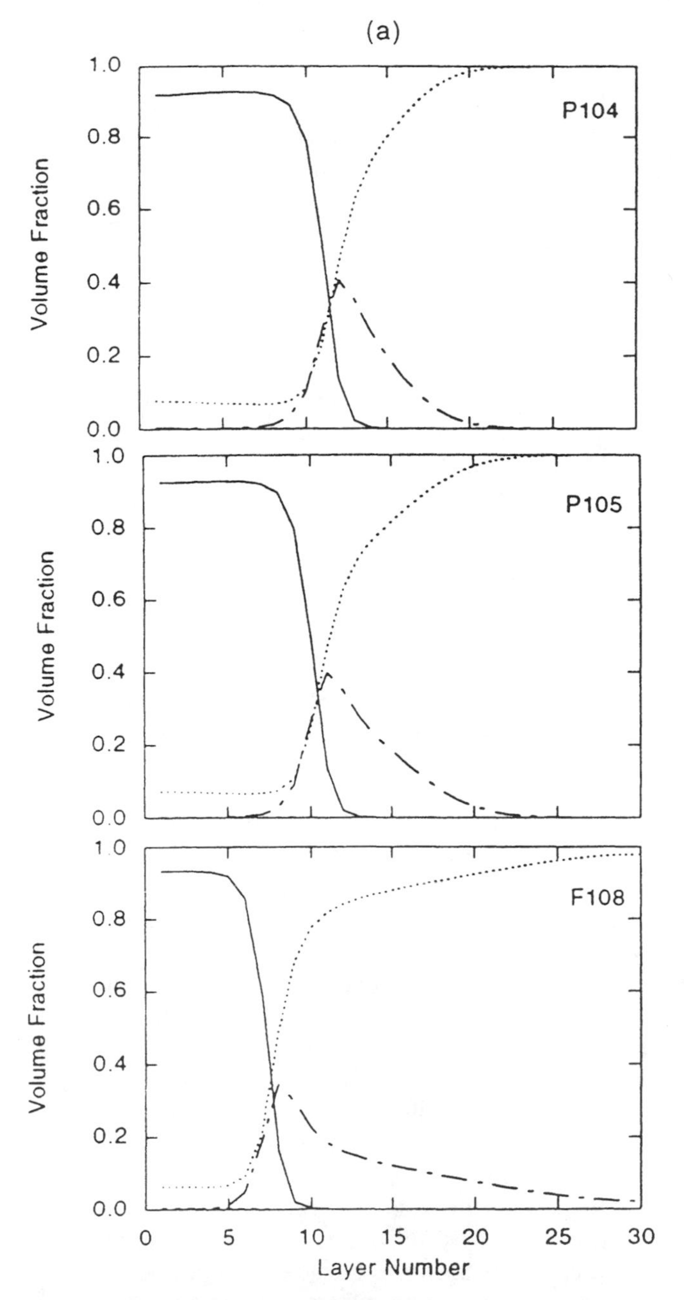

FIG. 20 Effect of PEO content on (a) segment density profiles in the micelles, and (b) micelle aggregation number and size for PEO-PPO-PEO copolymers having the same PPO block size. (From Ref. 71.)

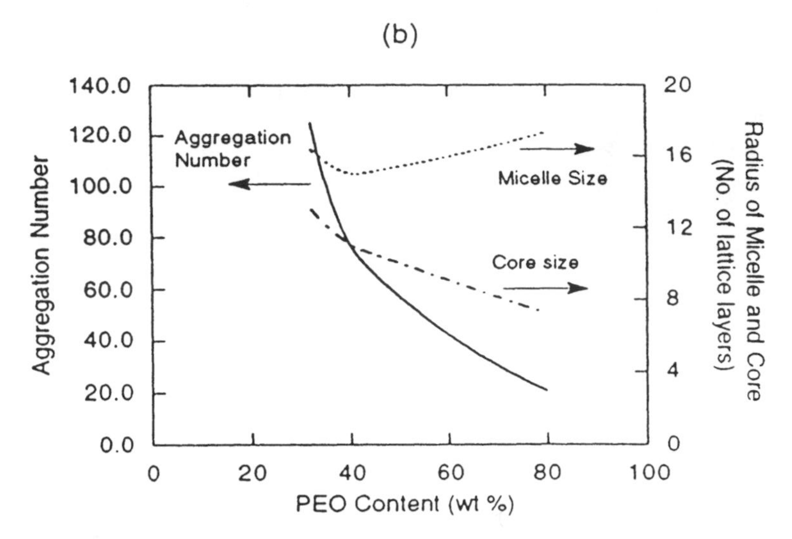

FIG. 20 (Continued)

with hexane uptake was greater than could be explained by the additional volume of the solute. They postulated that the increase in hydrodynamic radius resulted from a combination of an increase in the aggregation number and the incorporation of solute.

The presence of free polymer in the bulk solution enhances the solubility of naphthalene in this region, indicating that the copolymer associates with the naphthalene to some degree [71, 72]. The concentration of polymer in the bulk increases with increasing PEO content of the polymer, so that the naphthalene bulk concentration is higher for the polymers with higher PEO content, despite the fact that naphthalene is more compatible with PPO than with PEO. The polymer bulk (unmicellized) concentration decreases with increasing solute concentration, probably because the addition of a hydrophobic solute increases the tendency of the copolymer molecules to aggregate, resulting in a higher aggregation number and a lower concentration of polymer in the bulk solution. The fact that the polymer bulk concentration changes with solute concentration is one of the contributing factors leading to the nonlinear relationship between the overall solute concentration and the equilibrium concentration of naphthalene in water. This nonlinear relationship indicates that the partition coefficient between water and the micelles changes with the solute concentration, an aspect of the problem that has yet to be investigated experimentally. Again, the theory reproduces the experimental finding that there is a strong correlation between the PPO content of the polymer and its propensity to solubilize naphthalene, although it overpredicts the magnitude of the PPO effect. The agreement could possibly be improved by taking such factors as the statistical segment length and the difference in molecular volumes of the individual segments into account.

To summarize, the theory of Hurter et al. predicts that the partition coefficient of naphthalene increases with an increase in solute concentration, owing to the fact that the uptake of solute changes the structure of the micelle. The polymer aggregation number increases, and a more hydrophobic core is formed, with a lower water concentration and higher concentration of PPO segments. The concentration of polymer in the bulk also decreases upon solute addition. On the other hand, an increase in the polymer concentration results in more micelles being formed, with the same aggregation number, so that for a given bulk concentration of naphthalene, the partition coefficient remains constant with an increase in polymer concentration. These conclusions are illustrated by comparing the change in the micellar profiles upon solute addition and upon changing the polymer concentration. The addition of solute causes the micelles to become larger, as a result of the swelling due to solute addition, as well as an increase in the polymer aggregation number. Water is excluded from the core of the saturated micelles, and the concentration of solute in the core increases. The micellar profiles for naphthalene solubilized in 1.5% P105 and in 10% P105 solutions are essentially indistinguishable.

D. Effect of Copolymer Molecular Weight on Solubilization

It was shown experimentally that the molecular weight of the polymer has a significant effect on the partitioning behavior of naphthalene in Pluronic and Tetronic PEO-PPO copolymers. The lattice theory for flexible chain molecules [71] significantly underpredicted the magnitude of the molecular weight effect. The modified lattice theory [72] improves upon this prediction, as is illustrated in Fig. 21, where the theoretical predictions are compared to the experimental results for Pluronic polymers. The theory overestimates the molecular weight effect somewhat, though the agreement is reasonable. In addition to showing that naphthalene solubilization increases with polymer molecular weight, the theory is able to reproduce the nonlinear relationship between the naphthalene solubilized and the polymer concentration for low molecular weight polymers that was observed experimentally in both Pluronic and Tetronic copolymers.

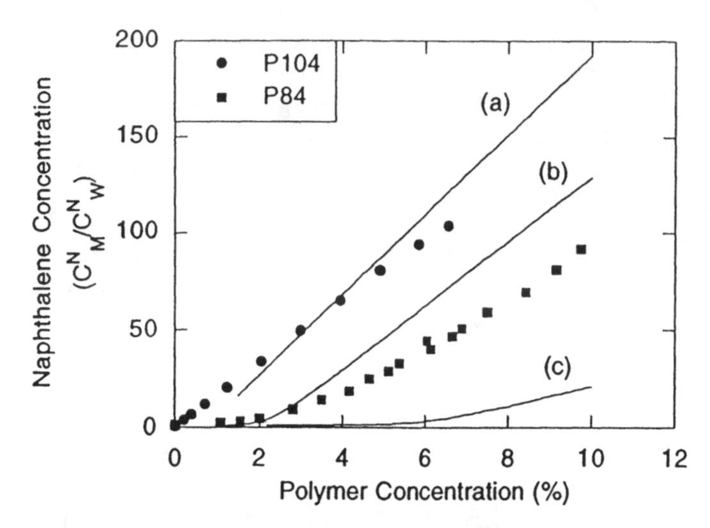

FIG. 21 Experimental results on naphthalene concentration in Pluronic P84 and P104 micellar solutions compared to the theoretical predictions for (a) EO_{30}-$PO_{61}EO_{30}$ (which corresponds to P104), (b) $EO_{25}PO_{51}EO_{25}$, and (c) $EO_{20}PO_{41}EO_{20}$ (which corresponds to P84). (From Ref. 5.)

An analysis of the theoretical results shows that the polymer bulk concentration is not constant for the low molecular weight polymers. For P94, the bulk concentration initially increases but then becomes constant above ~4% polymer. For low molecular weight polymers, micelles form at a higher critical concentration, and micelle properties change near the critical micelle concentration but are essentially constant far from it. The bulk or unmicellized polymer concentration is higher for lower molecular weight polymers, which means less of the polymer aggregates to form micelles, which decreases the amount of naphthalene solubilized and provides a partial explanation for the molecular weight effect. Fig. 22 compares the micellar profiles for naphthalene solubilized in Pluronics P104 and P84. The lower molecular weight polymer forms a smaller micelle, with a large interfacial region, and a heterogeneous core with a high PEO and water content. The core environment is thus much less hydrophobic for the lower molecular weight polymer, resulting in significantly less solubilization of naphthalene. This figure also shows the higher concentration of polymer in the bulk solution, which results in less polymer being available to form micelles.

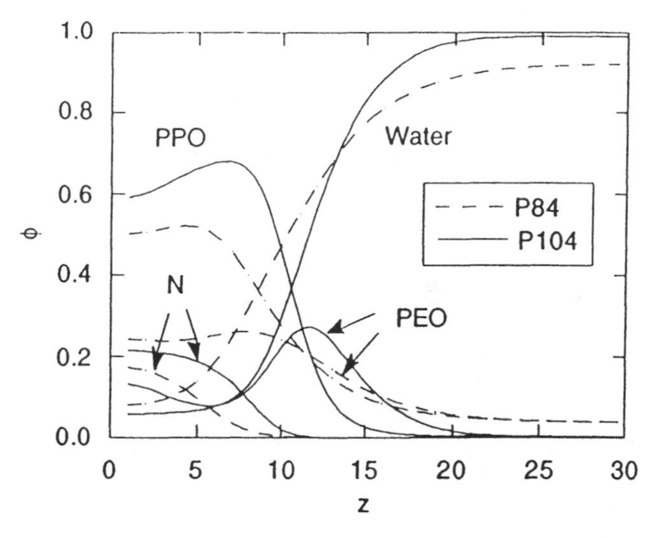

FIG. 22 Comparison of the density profiles of naphthalene solubilized in P84 and P104, Pluronics which have the same PEO/PPO composition but different molecular weights. (From Ref. 5.)

IX. CONCLUSIONS

The solubilization of hydrophobic solutes in block copolymer micelles depends strongly on the structure of the micelle-forming polymers, whether they are linear or branched, the chemical nature of the constituents, and on the hydrophobic to hydrophilic balance in the copolymer. Linear triblock copolymers are more accommodating than branched polymers, which is attributed to their ability to form tighter micelles. It is believed that the configurational constraints on branched polymers in micellar solutions leads to the formation of looser aggregates which do not provide as apolar an environment in the core as found in the case of the linear copolymers. As could be expected, the partitioning of hydrophobic solutes to the micelles is enhanced as the hydrophobic content of the block copolymers increases. Polymer molecular weight is important, with higher molecular weight polymers forming larger micelles and having a higher solubilizing capacity. The partitioning behavior also correlates well with the octanol-water partition coefficient of the solute, K_{ow}. Varying solution conditions such as ionic strength, pH, and temperature all have a significant effect on the micelle structure and, consequently, affect the solubilization behavior.

A self-consistent mean-field theory [5, 71] was able to elucidate the solubilization behavior of hydrophobic solutes in block copolymer micelles by providing detailed information on the microstructure of the micelles formed by linear and branched copolymers. The calculations showed qualitative agreement with experimental predictions on the effect of hydrophobicity and molecular weight of the polymer, and quantitative agreement was remarkably good considering that all parameters were obtained from independent experiments. Modifying the simple lattice theory for flexible chain molecules to allow for polar and nonpolar conformations of the PEO and PPO chains [72] enables the theory to successfully reproduce the anomalous phase behavior of these polymers and improves the ability of the theory to predict more subtle effects such as the molecular weight dependence of naphthalene solubilization in block copolymer micelles. This improvement results from the sensitivity of the equilibrium chain conformations to both solution composition and temperature.

The theory shows that as the temperature is increased the fraction of segments in the nonpolar conformation increases, thus increasing the hydrophobicity of the polymer. It was shown that this leads to exclusion of water from the core of the micelle as the temperature is raised, creating a more hydrophobic environment which is attractive to organic solutes. This is in agreement with the experimental results, which show an increase in solubilization of naphthalene with temperature. It also supports the observations of Al-Saden et al. [11], who found that raising the temperature leads to a decrease in the hydrodynamic radius of the micelle, which they postulated was due to the expulsion of water at higher temperatures.

Comparing the segment density profiles of naphthalene-saturated micelles formed by block copolymers of differing composition shows that the micelle structure is strongly dependent on the composition of the polymer. Polymers with a higher proportion of the hydrophilic block, PEO, tend to form smaller micelles, with a lower concentration of PPO and a higher concentration of water in the core, which results in less naphthalene being solubilized. The amount of naphthalene solubilized is thus not simply dependent on the amount of PPO in the solution but depends on the polymer structure and composition. Similarly, polymers of low molecular weight form micelles with a less hydrophobic core environment and are not as efficient at solubilizing naphthalene. Increasing the PEO content, or decreasing the molecular weight of the polymer, also tends to raise the concentration of free polymer in the bulk solution. This polymer is essentially wasted since it enhances the solubility of naphthalene in the bulk solution to some extent, but to a much lesser degree than the aggregated polymer.

The lattice theory has thus been successful in elucidating the underlying structural effects which lead to the complex solubilization behavior of polycyclic aromatic hydrocarbons in block copolymer micelles. The modified lattice theory has also shown that the micelle properties are strongly dependent on the solute concentration, whereas the only experimental results that were available for comparison were for saturated micelles. It would be interesting to compare the theoretical predictions to solubilization results in the low solute concentration region and to perform light-scattering and neutron-scattering experiments, which would provide more detailed information about the structure of the micelles. These interesting results on solubilization in block copolymer micelles show that the solubilization is affected by the detailed microstructure of the aggregates, which in turn depends on both polymer and solute structure and solution conditions. This allows a great deal of flexibility in designing novel solvents which can be used as extractants and solubilizers in a variety of pharmaceutical, biological, and industrial applications.

REFERENCES

1. S.-Y. Lin and Y. Kawashima, *Pharm. Acta Helv. 60*: 339 (1985).
2. J. H. Collett and E. A. Tobin, *J. Pharm. Pharmacol. 31*: 174 (1979).
3. M. Nowakowska, B. White, and J. E. Guillet, *Macromolecules 21*: 3430 (1988).
4. R. Nagarajan, M. Barry, and E. Ruckenstein, *Langmuir 2*: 210 (1986).
5. P. N. Hurter, Ph.D. thesis, Massachusetts Institute of Technology, Cambridge, MA, 1992.
6. I. R. Schmolka and A. J. Raymond, *J. Am. Oil Chem. Soc. 42*: 1088 (1965).
7. I. R. Schmolka, *J. Am. Oil Chem. Soc. 54*: 110 (1977).
8. R. A. Anderson, *Pharm. Acta Helv. 47*: 304 (1972).
9. W. Saski and S. G. Shah, *J. Pharm. Sci. 54*: 71 (1965).
10. K. N. Prasad, T. T. Luong, A. T. Florence, J. Paris, C. Vaution, M. Seiller, and F. Puisieux, *J. Colloid Interface Sci. 69*: 225 (1979).
11. A. A. Al-Saden, T. L. Whateley, and A. T. Florence, *J. Colloid Interface Sci. 90*: 303 (1982).
12. W. Brown, K. Schillen, M. Almgren, S. Hvidt, and P. Bahadur, *J. Phys. Chem. 95*: 1850 (1991).
13. Z. Zhou and B. Chu, *J. Colloid Interface Sci. 126*: 171 (1988).
14. G. Wanka, H. Hoffmann, and W. Ulbricht, *Colloid Polym. Sci. 266*: 101 (1990).
15. N. J. Turro and C.-J. Chung, *Macromolecules 17*: 2123 (1984).
16. P. Alexandridis, J. F. Holzwarth, and T. A. Hatton, *Macromolecules 27*: 2414 (1994).
17. P. Bahadur and N. V. Sastry, *Eur. Polym. J. 24*: 285 1987.
18. G. Riess and D. Rogez, *ACS Polym. Prepr. 23*: 19 (1982).
19. R. Xu, M. A. Winnik, F. R. Hallett, G. Riess, and M. D. Croucher, *Macromolecules 24*: 87 (1991).

20. R. Xu, M. A. Winnik, G. Riess, B. Chu, and M. D. Croucher, *Macromolecules 25*: 644 (1992).
21. M. Wilhelm, C.-L. Zhao, Y. Wang, R. Xu, M. A. Winnik, J.-L. Mura, G. Riess, and M. D. Croucher, *Macromolecules 24*: 1033 (1991).
22. Y. Gallot, J. Selb, P. Marie, and A. Rameau, *ACS Polym. Prepr. 23*: 16 (1982).
23. P. L. Valint Jr. and J. Bock, *Macromolecules 21*: 175 (1988).
24. T. Cao, P. Munk, C. Ramireddy, Z. Tuzar, and S. E. Webber, *Macromolecules 24*: 6300 (1991).
25. K. Prochazka, D. Kiserow, C. Ramireddy, Z. Tuzar, P. Munk, and S. E. Webber, *Macromolecules 25*: 454 (1992).
26. Z. Tuzar, S. E. Webber, C. Ramireddy, and P. Munk, *ACS Polym. Prepr. 32*: 525 (1991).
27. M. Ikemi, N. Odagiri, S. Tanaka, I. Shinohara, and A. Chiba, *Macromolecules 14*: 34 (1981).
28. M. Ikemi, N. Odagiri, S. Tanaka, I. Shinohara, and A. Chiba, *Macromolecules 15*: 281 (1982).
29. A. Tontisakis, R. Hilfiker, and B. Chu, *J. Colloid Interface Sci. 135*: 427 (1990).
30. P. N. Hurter and T. A. Hatton, *Langmuir 8*: 1291 (1992).
31. P. N. Hurter, L. A. Anger, L. J. Vojdovich, C. A. Kelley, R. E. Cohen, and T. A. Hatton, in *Solvent Extraction in the Process Industries* (D. H. Logsdail and M. J. Slater, eds.), Elsevier Applied Science, London, 1993, Vol. 3, p. 1663.
32. P. Alexandridis, J. F. Holzwarth, and T. A. Hatton, manuscript in preparation, 1994. P. Alexandridis, Ph.D. thesis, Massachusetts Institute of Technology, Cambridge, MA, 1994.
33. K. Pandya, P. Bahadur, T. N. Nagar, and A. Bahadur, *Colloids Surf. A 70*: 219 (1993).
34. A. V. Kabanov, E. V. Batrakova, N. S. Melik-Nubarov, N. A. Fedoseev, T. Yu. Dorodnich, V. Yu. Alakhov, V. P. Chekhonin, I. R. Nazalova, and V. A. Kabanov, *J. Controlled Release 22*: 141 (1992).
35. I. N. Topchieva, S. V. Osipova, and V. A. Polyakov, *Colloid J. USSR 52*: 347 (1990).
36. W. R. Haulbrook, J. L. Feerer, T. A. Hatton, and J. W. Tester, *Environ, Sci. Technol. 27*: 2783 (1993).
37. M. Nowakowska, B. White, and J. E. Guillet, *Macromolecules 22*: 3903 (1989).
38. M. Nowakowska, B. White, and J. E. Guillet, *Macromolecules 23*: 3375 (1990).
39. M. Nowakowska and J. E. Guillet, *Macromolecules 24*: 474 (1991).
40. E. Sustar, M. Nowakowska, and J. E. Guillet, *J. Photochem. Photobiol. A: Chemistry*, *53*: 233 (1990).
41. D. Kiserow, K. Prochazka, C. Ramireddy, Z. Tuzar, P. Munk, and S. E. Webber, *Macromolecules 25*: 461 (1992).
42. L. Leibler, H. Orland, and J. C. Wheeler, *J. Chem. Phys. 79*: 3550 (1983).

43. P. Flory, *Principles of Polymer Chemistry*, Cornell University, Ithaca, NY, 1953.
44. R. Nagarajan and K. Ganesh, *J. Chem. Phys. 90*: 5843 (1989).
45. J. Noolandi and K. M. Hong, *Macromolecules 16*: 1443 (1983).
46. D. A. Edwards, R. G. Luthy, and Z. Liu, *Environ. Sci. Technol. 25*: 127 (1991).
47. K. T. Valsaraj and L. J. Thibodeaux, *Water Res. 23*: 183 (1989).
48. C. McDonald and C. K. Wong, *J. Pharm. Pharmac. 26*: 556 (1974).
49. G. N. Malcolm and J. S. Rowlinson, *Trans. Faraday Soc. 53*: 921 (1957).
50. G. Karlstrom, *J. Phys. Chem. 89*: 4962 (1985).
51. A. A. Samii, D. Karlstrom, and B. Lindman, *Langmuir 7*: 1065 (1991).
52. M. Nowakowska, B. White, and J. E. Guillet, *Macromolecules 22*: 2317 (1989).
53. M. R. Munch and A. P. Gast, *Macromolecules 21*: 1360 (1988).
54. R. Nagarajan and K. Ganesh, *Macromolecules 22*: 4312 (1989).
55. A. Halperin, *Macromolecules 20*: 2943 (1987).
56. C. M. Marques, *Macromolecules 21*: 1051 (1988). D. Izzo and C. M. Marques, *Macromolecules 26*: 7189 (1993).
57. A. N. Semenov, *Sov. Phys. JETP 61*: 733 (1985).
58. Y. B. Zhulina and T. M. Birshtein, *Polym. Sci. USSR* (*Engl. transl.*) *12*: 2880 (1986).
59. B. van Lent and J. M. H. M. Scheutjens, *Macromolecules 22*: 1931 (1989).
60. Y. Wang, W. L. Mattice, and D. H. Napper, *Langmuir 9*: 66 (1993).
61. J. M. H. M. Scheutjens and G. J. Fleer, *J. Phys. Chem. 83*: 1619 (1979).
62. F. A. M. Leermakers, P. P. A. M. van der Schoot, J. M. H. M. Scheutjens, and J. Lyklema, in *The Equilibrium Structure of Micelles* (K. L. Mittal, ed.), Plenum, 1989.
63. F. A. M. Leermakers and J. M. H. M. Scheutjens, *J. Phys. Chem. 93*: 7417 (1989).
64. D. G. Hall and B. A. Pethica, Chapter 16 in *Nonionic Surfactants* (M. J. Schick, ed.), Marcel Dekker, 1967.
65. T. L. Hill, *Thermodynamics of Small Systems*, Vol. 1, Benjamin, 1963.
66. T. L. Hill, *Thermodynamics of Small Systems*, Vol. 2, Benjamin, 1964.
67. O. A. Evers, J. M. H. M. Scheutjens, and G. J. Fleer, *Macromolecules 23*: 5221 (1990).
68. J. M. H. M. Scheutjens and G. J. Fleer, *Macromolecules 18*: 1882 (1985).
69. J. M. H. M. Scheutjens, F. A. M. Leermakers, N. A. M. Besseling, and J. Lyklema, in *Surfactants in Solution*, Vol. 7 (K. L. Mittal, ed.), Plenum, 1989.
70. K. A. Cogan, F. A. M. Leermakers, and A. P. Gast, *Langmuir 8*: 429 (1992).
71. P. N. Hurter, J. M. H. M. Scheutjens, and T. A. Hatton, *Macromolecules 26*: 5592 (1993).
72. P. N. Hurter, J. M. H. M. Scheutjens, and T. A. Hatton, *Macromolecules 26*: 5030 (1993).
73. F. A. M. Leermakers, Ph.D. thesis, Wagenigen Agricultural University, Wagenigen, 1988.

7

Kinetics of Solubilization in Surfactant-Based Systems

ANTHONY J. WARD Department of Chemistry, Clarkson University, Potsdam, New York

SYNOPSIS

Interfacial mass transfer kinetic measurements require the simultaneous monitoring of interfacial area and volume as a function of time. Few experimental methods that have been applied to the study of solubilization kinetics have been well-defined in these terms. The drop-on-fiber and rotating liquid disk methods are described, which provide such definition for dilute systems in the former case, and more concentrated in the latter (concentration being defined in terms of the amount of oil solubilized per unit volume of surfactant). In the case of highly dilute alkane systems, the mechanistic information indicates a concerted surfactant monomer adsorption-desorption step being rate determining for solubilization. A complex situation dependent upon the rheology and total amount of oil solubilized is found for systems where the amount of solubilized oil in the surfactant solution is high.

I. INTRODUCTION

Solubilization of nominally insoluble material is a necessary process occurring in a wide range of phenomena found in pharmaceutical delivery systems, animal digestive processes, and the sensory processes of smell and taste. Much work done on the fundamental equilibrium aspects of solubilization has been limited to determinations of the equilibrium or saturation solubilization capacities of systems and to the effects of system variables on these quantities. In contrast, comparatively little attention has been focused upon the kinetics of the process with consequently little being known definitively about its mechanism. In these areas and others such as topical drug delivery, detergency, and enhanced oil recovery, a knowledge of the kinetics of the solubilization process is of great importance. Here a knowledge of the time scale of the process is often more important than the final equilibrium state of the system; i.e., it is not necessarily the total amount, determined by the ultimate thermodynamic equilibrium state of the system, that is important but the rate of change of composition in the system. This lack of available kinetic data arose because of the relatively slow nature of the process and the experimental difficulties in following time evolution of the requisite volume and interfacial parameters.

A great deal of interest concerning solubilization phenomena relates to the development of formulations using surfactants as vehicles for transdermal drug delivery. Much of the attraction of such surfactant-based systems lies in the acceptable rheological and wide-ranging physicochemical properties which can be obtained. The objects of system design are to achieve a suitable reservoir dosage and controlled release via percutaneous absorption. Contact between such vehicles and the skin inevitably entails modification of the skin's barrier function either by hydration changes or structural changes in the molecular arrangements in the stratum corneum. This is a result of absorption of components either from vehicle into the skin, or vice versa. Although thermodynamic equilibrium properties of the system will control the ultimate state of the system, the time scales of the various kinetic processes are usually of more importance for consideration in practical applications. Such systems focus attention on the kinetics of the transfer of components across oil/water interfaces in systems containing surfactant aggregation structures. The amount of available data is small when compared to the corresponding literature relating to equilibrium properties of surfactant phases, but an understanding of the mechanisms of mass transfer across interfaces in such systems is becoming clearer.

Despite the obvious importance of mechanistic information relating to the kinetics of the solubilization process, there have been relatively few studies which have been published. Some studies using rotating discs [1, 2] have been made of fatty acid solubilization in aqueous surfactant solutions and the role of interfacial liquid crystal formation has been established. Turbidity measurements [3, 4] have been used to study effects such as those due to the molecular architecture of the surfactants. The lack of available information owes, in part, to the difficulties in performing unambiguous and well-defined experiments which establish the time dependencies of both the phase volume and interfacial area. Some of these problems have been overcome with use of the captive drop-on-fiber technique [5, 6], which has been applied [6–11] to the study of oil solubilization in aqueous surfactant solutions.

The advances that have been made in recent years in understanding the mechanisms of oil solubilization in aqueous solutions of surfactants has been partly a result of the captive drop-on-fiber technique [5, 11] which allows observation in a well-defined high–surface area/volume system. The method overcomes the problems associated with the conventional methods using emulsions and the need to measure simultaneously the volume and interfacial area of the system as a function of time, which is a tedious and time-consuming process—a situation responsible for the almost complete lack of mechanistic information about the solubilization process. Some attempts to obtain kinetic information have been made [3, 4] using observations of turbidity; however, these methods usually cannot be reliably related to the solubilization process quantitatively because phenomena relating to particle size and coalescence also affect the value of turbidity.

The purpose of this chapter is to review the available information in the context of what it tells us about processes such as solubilization, interfacial liquid crystal formation, and diffusion which control the overall kinetic behavior of systems containing surfactant aggregates.

II. EXPERIMENTAL BACKGROUND

A. Passive Drop-on-Fiber

One of the problems of quantitatively examining systems such as emulsions to obtain kinetic data and to infer mechanistic information has been that of simultaneously defining the volume and interfacial area of the system as a function of time. Studies of stability in formulations to determine the factors involved in their time evolution are usually somewhat qualitative and system specific. In an emulsion, for example, coalescence and flocculation processes may contribute to interfacial area decrease in addition to any solubilization processes. An attempt to overcome some of the inherent difficulties in such measurements has been the development of the passive *drop-on-fiber* [5, 6] technique. This defines the system by fixing in space one droplet of emulsion size using an inert cylindrical fiber. Provided the distortions from gravity are negligible, the shape of the axisymmetric drop (Fig. 1) is purely determined by capillary forces. An analysis of this type of system [5] shows that the rate of solubilization into solution is related in a simple fashion to the relative dimensions of the droplet and fiber. Both the volume, V, of the drop and the interfacial area, A, can be obtained and used to calculate the rate from the general relationsip:

$$\text{Rate} = \mathrm{A}\frac{dV}{dt} \tag{1}$$

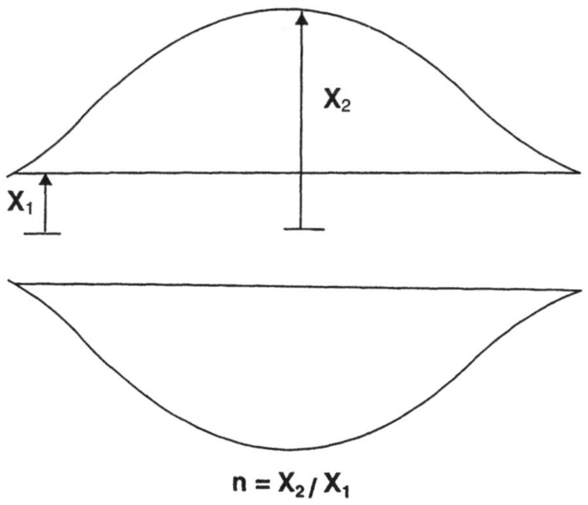

X_2

X_1

$n = X_2 / X_1$

FIG. 1 Profile of the drop-on-fiber system with characteristic parameters.

The way in which the experiment has been performed has mainly been concerned with systems where the amount of oil in the drop is extremely small compared to the amount required to saturate the micelles. In this respect, the data obtained are mainly concerned with the *initial kinetic* process so that contributions from micelles containing oil can be neglected.

III. RESULTS AND DISCUSSION

The results obtained using this technique can primarily be broadly discussed for systems characterized in terms of solubility in water of the solubilizate.

A. Insoluble Oil/Aqueous Micellar Surfactant

Application of the drop-on-fiber technique [5–8] to the study of initial solubilization kinetics of highly water insoluble oils into aqueous surfactant solutions has yielded consistent mechanistic descriptions. In this context, "insoluble" may arbitrarily be gauged as oil solubilities less than 10^{-6} wt.% in water. Pseudozero (Fig. 2) order kinetics were observed [5–8] for single-component oils solubilizing into different surfactants above their critical micellization concentrations (CMC), whereas the rates were not observable below the CMC (Fig. 3). This behavior was considered [5] in terms of the following possible diffusion mechanisms:

1. Diffusion of oil into water is related to the water solubility of the oil $C_{o,w}$ as

$$\text{Rate} = \frac{D_{o,w}C_{o,w}}{\delta} \tag{2}$$

   where $D_{o,w}$ is the diffiusion coefficient of the oil and δ is the diffusion layer thickness (Fig. 4). This mechanism was shown [3] to be significant only for oils with solubilities in water greater than ca. 10^{-6} wt.%.
2. A process limited by the rate of micellar diffusion to the oil/water interface with mass transfer of oil to micelle.

The rates of solubilization predicted by this mechanism are also too high compared to the experimentally observed values for water-insoluble oils. It was argued [5] that this was because the dissociation of the micelle, being a precursor to adsorption with consequent desorption of the mixed micelle containing oil, did not occur for every excursion of micelles into the interfacial region. In order to take this into account, a mechanism for the solubilization of sparingly soluble solubilizates was proposed [6]. In

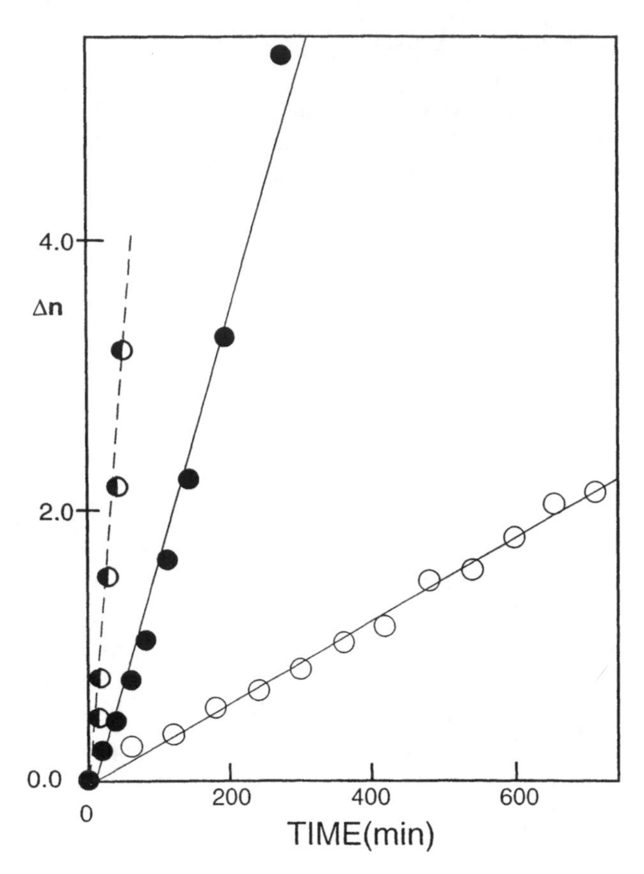

FIG. 2 Solubilization stage of oil droplets at 298 K as a function of time in 1% (w/w) surfactant solutions: ◐, dodecane in $C_{12}EO_6$; ●, nonane, ○, dodecane in *n*-dodecyldimethylammoniopropane sulfonate (DDAPS).

this, micelles diffuse to the interface where they either dissociate and adsorb or simply diffuse away intact (for micelles, not being surface active, are not expected to adsorb as entities) (Fig. 5). Those that dissociate and adsorb (as monomers) trigger by their adsorption a concerted desorption from the interface of an equivalent number of monomers in the form of micelles containing a certain quantity of the solubilizate. As only the diffusion of micelles *not* containing solubilized material to the oil-water interface is considered, and the model is pertinent only to initial rates, the experimental conditions described above were so arranged that it is this stage which was studied.

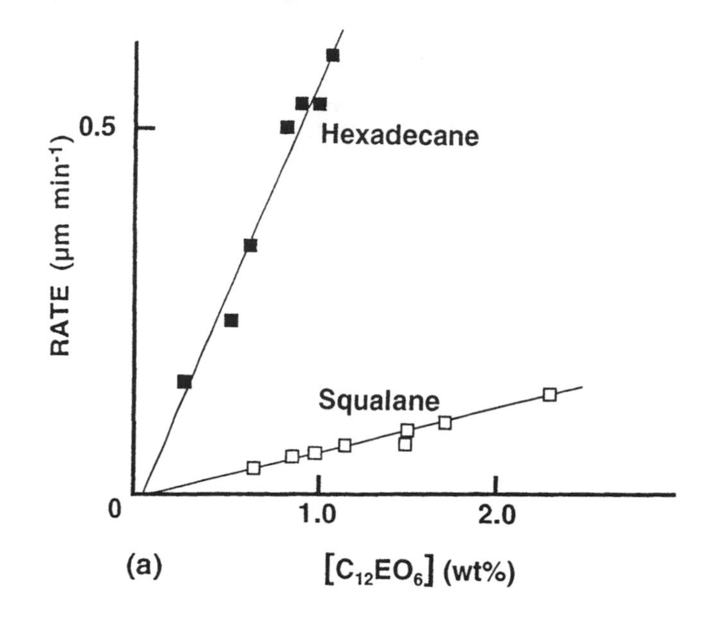

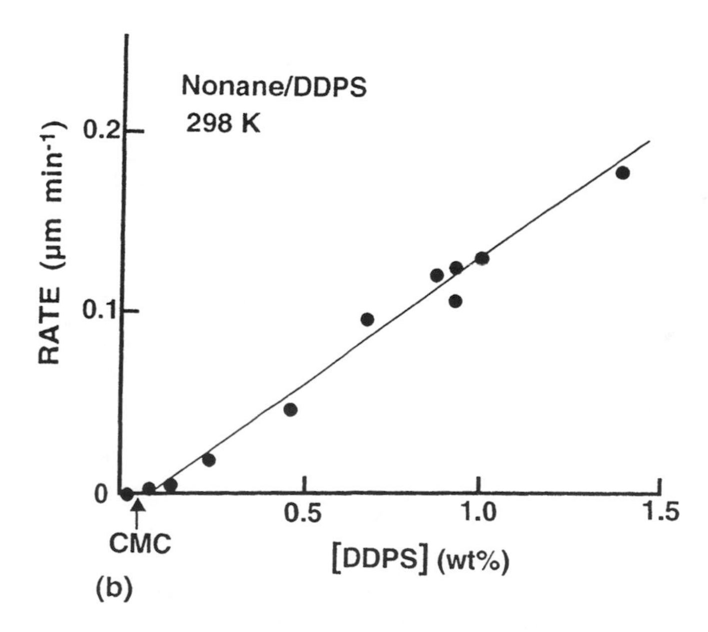

FIG. 3 Solubilization rate as a function of surfactant concentration for insoluble oils: (a) hexadecane and squalane solubilizing in $C_{12}E_6$; (b) nonane solubilizing in DDAPS.

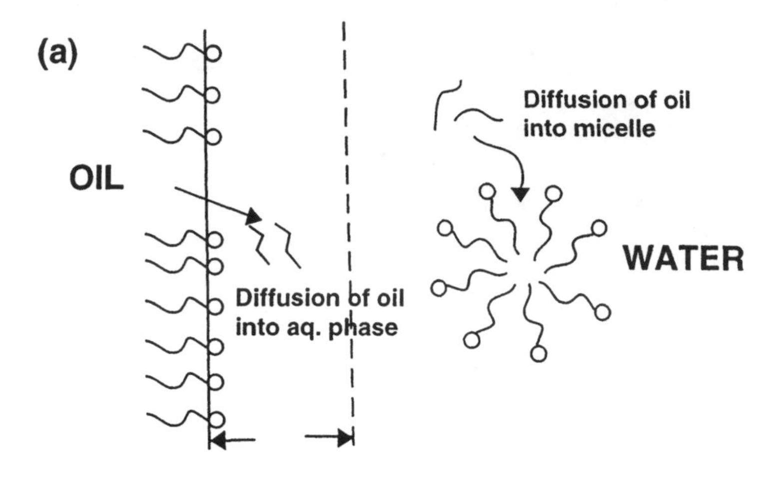

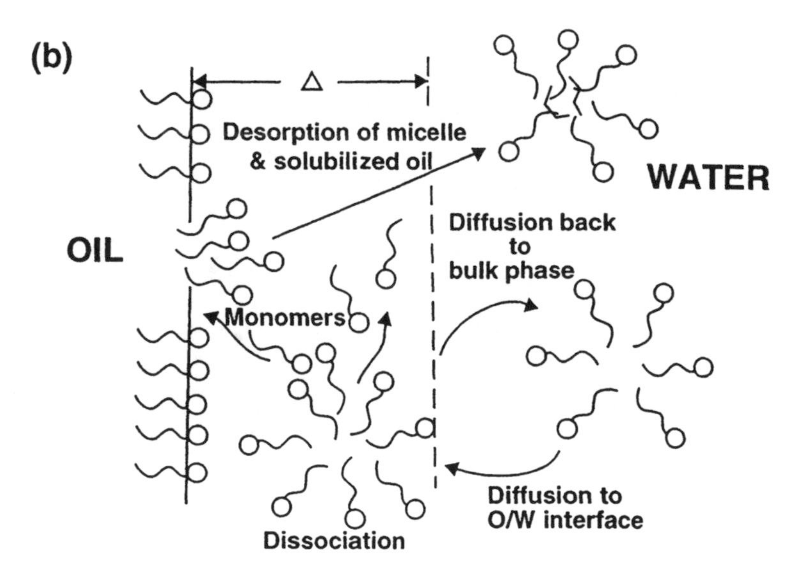

FIG. 4 Possible schemes for interfacial mass transfer across an oil/water interface: (a) passive diffusion of oils with significant water solubility; (b) micellar dissociation-association near to the oil/water interface in systems of water insoluble oils ($<10^{-6}$ wt.%).

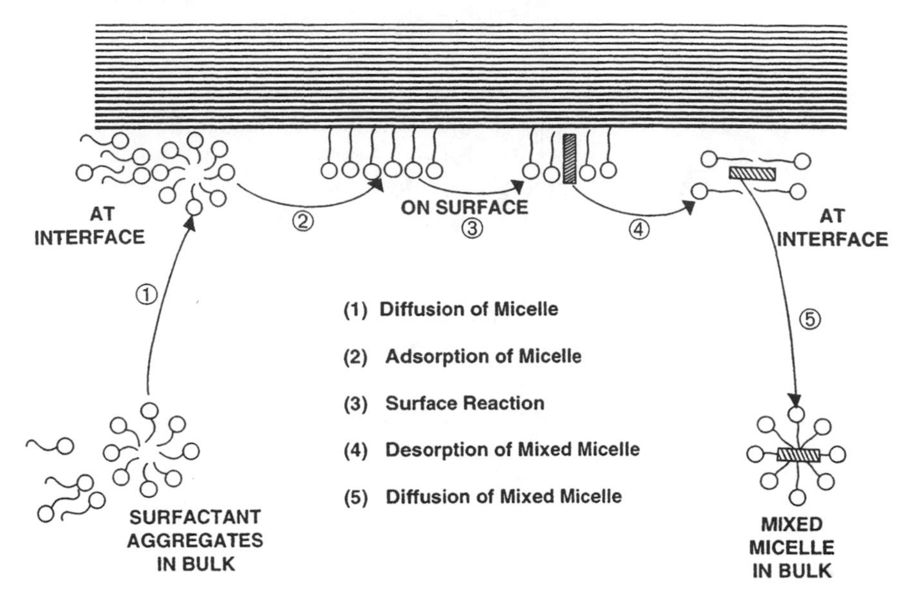

FIG. 5 Scheme for mass transfer at the aqueous micellar surfactant/oil interface.

The argument behind this mechanistic picture is based on consideration of the fluxes in the context of the micellar properties. Thus, the rate of solubilization can be written as the product of the flux, J_{out}, of micelles desorbing from the interface in unit time and the number of solubilizate molecules, b, carried by each of these micelles. J_{out} is related to J_{in}, the flux of micelles which actually adsorb, by the relation of $aJ_{out} = a_o J_{in}$, where the number of monomers in the desorbing and adsorbing entities are respectively a and a_o. The flux J_{in} can be expressed in terms of a diffusion layer thickness Δ, of order of micelle size, the bulk micelle concentration $(c - c_o)/a_o$ (where c_o is the critical micelle concentration) and the fraction of micelles which dissociate while in the interfacial region, which depends on the micelle decay constant τ_2. J_{in} has the final form [6]

$$J_{in} = \frac{V_m \Delta (c - c_o)}{a_o \tau_2} \tag{3a}$$

so that J_{out} has the form

$$J_{out} = V_m \Delta (c - c_o) a \tau_2 \tag{3b}$$

where V_m is the molar volume of the solubilizate.

The assumption is made that the amount of solubilizate, b, associated with a desorbing micelle is proportional to the equilibrium solubilization capacity, with the same proportionality constant for different oils. However, this is not necessarily true: the quantity b calculated in this way may differ from the equilibrium quantity for two reasons; first, the aggregation number, and hence the b value [12] of the desorbing micelle may differ from that of the fully equilibrated species and, second, the shape of the latter is certain to be quite different from the shape of the desorbing micelle at the stage where it parts company with the interface. The kinetic b is thus related to the equilibrium value b_s by a relation of the form $b = b_sG(p, q)H(s)$, where G is a function of the aggregate numbers a_o, a and a_s of the "empty," desorbing and equilibrated micellar species ($p = a/a_o$; $q = a_s/a_o$) and H is a function of the shapes of the micelle at the desorption stage. The overall rate of the process, the flux of oil into solution, is thus

$$F = b_sG(p, q)H(s)J_{out} \tag{4}$$

which becomes on use of the relations $a_oJ_{in} = aJ_{out}$ and $b_s = a_sB_s$, where B_s is the equilibrium capacity expressed in terms of molecules of oil per molecule of active,

$$F = \frac{a_oa_s}{a} B_sG(p, q)H(s)J_{in} \tag{5}$$

$$= F_s \frac{a_s}{a} G(p, q)H(s) = F_sG'(p, q)H(s) \tag{6}$$

where F_s is defined as the "standard" flux prevailing when $a = a_s$, i.e., $F_s = a_oB_sJ_{in}$; and $G'(p, q) = (a_s/a)G(p, q)$. Thus under the conditions where the oil flux is F_s it is expected that for two oils the ratio of the measured rates should be in the ratio of the respective B_s values, or that the slopes of the lines in Fig. 2 be proportional to this quantity at the same temperature. The explanation for deviations from this rule lies in deviations of the product of the correction factors, $G'(p, q)H(s)$, from unity.

An estimate of the shape factor, $G(p, q)$, can be made using a simple geometric argument. Let the aggregation number of the "empty" micelle be a_o, that of the desorbing micelle be a and that of the fully equilibrated micelle be a_s. If α is the area per headgroup of active while in the micelle, the surface area of the micelle is αa_o for "empty" (adsorbing) micelles, with similar expressions for the other two cases. The micelle volume is thus $ka_o^{3/2}$ (or $ka^{3/2}$, or $ka_o^{3/2}$) in the case of spherical micelles. If it is assumed that the amount of oil, b, associated with a micelle is related to the difference in volume between this and the "empty" micelle, then

$$b = k'(a^{3/2} - a_o^{3/2}) \tag{7}$$

If b_s refers to the saturation value of b ($a = a_s$), it is clear that

$$b = \frac{b_s(a^{3/2} - a_o^{3/2})}{a_s^{3/2} - a_o^{3/2}} \tag{8}$$

$$= b_s(p^{3/2} - 1)(q^{3/2} - 1) \tag{9}$$

where $p = a/a_o$ and $q = a_s a_o$. The factor $G(p, q)$ is evidently the coefficient of b_s in Eq. (9) and the factor $G'(p, q)$ has the form

$$G'(p, q) = \frac{q}{p}\frac{p^{3/2} - 1}{q^{3/2} - 1} = \frac{p^{1/2} - p^{-1}}{q^{1/2} - q^{-1}} \tag{10}$$

The ratio of these factors for two oils is thus

$$\frac{G_1}{G_2} = \left(\frac{p_1^{1/2} - p_1^{-1}}{p_2^{1/2} - p^{-1}}\right)\left(\frac{q_2^{1/2} - q_2^{-1}}{q_1^{1/2} - q_1^{-1}}\right) \tag{11}$$

The function $x^{1/2} - x^{-1}$ is easily tabulated for a series of values of x (= p or q) in the range pertinent to the problem (say $x = 1$ to 5). The parameter a_o is the same for both oils so that corresponding values of a and a_s are readily evaluated.

Given the conditions $a_o^1 = a_o^2$, $a_s^1 > a_s^2$ and $a^1 > a^2$, it is possible to obtain ratios greater than unity from Eq. (11) in two ways, which are respectively when $a_o < a < a_s$ and when $a_o < a_s < a$. In the former case $G' < 1$; in the latter case $G' > 1$. In the latter case supersaturation of the micelles occurs; such a case has been discussed [13].

Some experiments on the solubilization of hexadecane and squalane into a nonionic surfactant micellar system [7] indicated the ratio of the terms $G'(p, q)H(s)$ for the two oils to be about 2. It can be seen from the above that $G'(p, q)$ by itself could account for the factor of 2 for reasonable combinations of the aggregation numbers of the micelle species present.

The shape factor H is much more difficult to quantify, but an indication of its magnitude was provided by comparison of the B_s data obtained for hexadecane and squalane solubilization in micelles with similar data for the uptake of the same two oils by planar black films of glyceryl monooleate reported by [14]. For the micellar case, the ratio of hexadecane to HMT is about 7.5:1, whereas in the latter systems this ratio is 25:1. It is therefore easy to believe that the factor H may differ appreciably from unity for a micelle forming at the oil/water interface.

Overall, however, the product of the two terms G and H, although probably never unity, is nevertheless not expected to alter the theoretical rate by as much as one order of magnitude. It is therefore possible, by using the expression already given for J_{in} and the various experimental data, to obtain an order of magnitude estimate of the quantity Δ, which is given by the expression

$$\Delta = \frac{2a\tau_2}{bV_m}\frac{dF}{dc} \tag{12}$$

Rearrangement of Eq. (12) to show the factor describing the probability of the micelle dissociating within the vicinity of the oil/water interface leading to solubilization has to be included [6], giving the relation

$$\text{Rate} = \frac{\Delta}{2\tau}\frac{V_m b(C - \text{CMC})}{a} \tag{13}$$

where Δ is the thickness of the interfacial region (Fig. 5) which has dimensions of the order of the micelle diameter, τ is the time interval between micellar dissociations which lead to adsorption and concerted solubilization steps; V_m is the molar volume of the solubilizate, and b/a is micellar capacity for the solubilizate (i.e., moles oil/mole surfactant). The term $\Delta/2\tau$ represents the probability of a micellar dissociation within the interfacial region leading to monomer adsorption with a concerted desorption transporting oil molecules into a mixed micelle.

Equation (13) has been used (6–11) to describe the behavior of alkanes solubilizing into aqueous micellar solutions of various surfactant types. The term b/a is found to be proportional to the equilibrium solubilization capacity of the micelle for the homologous series of n-alkanes [10] containing 8 to 16 carbon atoms. Comparisons between oils of differing architecture [8], on the other hand, indicated the constant of proportionality between the equilibrium and kinetic values of the solubilization capacity to be different, possibly reflecting differences in the oil packing in the micelle.

For all cases of solubilization of single-component oils into aqueous surfactant solutions that have so far been studied, the value of n varies linearly with time (Fig. 2). Solubilization rates were derived [5] from the slopes of the Δn ($= n(t) - n(0)$) against time plots. In the case of oils with negligible water solubility, no rates were measurable below the critical micellization concentration (CMC) with approximately linearly increasing values above this concentration (Fig. 3). These observations are consistent with Eq. (13) and add support to its validity; similarly, decreasing the molecular size of the oil leads to an increase in solubilization rate (Fig. 6) at a fixed concentration of micellar surfactant. A detailed examination of the rate behavior for the homologous series of n-alkanes (C_nH_{2n+2}) for $n \geq 6$ showed the rate of increase becomes more rapid as the number of

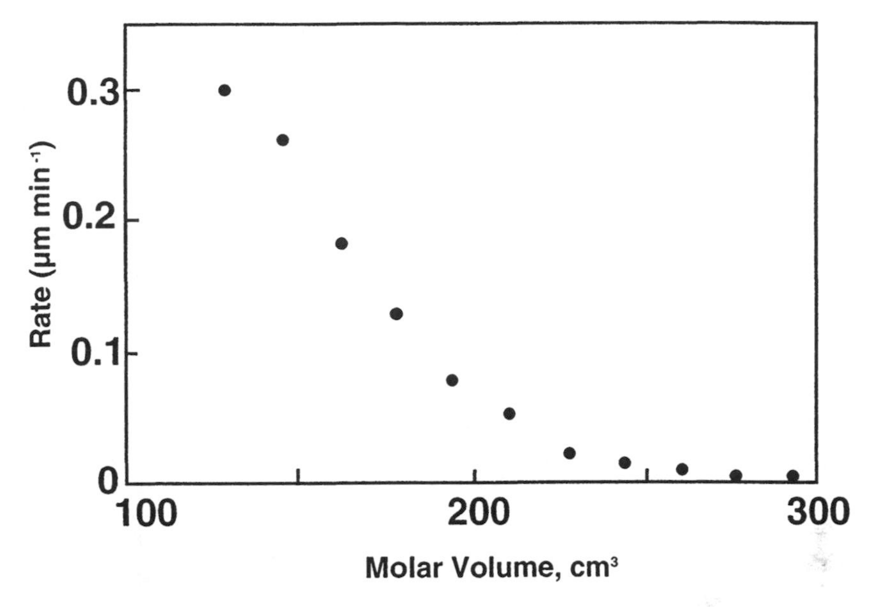

FIG. 6 Solubilization rate as a function of the molar volume of solubilizate in 1% DDAPS at 298 K.

carbon atoms is reduced below 12 in the case of the *n*-alkanes. Values for the rate were calculated assuming the kinetic values of b/a to be proportional to $(b/a)_{eq}$ using Eq. (13) and normalizing to the $\Delta/2\tau$ value derived from the rate of *n*-tetradecane solubilization. Good agreement (Fig. 7) was observed between the experimental and calculated values for the homologs in the range $10 \leq n \leq 16$.

This agreement does not mean that the desorbing mixed micelles necessarily contain the equilibrium amounts of solubilized oil but that the constant of proportionality relating $(b/a)_{kinetic}$ to $(b/a)_{eq}$ is the same for each member of the series. The steric factors determining the packing of the solubilizate molecules in the micelle interior are likely to be similar for such a homologous series and is implied by the analysis. In the previous case of squalane and hexadecane a different packing requirement would be expected for squalane which is highly branched. Thus, although it is the effective micellar volume available to the solubilizate, as also inferred by Chiu et al. [3], which determines the observed rate, the details of the actual mixing have to be considered when comparing different oils. Ultimately, this will be expressed in terms of the ideality or nonideality of the solubilizate mixing in the micelle.

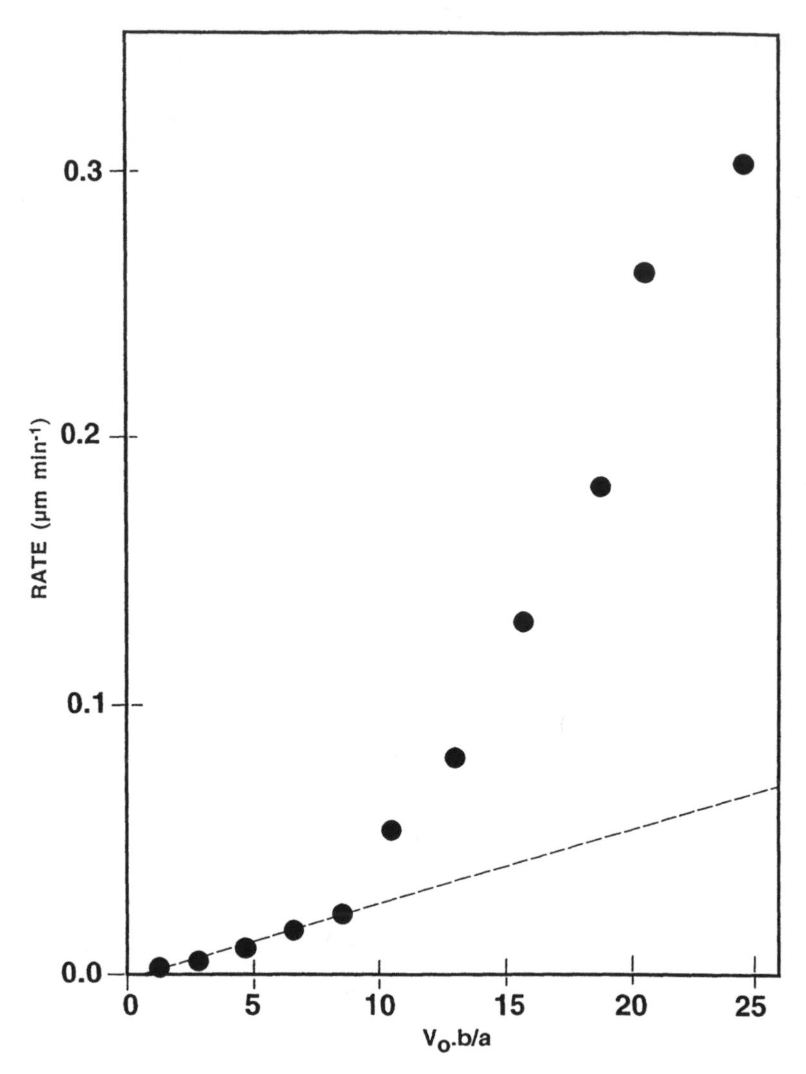

FIG. 7 Solubilization rate plotted as a function of the product of the equilibrium solubilization capacity (moles of oil/mole of micellized surfactant) and the molar volume of the oil for *n*-alkanes solubilizing in DDAPS.

The calculated values for the rates were found to be slightly lower than those obtained experimentally for chain lengths with $n \leq 10$. It is probable that for these shorter chains there is a contribution from a diffusion mechanism involving micelles and oil in the aqueous medium [6] arising from the higher water solubility of the shorter alkanes. Increased rates are also found (Fig. 8) for alkanes solubilizing into micellar solutions of nonionic surfactants as the temperature is increased into the region of the cloud temperature.

One conseqence of the proposed mechanism is that the rate is essentially independent of stirring in the system, unlike the passive diffusion process described by Eq. (2). Here the value of δ is decreased with the consequent rate increased by increased stirring. A similar conclusion has been reached more recently [15, 16] from the description of interfacial kinetics in experiments involving contacting phases with different states of surfactant aggregation. A linear increase in the aqueous micellar phase with time in contact with a lamellar phase was found. If the rate of mass transfer from the micellar phase had been diffusion limited, a square root of time dependence of the layer thickness would have been expected. In this case, the restructuring to produce surfactant monomers from the

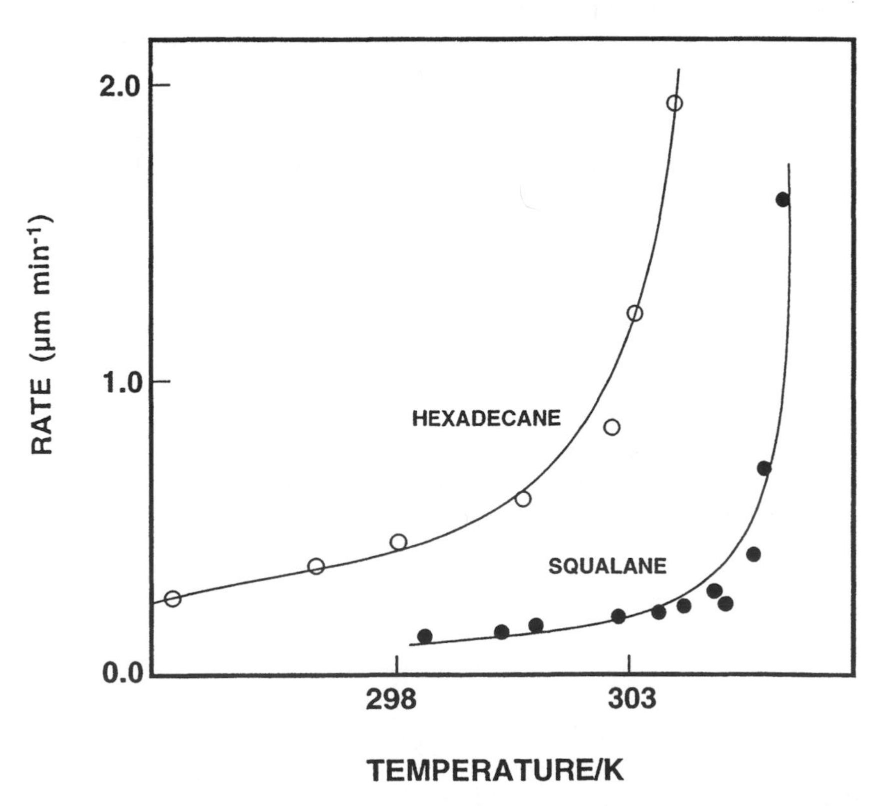

FIG. 8 Temperature dependence of solubilization rate: hexadecane and squalane solubilizing in $C_{12}EO_6$.

micellar dissociation in passage into the lamellar phase was regarded as being the rated limiting step.

B. Soluble Oil/Aqueous Micellar Surfactant

Oils which have sufficient water solubility to produce a significant mass transfer across the oil/water interface are more likely to produce kinetics dependent upon mechanisms in which a diffusion step is rate determining. Preliminary studies [9, 17] of systems containing benzene as either a single component or a component of a binary mixture have indicated this to be the case. Benzene is soluble to the extent of ca. 1500 ppm in water and any oil which has a water solubility greater than ca. 10^{-5} wt.% may be regarded as soluble in this context. Oil transport across the liquid/liquid interface is that essentially of solubility and is governed by the passive diffusion Eq. (2). Rates determined by the drop-on-fiber experiment are for systems under conditions of minimal stirring and therefore do not represent rates which are at the diffusion limit; i.e., the diffusion layer thickness, δ, is on the order of the dimensions of the oil molecule. Some rates typically found for soluble oils dissolving in water are presented in Table 1. It is interesting to note that the diffusion layer thicknesses derived in these experiments are similar to those found [18] in membrane processs *in vivo*. Further the ratios of the rates is the same as the ratio of the oil solubilities in water as required by Eq. (2).

Equation (2) shows the rate to be independent of the surfactant concentration inasmuch as it does not affect the activity of the oil in the aqueous phase. A surfactant concentration scan of the rate of benzene in contact with aqueous solutions of sodium dodecyl sulfate (Fig. 9) shows no increase until the total surfactant concentration is above the CMC. This observation may be understood in terms of the relative sizes of the oil "sinks" provided by the bulk water and the micelles; thus only when the volume of micelles available for solubilization is comparable to or greater than that from the water solubility will there be an observable effect from

TABLE 1

Oil	Rate (m min^{-1})	Diffusion layer thickness (μm)
Benzene	2.4×10^{-6}	50
Toluene	0.7×10^{-6}	55
p-Xylene	0.18×10^{-6}	55

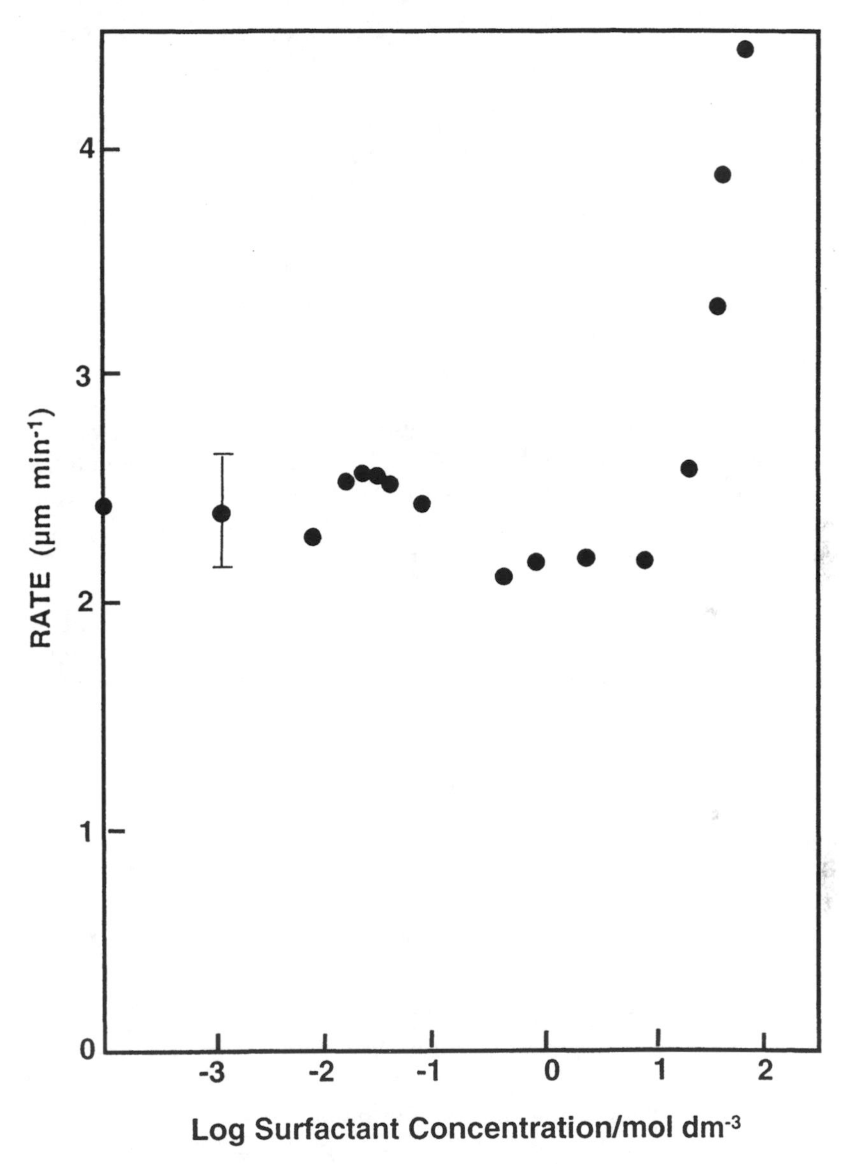

FIG. 9 Solubilization rate of benzene as a function of sodium dodecylsulfate (SDS) concentration (normalized to the CMC) at 298 K.

the surfactant on the observed rate. The concentration at which this occurs will obviously depend on values of the water solubility, CMC and micelle volume.

C. Solubilization in Mixed Surfactant Systems

Most formulations based on surfactants used in the detergency, cosmetic and pharmaceutical areas consist of mixtures of different surfactant types. In this way, the range of utility is enlarged over that obtained from pure single-surfactant systems. A particular aspect which has been of interest is that of solubilization which depends on the concentration of micelles present. In systems of mixtures of surfactants large deviations from ideality in the micellar mixing process has been observed [19–25] which have been discussed in terms of regular solution theory approximations [19, 21] or a phase separation approach [23, 25]. A result of such deviations from ideal mixing is that there exists a range of total surfactant concentrations where the CMC of the mixed system is lower than the CMC values of either pure components. This phenomenon, known as synergism, can be large under the correct conditions of mixture composition and electrolyte concentration giving CMC values from the mixtures which are two to three orders of magnitude smaller than either of the pure constituents.

A further development relating to the packing requirements of the micelle/solubilizate aggregate has been made [26, 27] in the consideration of solubilization from binary mixtures of aliphatic oils. It was shown [28], within the restriction of no preference in the solubilization of the oil components (i.e., the composition of the solubilizate mixture remaining the same as the bulk contacting oil phase), that relationships of the following form should apply:

$$b_1 = \kappa_1\{(\kappa^{o}_{1b} + \kappa^{o}_{2b}) - \kappa_1\kappa_2 P\} \tag{14a}$$

$$b_2 = \kappa_2\{(\kappa^{o}_{1b} + \kappa^{o}_{2b}) - \kappa_1\kappa_2 P\} \tag{14b}$$

where b_i^o are the solubilization capacities of the pure oils and κ_i are the oil mole fractions in the binary mixtures. The value of the factor P can vary between zero and $(1 + [b_1^o/b_2^o])$, the former value representing ideal mixing of the solubilizate in the micelle interior. Comparison of data determined from kinetic experiments with values determined by equilibrium [8] show agreement in the value of P required to fit the data in some cases, whereas in others different values are needed. The solubilization of oil from dodecane/dexadecane mixtures into micelles of the nonionic surfactant *n*-dodecyl hexaoxyethylene glycol ether, $C_{12}EO_6$, [28] shows good agreement (Fig. 10) with a P value of 1.3, i.e., very close to ideal mixing. In contrast, solubilization from similar mixtures into a micellar zwitterionic

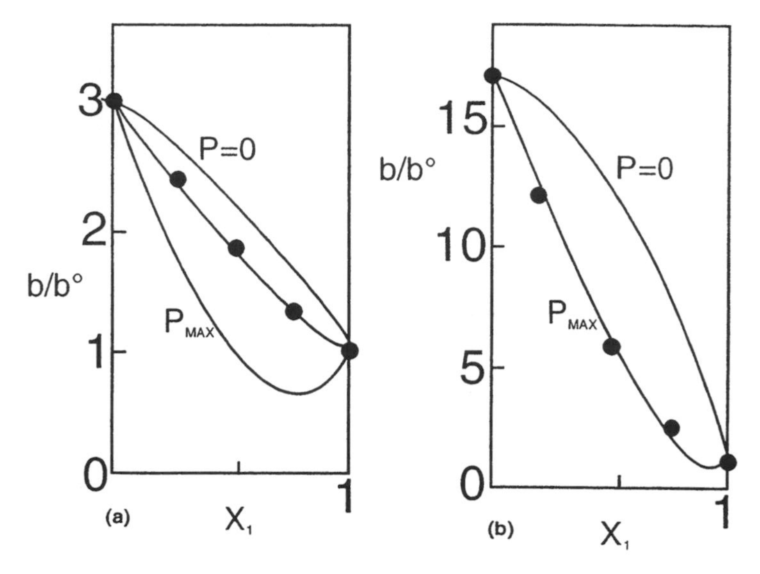

FIG. 10 Composition dependence of solubilization for dodecane:hexadecane mixtures at 298 K in (a) $C_{12}EO_6$ and (b) DDAPS.

surfactant, *n*-dodecyl dimethylammonium propane sulfonate, shows (Fig. 10) that while the equilibrium data require $P = 0$ (zero) the kinetic data are only fitted if the maximum possible value of P is used. It may be suggested that this discrepancy may reflect micelle shape changes in the first stages of the solubilization process which are not considered in the derivation of Eq. (14). A further manifestation of micellar shape changes occurring in the initial stages of solubilization is the nonlinear temperature dependencies observed [7, 8] for the rates in the region of the phase inversion temperature of some nonionic surfactants.

D. Solubilization in Binary Surfactant Mixtures

As part of a program of investigating the kinetics of solubilization to derive associated mechanisms, a study of alkane solubilization into anionic/nonionic and cationic/nonionic surfactant mixed micellar solutions has been carried out. Some preliminary results have been presented representing, to the best of our knowledge, the first kinetic study of a mixed surfactant system. The surfactants sodium dodecylsulfate and dodecyltrimethylammonium bromide were chosen as the ionic component of binary mixtures with the nonionic surfactant *n*-dodecyl hexaoxyethylene glycol ether because of their similar micellar aggregation numbers [29, 30] of 62 and 50 monomers/micelle, respectively. Drop size-time profiles were measured for nonane solubilizing into micellar solutions of SDS and DDTAB (Fig. 11) and were found to be linear. The rates derived from these plots were studied as a function of the mixture composition at fixed total surfactant concentration (Fig. 12). Initial addition of the ionic component to pure

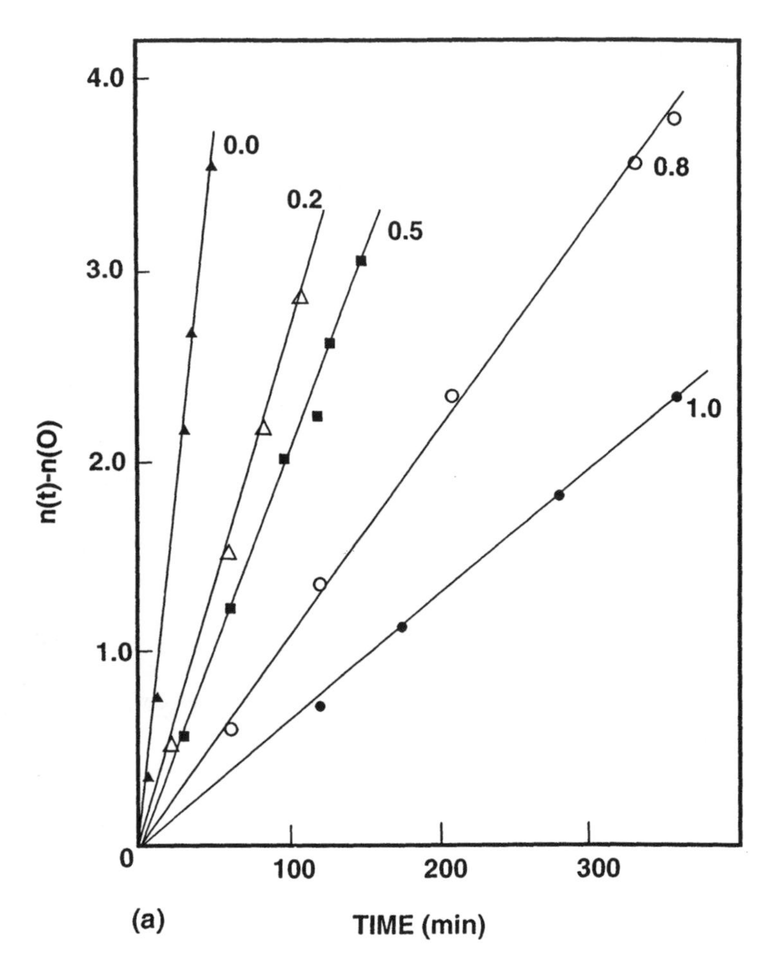

FIG. 11 Time evolution of $\{n(t) - n(0)\}$ for nonane solubilizing in (a) SDS/$C_{12}EO_6$ and (b) *n*-dodecyltrimethylammonium bromide (DDTAB)/$C_{12}EO_6$ mixtures at a total surfactant concentration of 0.025 M. (Figures on curves represent the mole fraction of ionic surfactant in the mixture.)

nonionic micelles leads to a rapid decrease in the rate. Above mole fractions of the ionic component (X_{ionic}) of 0.2–0.3 there is little further decrease in the rate. The behavior is reflected also in the variation of the micellar solubilization capacities of the systems (Fig. 13) with surfactant composition. An expression analogous to Eq. (13) may be written as

$$(\text{Rate})_{\text{mix}} = \left(\frac{\Delta}{2\tau}\right)_{\text{mix}} (C - \text{CMC})_{\text{mix}} \tag{15}$$

where the subscripts refer to the mixed surfactant systems. Values for

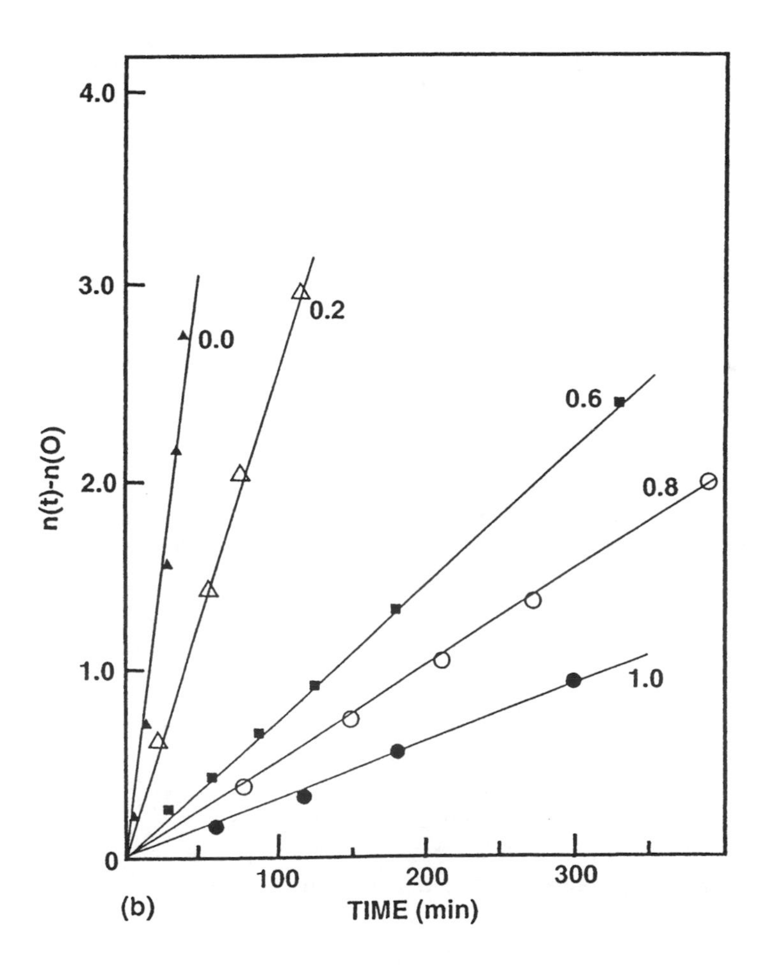

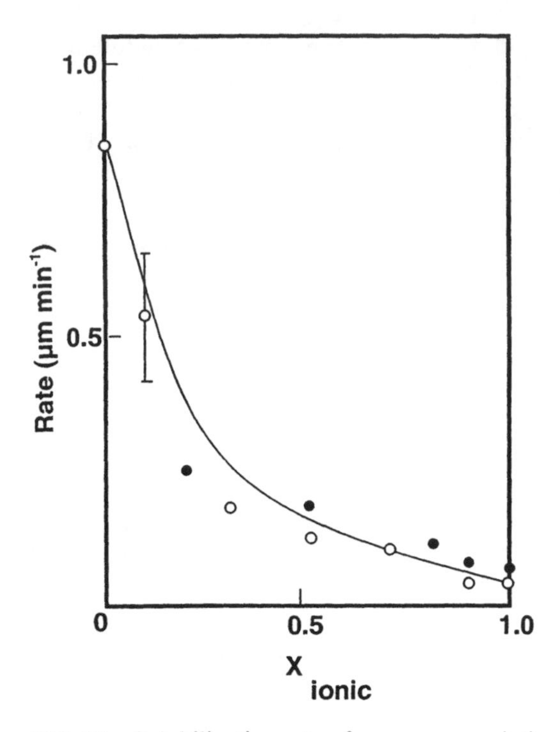

FIG. 12 Solubilization rate of *n*-nonane; variation with the composition of surfactant mixture at fixed total surfactant concentration (0.025 M) at 298 K: ●, DDTAB/$C_{12}EO_6$; ○, SDS/$C_{12}EO_6$; ——, calculated rate (Eqs. (15) and (16)).

$(ESC)_{mix}$ were measured at equilibrium and were used making the assumption that the kinetic value, i.e., the value under the conditions of the initial kinetics, was proportional to the equilibrium value. As discussed previously, this has been indicated [7–10] to be the case for simple alkane solubilizates. In the case of $\Delta/2\tau$ the assumption of a weighted average of the two pure values, $\Delta/2\tau_{ionic}$ and $\Delta/2\tau_{nonionic}$, was made in a derivation using Eq. (12) as follows:

$$\left(\frac{\Delta}{2\tau}\right)_{mix} = X_{ionic}\left(\frac{\Delta}{2\tau}\right)_{ionic} + (1 - X_{ionic})\left(\frac{\Delta}{2\tau}\right)_{nonionic} \tag{16}$$

Good agreement between rates estimated using Eqs. (15) and (16) is seen (Fig. 12) with the observed experimental values.

Both Δ and τ are not immediately accessible from experiment; however, an estimate of the Δ value can be made using the observed solubilization

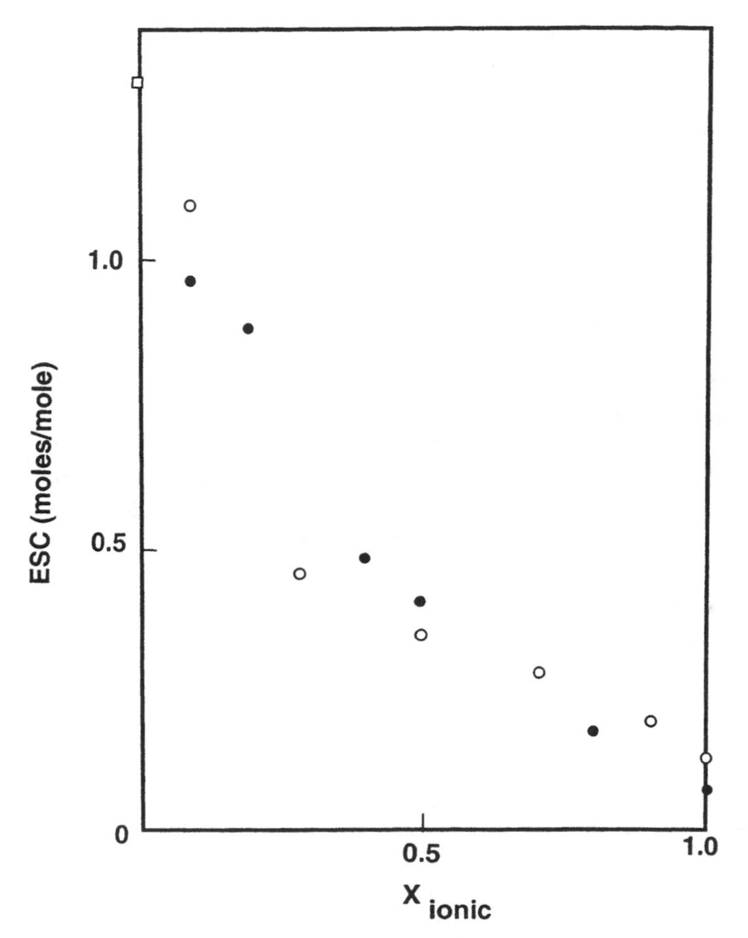

FIG. 13 Equilibrium solubilization capacity for *n*-nonane as a function of mixed surfactant composition: ○, DDTAB/$C_{12}EO_6$; ●, SDS/$C_{12}EO_6$.

capacity data (Fig. 13). It has been shown [31, 32] that a two-state model can be used to describe solubilization of molecules, such as *n*-alkanes, which locate in the micellar interior [33]. The mixing process of the solubilizate in the micelle may be regarded as ideal so that the excess pressure it produces, P, may be written as

$$PV = RT \ln X \tag{17}$$

where X is the mole fraction of solubilizate in the micelle and V is identified with the molar volume of the oil. If it is now assumed that the micelle interior can be described as a spherical droplet of radius, r, enclosed by an interface of tension, σ, this excess pressure acting across the interface is given by the Laplace equation; i.e.,

$$p = \frac{2\sigma}{r} \tag{18}$$

The value of Δ can then be equated with twice the value of r. Δ values of 2.5 nm and 9.2 nm were obtained for SDS and $C_{12}EO_6$ pure micelles, respectively, using a value or 18 mN m^{-1} as previously suggested [32]. These compare favorably with values of 3.4 nm and 15 nm derived [34] from light scattering measurements on these systems. Alternatively, if the data are normalized to give a value of 3.4 nm for SDS micelles, a σ value of 24.5 mN m^{-1} must be used. This yields a value of 12.5 nm for the estimated diameter of $C_{12}EO_6$ micelles and the values for the mixed surfactant systems are shown in Fig. 14. The agreement between the estimated and light-scattering values is remarkably good, but it should be remembered that micelles of this size are not spherical and this degree of agreement, therefore, must be coincidental. Similar micellar sizes are indicated in the composition range $0 < X_{\text{ionic}} < 0.3$ for both SDS and DDTAB mixtures with $C_{12}EO_6$, whereas at $X_{\text{ionic}} > 0.3$ the mixed systems with the cationic are consistently lower values in line with previous [31, 35] light-scattering results.

Values for the average time, τ, between successful micellar dissociations within the interfacial region described by Δ (i.e., those which lead to monomer adsorption at the oil/water interface with resulting concerted desorption of mixed oil and surfactant) calculated from the experimental rates and the estimated Δ values are given in Fig. 15. A value of ca. 1×10^{-3} s results for the case of pure SDS micelles, 2×10^{-3} s for $C_{12}EO_6$ and 3×10^{-4} s for DDTAB. The value for SDS can be compared to the relaxation times for micellar processes derived [36] from kinetic spectroscopy which show a slow relaxation time, τ_2, of 1.8×10^{-3} s and fast relaxation time, τ_1, of 1×10^{-4} s. The slow process is associated with whole micelle fragmentation and coagulation [35] and the fast process is related to single monomer excursions into and out of micelles. Again the τ values obtained show remarkable consistency with the time scales of these processes as might be expected, since the minimum possible value for τ would be ca. τ_1, while, in principle, there is no upper limit.

The variation of τ with mixed surfactant composition shows an initial increase from the value of the pure nonionic surfactant reaching a maximum value of X_{ionic} of approximately 0.2–0.3; thereafter, the values continuously decrease down to the pure ionic surfactant case. In the composi-

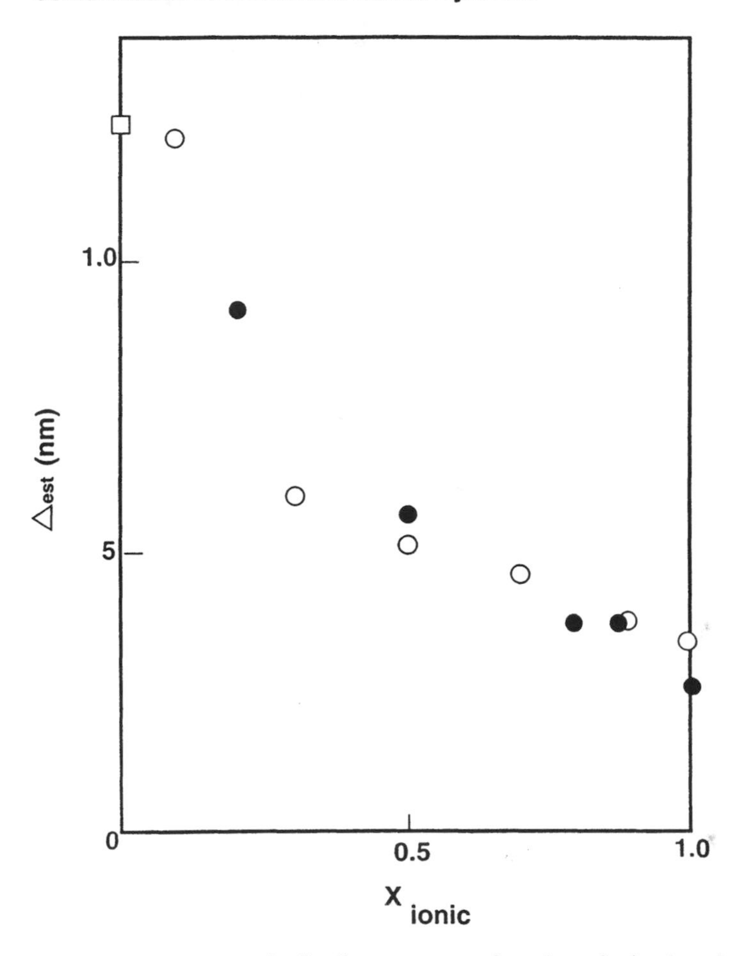

FIG. 14 Estimated micelle diameters as a function of mixed surfactant composition. ○, DDTAB/$C_{12}EO_6$; ●, SDS/$C_{12}EO_6$.

tion range of the maximal values there is no significant difference between the mixtures containing anionic surfactant and those containing cationic surfactant; however, in the ionic surfactant-rich mixtures those containing cationic surfactant are consistently lower than those with SDS. This behavior may be a result of the degree of counterion binding to the surfactant ions, i.e., dodecylsulfate or dodecyltrimethylammonium ions. It has been shown [36, 37] for mixed micelles of SDS and $C_{12}EO_7$ or $C_{12}EO_5$ that at $X_{ionic} < 0.5$ the sodium ions were more dissociated, indicating shielding

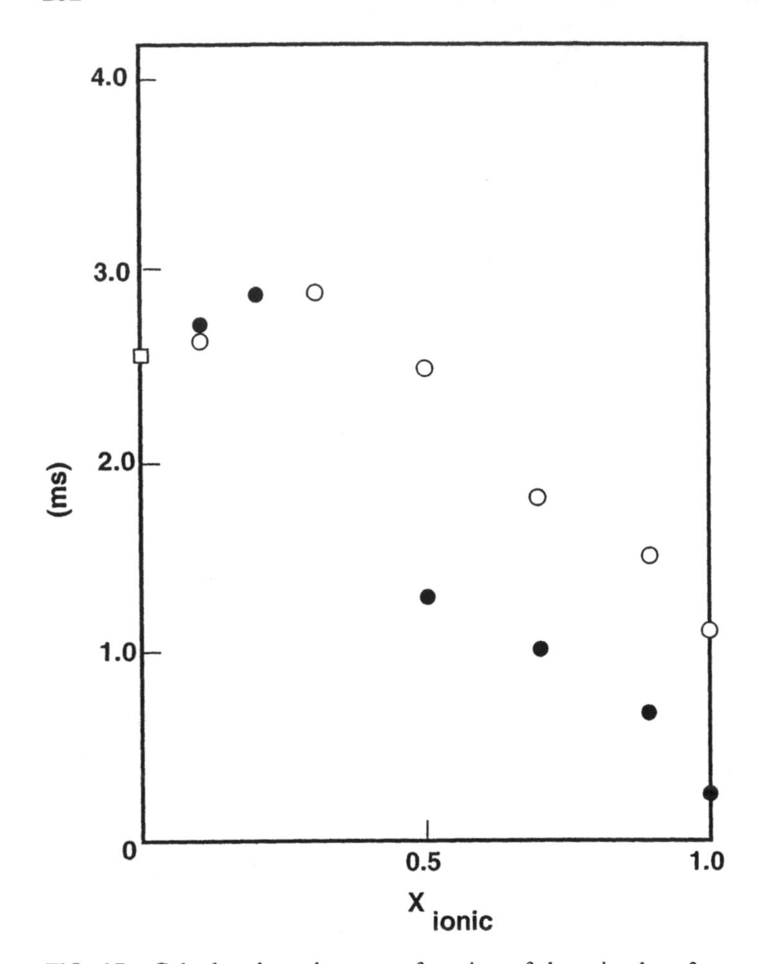

FIG. 15 Calculated τ values as a function of the mixed surfactant composition: ●, DDTAB/$C_{12}EO_6$; ○, SDS/$C_{12}EO_6$.

of the dodecylsulfate groups by the EO chains. For $X_{ionic} > 0.5$, the EO chains are loosely packed and considerable ion binding takes place between the sodium ions and the dodecylsulfate groups. The differences at the high ionic content, therefore, may be related to the differing ability of the sodium and bromide counterions to bind to the ionic surfactant component headgroups. Similarly, the absence of any significant differences at the low ionic concentrations in the mixtures is indicative of the shielding provided by the EO chains at the micelle surface.

The decrease in solubilization rate as X_{ionic} is increased may be explained as a result of the decreased micellar capacity. That this is the case is borne out by the relatively good fit to the experimental data obtained using Eqs. 15 and 16. The effect of increased surface charge density upon the solubilization rate of a nonpolar molecule like nonane is of secondary importance to the primary effect of changing micelle size. The ideal nature of the kinetic process inferred by Eq. (4) is consistent with the observed [40] phase behavior. Mixtures of pairs of strongly hydrophilic surfactants showed no strong intermolecular interactions, whereas combinations of a more hydrophobic and a hydrophilic counterpart, e.g., SDS and $C_{12}EO_2$, showed interesting interactions.

E. Differential Solubilization Effects

The potential of selection in the solubilization properties of surfactant solutions in contact with mixtures of oils in areas such as oil recovery and drug delivery systems warrants investigation. Some attention has been focused [26, 27, 39, 40] upon solubilization in aqueous surfactant solutions from binary oil mixtures. These studies have been concerned with the equilibrium solubilization properties of binary oil mixtures generally comprising two oils of significantly different polarity. A noticeable selectivity is observed in these cases; i.e., the composition of the oil solubilized contained a larger mole fraction of one of the components than in the original oil mixture contacting the surfactant solutions. The degree of selectivity was found to be related to the molar solubilization capacities of the individual oils. Generally the oil of higher polarity was found to be solubilized in preference to the other component. This may be accounted for, in part, by a difference in the solubilization sites available to the different components; thus, for example, in the case of benzene-cyclohexane and benzene-*n*-hexane mixtures studied by Nagarajan *et al.* [26, 27], the selection must arise partly as a result of the tendency of the benzene component to be adsorbed at the micelle/solution interface; such a tendency for aromatic molecules to accumulate at the micelle surface, in both ionic and nonionic surfactant systems is well known [41–43] and results from charge-transfer-type interactions. In contrast, as discussed in Sec. III.C, no selection is found for systems where the oils are of similar polarity, although sterically different [7], and where the solubilization site is the same for both oils. This may be because the polarity difference results in the selection of different locations of the oil components in the micelle. One of the aims of this work was to investigate the relative roles of the different factors in the mechanisms of solubilization.

An example of the differential extraction of oil components from the skin has been demonstrated [44]. Molecules interacting with the stratum corneum lipids primarily through hydrophobic forces and located within the lipid bilayers were extracted the most rapidly. Those molecules which were more amphiphilic in nature and interacted more intimately with the bilayer took much longer to be removed. This implies that the nature of the location of the oil component within the surfactant association structure is also of importance in the solubilization and related processes. Demonstration of such preference in the solubilization process is found in extraction from binary mixtures of benzene and cyclohexane [9] into either water or surfactant solutions. The kinetic profile derived from a drop-on-fiber experiment (Fig. 16) is no longer linear, implying that the composition of the oil drop is also time dependent. A bulk scale experiment in which the composition of the oil in the micellar phase was monitored by gas chromatography [45] showed the composition to increase to a ratio of ca. 9:1 from an initial 1:1 bulk oil composition within the first hour of contact between the bulk oil and micellar phases. In this case the process is determined by the diffusion rates into water which are in a ratio of approximately 30:1 in favor of benzene. Similar effects have recently been found [46] in the initial solubilization kinetics from binary mixtures of *n*-alkanes where the mechanism is that discussed in Sec. I.

Support for the idea that the benzene component is preferentially solubilized is found in the effect of changing the relative composition of the oil mixture on the Δn-τ curves (Fig. 17). Increasing the fraction of benzene in the mixture increases the fraction of the curve described by a rate of magnitude similar to that expected for benzene. A measure of the degree of selectivity may be made by comparing the observed $n(t)$-t curves with those calculated assuming a weighted volume fraction average (Fig. 17). Although it has been argued [26] that some fraction of the benzene may also reside in the micelle interior, it may be speculated that the absorption of benzene molecules at the micelle/water interface would retard the solubilization of cyclohexane in the initial stages as investigated here. This effect would be expected at the oil/water interface at concentrations below the critical micelle concentration. Preliminary measurements showing (Fig. 9) a reduction at surfactant concentrations where the surface excess and monolayer formation has been established just below the CMC may be indicative of such an effect.

1. Other Measurements of Solubilization Kinetics

The rotating disk method can be applied to systems where the solubilizate is in solid form. It has the advantage that the rate data may be analyzed in terms of the rheological behavior at the interface. This has been demonstrated [2] in a study of fatty acid solubilization into micellar solutions.

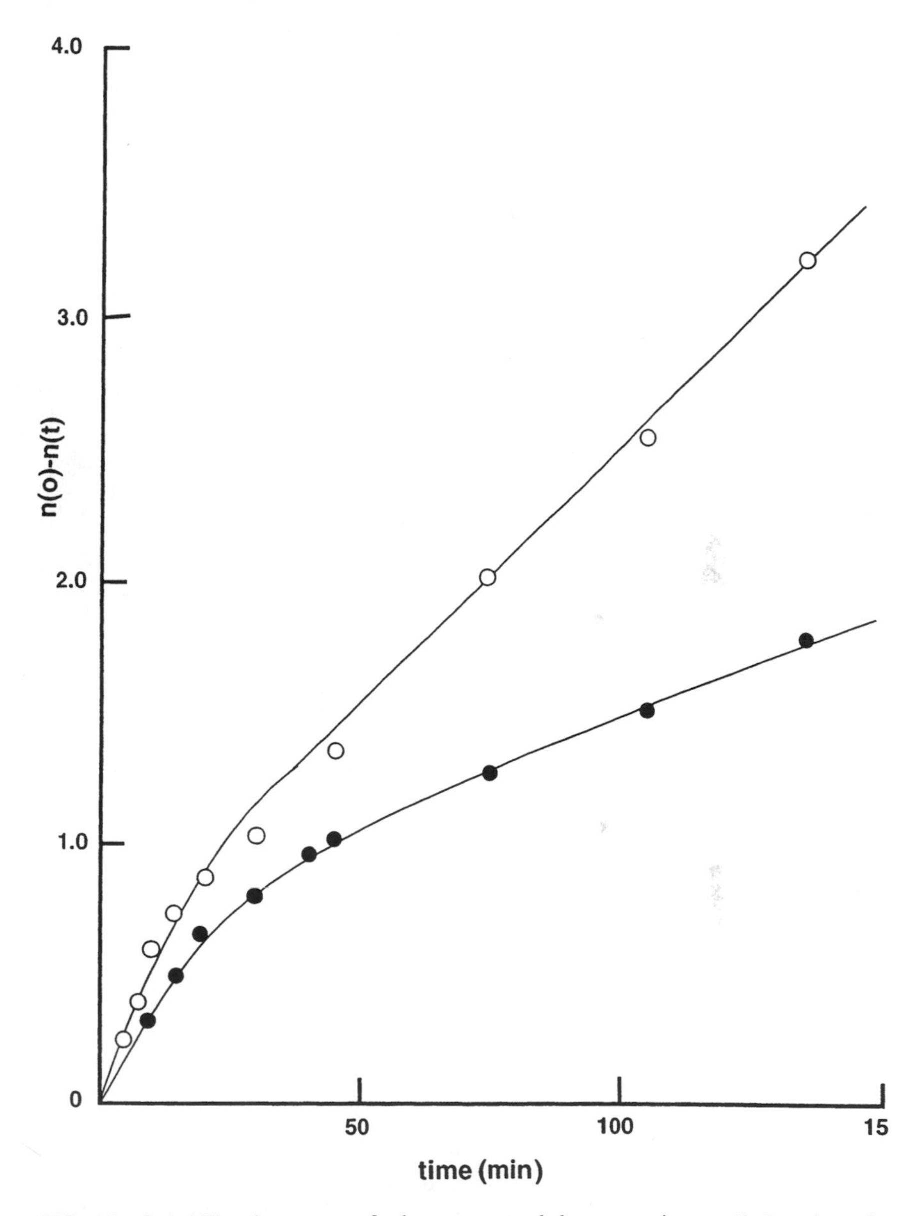

FIG. 16 Solubilization stage of a benzene:cyclohexane mixture (1:1 mole ratio) into (○) 0.11 M DDTAB and (●) pure water.

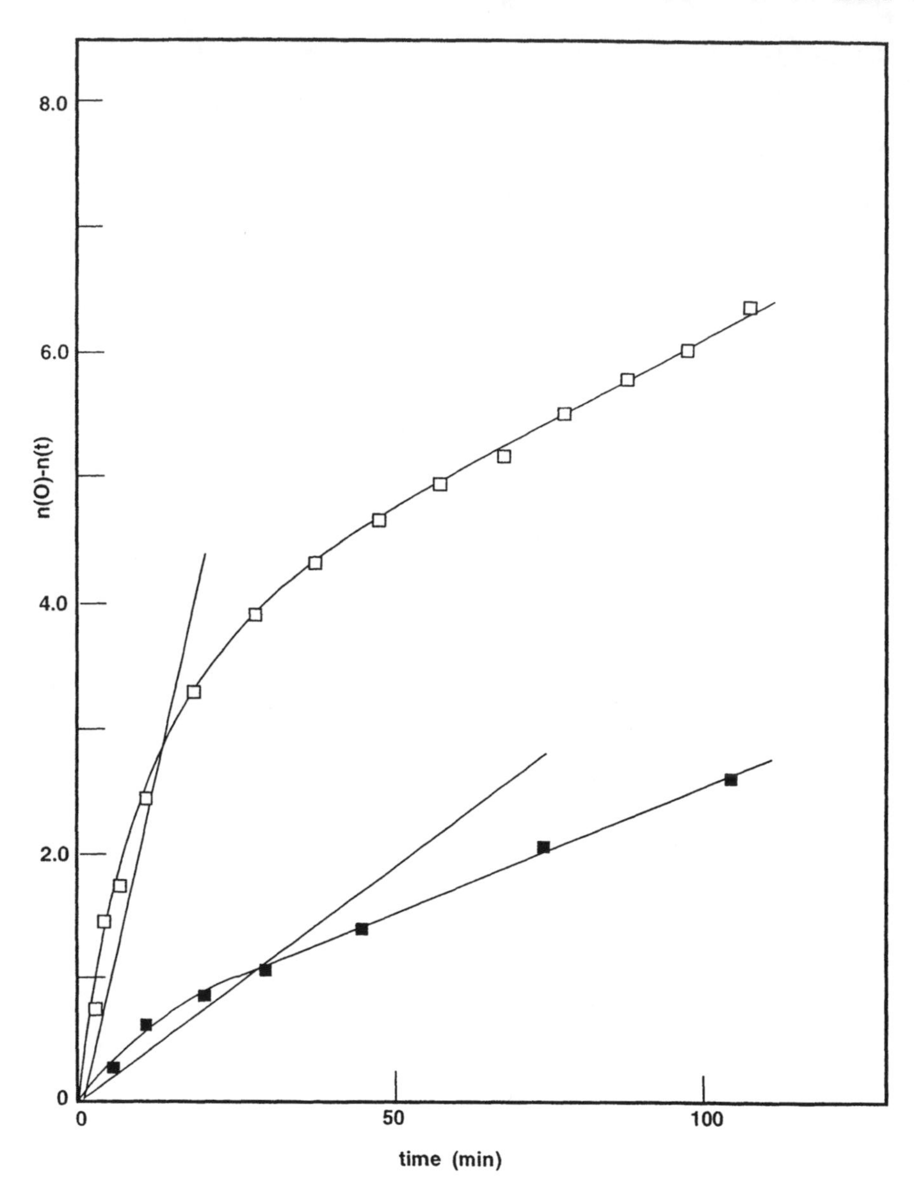

FIG. 17 Solubilization stage of benzene:cyclohexane droplets at 298 K as a function of time in 0.143 M DDTAB solutions for (□) 4:1 and (■) 1:1 mole ratios of benzene:cyclohexane; —— weighted average plots.

2. Rotating Disk Experiments

The use of a system with a well-defined solid/liquid or liquid/liquid geometrical interface allows the mass transfer process to be examined as a function of its rheology. This has been shown [2] to yield detailed mechanistic information in studies of fatty acid solubilization into surfactant or bile salt micelles.

The initial experiments used a solid fatty acid solubilizate that could be manufactured into a disk of well-defined geometry (Fig. 18) and doped with radiolabeled molecules. Solubilization rates were measured from the rate of appearance of the radiolabel in the receiving surfactant solution. An extensive analysis of the five-stage mechanism proposed for the solubilization process (Fig. 5) showed differing dependencies on rheology and active concentration for the various possible rate-determining schemes. These results obtained for the different possible situations have been analyzed in detail [48].

For the cases of long-chain fatty acids (C_n; $n \geq 12$) solubilizing into a variety of different surfactant micellar systems, the surfactant concentration and spinning rate dependence led to the conclusion that step 5 (Fig. 5) was rate determining, i.e., the step involving desorption of mixed micelles from the interface.

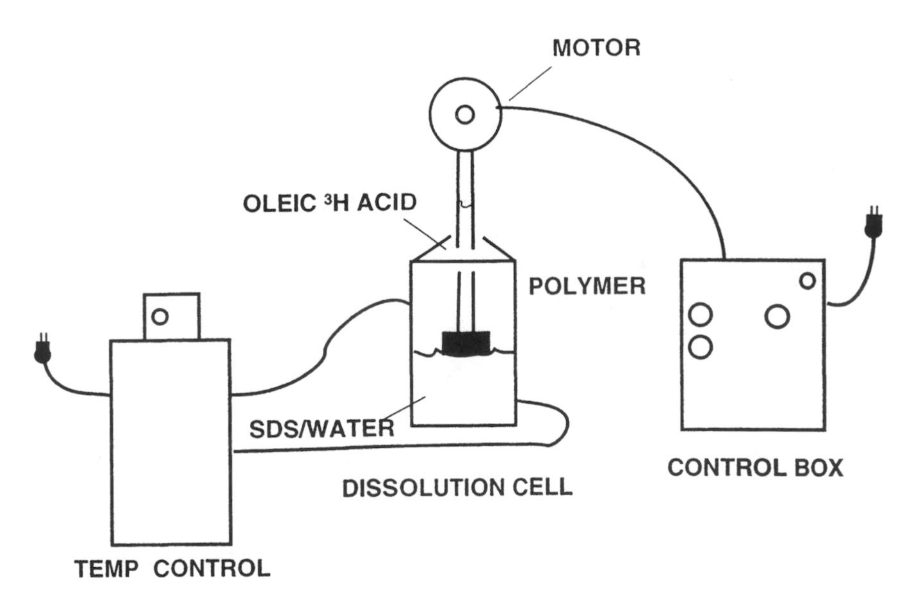

FIG. 18 Schematic of the rotating disk configuration.

Attempts were made to extend this technique to include liquid-state solubilizates by the use of a retaining porous membrane. A hollow metal cylinder was attached to the motor with the end sealed using a celaphor membrane. Studies of oleic acid solubilizing into the bile salt, sodium tauroxycholate indicated interfacial reactions (steps 3 and 4) to be rate determining. However, it was not possible to unambiguously determine if the observed data was unaffected by effects caused by the membrane itself. Indeed, it was thought likely that the diffusion gradients produced at the liquid/polymer membrane/liquid were dominating the observed kinetics.

Recently, in an attempt to overcome the problems associated with membranes used to contain a liquid solubilizate, a novel aerogel has been utilized [50]. Block copolymers of styrene and butadiene were synthesized in emulsion form of extremely high internal phase volume (HIPES). Such materials have been studied extensively [51–53] as substrates to contain fusion reactions on filters in cosmic ray studies.

These materials can be made with a distribution of cell radii from ca. 1 to 150 μ and are of extremely low density (<0.02 g cm^{-3}) while retaining good structural integrity. They are open-celled structures which can absorb liquids up to 95–98% of their geometrical volume, allowing the creation of a body that is greater than 95% liquid but has a defined geometrical shape. Preliminary studies of oleic acid solubilizing into micellar solutions of SDS have been made [47] using HIPES with different open cell sizes. In this way, any effects of the polymer matrix may be seen on the apparent kinetics. Fluxes obtained from observing the transport of radiolabeled molecules of oleic acid were found to be independent of the cell size of the HIPES material used over the experimentally accessible cell size range of 5–100 μ (Fig. 19). From the combination of the way in which the flux varied with surfactant concentration and spinning rate (Figs. 20, 21), it was deduced that the rate-determining step was desorption of mixed micelles (step 5); i.e., similar to that found previously for the similar system using solid fatty acid solubilizates.

F. Interfacial Liquid Crystal Formation

Early work [2] on systems with a fatty acid or soap constituent indicated the importance of interfacial liquid crystal formation in solubilization and associated detergency processes. This was further emphasized in studies [1, 2] of solubilization kinetics in such systems using a rotating disk apparatus. The mechanisms found in these situations involved both diffusion of components in and out of the interfacial layer and the degree of stirring. Experiments to determine the pathways of phase formation in surfactant/

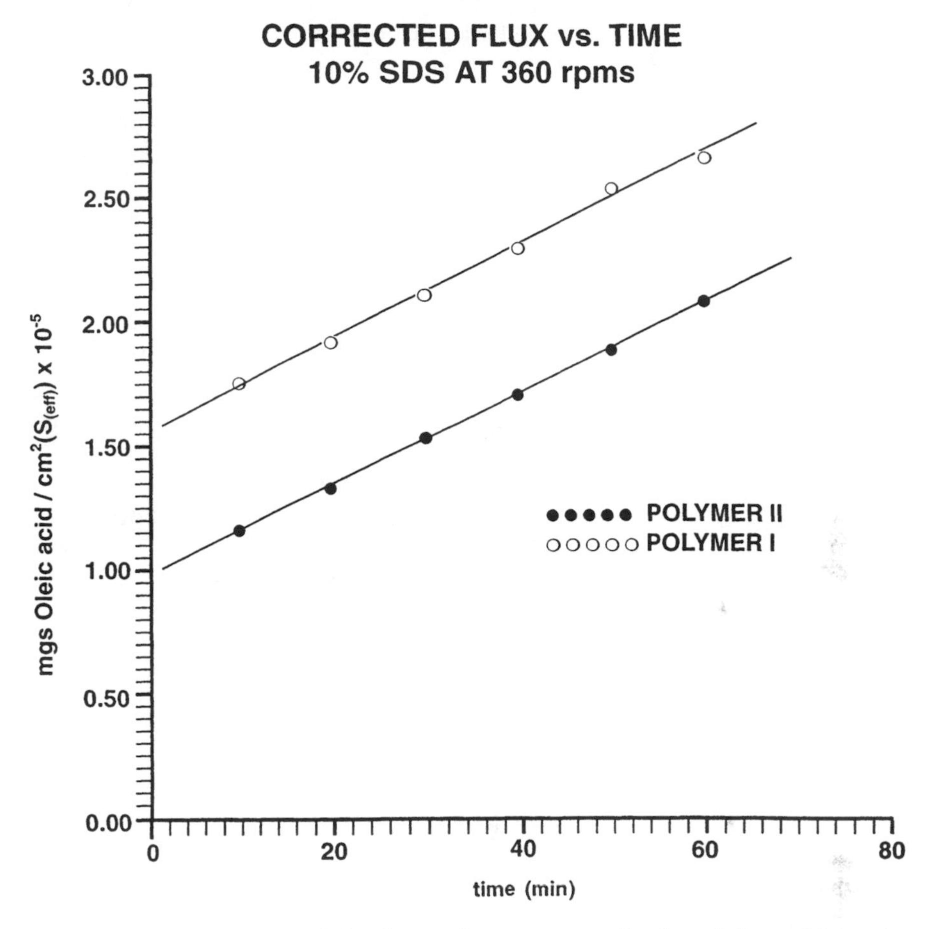

FIG. 19 Fluxes corrected for the surface area contribution of the solid matrix (mean cell diameter: ○, 10 μm; ●, 100 μm). The curves have been offset to allow them to be distinguished.

oil/water systems have shown [15, 16] the important role played by the formation of lamellar phases in interfacial mass transfer processes. Furthermore, the rate at which components diffuse in and out of this interfacial layer, when considered in the context of the phase behavior of the system, determines the overall observed kinetics. An example has been given [16] where the formation of a layer of interfacial water effectively stopped the transport of the oil component in the system. In this case, although the bulk composition of the total components indicated that a

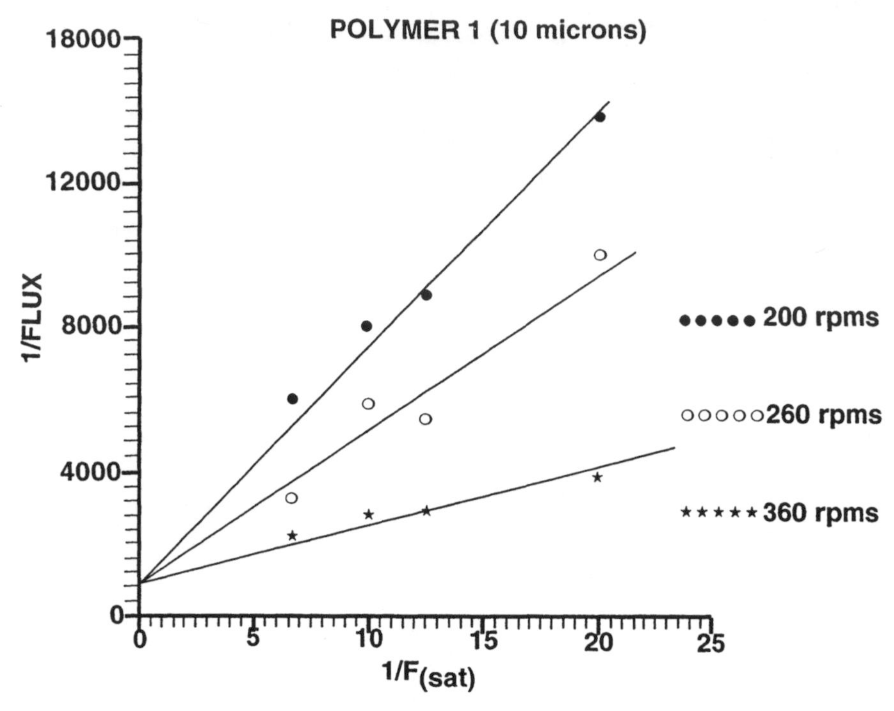

FIG. 20 Variation of flux of oleic acid solubilization with SDS concentration (F_{sat}) from a HIPES disk (10-μm cell diameter).

micellar phase should be formed, this state, in fact, was never reached because the kinetic pathway involved an effective barrier to transport in the unstirred system, namely, the formation of a water layer. These considerations must be taken into account in situations where multicomponent systems are involved and interfaces formed.

IV. SUMMARY

The work outlined shows some of the insights to be obtained from a controlled and well-defined experimental approach to the study of solubilization kinetics. The importance of interfacial surfactant association structures and their rearrangements has been demonstrated. Such considerations are important in many processes such as liquid chromotography, potential low-energy-cost oil separations and multivarious physiological processes.

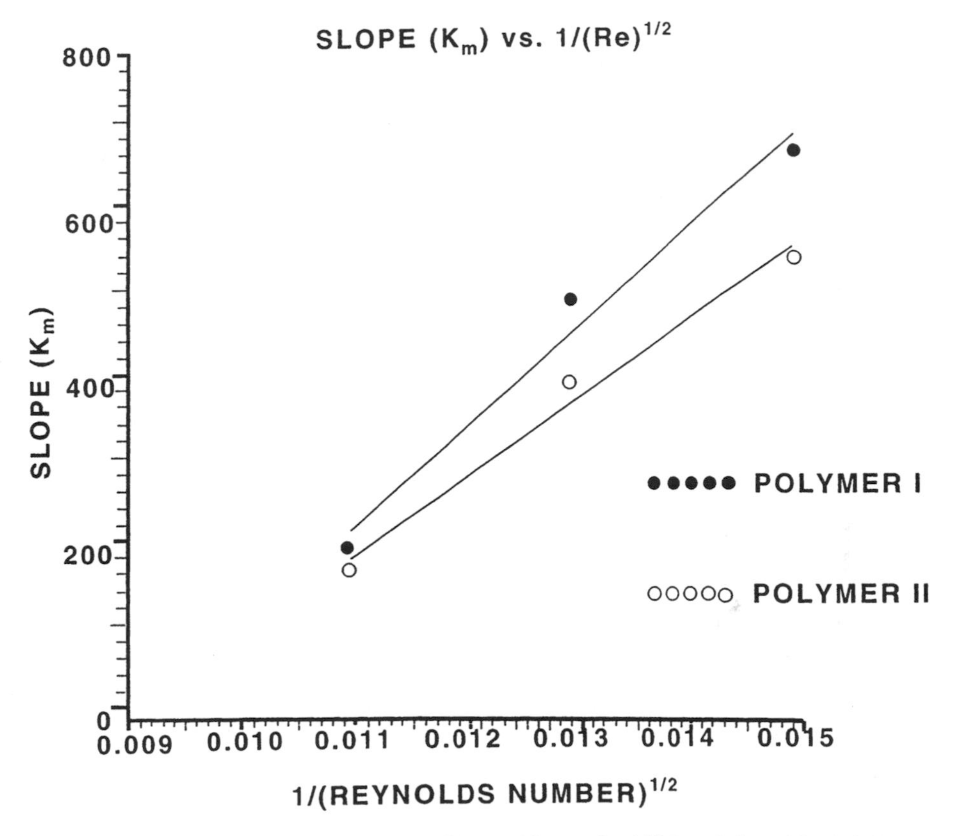

FIG. 21 Plot of rate constant versus (Reynolds number)$^{1/2}$ for oleic acid solubilizing in aqueous SDS solution (10%); HIPES average cell diameter: ○, 10 μm; ●, 100 μm.

Importance of the phase behavior in determining kinetic behavior has been demonstrated. It should be noted that a combination of local adsorption phenomena and development of interfacial concentrations dependent on the relative rates of ingress or diffusion away can lead to interfacial phase formation which is unexpected from the total concentrations of the components in the system. Thus, for example, interfacial liquid crystal formation can be observed at the oil/water interface in systems where the total oil is $\ll$1% of the amount required to saturate the surfactant micelles with which the oil is in contact in initial kinetics experiments.

A recent computer simulation of the time evolution of a water/oil/surfactant system [49] has been shown to be useful in following aspects of the solubilization process in micellar systems. Analysis of the time evolution of the system indicated concerted transfer processes involving the oil/water interface that are analogous to those suggested by the *drop-on-fiber* results obtained for insoluble oil solubilization. This approach, in parallel, offers a significant method to enhance the current experimental approaches to understanding the kinetic processes associated with solubilization phenomena.

ACKNOWLEDGMENT

The author would like to specially thank Dr. B. J. Carroll for his help in many stimulating discussions and for his support in the early years of developing the *drop-on-fiber method*.

REFERENCES

1. Shaeiwitz, J. A., Chan, A. F., Cussler, E. L. and Evans, D. F., *J. Colloid Interface Sci. 84*: 47 (1981).
2. Huang, C., Evans, D. F. and Cussler, E. L., *J. Colloid Interface Sci. 82*: 499 (1981).
3. Chiu, Y. C., Han, Y. C. and Cheng, H. M., in *Structure/Performance Relationships in Surfactants* (M. J. Rosen, ed.), A.C.S. Symposium Ser., 253 (1984).
4. Bolsman, T. A. B. M., Veltmaat, F. T. G. and van Os, N. M., *J. Am. Oil Chem. Soc. 65*: 280 (1988).
5. Carroll, B. J., *J. Colloid Interface Sci. 57*: 488 (1976).
6. Carroll, B. J., *J. Colloid Interface Sci. 79*: 126 (1981).
7. Carroll, B. J., O'Rourke, B. G. C., and Ward, A. J. I., *J. Pharm. Pharmacol. 34*: 287 (1982).
8. O'Rourke, B. G. C., Ward, A. J. I. and Carroll, B. J., *J. Pharm. Pharmacol. 39*: 865 (1987).
9. Ward, A. J. I., Carr, M. C. and Crudden, J., *J. Colloid Interface Sci. 106*: 558 (1985).
10. Donegan, A. C. and Ward, A. J. I., *J. Pharm. Pharmacol. 39*: 45 (1987).
11. Faulkner, P. G., Ph.D. thesis, National University of Ireland (1988).
12. Nakagawa, T., Kuriyama, K., and Inoue, H., *J. Colloid Interface Sci. 15*: 268 (1960).
13. Tondre, C., and Zana, R., *J. Disp. Sci. Technol. 1*: 179 (1980).
14. Gruen, D. W. R., and Haydon, D. A., *Pure Appl. Chem. 52*: 1229 (1980).
15. Mortensen, M., Friberg, S. E. and Neogi, P., *Sep. Sci. Technol. 20*: 285 (1985).
16. Neogi, P., Kim, M., and Friberg, S. E., *Sep. Sci. Technol. 20*: 613 (1985).
17. Ward, A. J. I., unpublished results.
18. Hadgraft, J., Williams, W. G. and Allan, G., in *Pharmaceutical Skin Penetration Enchancement* (Walters, K. A. and Hadgraft, J., eds.), Marcel Dekker, New York, 1993 Chap. 7.

19. Rubingh, D. N., in *Solution Chemistry of Surfactants*, Vol. 1 (Mittal, K. L., ed.), Plenum, New York, 1979, p. 337.
20. Hua, X. Y. and Rosen, M. J., *J. Colloid Interface Sci. 87*: 469 (1982).
21. Holland, P. M. and Rubingh, D. N., *J. Phys. Chem. 87*: 1984 (1983).
22. Zhu, B. Y. and Rosen, M. J., *J. Colloid Interface Sci. 99*: 435 (1984).
23. Bourrel, M., Bernard, D. and Gracia, A., *Tenside Deterg. 21*: 311 (1984).
24. Scamehorn, J. F., in *Phenomena of Mixed Surfactant Systems*, A.C.S. Symposium Ser. 311, Washington, 1986, p. 1.
25. Rathman, J. and Scamehorn, J. F., *Langmuir 2*: 354 (1986).
26. Nagarajan, R., and Ruckenstein, E., in *Surfactants in Solution* (Mittal, K. L., ed.), Plenum, New York, 1983.
27. Chaiko, M. A., Nagarajan, R. and Ruckenstein, E., *J. Colloid Interface Sci. 99*: 168 (1984).
28. Faulkner, P. G., M. Sc. thesis, University College, Dublin, 1985.
29. Fendler, J. F. and Fendler, E. J., in *Catalysis in Micellar and Microemulsion Systems*, Academic Press, New York, 1975.
30. Mukerjee, P. and Cardinal, J. R., *J. Phys. Chem. 82*: 1620 (1978).
31. Ericksson, J. and Gilberg, G., *Acta Chem. Scand. 20*: 2019 (1966).
32. Mukerjee, P. and Cardinal, J. R., *J. Phys. Chem. 82*: 1620 (1978).
33. Mysels, K. J. and Princen, L. H., *J. Phys. Chem. 63*: 1699 (1959).
34. Aniansson, E. A. G. and Wall, S. N., *J. Phys. Chem. 78*: 1024 (1974); *79*, 857 (1975).
35. Kahlweit, M., *Pure Appl. Chem. 53*: 2060 (1981); *J. Colloid Interface Sci. 90*: 92 (1982).
36. Meguro, K., Akasu, H. and Ueno, M., *J. Am. Chem. Soc. 53*: 145 (1976).
37. Guering, P., Nilson, P. G. and Lindman, B. J., *J. Colloid Interface Sci. 105*: 41 (1985).
38. Sagitani, H. and Friberg, S. E., *Bull. Chem. Soc., Jpn. 56*: 31 (1983).
39. Thomas, D. C. and Christian, S. D., *J. Colloid Interface Sci. 82*: 430 (1981).
40. Caceres, T. and Lissi, E. A., *J. Colloid Interface Sci. 97*: 298 (1984).
41. Henriksson, U., Klason, T., Odberg, L. and Eriksson, J. C., *Chem. Phys. Lett. 52*: 554 (1977).
42. Christenson, H. and Friberg, S. E., *J. Colloid Interface Sci. 75*: 276 (1980).
43. Ward, A. J. I., Rananavare, S. B., Larsen, D. W., and Friberg, S. E., *Langmuir 2*: 373 (1986).
44. Imokawa, G. and Hattori, M., *J. Invest. Dermatol. 84*: 282 (1985).
45. Crudden, J., Ph.D. thesis, National University of Ireland, 1987.
46. Faulkner, P. G., Ph.D. thesis, National University of Ireland, 1988.
47. Ward, A. J. I., unpublished results.
48. Cussler, E. L., Ph.D. dissertation, University of Minnesota, 1977.
49. Karborni, S., van Os, N. M., Esselink, K. and Hilbers, P. A. J., *Langmuir 9*: 1175 (1993).
50. Belleau, D., M. S. dissertation, Clarkson University, 1991.
51. Williams, J. M., *Langmuir, 4*, 44 (1988).
52. Williams, J. M. and Wrobleski, D. A., *Langmuir 4*: 656 (1988).
53. Williams, J. M., Gray, A. J., Wilkerson, M. H., *Langmuir 6*: 437 (1990).

III
Solubilization in Nonmicellar Surfactant Aggregates

8

Adsolubilization

JOHN H. O'HAVER and LANCE L. LOBBAN School of Chemical Engineering and Materials Science, University of Oklahoma, Norman, Oklahoma

JEFFREY H. HARWELL and EDGAR A. O'REAR III School of Chemical Engineering and Materials Science and Institute for Applied Surfactant Research, University of Oklahoma, Norman, Oklahoma

SYNOPSIS

Surfactants adsorbed at the solid/liquid interface can form aggregates that are much like micelles. These micelles can be used to solubilize organic

---

Financial support for this work was provided by E. I. DuPont de Nemours and Co., Kerr-McGee Corporation, Union Carbide Co., Sandoz Co., Dow Chemical Co., ICI Americas, Hitachi Research Laboratory, Hitachi-Shi, Japan, Oklahoma Center for Advancement of Science and Technology Award No. ARO-075, and National Science Foundation Grant Nos. CPE-8318864, CTS-9812806, and CTS-9009246.

molecules in the same manner that micelles are used. This phenomenon, called adsolubilization, has not been studied extensively and the first new technologies based on it are just beginning to emerge. The initial research on adsolubilization indicates that at high adsorption densities admicelles have adsolubilization capacities similar to micelles but with greater selectivity than micelles, probably because of the greater packing density which can be obtained in admicelles. At lower adsorption densities the capacity of admicelles may be many times the capacity of micelles for amphiphilic molecules.

I. INTRODUCTION

A. Definition

Adsolubilization is the surface analogue of solubilization, with aggregates of adsorbed surfactant playing the role of micelles. To the knowledge of the authors, this chapter is the first attempt to review the literature on the phenomenon of adsolubilization. In fact, while the phenomenon of solubilization has been extensively studied for the better part of this century and has important applications in numerous technologies, adsolubilization is hardly recognized as an important phenomenon at all: Almost all the studies of it have occurred since the mid-eighties, and the technological utilization of the phenomenon is only beginning. The authors feel, however, that adsolubilization may eventually be seen to be as significant and useful a phenomenon as solubilization.

B. Background

1. Surfactant Adsorption

Just as it would not be possible to understand solubilization without an understanding of micelle formation, it is not possible to understand adsolubilization without understanding the formation of admicelles. Figure 1 is a schematic of a typical surfactant adsorption isotherm. Such isotherms are commonly seen for ionic surfactants adsorbing from aqueous solution on surfaces with a charge opposite to the charge on the surfactant. Examples include anionic surfactants such as SDS on alumina at low or neutral pH or cationic surfactants such as cetyltrimethylammonium bromide on silica at pH values above 3. Similar isotherms are also sometimes—but not always—observed for nonionic surfactants. For example, an ethoxylated nonylphenol will exhibit a similar isotherm on silica, but will not adsorb sufficiently strongly to show all the details of such an isotherm on alumina.

Isotherms like those in Fig. 1 are commonly divided into four regions [1, 2]. Region 1 is a region of low adsorption densities and is sometimes

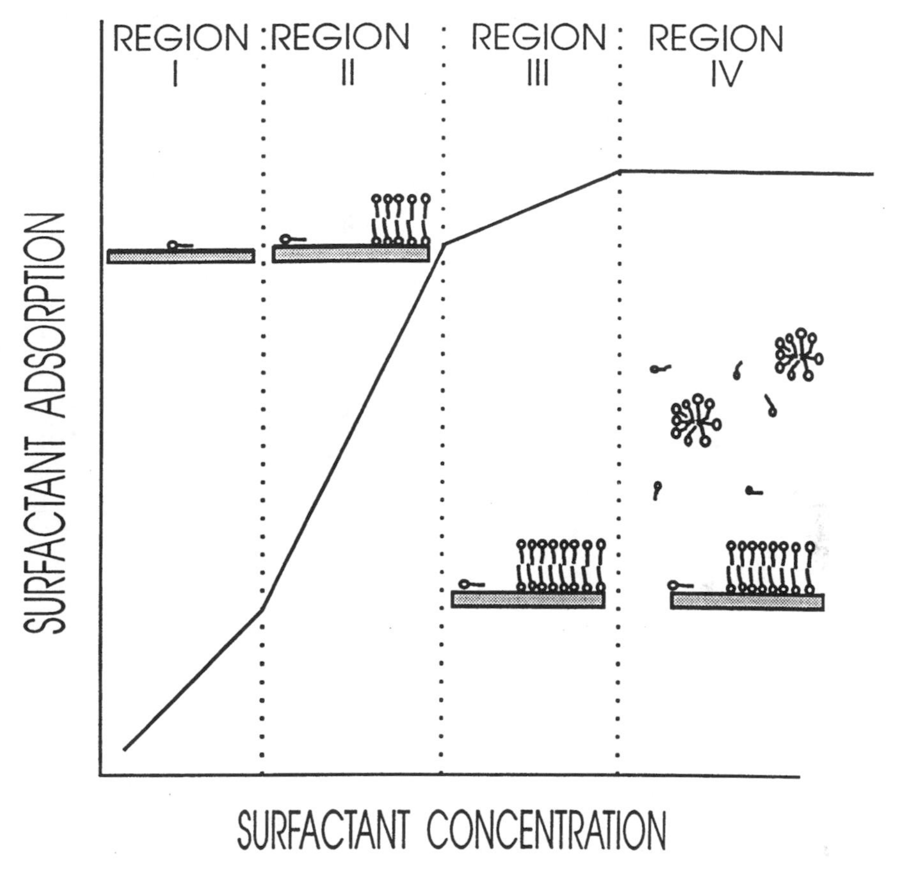

FIG. 1 Typical surfactant adsorption isotherm.

referred to as the Henry's law region. In region 1 the surfactants are adsorbed as monomers and do not interact with one another. For nonionic surfactants and for ionic surfactants with some added electrolyte, the slope of the isotherm is 1 in this region when the data is plotted on a log-log plot. For ionic surfactants on a surface of moderate charge density with no added electrolyte, the slope of the isotherm may be less than 1 because the charge on the surface is being reversed by the adsorption of the charged surfactant ions. The most important point, though, is that there are no aggregates of adsorbed surfactants in this region. In many systems where conditions are favorable to surfactant adsorption, this region may not even be detectable because it occurs at such low surfactant concentrations.

Region 2 is indicated by the sharp *increase* in the slope of the isotherm that occurs at the transition between region 1 and region 2. All investigators of surfactant adsorption have attributed this increase in slope to the formation of micelle-like aggregates of adsorbed surfactants [1–12]. The transition has been given designations analogous to the critical micelle concentration (CMC) such as critical admicelle concentration (CAC) or hemimicelle concentration (HMC) [5, 11]. The differences in terminology emphasize different views of the morphology of the aggregates at the transition (monolayered or bilayered), but they also emphasize the consensus that the aggregation phenomenon which occurs at the solid/liquid interface is analogous to micelle formation [9]. This conclusion, which was initially based primarily on the shape of the isotherm, has been reinforced by modern studies using fluorescence and spin probes; these studies indicate that the aggregates that form in region 2 have micelle-like microviscosities and micelle-like internal environments [13–17]. It is important to note that these micelle-like aggregates (we will call them admicelles for the rest of the chapter) are present even though the total adsorption density of the surfactant may be less than $\frac{1}{100}$th of a monolayer. Cases and co-workers have presented evidence which they interpret to indicate that the reason for the formation of these aggregates locally or patchwise at the interface is due to the heterogeneity of the surface [6].

The region 2/region 3 transition is marked by a decrease in the slope of the isotherm. In some systems, the change in slope can be quite dramatic. While there is widespread agreement about the mechanism for the change in slope from region 1 to region 2, there is little agreement about the reason for the change in slope from region 2 to region 3. The most commonly given explanation is that the aggregates forming in region 2 are monolayers adsorbing head down on an oppositely charged surface and that the change in slope coincides with the cancellation of the charge on the solid surface by the charge on the adsorbed aggregates so that subsequent surfactants are adsorbing onto a like-charged surface [5]. Sometimes it is proposed that a second layer on the aggregates does not begin to form until region 3 [13–15]. Other workers have presented data to indicate that the second layer forms suddenly at the CMC of the system and that only monolayers are present below the CMC [6]. Other workers have attributed the change in slope to the distribution of patch energies on the surface [2, 11]. It may be that a critical component of the change in slope is a Langmuir-like competition between aggregates for the remaining surface area [16, 17]. The region 2/region 3 transition does not always correspond to a change in the charge of the surface, however, and it is also observed in systems of nonionic surfactants adsorbed on silica [9, 10, 16–18]. Further, it is not clear how a change in the morphology of the aggregates between region 2 and region 3 can be reconciled with the results of fluorescence and spin probe studies which clearly show that both the microviscosity and the polarity of the aggregates is constant from the region 1/region 2 transition through region 3 [19].

A recent study by Kung and Hayes [20] sheds additional light on the morphology of the admicelle which may be important in future applications of adsolubilization. In a comparison of the FTIR spectra of cetyltrimethylammonium bromide micelles, crystals, and adsorbed layers, it was observed that surfactant in the adsorbed layer has a micelle-like spectrum in regions 2 and 3 with no obvious change from region 2 to region 3, which suggests that local bilayers are formed in this system beginning with the region 1/region 2 transition. When, however, the adsorbed layer spectra were compared to the spectra obtained upon drying of the adsorbed samples, it was found that the difference between the adsorbed and the dry samples decreased with increasing surface coverage; this was interpreted by the authors to indicate that the structure inside the adsorbed layer was becoming more solidlike at higher coverages. If this interpretation of the FTIR spectra is correct, there may be significant changes in the location of adsolubilizates and ability of the admicelle to distinguish between adsolubilizates when region 2 adsolubilization is compared to region 3 adsolubilization.

Region 4 is called the plateau adsorption region [12]. In most systems the region 3/region 4 transition occurs near the CMC of the surfactant. When the surfactant is monoisomeric (or nearly so, as in the case of some commercial polyethoxylated alcohols or phenols) the adsorption becomes nearly constant above the CMC. This phenomenon is easily understood in terms of the pseudophase separation model of micelle formation: micelles that form at the CMC have the same chemical potential as the monomer at the CMC. As the concentration of surfactant increases, the surfactant goes into the micellar pseudophase at almost the same chemical potential as the surfactant in the first micelles to form and so the chemical potential of the surfactant does not increase as dramatically with surfactant concentration above the CMC as it would in the absence of micelles. Since the admicelles are also in equilibrium with the same monomer phase, the last admicelle to form is also at the same chemical potential as the monomer at the CMC. The only way more admicelles can form is for the chemical potential of the surfactant to increase, but this occurs only slowly because the surfactant added above the CMC forms more micelles at the nearly the same chemical potential as that of the micelles that formed near the CMC. The pseudophase separation model is certainly only an approximation, but in general, the better a surfactant is at forming micelles, the more pronounced will be the region 3/region 4 transition, the closer that transition will be to the CMC, and the less the adsorption will increase above the CMC.

2. Factors Affecting Admicelle Formation

Though it should be very clear from these observations that admicelles are very much like micelles and so can be expected to exhibit something closely analogous to solubilization, the analogy can be reinforced by the following results: When a surfactant system is modified in a way that affects the CMC of the system, the CAC and the region 3/region 4 transition vary in the same direction and at about the same rate as the CMC. For example, addition of electrolyte causes a decrease in the CMC of an ionic surfactant; however, the plateau adsorption level of an ionic surfactant will vary little with the addition of electrolyte because the CAC (the region 2/region 3 transition) and the region 3/region 4 transition will vary at about the same rate and in the same direction as the CMC [10]. Similar results have been observed for changes in the length and degree of branching of the hydrophobic moiety of the surfactant [21].

There is one dramatic difference in the factors which affect micelle formation and the factors influencing admicelle formation: the concentration of potential-determining ions for the solid [2, 10]. This can be illustrated for the case of alumina. In an aqueous suspension of alumina, the charge density on the surface of the alumina will be found to vary dramatically with the pH of the suspension. At pH values below about 9.1 (depending somewhat on the history of surface) the surface of the alumina will be positively charged. Above pH 9.1, the surface will be negatively charged. At pH 9.1, the surface will be uncharged. For this reason, hydrogen and hydroxyl ions are called potential-determining ions for alumina. While pH will not greatly affect the CMC of most ionic surfactants, SDS (for example) will hardly adsorb at all on alumina at pH values above 9.1 but will adsorb to the extent of forming a complete bilayer on the surface at a pH of about 4 or below. Lithium ions have been shown to have a similar effect on SDS adsorption on alumina, probably because they are small enough to become incorporated into lattice sites on the surface of the alumina [10]. Potential-determining ions are a very powerful tool for manipulating the extent of surfactant adsorption and therefore the concentration of admicelles at a solid/liquid interface.

One final effect should be mentioned. Just as the addition of a small concentration of a low CMC nonionic can dramatically lower the CMC of an anionic surfactant system, so mixed admicelles can be formed; as the use of a mixture lowers the CMC of a system, so the use of mixtures is found to lower CAC values as well [18–19, 21–24]. This phenomenon has been studied in some detail for mixtures of anionic and nonionic surfactants adsorbed on alumina by Roberts et al. [22] and Lopata et al. [23]. The most striking observation is that while the nonionic polyethoxylated nonylphenols will hardly adsorb at all on the alumina by themselves, the addition of only 10 mole % of anionic surfactant leads to the formation of admicelles which are predominately nonionic in composition.

C. Potential Applications of Adsolubilization

As stated in the Introduction, the possible applications of adsolubilization are only now beginning to be studied. Several promising areas have already begun to develop, however. One is the use of adsolubilization in the formation of thin polymer films on solid surfaces [25, 26]. Wu et al. developed a three-step process for forming ultrathin (3–4 nm) polymer films on a solid surface [24]. The first step in the process is formation of an adsorbed surfactant layer, the second step is the adsolubilization of a reactive monomer in the admicelle layer, and the third step is the initiation of a polymerization reaction within that layer. This process has now been studied for the formation of new ultrathin solid lubricating layers, for the compatibilization of filler and reinforcer surfaces for polymer composites, and for the formation of reverse-phase chromatography packings [27].

Adsolubilization has also been proposed as the basis of a new surfactant-based separation process [28]. One particularly promising variation of this is the use of chiral surfactants for the resolution of mixtures of optical isomers [29]. More recently Lobban and co-workers [30] have begun to investigate the use of adsolubilization for admicellar catalysis, a surface analogue of micellar catalysis which may give greater stereoselectivity than micellar catalysis and may be more amenable to development into industrial scale processes because it can be employed in a fixed-bed mode. Though the possibilities of use of the phenomenon in drug release applications seem obvious, to our knowledge not a single such study has been reported.

II. STUDIES OF ADSOLUBILIZATION

The phenomenon of adsolubilization has been studied by researchers in the past two decades for a variety of reasons. The different studies can be divided into one of four categories based upon their purpose: (1) the use of adsolubilized molecules to help characterize the adsorbed surfactant layer; (2) the use of adsolubilization in separation processes; (3) the use of adsolubilization in the formation of ultrathin films; (4) the use of adsolubilization in admicellar catalysis. Research in each of these areas will now be described.

A. Characterization of Admicelles

Most studies of adsolubilization have focused on what the process of adsolubilization would reveal about the structure and nature of an absorbed surfactant layer on various substrates. In an early study, Koganovskii, Klimenko, and Tryasorukova [31] investigated the coadsorption of nonionic surfactants and solubilized naphthalene from aqueous solutions onto acetylene black. Three nonionic surfactants, $C_{10}H_{21}O(C_2H_4O)_{17}H$, $C_{12}H_{25}O(C_2H_4O)_{17}H$, and $C_{12}H_{25}O(C_2H_4O)_{23}H$, were used in the study. They found that the amount of naphthalene adsolubilized in an adsorbed layer of a given surfactant was greater than the amount solubilized in the micelles of that surfactant. They concluded that the naphthalene adsolubilized both into the interior of the admicelle as well as coadsorbed with the surfactant onto the surface of the acetylene black.

Nunn et al. [32], gathered evidence about the nature of the adsorbed layer by adsolubilizing the dye pinacyanol chloride in an adsorbed layer of sodium *p*-(1-propylnonyl) benzenesulfonate on γ-alumina. Pinacyanol is red in aqueous media and blue in organic surroundings. They found that when pinacyanol was added to a solution of water and alumina, the alumina remained white while the solution was red. When the solution also contained surfactant at an equilibrium concentration in the bulk solution below the CMC, the solution was clear and the alumina became blue. When the concentration of the surfactant was raised so that the equilibrium concentration was above the CMC, both the alumina and the solution were blue. This evidence led to the conclusion that the adsorbed surfactant layer was micelle-like in structure, having a hydrocarbon core.

A study of the structure of the adsorbed surfactant bilayer of Triton X-100 (octylphenoxyethanol with an average of 9 to 10 oxyethylene units) on silica using adsolubilized pyrene was done by Levitz, Van Damme, and Keravis [16, 17]. This study also supported the existence of condensed assemblies of surfactant molecules on the surface of the silica. These assemblies gave rise, by the interaction of the aliphatic tails, to a hydrophobic, organic environment in their core. The environment of the admicellar layer for pyrene was found to be similar to that of pyrene in Triton X-100 micelles, regardless of the extent of coverage of the surfactant layer on the silica surface. By studying the decay of the pyrene at different surfactant coverages and different pyrene concentrations, some interesting results were surmised about the extent of surfactant bilayer coverage. For low pyrene concentrations, the decay was a single exponential, indicating a single pyrene monomer per adsorbed area, independent of surfactant surface coverage. For high pyrene concentrations, with low surface coverage (less than $\frac{1}{2}$ of the maximum plateau adsorption), the decay was similar to that of a micellar medium, suggesting a fragmented structure on the surface. When the surfactant surface coverage was increased to above 80% of the maximum, the decay became very similar to that of pyrene monomer, suggesting a continuous monomer phase within the adsorbed surfactant bilayer. This study also dealt with the adsorbed aggregate size, finding increasing aggregate size with increasing surfactant coverage up to about 80% of the maximum where the coverage appears to be nearly continuous on the silica surface.

Chandar et al. [15] used pyrene and dinaphthylpropane (DNP) as fluorescence probes to study the structure of adsorbed sodium dodecyl sulfate bilayers on alumina. Fluorescent probes had been found to have responses that are highly dependent on the environment. These responses include the excitation and emission spectra, decay rates, quantum yields of emission, fluorescence polarization, and quenching. These properties had been related to the micellar properties of polarity, viscosity, diffusion, solute partitioning, and aggregate number. This study looked at the polarity, microviscosity, and aggregate size of the adsorbed surfactant layer. Measurements of the polarity of the environment gave results similar to those of pyrene in SDS micelles, again giving evidence to the admicelle being the surface analogue of a micelle. Microviscosity measurements found that the mobility of the probes was more restricted in the adsorbed layer than in a micelle. This suggests that the interior of the admicellar layer is more highly structured and rigid than is the interior of a micelle. Little change was found to occur in the microviscosity with increasing layer coverage, indicating that the structure of the adsorbed layer changes little with increasing coverage. Finally, aggregate size was found to increase with increasing equilibrium surfactant concentration in the bulk solution up to the CMC.

Waterman et al. [13] used ESR spectroscopy and the nitroxide spin probe 16-doxylstearic acid to study sodium dodecyl sulfate, SDS, adsorbed onto alumina. They found that at low equilibrium SDS concentrations in the bulk solution, below 150 *M*, the probe tended to aggregate onto the surface of the alumina along with the SDS. When SDS concentrations in the bulk exceeded 10 mM, giving bilayer coverage on the surface of the alumina, the probe aggregates were broken up. This is consistent with the probe being adsolubilized into the interior of the bilayer.

The study was expanded by Chandar et al. [14] by using three nitroxide spin probes adsolubilized in different locations in the admicelle; 5-, 12-, and 16-doxylstearic acid were used to investigate the microviscosity of the SDS-admicellar environment at various distances from the alumina surface. The rotational correlations of each spin probe were first measured in SDS micelles, where it was found that the viscosity measured with each probe was effectively the same. The same measurements were then taken in the adsorbed surfactant layer. The microviscosity of the admicellar layer varied inversely with increasing distance of the probe from the SDS headgroups of the admicelle. Thus, the admicelle appears to be more rigid or structured near the headgroups but more flexible in the interior. This is consistent with the headgroups being fairly firmly attached electrostatically to the solid surface while the tails are able to move somewhat in the hydrocarbon core. Similar observations have been made for use of the same probes in lipid bilayers.

Zhu, Zhao, and Gu [34] have investigated the solubilization and adsolubilization of Chrome Azurol S (CAS), dye into the micelles of three different surfactants and adsorbed bilayers of each on two different silica gels. The surfactants chosen were decylmethyl sulfoxide (DEMS), Triton X-100 (PEO 10), and Triton X-305 (PEO 30). One silica gel had a BET surface area of 417 m^2/g and an average pore diameter of 9.2 nm. The second silica gel had a BET surface area of 47.1 m^2/g and an average pore diameter of 170 nm. CAS was first analyzed in pure water, and then in surfactant solutions both below and above their respective CMCs. In water, CAS exhibits a yellow color. In each of surfactant solutions when the concentrations were below the CMCs, the CAS also exhibited a yellow color. However, in the solutions where the surfactants were present above the CMCs of the surfactants, the solution became brown-red. This indicates the solubilization of the CAS into the organic interior of the micelles.

CAS was not found to adsorb from aqueous solution onto either of the silica gels. However, as the equilibrium bulk surfactant concentration was increased and therefore the amount of surfactant adsorbed onto the surface of the silica gels was increased, the amount of CAS which adsolubilized also increased. The color exhibited by both of the silica gels went from light red to deep red as the amount of adsolubilized CAS increased. The ratio of CAS to adsorbed surfactant molecules was measured at each data point. The maximum ratios of CAS to surfactant were 1:63 in the DEMS layer, 1:135 in the Triton X-100 layer, and 1:34 in the Triton X-305 layer. The demonstrated order of solubilization/adsolubilization (Triton X-305 > DEMS > Triton X-100) is the same for both micelles and admicelles.

Another study of the adsolubilization of dyes into admicelles was carried out by Esumi et al. [35]. They investigated the adsolubilization of the dyes yellow OB and azobenzene in mixed admicelles containing anionic and nonionic surfactants on γ-alumina. The surfactants used were lithium perfluorooctanesulfonate (LiFOS), lithium dodecyl sulfate (LDS), and two polyoxyethylene nonylphenols (NP7.5 and NP20). As in previous studies, the amount of adsolubilized dye increased with increasing concentration of the dye in the bulk solution. In a system with a mixed bilayer of LiDS and NP7.5 or NP20, the maximum adsolubilization of yellow OB was found to be greater than in a mixed bilayer of LiFOS and NP7.5 or NP20. This was somewhat expected as yellow OB is not found to adsolubilize into a bilayer of LiFOS alone. The amount of adsolubilized yellow OB was also found to be greater in a mixed bilayer of LiDS or LiFOS with NP20 than with NP7.5. Thus, adsolubilization appears to increase with increasing poly(oxyethylene) chain length, indicating that the dye is primarily adsolubilized by its interaction with the poly(oxyethylene) chain. The amount of adsolubilized yellow OB was found to be greater in a mixed admicelle as compared to a mixed micelle of the same surfactants.

The results of the azobenzene tests showed that the amount of adsolubilized azobenzene decreased with increasing poly(oxyethylene) chain length in mixed surfactant systems. This is consistent with the greater solubilization of azobenzene in NP7.5 micelles compared to NP20 micelles. As with yellow OB, the amount of adsolubilized azobenzene in the mixed admicelles was found to be greater than its solubilization in mixed micelles of the same surfactants.

Esumi, Shibayama, and Meguro [36] studied the adsolubilization of hexanol and 2,2,3,3,4,4,4-heptafluorobutanol (HFB) in the adsorbed bilayers of lithium dodecyl sulfate (LDS) and lithium perfluorooctanesulfonate (LiFOS) on alumina. Using pyrene as a fluorescence spin probe, they also investigated the effect of the adsolubilized alcohol on the polarity of the bilayer. Because hexanol and HFB are only slightly soluble in water, they tend to adsolubilize into the admicelle. The amount of adsolubilized hexanol in LDS was found to increase with increasing hexanol concentration in the bulk solution, reaching a maximum of about 5×10^{-4} (mols hexanol /g of alumina). The polarity of the admicelle was found to decrease as the amount of adsolubilized hexanol increased until the polarity approached that of pyrene in pure hexanol. The amount of hexanol which adsolubilized in the LiFOS bilayer likewise increased with increasing hexanol concentration, reaching a maximum of about 1×10^{-4} (mols hexanol /g of alumina). The polarity of the LiFOS admicelle was found to decrease as the amount of adsolubilized hexanol increased. The results of the adsolubilization of HFB in LDS and LiFOS admicelles showed similar trends, although the adsolubilization of HFB in LiFOS did not have an appreciable effect on the polarity of the LiFOS admicelle. Additionally, the polarity of the LDS bilayer containing adsolubilized HFB, as indicated by the pyrene, was nearly the same as that of pyrene in LDS micelles, indicating that the pyrene adsolubilizes in the LDS, and not the HFB.

A study by Esumi et al. [37] on the characterization of an adsorbed cationic surfactant bilayer on silica used electron spin resonance and fluorescence spectra to investigate the micropolarity and microviscosity of admicelles of dodecyl pyridinium chloride (DPCI) and cetyl pyridinium chloride (CPCI). As was the case with the anionic and nonionic surfactants, the admicelles of these surfactants exhibited a lower micropolarity and higher microviscosity than did their respective micelles.

Esumi and Yamada [38], recently investigated the characteristics of phospholipid layers adsorbed onto silica. L-distearoylphosphatidylcholine (DSPC), DL-tocopherol acetate (VEA), octyl methoxycinnamate (PARSOL MCX), and butyl methoxydibenzoylmethane (PARSOL 1789) were used in the study. The adsorption of DSPC was found to follow the same trends of previously studied surfactants, showing increasing adsorption with increasing bulk solution concentration. The DSPC gave evidence of first forming adsorbed monolayers and then bilayers at saturation. The adsolubilization of VEA, PARSOL MCX, and PARSOL 1789 were studied and showed that adsolubilization decreased with increasing molecular weight. The order of adsolubilization was PARSOL MCX > VEA > PARSOL 1789.

B. Separation Processes

The second area involving adsolubilization is its use in separation procedures. This seems to follow as a natural extension of the demonstrated differences in the solubilities of various compounds into admicelles. These new, low-temperature processes could be useful in biotechnology and other applications which are now limited by current separations technology. Lee et al. [39] investigated the adsolubilization of a series of normal alcohols in the admicelles of sodium dodecyl sulfate (SDS) on alumina. Somewhat unexpected results were obtained: First, very high ratios of alcohol to surfactant in the admicelles were found at very low levels of surfactant adsorption, but these ratios decreased with increasing surfactant adsorption to a value similar to that in micelles of SDS. Second, in the presence of the alcohols there was a remarkable increase in surfactant adsorption below the CMC but a slight decrease in the plateau adsorption of the surfactant. In order to explain these results, a two-site model of alcohol adsolubilization was presented, as illustrated in Fig. 2. It was proposed that alcohols, because of their polar end group, could adsolubilize both between the surfactant headgroups in the admicelle and at the perimeter of a disklike, bilayered admicelle. This would explain the high adsolubilization ratios at low surfactant concentrations, with the effect becoming less pronounced as the surfactant coverage increases; at low coverage, the ratio of the adsolubilization sites around the perimeter to the sites between the headgroups decreases with increasing surfactant

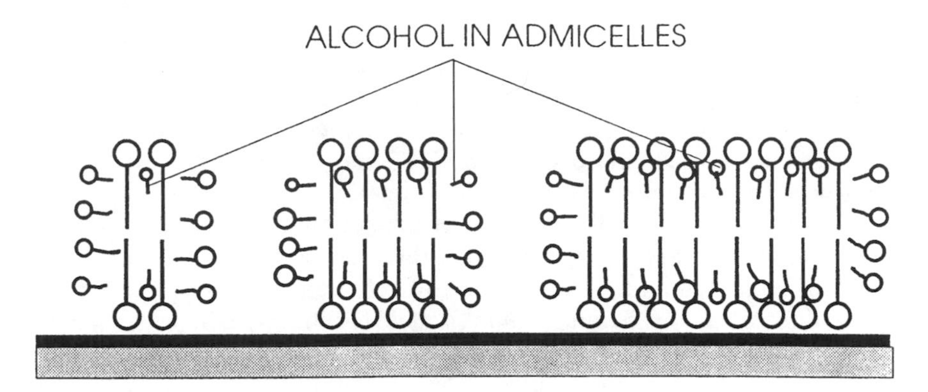

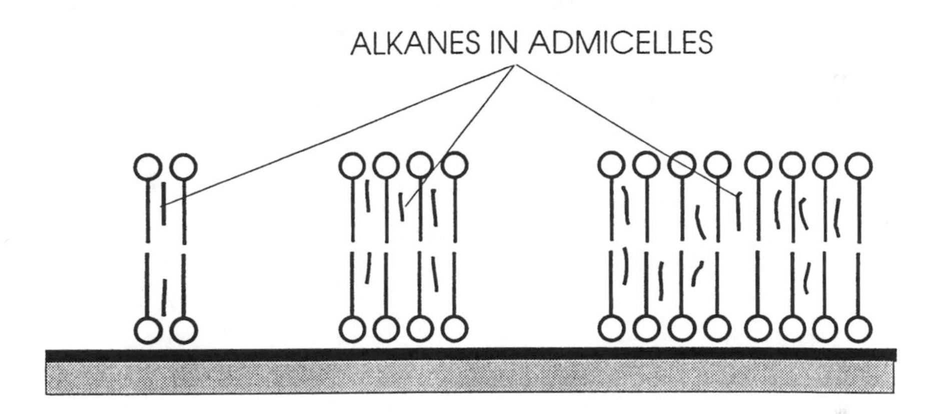

FIG. 2 Location of adsolubilizates in admicelles at increasing levels of surfactant adsorption.

coverage. This also explains the dramatic increase in the stability of the aggregates at very low coverages combined with the exclusion of surfactant from the bilayer at very high coverages.

The study also investigated the effect of alcohol chain length on adsolubilization as well as aggregate size. The results indicated that as the alkane chain length increased, the adsorption of the alcohol increased. It was suggested that this is due to an increase in the hydrophobic contribution of the chain. Aggregate size at $\frac{1}{100}$th of bilayer coverage was calculated to be about 60, very similar in size to an SDS micelle. As the bulk solution concentration of SDS approached the CMC, the aggregate size increased, approaching a maximum value of approximately 25,000.

Barton et al. [28] studied the use of admicellar chromatography as a means of separating and concentrating three of the isomers of heptanol. Alumina was used as the particulate to pack a chromatography column while the anionic surfactant sodium dodecyl sulfate (SDS) was used to form the admicelles. The alcohols *n*-heptanol, 3-heptanol, and 2-methyl-2-hexanol were used in the study. The adsolubilization of each alcohol was first studied, showing the following pattern: *n*-heptanol > 3-heptanol > 2-methyl-2-hexanol. As can be seen from the adsorption isotherms (Figs. 3–4), *n*-heptanol could be concentrated from either of the other isomers at the low end of the isotherm. At the upper end, it appears that 2-methyl-2-hexanol could be separated from the other isomers. Predictions based upon the single-component isotherms were then tested using a 50:50 mix of *n*-heptanol and 2-methyl-2-hexanol at various points along the ad-

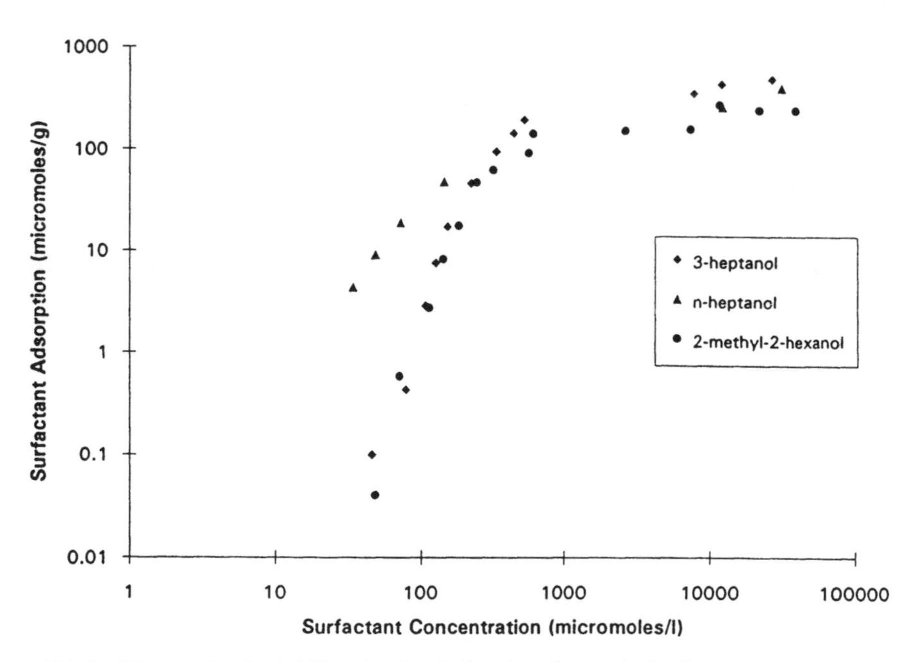

FIG. 3 Heptanol adsolubilization in dodecyl sulfate admicelles.

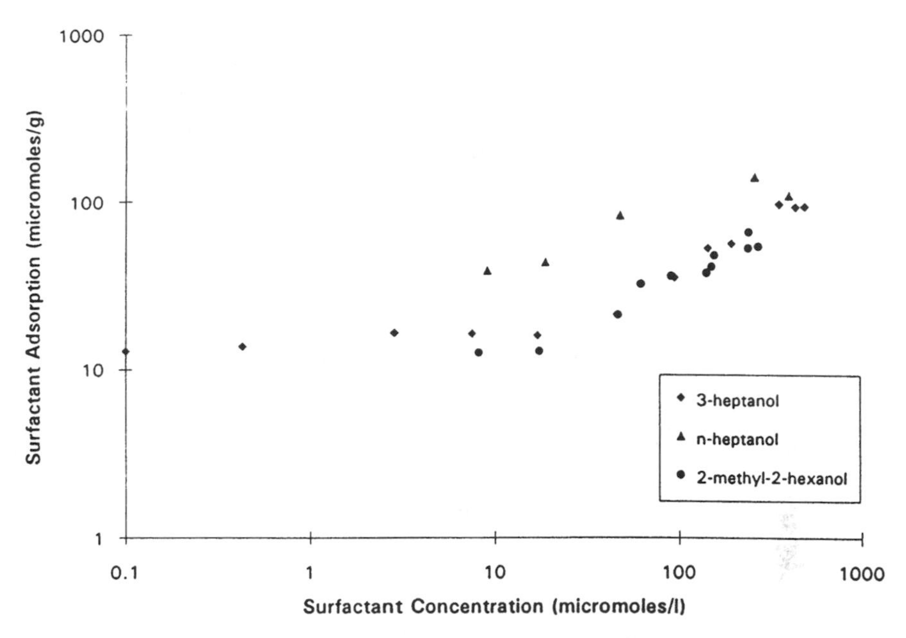

FIG. 4 Dodecyl sulfate adsorption isotherms in the presence of different isomers of heptanol.

sorption isotherm. Separation factors as large as 4 were obtained for the system.

C. Thin-Film Formation

A third use of adsolubilization is in the surface modification of particulates by in situ polymerization of adsolubilized monomers within the surfactant bilayer. This technology offers several possible applications: (1) the potential of modifying materials such as titanium dioxide, alumina, and silica, which are used as fillers, pigments, and reinforcers in a variety of applications, to improve their performance in these composites, (2) a new method of producing various polymers, (3) a methodä of modifying surfaces for biotechnology and separation applications, (4) a method of binding a variety of compounds to different substrates for a variety of applications.

The first work in this area was done by Wu et al. [25–27], who studied the application of an ultrathin polystyrene film onto the surface of alumina using the in situ polymerization of styrene monomers adsolubilized within sodium dodecyl sulfate (SDS) admicelles. Styrene was found to adsolubilize into SDS admicelles to a maximum value of one styrene monomer to approximately two adsorbed SDS molecules over a range of styrene concentrations. After washing, the modified alumina was found to be very hydrophobic.

Mangaokar [40] studied the application of polystyrene to silica using a similar process. This study used both a cationic surfactant, cetyl trimethyl ammonium bromide (CTAB), and a nonionic surfactant, Triton X-100. The admicelles were formed on two different silicas, one fused and one precipitated. The process produced ultrathin polystyrene films which were then extracted and analyzed.

The first use of a gaseous monomer was made by Lai [41], who studied the adsolubilization and polymerization of tetrafluoroethylene (TFE) in perfluorinated surfactant admicelles on powdered alumina and thin plates of alumina. Tridecafluoroheptanoic acid sodium salt was used as the surfactant. The amount of adsolubilized TFE was found to increase with increasing initial pressure of TFE above the solution. The partition coefficient for adsolubilization, K_{ads}, reached a maximum value in the range of 670 to 800 M^{-1}; K_{ads} is defined the ratio of adsolubilized TFE molecules to adsorbed surfactant molecules divided by the concentration of TFE in the supernatant. After washing with water, the modified surface of the alumina were found to be highly hydrophobic. The polymer formed was analyzed by FTIR and other methods. Chen [42] extended the process to include ultrathin polystyrene films on titanium dioxide substrate using admicelles of sodium dodecyl sulfate.

The second use of a gaseous monomer, and the first copolymer work, was done by O'Haver [43], forming styrene-butadiene copolymer on alumina. Sodium dodecyl sulfate was used as the surfactant, and the polymerization was carried out at 5°C using a redox initiation system similar to those used in commercial emulsion polymerization. The polymer was extracted and analyzed by FTIR, NMR, and other methods.

D. Studies of Adsolubilization in Admicellar Catalysis

Surfactant aggregates (micelles) have been studied extensively for the catalysis of chemical reactions. Rate enhancement occurs primarily by concentration enhancement of organic substrates in the hydrophobic core and of ionic reactants in the Stern layer. Recently it has been demonstrated that surfactant aggregates adsorbed on high surface area supports also catalyze chemical reactions [30]. In this work, the hydrolysis of trimethyl orthobenzoate (TOB) was catalyzed by SDS admicelles on porous alumina. Under conditions similar to that used for the micellar catalysis of the same reaction [44], comparable rate enhancement was achieved. Itä has theorized that this catalysis is also primarily due to concentration enhancement. Application of a modified pseudophase ion exchange model [45–47] requires adsolubilization constants for TOB and for the product methyl benzoate (MB) which is also adsolubilized in the admicelle interior. The adsolubilization constant K_T has recently been measured to be approximately 45 M^{-1} for surface coverages of 200 mols/g alumina and a solution pH of 8.4. The adsolubilization constant is defined by

$$K_T = \frac{[T]_m}{[T]_W[C]_D}$$

where $[T]_m$ is the concentration of TOB in the admicelle, $[T]_W$ is the equilibrium aqueous phase concentration, and $[C]_D$ is the concentration of adsorbed surfactant in aggregate form. The definition of $[C]_D$ for the admicellar system differs from that of the micellar system since, in general, the monomeric surfactant concentration is below the CMC. For application of the PIE model to admicellar catalysis, $[C]_D = [C]_t - [C]_W$, where $[C]_t$ is the total surfactant concentration and $[C]_W$ is the concentration of monomeric surfactant in the aqueous phase. The adsolubilization constant for MB was measured at pH 5 and a surfactant coverage of 480 mols/g alumina to be 27 M^{-1}. These values are somewhat less than the value of 73 M^{-1} reported in [44]. However, Yu et al. [30] found evidence that the catalytic activity of the admicelle increased with surface coverage. This may be due to increased adsolubilization upon bilayer formation or upon tighter surfactant packing on the surface, and the K_T for the admicelle system may show significant increase at higher surface coverages.

III. SUMMARY AND CONCLUSIONS

Given the technological potential of the adsolubilization phenomenon it is remarkable that it has not been seriously studied until recently. In fact, most of the research that has involved adsolubilization has not actually focused on either adsolubilization or a use of adsolubilization, but rather on the use of an adsolubilized probe to characterize the adsorbed surfactant layer. One of the few true studies of adsolubilization itself, that of Lee et al [39] produced quite unexpected results. This is encouraging; though there are many similarities between solubilization and adsolubilization, we can expect that those differences will produce still further unexpected behaviors which will open the doors to future technologies.

REFERENCES

1. P. Somasundaran and D. W. Fuerstenau, *J. Phys. Chem. 79*: 90 (1982).
2. J. F. Scamehorn, R. S. Schechter, and W. H. Wade, *J. Colloid Interface Sci. 85*: 463 (1982).
3. A. M. Gaudin and D. W. Fuerstenau, *Min. Eng. 7*: 958 (1955).
4. A. M. Gaudin and D. W. Fuerstenau, *Trans. AIME 202*: 66 (1966).
5. P. Somasundarum, T. W. Healy, and D. W. Fuerstenau, *J. Phys. Chem. 68*: 3562 (1964).
6. J. M. Cases and F. Villieras, *Langmuir 8*: 1251 (1992).
7. M. R. Böhmer and L. K. Koopal, *Langmuir 8*: 1594 (1992).
8. M. R. Böhmer, L. K. Koopal, R. Janssen, E. M. Lee, R. K. Thomas, and A. R. Remie, *Langmuir 8*: 228 (1992).
9. M. A. Yeskie and J. H. Harwell, *J. Phys. Chem. 92*: 2346 (1988).
10. D. Bitting and J. H. Harwell, *Langmuir 3*: 531 (1987).
11. J. H. Harwell, J. Hoskins, R. S. Schechter, and W. H. Wade, *Langmuir 1*: 251 (1985).
12. F. J. Trogus, R. S. Schechter, and W. H. Wade, *J. Colloid Interface Sci. 70*: 293 (1979).
13. K. C. Waterman, N. J. Turro, P. Chandar, and P. Somasundaran, *J. Phys. Chem. 90*: 6828 (1986).
14. P. Chandar, P. Somasundaran, K. C. Waterman, and N. J. Turro, *J. Phys. Chem. 91*: 148 (1987).
15. P. Chandar, P. Somasundaran, and N. J. Turro, *J. Colloid Interface Sci. 117*: 31 (1987).
16. P. Levitz, H. Van Damme, D. Keravis, *J. Phys. Chem. 88*: 2228 (1984).
17. P. Levitz and H. Van Damme, *J. Phys. Chem. 90*: 1302 (1986).
18. J. H. Harwell and J. F. Scamehorn, in *Mixed Surfactant Systems* (K. Ogini and M. Abe, eds.), Marcel Dekker, New York, 1993, p. 263.
19. J. H. Harwell and M. A. Yeskie, *J. Phys, Chem. 93*: 3372 (1989).
20. K. S. Kung and K. F. Hayes, *Langmuir 9*: 263 (1993).
21. T. Wakamatsu and D. W. Fuerstenau, *Adv. Chem. Ser. 79*: 161 (1968).
22. B. L. Roberts, J. F. Scamehorn, and J. H. Harwell, in *Phenomena in Mixed Surfactant Systems* (J. F. Scamehorn, ed.), ACS Symp. Ser. 311, Amer. Chem. Soc., Washington, DC, 1986, p. 200.
23. J. J. Lopata, J. H. Harwell, and J. F. Scamehorn, in *Surfactant-Based Mobility Control* (D. H. Smith, ed.), ACS Symp. Ser. 373, Amer. Chem. Soc., Washington, DC, 1988, p. 205.
24. P. Somasundaran, E. Fu, and Qun Xu, *Langmuir 8*: 1065 (1992).
25. J. Wu, J. H. Harwell, and E. A. O'Rear, *J. Phys. Chem. 91*: 623 (1987).
26. J. Wu, J. H. Harwell, and E. A. O'Rear, *Langmuir 3*: 531 (1987).
27. J. Wu, J. H. Harwell, E. A. O'Rear, and S. D. Christian, *AIChE J. 34*: 1511 (1988).
28. J. W. Barton, T. P. Fitzgerald, C. Lee, E. A. O'Rear, and J. H. Harwell, *Sep. Sci. Tech. 23*: 637 (1988).
29. C. Lee, E. A. O'Rear, J. H. Harwell, and J. A. Sheffield, *J. Colloid Interface Sci. 137*: 296 (1990).

30. C.-C. Yu, D. W. Wong, and L. L. Lobban, *Langmuir 8*: 2582 (1992).
31. A. M. Koganovskii, N. A. Klimenko, and A. A. Tryasorakova, *Kolloid. Zh. 36*: 165 (1974).
32. C. C. Nunn, R. S. Schechter, and W. H. Wade, *J. Phys. Chem. 86*: 3271 (1982).
33. W. L. Hubbell and H. M. McConnell, *Proc. Nat. Acad. Sci. U.S.A. 64*: 20 (1969).
34. B. Zhu, X. Zhao, and T. Gu, *J. Chem. Soc., Faraday Trans. 1 84*: 3951 (1988).
35. K. Esumi, Y. Sakamoto, T. Nagahama, and K. Meguro, *Bull. Chem. Soc. Jpn. 62*: 2502 (1989).
36. K. Esumi, M. Shibayama, and K. Meguro, *Langmuir 6*: 826 (1990).
37. K. Esumi, T. Nagahama, and K. Meguro, *Colloids Surf. 57*: 149 (1991).
38. K. Esumi and T. Yamada, submitted to *J. Colloid Interface Sci.*
39. C. Lee, M. A. Yeskie, J. H. Harwell, and E. A. O'Rear, *Langmuir, 6*: 1758 (1990).
40. A. S. Mangaokar, M. S. thesis, University of Oklahoma, 1991.
41. C. L. Lai, M. S. thesis, University of Oklahoma, 1992.
42. H. Y. Chen, M. S. thesis, University of Oklahoma, 1992.
43. J. H. O'Haver, Ph.D. dissertation, University of Oklahoma, in preparation.
44. R. B. Dunlap and E. H. Cordes, *J. Am. Chem. Soc. 90*: 4395 (1968).
45. L. S. Romsted, in *Micellization, Solubilization, and Microemulsions* (K. L. Mittal, ed.), Plenum Press, 1977, Vol. 2, p. 509.
46. L. S. Romsted, in *Surfactants in Solution* (K. L. Mittal and B. Lindman, eds.), Plenum Press, 1984, Vol. 2, p. 1015.
47. F. H. Quina and H. Chaimovich, *J. Phys. Chem. 83*: 1844 (1979).

9

Solubilization in Micelles and Vesicles Studied by Fluorescence Techniques: Interplay Between the Microproperties of the Aggregates and the Locus and Extent of Solubilization

EDUARDO LISSI and ELSA ABUIN Departamento de Quimica, Facultad de Ciencia, Universidad de Santiago de Chile, Santiago, Chile

SYNOPSIS

An overview of fluorescence techniques and methodologies that are commonly employed in studies of solubilization in micellar and vesicular solutions is presented in this chapter. Three aspects of solubilization are the main topics to be treated in detail: the solubilization extent, the solubilization loci, and the effect that the solubilizate can exert upon properties of the microaggregates. The first part deals with the evaluation of partition coefficients. Methods useful to obtain partitioning data pertaining to fluorophores, fluorescence quenchers, and additives that are nonquenchers but modify a fluorescence characteristic are described and their potential and limitations are discussed. The second part is devoted to analyzing how fluorescence measurements can be employed to monitor the characteristics of the microenvironment of solutes incorporated to micelles and vesicles and how the properties of the aggregate are modified by the presence of the solute. The interplay between solubilization and microaggregate properties is discussed, with special emphasis in cosolubilization extent, dependence of partition coefficients with solute concentration, and the changes in the fluorescence characteristics of fluorophores elicited by cosolute incorporation.

I. INTRODUCTION

During the last two decades, fluorescence techniques have been extensively employed to study several aspects of solubilization in micellar (mostly) and vesicular solutions. The field is currently very active and has received extensive coverage in several recent textbooks and review articles [1–10]. This chapter does not attempt an exhaustive description of all the aspects of solubilization in these types of solutions that can be studied by means of fluorescence techniques. Our aim will be to focus on the extent of solubilization, the loci of solubilization, and the effects that the solubilizate can exert upon properties of the microaggregates in relation to fluorescence. The emphasis will be in trying to put in correct terms the methodologies and in discussing their potential and limitations. The interpretation of the data will be exemplified with a few selected cases rather than a comprehensive listing of results available in the literature.

A. Brief Outline of Fluorescence Parameters Used in Solubilization Studies

Fluorescence is the emission of light accompanying the transition between two electronic states of the same multiplicity. In this chapter, it refers to the transition from the first excited singlet state of a fluorophore (the light-emitting species) to the ground state. For a fluorophore to be effective in studies of solubilization in micellar or vesicular solutions, some parameter pertaining to fluorescence must be sensitive to changes in the environment and/or be modified in the presence of the solubilizate. The various fluorescence parameters used in this context are the position of the emission band, the intensity of the light emitted at a given wavelength (or the fluorescence quantum yield), the fluorescence lifetime, the distribution of lifetimes, and the polarization (or anisotropy) of the emission. For most fluorophores, all these parameters are sensitive to the surrounding environment, and, hence, their changes in the presence of a microphase can be (and have been) employed to obtain information on several aspects of the solubilization process, such as the fluorophore distribution, the solubilization loci, and the characteristics of the microenvironment where it is located. Furthermore, the fluorescence can be quenched (decreasing the intensity or the lifetime) or sensitized (increasing the intensity) in the presence of additives. Hence, the changes elicited by the presence of a solubilizate upon the fluorescence intensity (or lifetime) from bound (associated to the microphase) or free (located in the aqueous phase) fluorophores can be employed to obtain the distribution of the solute between the aqueous phase and the micellar (or vesicular) pseudophase.

B. Definitions of Partition Coefficients

There is yet no real consensus regarding the best form in which the association of solutes with micellar or vesicular aggregates can be quantified. Within the framework of the pseudophase model [11], the distribution of a solute (S) between the aqueous pseudophase (or aqueous phase) and the micellar or vesicular pseudophase can be described in terms of partition constants defined by

$$K = \frac{X_{agg}}{X_W} \tag{1}$$

or

$$K' = \frac{X_{agg}}{[S]_W} \tag{2}$$

where X is the solute mole fraction in the dispersed pseudophase (X_{agg}) or in the aqueous phase (X_W), and $[S]_W$ is its concentration in the aqueous phase. Evaluation of K or K' only requires a knowledge of the amount of S associated with the aggregates (or its concentration in the aqueous phase), and the concentration of micellized (or in-vesicle) surfactant. Other definitions of the partition constant such as

$$K'' = \frac{n}{[S]_W} \tag{3}$$

where n is the solute mean occupation number, or

$$K''' = \frac{[S]_{in}}{[S]_W} \tag{4}$$

where $[S]_{in}$ is the concentration of the solute inside the aggregates, require a knowledge of the micelle (or vesicle) aggregation number (K'') or an assumption of a volume for the dispersed pseudophase (K'''). Partition constants defined by Eq. (3) and (4) then involve a certain degree of arbitrariness, particularly in the presence of large amounts of incorporated additives.

The partition constants defined by Eq. (1) to (4) are expressed in terms of concentrations, not activities. They must then be considered as "pseudo" constants or partition coefficients and as such, they could depend upon the intra-aggregate solute concentration. This is particularly relevant since strong nonideal behavior can be expected in these systems.

The definitions of association constants given by Eq. (1) to (4) cover partition constants (within the pseudophase approximation) and binding constants, where it is considered that the aggregates have a fixed number of sites to bind the solute [12, 13]. The dependence of partition constants on the concentration of the solute inside the aggregates can then be explained either in terms of binding models and/or be ascribed to the nonideality of the system (and hence to changes in the "pseudo" partition constant with the intra-aggregate solute concentration).

The association of the solute with the aggregates, considered as a dynamic equilibrium, implies that

$$K'' = \frac{k_+}{k_-} \tag{5}$$

where k_+ is the bimolecular rate constant for the incorporation of the solute molecules into the aggregates (in terms of the molar concentration of aggregates) and k_- is the unimolecular rate constant for the exit of the solute molecules from the aggregates. Values of K'' can therefore be obtained from a kinetic analysis of time-resolved fluorescence data [14]. It is important to note that, when the data analysis is based on the entrance and exit rates of the fluorophore, the K'' values obtained are those corresponding to the excited molecules, which can be different than those obtained by other procedures that renders the distribution constant of the molecules in the ground state.

II. EVALUATION OF PARTITION COEFFICIENTS

The experimental procedures and the assumptions required to obtain partitioning data from fluorescence measurements can be divided in two groups: those pertaining to the distribution of the fluorophore and those relating to the distribution of additives that modify the fluorescence of the fluorophore. For each of these methods, any of the fluorescence properties already mentioned can be employed to evaluate partition constants, either by employing steady-state or time-resolved fluorescence measurements. The present chapter will emphasize methods based on steady-state fluorescence measurements.

A. Partition Coefficients for Fluorophores

For a fluorophore distributed between the dispersed pseudophase and the dispersium medium, three situations can be envisaged depending upon the relationship between the entrance and exit rates and the fluorophore lifetime τ:

(i) k_- and k_+ [aggregates] $\ll 1/\tau$ (static limit)
(ii) k_- and k_+ [aggregates] $\gg 1/\tau$ (dynamic limit)
(iii) k_- and k_+ [aggregates] $\approx 1/\tau$

Case (iii) is particularly useful to evaluate k_+ and k_- from time-resolved fluorescence measurements, but for measuring partition coefficients by employing steady-state fluorescence techniques, it is the more complex. We will refer mostly to case (i) since the dynamic limit is only attained for long-lived excited states (such as triplet states or singlet oxygen).

1. Evaluation of Partition Coefficients from Steady-State Fluorescence Measurements as a Function of the Surfactant Concentration

(a) Static Limit. Under conditions where the static limit holds and considering, for simplicity, that the absorbance of the sample at the working excitation wavelength is low, the fluorescence intensity I (in arbitrary units) is given by

$$I = J(\epsilon_{agg}\Phi_{agg}[S]_{agg} + \epsilon_W\Phi_W [S]_W) \quad (6)$$

where J is a constant proportional to the intensity of the incident light, $[S]_{agg}$ and $[S]_W$ are the analytical concentrations of the fluorophore associated to the aggregates and in the aqueous solution, and ϵ and Φ are the extinction coefficients and fluorescence quantum yields, respectively. The analytical concentrations are related to the concentration in the aqueous phase $[S]_W$ and to the intra-aggregate solute mol fraction through the relations

$$[S]_W = (1 - \alpha)[S]_W \tag{7}$$

and

$$X_{agg} = \frac{[S]_{agg}}{([D] - \text{cmc}) + [S]_{agg}} \tag{8}$$

where α is the volume fraction of the dispersed pseudophase, [D] is the analytical concentration of the surfactant, and cmc denotes the analytical free-surfactant concentration under the conditions employed. From Eq. (6) to (8), it can be derived that, at a fixed analytical concentration of the fluorophore,

$$\frac{I_W}{I - I_W} = \left[\frac{I_W}{I_{agg} - I_W}\right] + \left[\frac{I_W}{I_{agg} - I_W}\right]\left(\frac{1}{K'}\right)\frac{1 - \alpha}{[D] - \text{cmc}} \tag{9}$$

where

$$I_W = J\epsilon_W \Phi_W [S]_W \tag{10}$$

and

$$I_{agg} = J\epsilon_{agg}\Phi_{agg}[S]_{agg} \tag{11}$$

and I_W and I_{agg} correspond to the intensities expected if all the fluorophore were present in the aqueous (I_W) or dispersed (I_{agg}) pseudophases. A plot of the left-hand side of Eq. (9) against $(1 - \alpha)/([D] - \text{cmc})$ gives the value of K' as the quotient between the intercept and the slope. This treatment requires that, in the surfactant concentration range considered, K' must be independent of [D] and the intra-aggregate fluorophore concentration. Furthermore, the values of α and cmc required to apply Eq. (9) can be bound to some uncertainties due to their dependence with the surfactant and fluorophore concentrations. Equations very close to Eq. (9) have been employed to obtain partitioning data for a wide range of aromatic compounds [15, 16] and other fluorophores such as psoralens [17] and 1-benzyl-1,4-dihydronicotinamide [18]. A different treatment of the data must be carried out when the fluorophore has a very low fluorescence quantum yield in aqueous solution (e.g., 1-anilinonaphthalene-8-sulfonic acid (ANS) or diphenylhexatriene derivatives). In the limiting situation where $\Phi_W = 0$, Eq. (12) can be employed [19]:

$$\frac{1}{I} = \frac{1}{I_\infty} + \frac{1}{I_\infty}\left(\frac{1}{K'}\right)\frac{1 - \alpha}{[D] - \text{cmc}} \tag{12}$$

where I_∞ is the light intensity at "infinite" quencher concentration. A plot of the left-hand side of Eq. (12) against $(1 - \alpha)/([D] - \text{cmc})$ gives the fluorescence intensity of the bound fluorophore as the intercept, and K' from the intercept/slope ratio.

An interesting extension of the methods based on fluorescence intensity measurements at different surfactant concentrations has been developed by Lakowicz and Keating [20] based on the phase-sensitive fluorescence detection. By an appropriate choice of the detector phase angle, it is possible to monitor selectively the emission arising from the fluorophores associated to the microphase or that arising from the surrounding medium. This selectivity allows for a direct evaluation of the fraction incorporated (or excluded) from the aggregates, which is particularly useful when one of the fractions is small. One of the limitations of the method is that its application requires a knowledge of the lifetimes in both media. The authors have applied this procedure to evaluate the fraction of 11-(3-hexyl-1-indolyl)-undecyltrimethylammonium bound to cetyltrimethylammonium bromide (CTAB) micelles over a wide range of surfactant concentrations.

(b) Dynamic Limit. Under these conditions, the light absorbed (if ϵ_{agg} is not equal to ϵ_W) is determined by the fluorophore distribution in the ground state, while the fluorescence quantum yield depends on the distribution in the excited state. Working at an isosbestic point ($\epsilon_{agg} = \epsilon_W$), it can be easily shown that, if the data conform to Eq. (9), the intercept/slope ratio gives the value of the partition constant of the fluorophore in the excited state.

2. Procedures Based on Changes of the Position or Shape of the Emission Spectrum as a Function of the Surfactant Concentration

The fluorescence of the fluorophore inside the aggregates and in the aqueous phase can differ in the shape of the emission band and/or in the band position. This latter effect can be characterized by the wavelength of maximum fluorescence intensity or by other parameters (such as the generalized polarization [21]) amenable to quantification.

If A' is the value of a fluorescence property in the "bound" state and A^0 in the "free" state, the fraction of bound molecules β can directly be derived from the value of A at a given surfactant concentration by

$$\beta = \frac{A^0 - A}{A^0 - A'} \tag{13}$$

only if A is linearly related to the extent of binding, as previously discussed for the fluorescence intensity at low (or fixed) absorbances. However, this simple relationship cannot be employed in general for properties related to the band shape or position. Among the fluorophores whose band shape changes with the microenvironment is pyrene (Py), one of the compounds frequently employed in this type of studies. In particular, the I_1/I_3 ratio between the intensities of the first and third vibronic bands of the emission spectrum has been widely employed to characterize microenvironments [2, 3, 22, 23], and also for the evaluation of its distribution and/or association with microphases [24]. However, since the I_1/I_3 ratio is not directly related to the Py distribution, Eq. (13) cannot be employed to obtain its partition. In this case, the $(A^0 - A)/(A^0 - A')$ ratio is related to the fraction of "bound" Py by

$$\frac{A^0 - A}{A^0 - A'} = \frac{\beta}{\beta + (1 - \beta)\Phi_W \epsilon_W/\Phi_{agg}\epsilon_{agg}} \tag{14}$$

where the quantum yields are those measured at the wavelength corresponding to the third vibronic band. Equation (13) only applies when

$$\Phi_W\epsilon_W = \Phi_{agg}\epsilon_{agg} \tag{15}$$

a condition under which the I_1/I_3 ratio measures the changes in I_1 and then the procedure is equivalent to measurement of the band intensity (and not the band shape). These considerations are particularly important since changes in band shape are generally accompanied by changes in lifetimes (and hence in Φ). The errors introduced by the improper use of Eq. (13) can be significant. For example, a value of $(A^0 - A)/(A^0 - A') = 0.5$ corresponds to β values of 0.66, 0.5, or 0.33 when $\epsilon_{agg}\Phi_{agg}/\epsilon_W\Phi_W$ are 0.5, 1.0, or 2.0, respectively. Similar considerations preclude a direct evaluation of partition constants from the observed changes in band position [17, 25] or anisotropy [26] with surfactant concentration.

3. Procedures Based on Time-Resolved Fluorescence Measurements

The expressions for the decay of the fluorescence of a partitioned fluorophore in terms of K'', k_+, k_- and its lifetime inside and outside the aggregates have been extensively worked out [1, 3]. The simplest situations apply for the limiting cases (i) and (ii).

(a) Static Limit. The decay follows a biexponential function

$$I = A_W \exp\left(\frac{-t}{\tau_W}\right) + A_{agg} \exp\left(\frac{-t}{\tau_{agg}}\right) \tag{16}$$

where the coefficients A_W and A_{agg} measures the initial ($t = 0$) intensity of the light emitted from the fluorophores in the aqueous and dispersed pseudophases, respectively [27]. The partition constant is given by

$$K' = \frac{A_{\text{agg}}}{A_{\text{W}}} \left(\frac{\epsilon_{\text{W}}}{\epsilon_{\text{agg}}} \right) \frac{(k_f)_{\text{W}}/(k_f)_{\text{agg}}}{[\text{D}] - \text{cmc}} \tag{17}$$

where k_f is the fluorescence specific rate constant ($k_f = \Phi_f/\tau$). If the initial intensity at the same absorbance in the absence of surfactant $(A^0)_{\text{W}}$ is determined, the partition constant is simply given by

$$K' = (1 - \alpha) \frac{(A^0)_{\text{W}} - A_{\text{W}})}{A_{\text{W}}([\text{D}] - \text{cmc})} \tag{18}$$

(b) Dynamic Limit. In the pseudoequilibrium limit, the decay becomes monoexponential with

$$\frac{1}{\tau} = (1 - \beta) \frac{1}{\tau_{\text{W}}} + \beta \left(\frac{1}{\tau_{\text{agg}}} \right) \tag{19}$$

If the lifetimes of the fluorophore in the aqueous phase and inside the aggregates are known or can be determined independently, Eq. (19) allows the evaluation of β and from its value, the partition constant for the excited fluorophore.

(c) Case (iii). In this situation, the decay becomes complex. Fitting the data to simplified models [1, 3] or a global compartmental analysis treatment [28] allows the evaluation of K'', k_+, and k_-.

4. Procedures Based on Selective Quenching by Totally Incorporated or Totally Excluded Quenchers

The partitioning of a fluorescent compound in a microheterogeneous solution can be determined from methods based on the selective quenching of its luminescence in one of the pseudophases.

In solutions of aggregates formed by ionic surfactants, selective quenching of the fluorescence in the aqueous phase can be achieved by using small hydrophilic co-ions (i.e., ions having the same sign of charge as that of the surface) as quenchers. In fact, several studies indicate that ionic quenchers located in the aqueous phase are not able to quench the fluorescence of aromatic compounds incorporated in micelles when the ions and the micelle surface are like charged. For example, iodide ions do not have influence on the fluorescence of probes incorporated in sodium dodecylsulfate (SDS) micelles [29], naphthalene in SDS micelles is unquenched by Br^- ion [27], and anthracene in CTAB micelles is unquenched by pyridinium ion [30]. Similarly, Py and Py derivatives incorporated in large dioctadecyldimethylammonium chloride (DODAC) vesicles are completely protected from quenching by acrylamide [31]. Quina and

Toscano [32] were the first to propose a method based on the selective quenching of the fluorescence in the aqueous phase for the evaluation of the partition constants of a fluorescent compound in micellar solutions. Data for the quenching of pyrene butyrate (PBA) by I^- ions in micellar solutions of SDS were analyzed in terms of a two-state model in which only the aqueous fluorescer is quenchable. The method assumes that the quencher is excluded from the micelle, that the excited molecule does not enter or leave the micelle during its lifetime, and that the decay of the fluorescer in the aqueous phase follows Stern-Volmer kinetics with a Stern-Volmer constant defined in terms of the analytical concentration of the quencher. The two first assumptions can be verified experimentally, for example, by showing that the lifetime of the fluorescer in the micellar state is unaffected upon addition of the quencher [32]. The condition of free-surfactant kinetics for the quenching of the fluorescer in the aqueous state could be invalid if excluded volume effects (due to like-charge repulsions between the micelles and the quencher) were important. In this case, the effective concentration of the quencher in the aqueous phase will be different than the analytical concentration and dependent on the concentration of the surfactant and the ionic strength of the solution. With the above-mentioned assumptions, Eq. (20) can be derived:

$$\frac{I}{I^0 - I} = \left[\frac{[\mathrm{S}]_{\mathrm{agg}}}{[\mathrm{S}]_{\mathrm{W}}}\left(\frac{\Phi_{\mathrm{agg}}}{\Phi_{\mathrm{W}}}\right) + 1\right]\left[1 + \frac{1}{K_{\mathrm{SV}}}\left(\frac{1}{[\mathrm{S}]}\right)\right] \tag{20}$$

where K_{SV} is the Stern-Volmer constant for the quenching of the fluorescer in the aqueous phase. A plot of the left-hand side of Eq. (20) against the reciprocal of the quencher concentration $1/[S]$ gives the ratio $[S_{agg}]/[S_W]$ from the intercept if Φ_{agg}/Φ_W is known [32]. Finally, division of $[S_{agg}]/[S_W]$ by the corresponding concentration of micellized surfactant gives the partition constant. The quenching of the fluorescence of naphthalene by Br^- in micellar solutions of SDS has also been interpreted in terms of this model by Almgren et al. [16] to determine the partition constant of naphthalene. A slightly modified version of Eq. (20), which takes into account the possibility of different absorptivities in the micellar pseudophase and the aqueous phase, has been proposed by Abuin and Lissi [33] and applied to determine the partitioning of naphthalene [33] and a series of aromatic alcohols [34] in CTAB micellar solutions, by employing Ni^{2+} ions as selective quenchers of the luminescence in the aqueous phase. The method has also been applied in a study of the solvent capacity of SDS and CTAB micelles toward several naphthalene derivatives employing Cu^{2+} and Cs^+ as selective quenchers in the CTAB solutions, and I^- in SDS solutions [15].

When quenchers excluded from the aggregates are employed, extrapolation of the fluorescence intensity to "infinite" quencher concentration results in the fluorescence intensity originated from inside the aggregates (I_∞). If the intensity measured when the same concentration of fluorophore is present in aqueous solution is I_W, the value of K' can simply be derived from

$$K' = \frac{I_W + I_\infty - I^0}{(I^0 - I_\infty)([\mathrm{D}] - \mathrm{cmc})} \tag{21}$$

Equation (21) holds at low absorbances and/or if the excitation is carried out at a wavelength such that $\epsilon_{agg} = \epsilon_W$, and it only assumes that the fluorescence quantum yield of the fluorophore in the aqueous phase is not modified by the surfactant. The advantage of Eq. (21) is that it does not assume an independence of K_{SV} in the aqueous pseudophase with the surfactant concentration and it can be applied, at low absorbances, even if $\epsilon_W \neq \epsilon_{agg}$.

Selective quenching of the fluorescence from the dispersed pseudophase can also be employed to determine the fluorophore partition. This can be achieved either by employing highly hydrophobic quenchers or counterions that will concentrate at the surface of the aggregates. A drawback of this procedure, relative to that based on the selective quenching of the free probes, is that intra-aggregate quenching processes rarely follow a simple Stern-Volmer kinetics. Furthermore, its application requires a knowledge of the emission from totally incorporated probes. This situation can be reached by extrapolation of the measured intensity to "infinite" surfactant concentration. Under these conditions, changes in the size and/or shape of the aggregates could modify the fluorescence yields of the incorporated fluorophore.

Casarotto and Craik [35] have evaluated the distribution of the antidepressant drugs imipramine and amitriptyline in dodecyldimethylammonium chloride micelles employing carbon tetrachloride as a "selective" quencher of the fluorescence from the micelle-bound drugs. Even if it is accepted that the amount of carbon tetrachloride remaining in the aqueous phase negligibly quenches the fluorescence of the free drugs, extrapolation of the data to "infinite" quencher concentration will give the fraction of light arising from unbound molecules. Evaluation of the fraction of the free drugs requires an independent determination of the $\epsilon_{agg}\Phi_{agg}/\epsilon_W\Phi_W$ ratio.

B. Partition Coefficients for Fluorescence Quenchers

The partitioning of a large variety of compounds can be determined from fluorescence quenching techniques on the basis of the effect their addition has on the fluorescence of a micelle (or vesicle) incorporated fluorophore.

There exist at present several experimental approaches useful for determining the partitioning of quenchers, employing either time-resolved fluorescence [1, 3, 14] or steady-state fluorescence measurements [1, 3, 36, 37]. Frequently, both the value of the partition constant obtained and its interpretation depend upon the type of measurement being performed or the method selected to carry out the data analysis, casting doubts on the validity of these techniques to obtain reliable partitioning data [38]. The main limitation of the methods based on steady-state fluorescence quenching measurements is the assumption that the quenching follows a Stern-Volmer relationship, a condition that is only rarely fulfilled in microheterogeneous systems.

For a dynamic quenching that follows a Stern-Volmer relationship in terms of the analytical quencher concentration with an apparent quenching constant k_{ap}, Lakowicz et al. [39] have derived Eq. (22):

$$\frac{1}{k_{ap}} = \alpha\left[\frac{1}{k_{agg}} - \frac{1}{k_{agg}K'''}\right] + \frac{1}{k_{agg}K'''} \tag{22}$$

This allows an evaluation of the intra-aggregate bimolecular rate constant (k_{agg}) and the partition constant K''' by plotting the inverse of k_{ap} against the volume fraction of the dispersed pseudophase (α). This procedure has been applied to determine the partitioning of hexachlorocyclohexane (using carbazole-labeled phospholipids as a fluorophore) [39] and several ubiquinone homologs (using 12-(9-anthroyloxy) stearic acid) as the fluorophore) [40] in vesicles. When $K''' \gg 1$ (as in most systems that can be treated by these procedures), Eq. (22) reduces to

$$\frac{1}{k_{ap}} = \frac{\alpha}{k_{agg}} + \frac{1}{k_{agg}K'''} \tag{23}$$

Omann and Glaser [41] have extended this procedure to three-pseudophase systems comprising a fluorescently labeled vesicle, a nonlabeled vesicle, and the aqueous solution. The method allows, in the absence of interaction between vesicles and if the fluorophore remains in the original aggregates, an evaluation of K''' for the unlabeled bilayer.

A slightly modified procedure can be employed if the fluorophore is totally excluded from the dispersed pseudophase. Under these conditions, K' can be obtained from Eq. (24):

$$\frac{1}{k_{ap}} = \frac{1}{k_W} + \frac{(1/k_W)K'([D] - \text{cmc})}{1 - \alpha} \tag{24}$$

This equation has the advantage that k_W can be directly measured in the absence of surfactant and that the quenching, taking place in the aqueous pseudophase, usually follows simple Stern-Volmer kinetics, allowing the evaluation of k_{ap}. The use of Eq. (24) assumes that the quenching rate constant and fluorophore lifiteme in the aqueous pseudophase are independent of the surfactant concentration. This condition could not be fulfilled for ionic surfactants if the quenching process is sensitive to the presence of electrolytes, but it most likely applies for neutral or zwitterionic surfactants. The only condition required to apply Eq. (24) is that the fluorophore must be totally excluded (or totally emitting) from the aqueous pseudophase. The linearity of the Stern-Volmer plot can be considered as a proof that this condition is being fulfilled. Furthermore, even if the fluorophore is partitioned, a modified treatment of the data allows an evaluation of k_{ap} in the aqueous pseudophase [42] if the static condition limit is fulfilled. The main drawback of procedures based on the use of Eq. (24) is that it can only be applied to fluorophores having long lifetimes (ca. > 50 ns) and for quenchers whose quenching rate constants are larger than ca. 10^9 M^{-1} s^{-1}. Otherwise, at the surfactant concentrations usually accessible (10 to 100 mM) the partition constant can only be evaluated at high (ca. > 0.2) mole fractions of the incorporated additive.

Encinas and Lissi [43] have proposed a general method based on steady-state fluorescence intensity measurements that can be applied for the evaluation of partition constants irrespective of the quenching mechanism (static, dynamic, or mixed), the quencher distribution statistics, the distribution of the quencher and the donor inside the micelle (or vesicle), the relationship between the probe lifetime and the quencher exit and entrance rates, and the type of equation relating the quenching effects to the analytical concentration of the quencher. Furthermore, the method allows for the evaluation of partition constants as a function of the intra-aggregate concentration of the quencher. This procedure has been applied to several systems comprising quenchers such as amines [31, 44, 45], peroxides and hydroperoxides [43, 46], carbon tetrachloride [31, 47, 48], dimethyl aniline [31, 45], nitroxides [43], olefins [43, 49], chloroform [47], tryptophan [50], acetone [51], and *n*-doxyl stearates [52] both in micelles and vesicles. In this method, quenching curves are obtained at several surfactant concentrations, and the analytical concentrations of the quencher $[S]_i$ that leads to a given I^0/I value are plotted against the corresponding concentration of surfactant in the aggregates [43]. Since it can be considered that all the systems presenting the same amount of quenching (measured by the I^0/I value) must have the same intra-aggregate quencher concentration (and hence the same concentration in the aqueous phase), simple mass balance considerations lead to

$$[S]_i = [S]_{water} + R([D] - \text{cmc})_i \tag{25}$$

where R is the mean number of solute molecules in the aggregates per micellized surfactant. Plotting the quencher concentration $[S]_i$ values against the corresponding concentration of micellized surfactant $([D] - \text{cmc})_i$ makes it possible to infer the quencher concentration remaining in the aqueous phase at the selected I^0/I value from the intercept and the mean number of molecules of quencher in the aggregates per micellized (or in-vesicle) surfactant from the slope. From these data, K' is directly obtained by

$$K' = \frac{R/(1 + R)}{[S]_W} \tag{26}$$

allowing the determination of K' as a function of the intra-aggregate concentration of the solute (determined by the I^0/I value selected).

Equation (26) is given in terms of analytical concentrations and is only valid when the volume fraction (α) of the dispersed pseudophase is negligible. If α is considered, it can be rearranged to

$$(1 - \alpha_i)[S]_i = [S]_{water} + R(1 - \alpha_i)([D] - \text{cmc})_i \tag{27}$$

The method proposed originally by Encinas and Lissi [43], and extensions of the methods [36, 53], only require that

(i) For all [D], the fluorophore be totally incorporated into the microaggregates or totally excluded from them
(ii) The values of K′ be independent of [D] and only determined by the intra-aggregate solute concentration
(iii) The value of the measured property be only determined by the intra-aggregate additive concentration and, hence, through (ii), by its concentration in the aqueous phase

The above requirements can be considered to hold in most systems (at least over a limited range of surfactant concentrations) but, when they do not hold, the method can lead to erroneous conclusions. For example, the ability of acrylamide to quench the fluorescence of pyrene derivatives incorporated in SDS micelles decreases with the surfactant concentration [54]. Treatment of the data according to Eq. (25) allows an evaluation of the acrylamide partition constant. However, the value obtained is considerably larger than that obtained from ultrafiltration experiments, which show that the incorporation of acrylamide in the micelles is negligible [54] even at the highest surfactant concentration employed in the quenching experiments (0.15 M). The disaggreement indicates that assumptions (ii) and/or (iii) are not valid in the system. It is known that the surface of micelles becomes drier when the surfactant concentration increases, even at concentrations far below that of the sphere-rod transition [55]. The low degree of acrylamide incorporation in the micelles, its hydrophilicity (which can decrease the partition coefficient with [D]) and the high sensitivity of the fluorescence quenching by acrylamide to the polarity of the solvent [31] magnify the above-mentioned effect, leading to the observed decrease in k_{ap} with increasing SDS concentration, even under conditions of negligible quencher incorporation. Caution must be then exercized when the method is applied for the evaluation of partition constants of poorly incorporated substrates whose quenching rate constants are strongly dependent on the characteristics of the fluorophore microenvironment. In these systems, the values obtained must be cross-checked by other methods and/or by employing several fluorophores located at different positions in the aggregates. A partition constant can be considered as reliable when the value obtained is independent on the location of the fluorophore in the aggregates. Alternatively, a safer approach could be, in these systems, to employ fluorophores totally located in the aqueous solution.

C. Partition Coefficients for Additives That Are Not Quenchers but Modify a Fluorescence Characteristic

If an additive changes any of the characteristic parameters of the fluorescence of a fluorophore totally incorporated in the dispersed pseudophase, the changes elicited by its addition can be employed to evaluate the extent of its incorporation. This approach has been widely employed with fluorescent probes such as ANS, whose fluorescence quantum yield depends both on the extent of incorporation and the characteristics of the interface, either to evaluate the changes that the additive produces upon the interface [56] and/or the extent of additive incorporation [57]. However, a quantitative evaluation of the degree of incorporation of the additive from these data is difficult because the relationship between the observed effect and the extent of the additive incorporation is usually unknown. The method proposed originally by Encinas and Lissi for the determination of partition constants of fluorescence quenchers [43] has been extended to evaluate the distribution of any additive able to modify any characteristic of the fluorescence, irrespective of the functional relationship between the additive concentration and the effect produced by its incorporation [53]. This procedure can then be applied to any fluorescence property (fluorescence intensity [53, 58, 59], lifetime, generalized polarization [60], anisotropy [60], and excimer-to-monomer intensity ratio [61]), as well as to data obtained by other techniques (e.g., EPR) [61]. The method has the advantage that, with a minimum of assumptions, it directly provides the intra-aggregate solute concentration associated with a given value of the measured property.

Abuin and Lissi have measured the partition of *n*-alkanols in cationic, anionic, and neutral micelles, in anionic/neutral mixed micelles and in DODAC vesicles from the effect that their addition produces upon the intensity of pyrene fluorescence [36, 48, 53, 58, 62]. In all these systems, the effect of the alkanol upon the intensity of Py fluorescence is due to an increase in the quenching rate constant by oxygen. Sotomayor et al. [60] have also evaluated the partition of alkanols from the effect that they produce upon the anisotropy of DPH (1,6-diphenyl-1,3,5-hexatriene) or the generalized polarization of LAURDAN (6-dodecanoyl-2-dimethyl-aminonaphthalene) in DODAC vesicles. Lissi et al. [61] measured the partition of *n*-nonanol in egg phosphatidylcholine vesicles from changes in the excimer-to-monomer fluorescence intensity ratio of 3-palmitoyl-2-1(1-pyrene decanoyl)-L-*a*-phosphatidylcholine, and Abuin et al. [36] have measured the incorporation of *n*-hexanol to CTAB micelles from its effect upon the fluorescence of ANS.

In all of the above-mentioned systems a "property" is measured as a function of the additive concentration at different surfactant concentrations. If it is assumed that the property considered is only determined by the intra-aggregate additive concentration (and hence if a partition is established, by its concentration in the aqueous pseudophase), a plot of the concentration required to produce a given effect on the property considered against the surfactant concentration gives the concentration of the additive in the aqueous phase (as the intercept) and the intra-aggregate concentration (as the slope). The value of the partition coefficient, at the given intra-aggregate concentration, can be directly obtained from these data [53, 61]. Primary data obtained on the effect of *n*-hexanol upon ANS fluorescence intensity in CTAB solutions and $[S]_i$ values plotted according to Eq. (25) are shown in Figs. 1 and 2. Figure 2 shows that the data conform to the proposed linear relationship between $[S]_i$ values and the surfactant concentration, allowing the evaluation of the partition constant for the alkanol. Values of K obtained from the dilute region up to saturation are given in Fig. 3 and agree with those obtained employing ultrafiltration [63] or saturation measurements [64].

The data treatment proposed by Sotomayor et al. [60], based on steady-state anisotropy measurements, leads to the simultaneous determination of the solute partition constant and to the anisotropy produced by a given intra-aggregate solute concentration. Primary data on the effect of the analytical concentration of heptanol upon DPH anisotropy in DODAC giant vesicles at three surfactant concentrations are given in Fig. 4. Figure 5 gives $[S]_i$ values as a function of the analytical surfactant concentration.

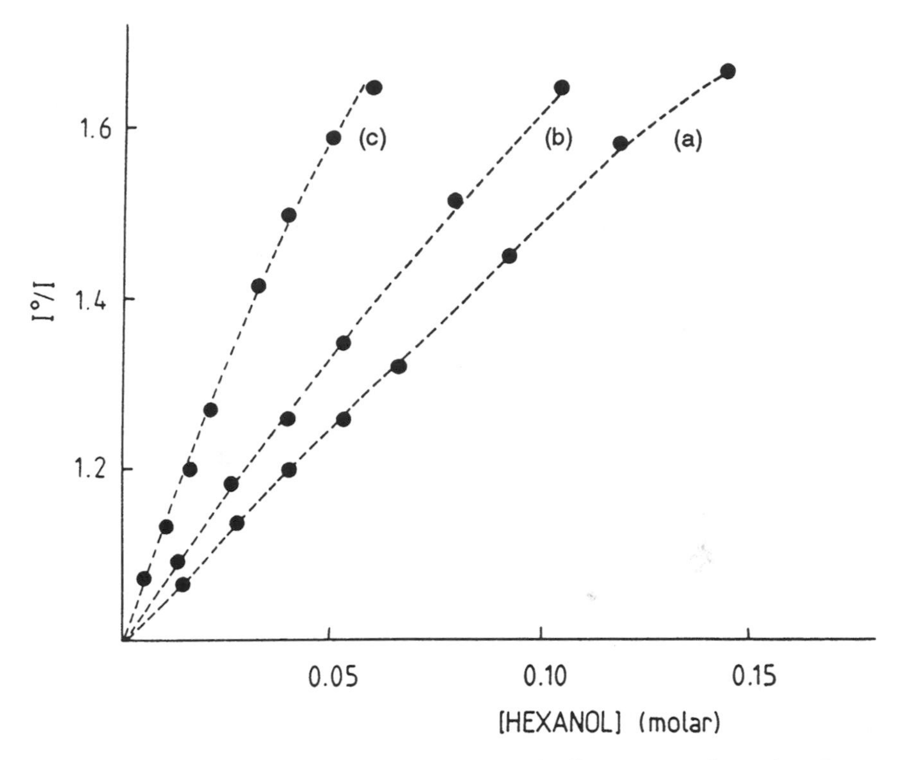

FIG. 1 Effect of *n*-hexanol concentration on the fluorescence intensity of ANS incorporated in CTAB micelles. Surfactant concentration: (a) 0.08 *M*; (b) 0.05 *M*; (c) 0.01 *M*. (Adapted from Ref. 36.)

The data conform to Eq. (25), allowing the evaluation of *K*. The values obtained by this procedure agree with those obtained from changes in the fluorescence intensity of Py [48] or LAURDAN generalized polarization [60], making the proposed methodology a versatile procedure to determine partition constants in rigid aggregates whose fluidity can be significantly modified by the incorporation of the additive.

A pertinent question that can be raised when determining the partitioning of a quencher or an additive that modifies a fluorescence characteristic is: to which aggregate subpopulation does the measured value correspond? This is a relevant question since (i) the system could comprise microaggregates of different characteristics (e.g., size) and, (ii) in the presence of an extrinsic fluorophore, at least two populations of the aggregates must be considered, those devoid of fluorophore and those bearing a fluor-

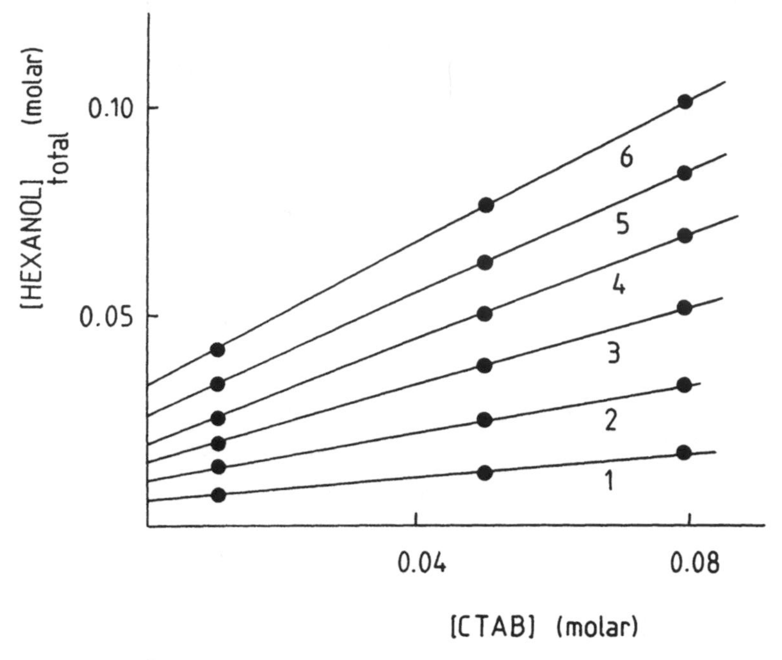

FIG. 2 Experimental data of Fig. 1 plotted according to Eq. (25). [*n*-hexanol]$_i$ values taken at I^0/I ratios of 1.1 (1); 1.2 (2); 1.3 (3); 1.4 (4); 1.5 (5) and 1.6 (6) (Adapted from Ref. 36.)

ophore (if it is considered, as in most of these systems, that the mean occupation number of the fluorophore is $\ll 1$). The answer to this question depends upon the procedure employed. Procedures based on an analysis of the extent of intramicellar quenching as a function of the analytical quencher concentration at a given surfactant concentration mostly measure the solute partitioning into fluorophore bearing aggregates. On the other hand, the procedure proposed by Encinas and Lissi [43] evaluates the capacity of the whole aggregate ensemble to extract solute from the aqueous phase and, if most of the aggregates are devoid of fluorophore, it measures the partition between the aqueous solution and the fluorophore-free aggregates. In agreement with those considerations, it has been found that the measured partition constant of the quencher is independent of the employed fluorophore [43, 45, 52]. On the other hand, it has been reported [45] that K''' values obtained by this procedure for the solubilization of dimethylaniline in neutral micelles depends on the fluorophore

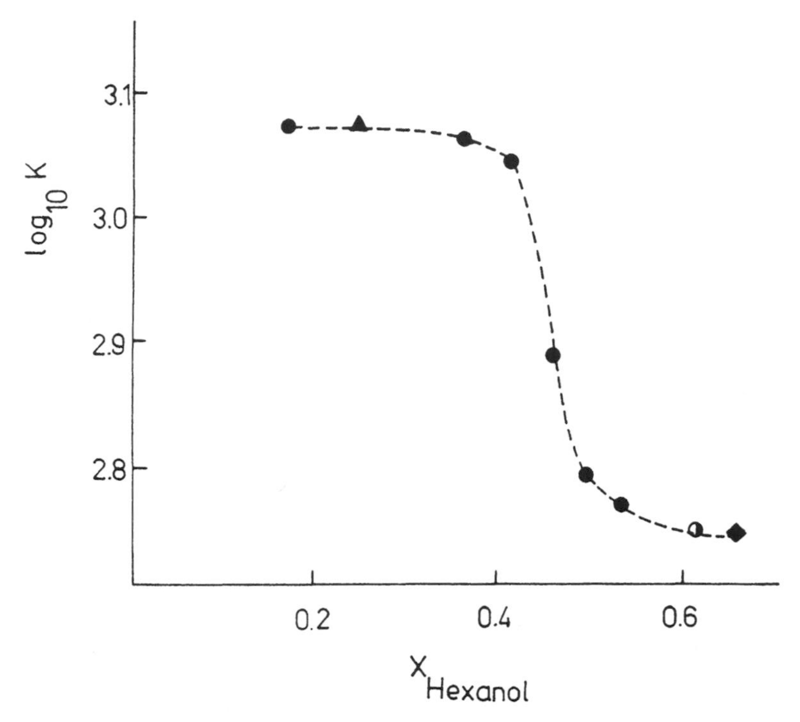

FIG. 3 Plot of the partition constant K' values as a function of the *n*-hexanol mole fraction in the micellar pseudophase. (●) Data from Fig. 2. (▲) Measured by ultrafiltration (Ref. 63). (◑) Data from Ref. 64; (◆) value obtained at saturation. (Taken from Ref. 36.)

employed in its determination. These results appear as anomalous and incompatible with the premises in which the method is based.

III. OTHER ASPECTS OF SOLUBILIZATION

A. Location of the Fluorophore or the Quencher Inside the Aggregates

The molecular properties of the amphiphile and the solubilizate affect the location of the solubilizate in the micelles or vesicle bilayers [65]. All the procedures discussed above give only a partition (or binding) constant within the framework of the pseudophase model and provide no informa-

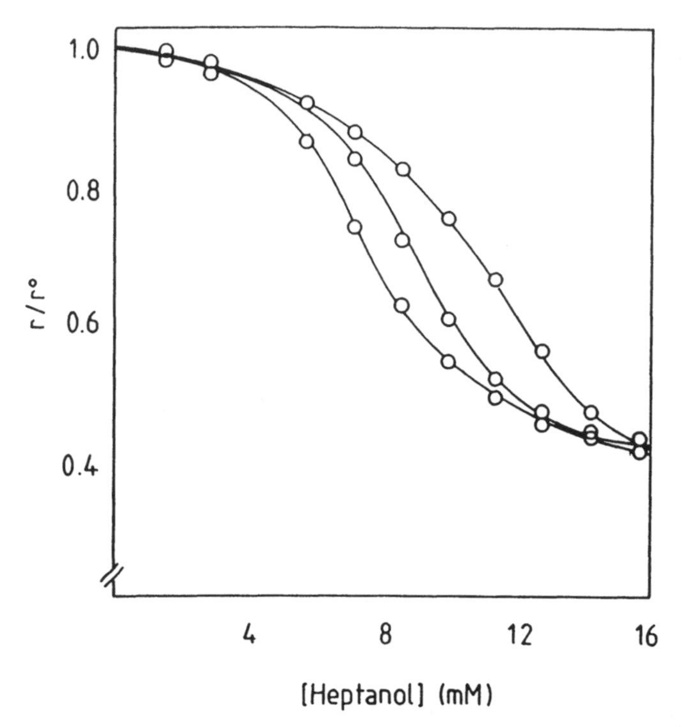

FIG. 4 Anisotropy of DPH in giant DODAC vesicles at 25°C (relative to the value obtained in the absence of added alkanol) as a function of the analytical concentration of *n*-heptanol. Concentration of DODAC, from bottom to top: 1 mM, 1.5 mM, and 2 mM. (Adapted from Ref. 60.)

tion regarding the location of the fluorophore or the additive inside the aggregates.

Procedures based on fluorescence quenching have been proposed for the evaluation of the local concentration of the quencher in the surroundings of fluorescent probes located at more or less specific sites inside the aggregates [52, 66, 67]. The extent of quenching can be expressed in terms of the bimolecular quenching constant and the local concentration of the quencher, determined by a local partition coefficient K_{local}, defined by

$$K_{local} = \frac{[S]_{local}}{[S]_W} \tag{28}$$

where $[S]_{local}$ is the intra-aggregate quencher concentration in the subvolume containing the fluorophore. It has been proposed [52, 66, 67] that

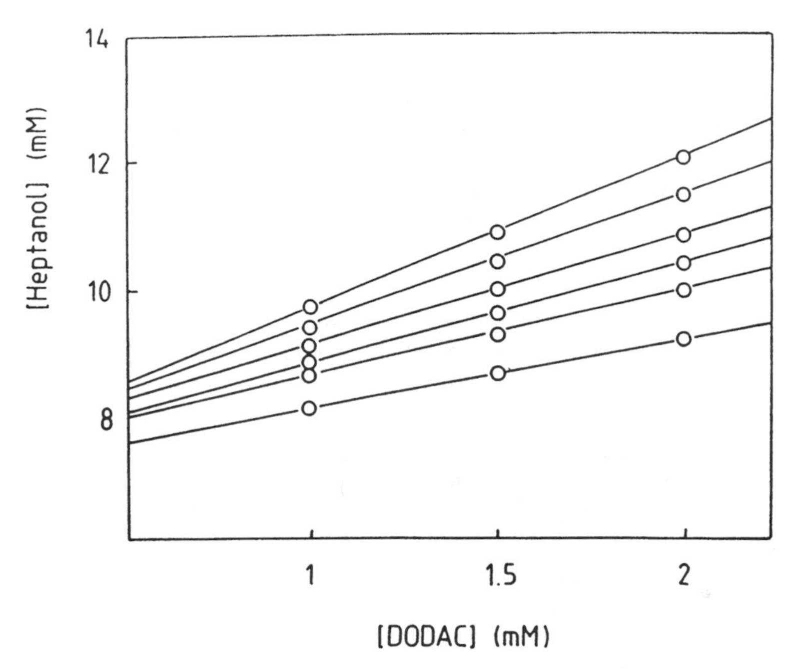

FIG. 5 Values of [n-heptanol]$_i$ from (Fig. 4) plotted according to Eq. (25). From bottom to top, data obtained at r/r^0 values of 0.85, 0.8, 0.75, 0.7, 0.65, and 0.55. (Adapted from Ref. 60.)

the intra-aggregate bimolecular quenching constant k_{agg} and K_{local} can be obtained from Eq. (29):

$$\frac{I}{I^0 - I}[Q] = \frac{1}{k_{agg}\tau K_{local}} + \left(\frac{1}{k_{agg}\tau}\right)\frac{V_L}{V} \tag{29}$$

However, this treatment can only be applied if both the probe and the quencher incorporated in the aggregates are totally confined to a known subvolume V_L of the dispersed pseudophase. If V_L corresponds to all the volume of the dispersed pseudophase [36], the proposed treatment only yields K''' with the disadvantage, with regard to the method proposed by Encinas and Lissi [43], that it assumes a linear Stern-Volmer relationship for the intra-aggregate quenching process.

Information regarding the location of solubilizates can be obtained, albeit in qualitative form and/or with a series of assumptions, from fluorescence data. Fluorescence characteristic parameters can be employed to evaluate the location of (or the microenvironment sensed by) the incorporated fluorophores. Lifetime [68], intensity [69, 70], polarization [69], band position [71–73], and band shape [2, 3, 22, 23, 74–76] measurements have been employed to characterize the location of the fluorophore and/or the properties of the microaggregates. However, it must be emphasized that the presence of the fluorophore can create local "defects" [52, 77], particularly in highly structured bilayers such as giant vesicles below their phase transition temperature, in which the properties sensed can be very different than those of the unperturbed bilayers. Furthermore, care must be exercised in interpolating parameters measured in microheterogeneous systems to scales defined in terms of measurements carried out in homogeneous solvents. In particular, it is difficult to evaluate how the orientation of the probe [78] and the peculiar properties of the microaggregate interface [79] (e.g., strong gradient of properties, strong electrical field, and high concentration of charges) can affect a measured property such as the position of an absorption or emission band, a fluorescence lifetime or the fluorescence band fine structure. In particular, it has been shown that pyrene groups (as well as other aromatics) strongly interact with tetraalkylammonium ions [75, 80, 81], casting doubts on the use of pyrene as a micropolarity probe in aggregates formed by surfactants bearing tetraalkylammonium heads. Furthermore, a quantitative evaluation of the polarity of the microenvironment (i.e., by assigning a number to the "polarity" of the microenvironment sensed by the probe) is generally difficult [73] since, as for other spectroscopic properties [82], the value obtained depends upon the solvents used in the calibration curve. This is particularly relevant for properties such as the Py I_1/I_3 fluorescence ratio, whose relationship with solvent properties (such as the E_{T30} parameter) that sense "polarity" is different for different types of solvents (i.e., protic or nonprotic) [83]. Similar considerations can be put forward when the intraaggregate viscosity is derived from the rate of Py excimer formation. In these type of studies, the uneven distribution of Py groups inside an aggregate (when intermolecular excimer formation is monitored) or the presence of preferred conformations of the molecule (when intramolecular excimer formation is considered) make difficult a quantitative evaluation of the microviscosity of the aggregate from the rate or extent of excimer formation. Furthermore, at high probe-to-surfactant ratios and particularly in vesicles, where phase separation can take place, the possible formation of ground state complexes can increase the excimer-to-monomer ratio [84], rendering erroneous intra-aggregate microviscosity values.

Fluorescence quenching experiments can also be employed to evaluate the location of the fluorophore or the quencher. Luisetti et al. [85] have concluded that for a series of Py derivatives whose heads are anchored to the surface of phosphatidylcholine bilayers, the chromophore is mainly located at a position that is determined by the length of the methylene chain that acts as spacer between the ionic head and the Py moiety. This conclusion is based on the observation that quenching of Py, pyrenebutyrate, and pyrenehexadecanoate by X-nitroxide stearates (X-5, 12, 16) is optimal when the spin label quencher is situated at the same chain length distance from the polar headgroup as is the Py moiety in the probes.

The degree of exposure of the fluorophore to the external solution has been estimated from the extent of quenching by water soluble (e.g., ionic) quenchers [77, 86–88]. These experiments are based on the premise that the rate of quenching must increase when the fluorophore moves toward the interface or to zones of high water penetration. In agreement with these considerations, the quenching by Cu^{2+} of several *n*-(9-antroyloxy) fatty acids in vesicles decreases when the number of carbon atoms between the fluorophore and the polar head increases [86]. However, quenching by I^- of *n*-(9-anthroyloxy)fatty acids in DODAC giant vesicles [89] or pyrene-labeled fatty acids in small unilamellar vesicles of phosphatidylcholine in its liquid crystalline state [90] do not show a clear trend with the length of the spacer alkyl chain, casting doubts on the chromophore location and/or on the general validity of these types of procedures to evaluate it.

Quenching experiments performed by using quenchers totally incorporated into the aggregates and localized at different depths in the bilayer can also be used to estimate the average location of mobile fluorophores. This procedure has been employed to estimate the average location of α-tocopherol in phospholipid bilayers employing a series of derivatives as quenchers [91]. Conversely, by localizing several closely related donors at different depths, the average location of quenchers in vesicle bilayers has been estimated on the basis that the faster quenching must take place when fluorophore and quencher intra-aggregate distributions are similar. This approach has been employed to evaluate the intravesicular distribution of diethylaniline [90], gramicidine A, and local anesthetics in lipid bilayers [92]. However, these procedures assume, when the quenching is not diffusion controlled, that the quenching rate constant (in terms of local quencher concentrations) are similar for all the loci inside the aggregates. This can be a poor assumption for quenching processes whose rates are extremely solvent dependent.

The relative proximity of several quenchers to a given donor can be evaluated even employing nonfixed donors such as DPH or pyrene. This approach has been employed to evaluate the relative location of DPH and ubiquinone-10 or ubiquinol-10 in DPPC vesicles [93]. In these types of experiments, it is considered that the quencher having the largest k_{agg}/k_Q ratio (where k_Q is the quenching rate in a reference solvent such as ethanol) is that whose intra-aggregate distribution best matches that of the donor. This procedure, although it can be employed to obtain a rough estimate of the quencher distribution (if that of the donor is considered as known), can be biased by the probe distribution and/or movement. For example, if the quenching rate increases with the polarity of the microenvironment, a quencher located near the interface could present a higher quenching rate constant in spite of a predominantly interior location of the donor. Furthermore, in order to ascertain the quencher location from relative intra-aggregates quenching constants, the quencher must be totally incorporated or the comparison must be carried out by employing intra-aggregate concentrations in the evaluation of the quenching rate constants [52]. If diffusion-controlled quenching processes are considered, the relative quenching rate of fluorophores localized at different positions inside the bilayer are not directly related to the quencher concentration around the probe but to its rate of access to the fluorophore locus [94]. These types of data have been extensively employed to evaluate the solubility/mobility of oxygen inside bilayers [95].

The photobehavior of exciplexes in micellar or vesicular aggregates has been employed to characterize the location of the donor and/or the quencher [71, 88, 96]. However, exciplex formation and decay are promoted by polar environments [97]. This can lead to exciplex behaviors similar to those observed in very polar (e.g., aqueous) environments even for systems with internal average locations of the donor and/or acceptor. Hence, the enhanced polarity sensed by the exciplex can result either from a faster formation rate in the more polar microenvironments, or from a displacement of the formed exciplex toward these locations. These considerations apply to both inter- and intramolecular exciplexes [71, 98].

A procedure for evaluating the probe (Py or DPH) and quencher (12-doxylstearic acid methyl ester) distribution in lipid bilayers has been developed by Duportail et al. [99], employing a model resulting from the theory of random walks in fractal structures to analyze the fluorescence decay profiles. This procedure also allows, in different time scales, an evaluation of the dimensionality of the movement of the quenchers inside the bilayer.

In vesicles, solid and fluid lipid domains can coexist near the phase transition temperature in the presence of additives or when formed by more than one surfactant. Sklar et al. [100], taking advantage of the fact that *cis*-parinaric acid partitions preferably in the fluid lipid while transparinaric acid prefers the solid lipid domains, have developed a methodology based on fluorescence polarization measurements to determine phospholipid phase separations and the partition of the probes between the coexisting phases. Similarly, employing LAURDAN as a fluorescence probe,

Parasassi et al. [101] have studied phase coexistence and phase interconversion in phospholipid vesicles.

In systems that do not involve phase separation but present a continuum of different microenvironments, the lifetime distribution [55, 102] or the width of the fluorescence band [73] can provide insight regarding the homogeneity of the probe population. A "two-site" solubilization model [103] is frequently employed to characterize the inhomogeneous fluorophore population from lifetime and/or polarization measurements [29, 31, 104].

In unilamellar vesicles, the outer and inner leaflets have been differentiated as two distinct solubilization sites. Selective quenching of the fluorescence arising from fluorophore molecules residing at the outer leaflet by nonpenetrating (or slowly-penetrating) quenchers can be employed to monitor the outer/inner leaflets distribution of the fluorophore. This procedure has been employed by Almgren [105] and by Abuin and Lissi [106] using iodide ions as quenchers of the fluorescence of pyrene in cationic unilamellar vesicles.

B. Dependence of the Partition Coefficient on the Solute Concentration

The dependence of the partition constant on X_{agg} in micelles is determined by the main localization of the solute at low intra-aggregate mole fractions. For polar (or amphiphatic) solutes (such as alkanols) partition constants decrease with X_{agg} (see for example Fig. 3) [36, 53, 58, 107]. This effect can be explained in terms of saturation of the micellar surface [53, 58, 108] and described in terms of a mixed incorporation model comprising both "binding" at the interface and "partition" into the core [13]. On the other hand, a moderate increase in the partition constant with X_{agg} has been reported for the solubilization of less polar solutes [109]. In terms of intramicellar activity coefficients, these results imply that, for hydrophobic solutes (e.g., cyclohexane, *n*-hexane, 2-hexene, or carbon tetrachloride), the activity coefficients decrease when the amount of solute incorporated increases, while for hydrophilic or amphiphilic solutes (e.g., *n*-hexanol or methylmethacrylate), activity coefficients increase when the intramicellar concentration of the solute increases [109, 110].

Abuin et al. [62] have measured the partition constant of *n*-hexanol in mixed micelles composed of SDS and poly(oxyethylene)lauryl ether (POE23) as a function of the composition of the micelles and the intramicellar concentration of the alcohol from the dilute region up to saturation. Most of the data were obtained from the effect of the alcohol upon the fluorescence intensity of Py incorporated to the micelles. The results obtained (Fig. 6) show a peculiar solubilization behavior, passing from positive (POE23-rich mixtures) to negative (SDS-rich mixtures) deviation from ideal mixing, irrespective of the intramicellar concentration of the alcohol (from 0.3 to 0.6 mole fraction in the micelles). This behavior has been explained in terms of the "two-site" model for the solubilization of the alcohol, together with synergistic/antagonistic effects promoted by surfactant mixing [62, 111, 112].

There exist few data dealing with the dependence of the partition constant on the solute concentration in vesicles. De Young and Dill [113]

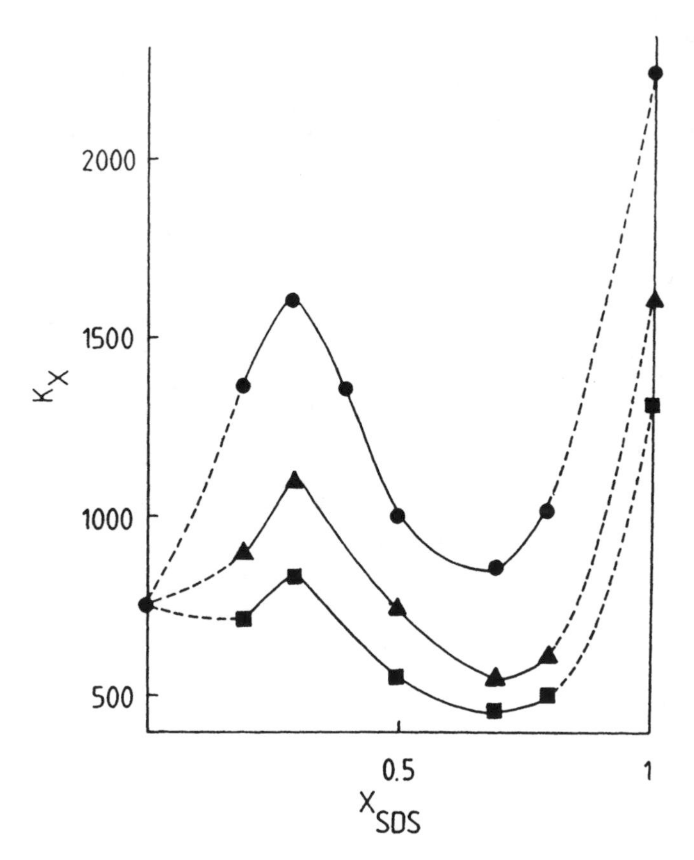

FIG. 6 Dependence of the partition constant of *n*-hexanol on the composition of mixed micelles of SDS and POE23 at different mol fractions of the alcohol in the micellar pseudophase. Mole fraction of the alcohol in the aggregates: (●) 0.3; (▲) 0.5; (■) 0.6. (Adapted from Ref. 62.)

have reported a small decrease in the partition constant with increasing mole fraction of hexane in dimyristoyl phosphatidylcholine vesicles at 25°C. On the other hand, the strong increase in solubility with X_{agg} observed for the incorporation of alkanols into giant DODAC vesicles below the phase transition temperature can be ascribed to a progressive disorganization of the bilayer [48]. The partition constant measured in small (sonicated) DODAC vesicles is higher than that measured in giant vesicles, both for alkanols [48] and anthracene derivatives [114], pointing to a relationship between solubility and the degree of order of the bilayer. In agreement with these considerations, Omann and Glaser [41] have reported that partitioning of 1,1-dichloro-2,2-bis(*p*-chlorophenyl ethylene) into gel phase lipids is at least 100-fold less than in fluid-phase lipids, and quenching data obtained in DODAC vesicles below and above the bilayer transition temperature also indicate a greater solubility of hydrophobic solutes in the more fluid bilayer [31].

C. Changes in the Properties of the Aggregates Elicited by the Additive

The incorporation of an additive into micelles or vesicles can produce noticeable changes in most properties of the aggregates: cmc depression [53, 115–118], decreased cloud point [119], changes in size [9, 116, 120, 121], shape [122], sphere-rod transition [123], fluidity [61, 104, 122, 124], counterion binding [115, 122, 125], and solute intermicellar migration rates [126], as well as modifications in the capacity of the aggregates to incorporate cosolutes.

1. Cosolubilization Extent

In a study of solubilization in cetyltrimethylammonium chloride and SDS from binary solvent mixtures, it has been found that, while the amount of each hydrocarbon solubilized from a *n*-heptane/*n*-hexane mixture is proportional to the mole fraction in the saturating organic mixture, a noticeable synergistic effect is observed in *n*-hexane/methylmethacrylate mixtures [107]. In micellar solutions, it has been proposed that those solutes competing for the same solubilization region (the "interface" or the "core") will mutually inhibit their incorporation into the aggregates, while those that are mainly solubilized in different sites will mutually favor their incorporation [127, 128]. The displacement of Py (as sensed by the I_1/I_3 ratio) toward less polar environments in the presence of polar (alkanols) additives [53, 124, 129–131] and toward similar or more exposed positions in the presence of alkanes [23] is fully compatible with the view that a solute located at the surface favors the incorporation of cosolutes in the core (by lowering the Laplace pressure), while those located at the core "open" the interface making it more active toward polar (or polarizable) solutes. Similarly, any factor (increasing surfactant concentration or salt addition) leading to "drier" (less open) micelles decreases the solubilization of those solutes (e.g., chloroform) that mostly localize at the surface and increases the solubilization of solutes (e.g., *n*-pentane) located at the micellar core [127]. For systems in chemical equilibrium inside the aggregates, the selective effect of the additive upon the solubility of reactants and products can also modify the position of the equilibrium inside the aggregate [132].

It can be expected that the capacity of vesicles to incorporate solutes could be strongly altered by the presence of additives that modify the degree of order (or packing) of the bilayer [113]. Changes in the fluorescence spectra of hematoporphyrin derivatives [133] and merocyanine 540 [134] in lecithin vesicles elicited by cholesterol addition have been interpreted as an indication of a reduced solubility of the dyes in the presence of the additive. This reduced solubility has been related to the increase in microviscosity of the liposomes.

2. Changes in the Fluorescence Characteristics of Extrinsic Fluorophores Elicited by Cosolute Incorporation

Several of the alterations produced in the aggregates by solute incorporation can be evaluated from changes in the fluorescence properties of incorporated fluorophores. When extrinsic fluorophores are employed, the response of their fluorescence properties to the additive can reflect both changes in the "solvent" composition around the probe (without displacing it to other locations) and/or its displacement from a location (e.g., the interface) to another (e.g., the inner core). Anchoring the fluorophore to more or less fixed positions (e.g., by employing probes bearing an ionic head) can reduce the second effect, and hence the observed changes will mostly sense the alterations in the characteristics of the aggregates at the site of location of the fluorophore. Extensive studies of this type have been carried out both in micelles and vesicles by sensing changes in fluorescence intensity, fluorescence quenching rates, excimer and exciplex formation rates, fluorescence band shape, and fluorescence polarization of incorporated fluorophores as a function of additive (mostly alkanols) incorporation. For example, the decrease in I_1/I_3 ratio observed for Py when alkanols are added to micellar [23, 53, 124, 128–131] or vesicular solutions [48] has been attributed to a displacement of the molecule toward less polar microenvironments, while the increase in the excimer/monomer fluorescence ratio observed for Py (intermolecular excimer) and dipyrenylpropane (intramolecular excimer) [8, 120, 123] has been related to a decrease of the effective microviscosity of the probe surroundings. The increase in Py lifetime and in the rate of quenching by oxygen upon the addition of *n*-alkanols (from heptanol to nonanol) in large DODAC vesicles [48] are also compatible with a displacement of Py toward less polar and more fluid environments. Russell and Whitten [136] have evaluated the effect of 1-heptanol addition upon the ground-state complexation of a hydrophilic quencher (methyl viologen) with several hydrophobic fluorescent probes. In SDS, this complexation is notably reduced by addition of 1-heptanol, indicating that the structure of the micelle changes from open (with the flurophore exposed) to more closed structures in which hydrophobic solubilizates becomes "isolated" from hydrophilic reagents bound to the surface. A similar explanation can be advanced to explain the decrease in the quenching efficiency of acrylamide toward methyl pyrene incorporated in SDS micelles elicited by *n*-octanol incorporation [54]. However, some care must be exercized in the interpretation of these data since electron spin-echo modulation experiments [137] and a kinetic study of the effect of alkanols on the acid denitrosation of *N*-nitroso-*N*-methyl-*p*-toluenesulfonamide failed to reveal a decreased polarity of the micellar surface in the presence of alkanols [138].

The lifetimes and fluorescence quantum yields of anthracene derivatives are extremely sensitive to the presence of water [139] and their values in microheterogeneous systems can be considered as an indication of the extent of water penetration in the locus of the anthracene moiety. However, the geometry of the molecules (particularly that of the *n*-(9-anthroyloxy) stearic acids, *n*-AS) is such that the extent of water penetration in the probe surroundings can be very different than that of the unperturbed aggregate at the depth supposedly sensed by the probe. This difference can be due to movements of the probe that push the chromophore toward the interface and/or to defects (e.g., channels) created by its incorporation. For example, it has been observed [89] that the lifetimes of 2-AS, 7-AS, and 12-AS are rather similar in DODAC vesicles below the phase transition temperature, and that they are almost equally quenched by iodide ions. These results are very difficult to interpret without invoking the defects generated by the probe and/or their movements toward the interface [88, 94].

Microenvironments sensed by anthracene and Py derivatives incorporated in vesicles change notably when the characteristics of the bilayers are modified either by changing the temperature or by the incorporation of additives. Figure 7 shows the effect of temperature and of *n*-octanol addition on the fluorescence intensity of 9,10-dimethylanthracene incorporated to DODAC vesicles [140]. These data can be interpreted as indi-

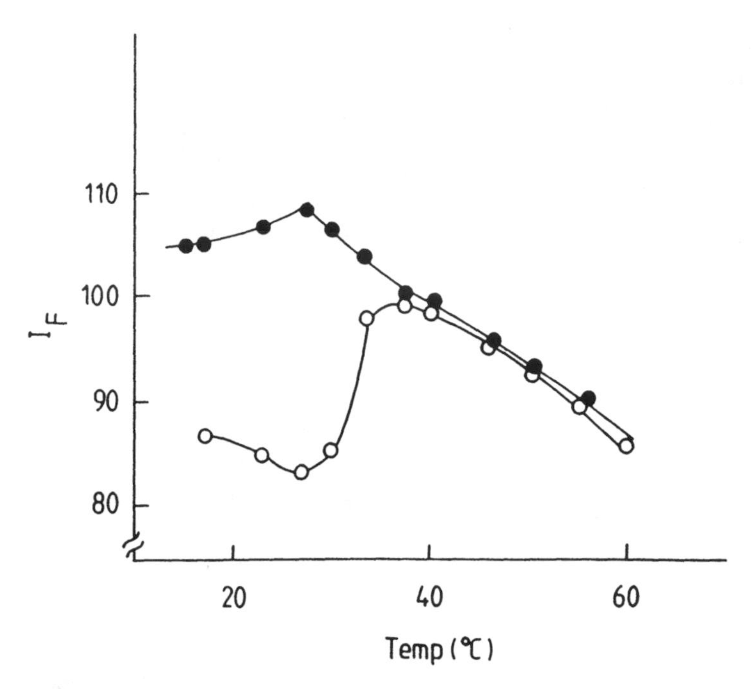

FIG. 7 Fluorescence intensity of 9,10-dimethylanthracene incorporated in giant DODAC vesicles as a function of temperature in the absence (○) and presence of 1.68 mM of *n*-octanol (●); DODAC concentration: 2 mM. (Adapted from Ref. 140.)

cating that the anthracene moiety moves to less polar (less water exposed) environments when the fluidity of the bilayer increases either by going through the phase transition or by *n*-alkanol incorporation. Similar results have been reported employing Py as fluorescent probe, where a significant decrease in the I_1/I_3 ratio is observed at high temperatures or in the presence of alkanols [48]. When probes anchored to the interface are employed (e.g., anthracene propionic acid), smaller changes are observed, suggesting that displacement of the probe toward the bilayer interior contributes to the observed effect when free probes (such as Py or 9-methyl anthracene) are employed as sensors.

Fluorescence anisotropy measurements have been extensively used to monitor intra-aggregate microviscosities [1–4, 7]. The changes caused by addition of an alkanol to a vesicle strongly depend on the properties of the alkanol (as measured by the length and branching of its alkyl chain), the state (liquid crystalline or gel) of the bilayer and the region of the vesicle (the surface or the bilayer interior) being sensed. These factors have been analyzed by Hitzemann [141], who studied the changes in the anisotropy of DPH and TMA-DPH (1-(4-trimethylammoniumphenyl)-6-phenyl-1,3,5-hexatriene, *p*-toluene-sulfonate) elicited by several alkanols in DMPC multilamellar liposomes at 18°C and 30°C. In the liquid crystalline state, all alkanols considered (from ethanol to hexanol) decrease the steady-state anisotropy of DPH, but only the most hydrophobic alkanols modify the anisotropy of TMA-DPH. In the gel state, the effect of different additives upon DPH anisotropy is more selective. These results are attributed to a deeper solute location in the gel phase. This conclusion is opposite to that reached for the relative position of aromatic compounds in DODAC vesicles [48, 140]. These differences could reflect the dependence of the location with the size and shape of the solutes. The influence of the shape of the solute upon its incorporation and location, as well as on the effect that it produces on the properties of the aggregates, has been very little addressed and appears as one of the more interesting fields to apply fluorescence-based techniques to solubilization studies.

ACKNOWLEDGMENT

Financial support by FONDECYT (project 0450/91) is appreciated.

REFERENCES

1. K. Kalyanasundaram, *Photochemistry in Microheterogeneous Systems*, Academic Press, New York, 1987.
2. K. Kalyanasundaram, in *Photochemistry in Organized and Constrained Media* (V. Ramamurthy, ed.), VCH, New York, 1991, Chapter 2.
3. C. Bohne, R. W. Redmond, and J. C. Scaiano, in *Photochemistry in Organized and Constrained Media* (V. Ramamurthy, ed.), VCH, New York, 1991, Chapter 3.
4. J. R. Lakowicz, *Principles of Fluorescence Spectroscopy*, Plenum Press, New York, 1983.
5. R. Zana, in *Surfactant Solutions: New Methods of Investigation* (R. Zana, ed.), Marcel Dekker, New York, 1987.
6. A. Malliaris, *Int. Rev. Phys. Chem. 7*: 95 (1988).
7. F. Grieser and C. L. Drummond, *J. Phys. Chem. 92*: 5580 (1988).
8. K. A. Zachariasse, in *Surfactants in Solution* (K. L. Mittal, ed.), Plenum Press, New York, 1989, Vol. 7, pp. 79–104.
9. A. Malliaris, J. Lang, and R. Zana, in *Surfactants in Solution* (K. L. Mittal, ed.), Plenum Press, New York, 1989, Vol. 7, pp. 125–139.
10. M. Van der Auweraer, E. Roelants, A. Verbeeck, and F. C. De Schryver, in *Surfactants in Solution* (K. L. Mittal, ed.), Plenum Press, New York, 1989, Vol. 7, pp. 141–157.

11. L. Sepulveda, E. Lissi, and F. Quina, *Adv. Colloid Interface Sci. 25*: 1 (1986).
12. M. Rotenberg and R. Margalit, *Biochim. Biophys. Acta 905*: 173 (1987).
13. E. Blatt and W. H. Sawyer, *Biochim. Biophys. Acta. 822*: 43 (1985).
14. (a) J-E. Lofroth and M. Almgren, *J. Phys. Chem. 86*: 1636 (1982); (b) S. Reekmans, N. Boens, M. Van der Auweraer, H. Luo, and F. C. De Schryver, *Langmuir 5*: 948 (1989); (b) M. Tachiya, *Chem. Phys. Lett. 33*: 289 (1975); (c) M. Tachiya, *J. Chem. Phys. 78*: 5282 (1983).
15. M. Gonzalez, J. Vera, E. B. Abuin, and E. Lissi, *J. Colloid Interface Sci., 98*: 152 (1984).
16. M. Almgren, F. Grieser, and J. K. Thomas, *J. Am. Chem. Soc. 101*: 279 (1979)
17. Y. H. Paik and S. Ch. Shim, *J. Photochem. Photobiol. A: Chem. 56*: 349 (1991)
18. K. Yamashita, H. Ishida, and K. Okubo, *J. Chem. Soc. Perkin Trans. II* 2091 (1989).
19. B. Ehrenberg, *J. Photochem. Photobiol. B: Biol. 14*: 383 (1992).
20. J. R. Lakowicz and S. Keating, *J. Biol. Chem. 258*: 5519 (1983).
21. T. Parasassi, G. De Stasio, A. d'Ubaldo, and E. Gratton, *Biophys. J. 57*: 1179 (1990).
22. (a) K. Kalyanasundaram and J. K. Thomas, *J. Am. Chem. Soc. 99*: 2039 (1977); (b) C. M. Paleos, C. I. Stassinopoulou, and A. Malliaris, *J. Phys. Chem. 87*: 251 (1983).
23. E. Lissi, A. Dattoli, and E. Abuin, *Bol. Soc. Chil. Quim. 30*: 37 (1985).
24. (a) A. Muñoz de la Peña, T. Ndou, J. B. Zung, and J. M. Warner, *J. Phys. Chem. 95*: 3330 (1991); (b) J. B. Zung, A. Muñoz de la Peña, T. Ndou, and I. M. Warner, *J. Phys. Chem. 95*: 670 (1991).
25. K. Tamori, Y. Watanabe, and K. Esumi, *Langmuir 8*: 2344 (1992).
26. S. Basu, *J. Photochem. Photobiol. A; Chem. 56*: 339 (1991).
27. R. R. Hautala, N. E. Schore, and N. J. Turro, *J. Am. Chem. Soc. 95*: 5508 (1973).
28. N. Boens, R. Andriessen, M. Ameloot, L. V. Dommelen, and F. C. De Schryver, *J. Phys. Chem. 96*: 6331 (1992).
29. M. Van Bosckstaelle, J. Gelan, H. Martens, J. Put, J. C. Dederen, N. Boens, and F. C. De Schryver, *Chem. Phys. Lett. 58*: 211 (1978).
30. H. J. Pownall and L. C. Smith, *Biochemistry 13*: 2594 (1974).
31. E. A. Lissi, S. Gallardo, and P. Sepulveda, *J. Colloid Interface Sci. 152*: 104 (1992).
32. F. H. Quina and V. G. Toscano, *J. Phys. Chem. 81*: 1750 (1977).
33. E. B. Abuin and E. A. Lissi, *J. Phys. Chem. 84*: 2605 (1980).
34. E. Abuin, E. Lissi, and A. M. Rocha, *J. Phys. Chem. 84*: 2406 (1980).
35. M. G. Casarotto and D. J. Craik, *J. Phys. Chem. 96*: 3146 (1992).
36. E. Abuin, M. V. Encinas, E. Lissi, and M. A. Rubio, *Bol. Soc. Chil. Quim. 35*: 113 (1990).
37. H. W. Ziemiecki, R. Holland, and W. R. Cherry, *Chem. Phys. Lett. 73*: 145 (1980).
38. (a) M. C. Antunes-Madeira and V. M. C. Madeira, *Biochim. Biophys. Acta*

820: 165 (1985); (b) O. T. Jones and A. G. Lee, *Biochim. Biophys. Acta 812*: 731 (1985); (c) O. T. Jones, R. J. Froud, and A. G. Lee, *Biochim. Biophys. Acta 812*: 740 (1985).
39. J. R. Lakowicz, D. Hogan, and G. Omann, *Biochim. Biophys. Acta 471*: 401 (1977).
40. R. Fato, M. Battino, G. Parenti Castelli, and G. Lenaz, *FEBS Lett. 179*: 238 (1985).
41. G. E. Omann and M. Glaser, *Biophys. J. 47*: 623 (1985).
42. S. L. Lehrer, *Biochemistry 10*: 3254 (1971).
43. M. V. Encinas and E. A. Lissi, *Chem. Phys. Lett. 91*: 55 (1982).
44. M. H. Gehlen, P. Berci Fo., and M. G. Neumann, *J. Photochem. Photobiol. A Chem. 59*: 335 (1991).
45. E. C. C. Melo and S. M. B. Costa, *J. Phys. Chem. 91*: 5635 (1987).
46. M. V. Encinas and E. A. Lissi, *Photochem. Photobiol. 37*: 125 (1983).
47. M. V. Encinas, M. A. Rubio, and E. A. Lissi, *Photochem. Photobiol. 37*: 125 (1983).
48. E. Abuin, E. Lissi, D. Aravena, and M. Macuer, *J. Colloid Interface Sci. 122*: 201 (1988).
49. M. V. Encinas, E. Guzman, and E. A. Lissi, *J. Phys. Chem. 87*: 4770 (1983).
50. M. V. Encinas and E. A. Lissi, *Photochem. Photobiol. 44*: 579 (1986).
51. J. C. Leigh and J. C. Scaiano, *J. Am. Chem. Soc. 105*: 5652 (1983).
52. E. Blatt, R. C. Chatelier, and W. H. Sawyer, *Photochem. Photobiol. 39*: 477 (1984).
53. E. B. Abuin and E. A. Lissi, *J. Colloid Interface Sci. 95*: 198 (1983).
54. M. A. Rubio and E. A. Lissi, *J. Photochem. Photobiol. A Chem. 71*: 175 (1993).
55. (a) W. R. Ware, in *Photochemistry in Organized and Constrained Media* (V. Ramamurthy, ed.), VCH, New York, 1991, Chapter 13; (b) A. Seret and A. Van der Vorst, *J. Phys. Chem. 94*: 5293 (1990); (c) A. Siemiarczuk, W. R. Ware, and K. D. Palme, Int. Conf. Luminescence, Lisboa, Portugal, 1990.
56. (a) J. Slavik, *Biochim. Biophys. Acta 694*: 1 (1982); (b) T. Ohyashiki, R. Adachi, and K. Matsui, *Chem. Pharm. Bull. 39*: 3235 (1991).
57. R. Mannhold, W. Voigt, and K. Dross, *Cell Biol. Int. Rep. 14*: 361 (1990).
58. A. Leon, E. Abuin, E. Lissi, L. Gargallo, and D. Radic, *J. Colloid Interface Sci., 115*: 529 (1987).
59. H. Chaimovich, M. Kawamuro, I. Cuccovia, E. Abuin, and E. Lissi, *J. Phys. Chem. 95*: 1458 (1991).
60. C. Sotomayor, M. Bagnara, M. Soto, E. Abuin, and E. Lissi, presented at the III Encuentro Latinoamericano de Fotoquimica y Fotobiologia, Buenos Aires. Argentina, 1991.
61. E. A. Lissi, M. L. Bianconi, A. T. do Amaral, E. de Paula, L. E. B. Blanch, and S. Schreier, *Biochim. Biophys. Acta 1021*: 46 (1990).
62. E. Abuin, E. Lissi, and A. M. Campos, *Bol. Soc. Chil. Quim. 37*: 169 (1992).
63. C. Gamboa, L. Sepulveda, and R. Soto, *J. Phys. Chem. 85*: 1429 (1981).
64. J. Gettins, D. Hall, P. L. Jobling, J. E. Rassing, and E. Wyn-Jones, *J. Chem. Soc. Faraday II 74*: 1957 (1978).

65. M. Aamodt, M. Landgren, and B. Jonsson, *J. Phys. Chem., 96*: 945 (1992).
66. K. A. Sikaris, K. R. Thulborn, and W. H. Sawyer, *Chem. Phys. Lipids 29*: 23 (1983).
67. R. C. Chatelier, E. Blatt, and W. H. Sawyer, *Chem. Phys. Lipids 36*: 131 (1984).
68. E. Blatt, K. P. Ghiggino, and W. H. Sawyer, *J. Phys. Chem. 86*: 4461 (1982).
69. M. T. Allen, L. Miola, D. M. Shin, B. R. Suddaby, and D. G. Whitten, *J. Membrane Sci. 33*: 201 (1987).
70. R. B. Suddaby, P. E. Brown, J. C. Russell, and D. G. Whitten, *J. Am. Chem. Soc. 107*: 5609 (1985).
71. (a) S. Lukac, *J. Am. Chem. Soc. 106*: 4386 (1984); (b) C. David, E. Szalai and D. Baeyens-Volant, *Ber. Bunsenges Phys. Chem. 86*: 710 (1982).
72. (a) K. Kalyanasundaram and J. K. Thomas, *J. Phys. Chem. 81*: 2176 (1977); (b) N. J. Turro and T. Okubo, *J. Phys. Chem. 86*: 159 (1982).
73. T. Handa, M. Nakagaki, and K. Miyajima, *J. Colloid Interface Sci. 137*: 253 (1990).
74. G. P. L'Heuereux and M. Fragata, *Biophys. Chem. 30*: 293 (1988).
75. P. Lianos, M-L Viriot, and R. Zana, *J. Phys. Chem. 88*: 1098 (1984).
76. P. Lianos and R. Zana, *J. Colloid Interface Sci., 84*: 100 (1981).
77. (a) M. T. Allen, L. Miola, and D. G. Whitten, *J. Am. Chem. Soc. 110*: 5198 (1988); (b) M. Langner and S. W. Hui, *Chem. Phys. Lipids 60*: 127 (1991).
78. T. Handa, K. Matusuzaki, and M. Nakagaki, *J. Colloid Interface Sci., 116*: 50 (1987).
79. C. Ramachandran, R. A. Pyter, and P. Mukerjee, *J. Phys. Chem. 86*: 3198 (1982).
80. E. B. Abuin and E. A. Lissi, *Bol. Soc. Chil. Quim. 34*: 59 (1989).
81. K. Viaene, A. Verbeeck, E. Gelade, and F. C. De Schryver, *Langmuir 2*: 456 (1986).
82. S. A. Simon, R. V. McDaniel, and T. J. Mc Intosh, *J. Phys. Chem. 86*: 1449 (1982).
83. D. C. Dong and M. A. Winnik, *Photochem. Photobiol. 35*: 17 (1982).
84. G. P. L'Heureux and M. Fragata, *J. Photochem. Photobiol. B Biol. 3*: 53 (1989).
85. J. Luisetti, H. Mohwald, and H. J. Galla, *Biochim. Biophys. Acta 552*: 519 (1979).
86. K. R. Thulborn and W. H. Sawyer, *Biochim. Biophys. Acta 511*: 125 (1978).
87. E. Blatt, K. P. Ghiggino, and W. H. Sawyer, *J. Phys. Chem. 86*: 4461 (1982).
88. M. Van Bockstaele, J. Gelan, H. Martens, J. Put, and F. C. De Schryver, *Chem. Phys. Lett. 70*: 605 (1980).
89. E. Lissi, C. Sotomayor, and P. Sepulveda, unpublished results.
90. Y. Barenholz, T. Cohen, R. Korenstein, and M. Ottolenghi, *Biophys. J. 59*: 110 (1991).
91. F. J. Aranda, *Biochim. Biophys. Acta 985*: 26 (1989).
92. (a) E. A. Haigh, K. R. Thaulborn, and W. H. Sawyer, *Biochemistry 18*:

3525 (1979); (b) K. A. Sikaris and W. H. Sawyer, *Biochem. Pharmacol.* *31*: 2625 (1982).
93. F. J. Aranda and J. C. Gomez-Fernandez, *Biochem. Int. 12*: 137 (1986).
94. H. Merkle, W. K. Subczynski, and A. Kusumi, *Biochim. Biophys. Acta* *897*: 238 (1987).
95. E. B. Abuin and E. A. Lissi, *Prog. Reaction Kinet. 16*: 1 (1991).
96. (a) S. S. Atik and J. K. Thomas, *J. Am. Chem. Soc. 103*: 3550 (1981); (b) Y. Waka, F. Tanaka, and N. Mataga, *Photochem Photobiol. 32*: 335 (1980); (b) K. Kano, H. Kawazumi, T. Ogawa, and J. Sunamoto, *J. Phys. Chem.* *85*: 2204 (1981).
97. N. Mataga and M. Ottolenghi, in *Molecular Association* (R. Foster, ed.), Academic Press, London, 1979.
98. (a) H. Masuhara, K. Kaji, and N. Mataga, *Bull. Chem. Soc. Jpn. 50*: 2084 (1977); (b) B. Katusin-Razem, M. Wong, and J. K. Thomas, *J. Am. Chem. Soc. 100*: 1679 (1978).
99. G. Duportail, J. C. Brochon, and P. Lianos, *J. Phys. Chem. 96*: 1460 (1992).
100. L. A. Sklar, G. P. Miljanich, and E. A. Dratz, *Biochemistry 18*: 1707 (1979).
101. T. Parasassi, G. De Stasio, G. Ravagnan, R. M. Rush, and E. Gratton, *Biophys. J. 60*: 179 (1991).
102. (a) R. Alcala, E. Gratton, and F. Prendergast, *Biophys. J. 51*: 587 (1987); (b) R. Alcala, E. Gratton, and F. Prendergast, *Biophys. J. 51*: 925 (1987).
103. (a) P. Mukerjee and J. R. Cardinal, *J. Phys. Chem. 82*: 1620 (1978); (b) P. Mukerjee, *Pure Appl. Chem. 52*: 1317 (1980).
104. (a) D. J. Jobe and R. E. Verrall, *Langmuir 6*: 1750 (1990); (b) E. Blatt, K. P. Ghiggino, and W. H. Sawyer, *J. Chem. Soc. Faraday Trans. I 77*: 2551 (1981); (c) E. Blatt, W. H. Sawyer, and K. P. Ghiggino, *Aust. J. Chem.* *36*: 1079 (1983).
105. M. Almgren, *J. Phys. Chem. 85*: 3599 (1981).
106. I. M. Cuccovia, H. Chaimovich, E. Lissi, and E. Abuin, *Langmuir 6*: 1601 (1990).
107. T. Caceres and E. A. Lissi, *J. Colloid Interface Sci. 97*: 298 (1984).
108. A. M. Blokhus, H. Hoiland, and S. Backlund, *J. Colloid Interface Sci. 114*: 9 (1986).
109. E. B. Abuin, E. Valenzuela, and E. A. Lissi, *J. Colloid Interface Sci. 101*: 401 (1984).
110. G. A. Smith, S. D. Christian, E. E. Tucker, and J. F. Scamehorn, *J. Colloid Interface Sci. 130*: 254 (1989).
111. (a) C. Treiner, A. A. Khodja, and M. Fromon, *Langmuir 3*: 729 (1987); (b) R. Bury, E. Souhalia, and C. Treiner, *J. Phys. Chem. 95*: 3824 (1991).
112. C. M. Nguyen, S. D. Christian, and J. F. Scamehorn, *Tenside Deterg. 25*: 328 (1988).
113. L. R. De Young and K. A. Dill, *J. Phys. Chem. 94*: 801 (1990).
114. M. V. Encinas, E. Lemp, and E. A. Lissi, *J. Photochem. Photobiol. B Biol. 3*: 113 (1989).
115. R. Zana, S. Yiv, C. Strazielle, and P. Lianos, *J. Colloid Interface Sci. 80*: 208 (1981).

116. A. Malliaris, *J. Photochem. Photobiol. A Chem. 40*: 79 (1987).
117. (a) M. Manabe and M. Koda, *Bull. Chem. Soc. Jpn. 51*: 1599 (1978); (b) E. Vikingstad and O. Kummen, *J. Colloid Interface Sci. 74*: 16 (1980); (c) F. Yamashita, G. Perron, J. E. Desnoyer, and J. C. T. Kwak, *J. Colloid Interface Sci. 114*: 548 (1986); (d) D. G. Marangoni and J. C. T. Kwak, *Langmuir 7*: 2083 (1991).
118. (a) V. Perez-Villar, V. Mosquera, M. Garcia, C. Rey, and D. Attwood, *Colloid Polym Sci. 258*: 965 (1990); (b) C. Treiner, *J. Colloid Interface Sci., 90*: 44 (1982); (c) K. Hayase, S. Hayano, and H. Tsubota, *J. Colloid Interface Sci. 101*: 336 (1984).
119. Y. Tokuoka, H. Uchiyama, M. Abe, and K. Ogino, *J. Colloid Interface Sci. 152*: 402 (1992).
120. G. Oradd, G. Lindblom, L. B.-A. Johansson, and G. Wikander, *J. Phys. Chem. 96*: 5170 (1992).
121. (a) M. Almgren and S. Swarup, *J. Phys. Chem. 86*: 4212 (1982); (b) M. Almgren and S. Swarup, *J. Colloid Interface Sci. 91*: 256 (1983).
122. P. Lianos, J. Lang, C. Strazielle, and R. Zana, *J. Phys. Chem. 86*: 1019 (1982).
123. (a) D. Nguyen and G. L. Beltrand, *J. Colloid Interface Sci. 150*: 143 (1992); (b) D. Nguyen and G. L. Beltrand, *J. Phys. Chem. 96*: 1994 (1992).
124. P. Lianos, J. Lang, and R. Zana, *J. Phys. Chem. 86*: 4809 (1982).
125. M. Manabe, M. Koda, and K. Shirayama, *J. Colloid Interface Sci. 77*: 189 (1984).
126. M. Malliaris, J. Lang, J. Sturm, and R. Zana, *J. Phys. Chem. 91*: 1475 (1987).
127. E. Valenzuela, E. B. Abuin, and E. A. Lissi, *J. Colloid Interface Sci. 102*: 46 (1984).
128. J. C. Hoskins and A. D. King, Jr., *J. Colloid Interface Sci. 82*: 260 (1981).
129. E. B. Abuin, E. Lissi, and H. Casal, *J. Photochem. Photobiol. A Chem. 57*: 343 (1991).
130. A. Malliaris, *Adv. Colloid Interface Sci. 27*: 153 (1987).
131. (a) P. Lianos and R. Zana, *J. Colloid Interface Sci. 101*: 587 (1984); (b) P. Lianos and R. Zana, *Chem. Phys. Lett. 72*: 171 (1980); (c) J. C. Russell, U. P. Wild, and D. G. Whitten, *J. Phys. Chem. 90*: 1319 (1986).
132. J. A. Zoltewicz and L. B. Bloom, *J. Phys. Chem. 96*: 2007 (1992).
133. E. Gross, Z. Malik, and B. Ehrenberg, *J. Membrane Biol. 97*: 215 (1987).
134. P. Williamson, K. Mattocks, and R. A. Schlegel, *Biochim. Biophys. Acta 732*: 387 (1983).
135. D. Miller, *Ber. Bunsenges Phys. Chem. 85*: 337 (1981).
136. J. C. Russell and D. G. Whitten, *J. Am. Chem. Soc. 104*: 5937 (1982).
137. P. Baglioni and L. Kevan, *J. Phys. Chem. 91*: 1516 (1987).
138. C. Bravo, J. R. Leis, and M. E. Peña, *J. Phys. Chem. 96*: 1957 (1992).
139. A. L. Macanita, F. P. Costa S. M. B. Costa E. C. Melo, and H. Santos, *J. Phys. Chem. 93*: 336 (1989).
140. M. A. Rubio, E. Lemp, M. V. Encinas, and E. A. Lissi, *Bol. Soc. Chil. Chim. 37*: 33 (1992).
141. R. J. Hitzemann, *Biochim. Biophys. Acta 983*: 205 (1989).

10

Solubilization of Organic Compounds by Vesicles

MASAHIKO ABE and KEIZO OGINO Faculty of Science and Technology, Science University of Tokyo, Noda, Chiba, Japan

HITOSHI YAMAUCHI Developmental Research Laboratories, Daiichi Pharmaceutical Co., Ltd., Edogawa-ku, Tokyo, Japan

SYNOPSIS

When phospholipids and synthetic double-chained surfactants are added into aqueous solutions during sonicating, they can form liposomes and/or vesicles consisting of a lipid bilayer (hydrophobic portions) and an inner water phase (hydrophilic portions). Liposomes and/or vesicles can solubilize organic compounds and water-soluble compounds simultaneously. In this study, the incorporation of amphoteric compounds, lipophilic compounds, and/or alcoholic compounds into the lipid bilayers (hydrophobic portions) of liposomes and/or vesicles were examined. Large amounts of amphoteric compounds and lipophilic compounds which penetrate into the lipid bilayers will result in destruction of the bilayer structure and an increase in water-soluble compound leakage. Moreover, an alcoholic compound is able to solubilize more strongly by vesicles than by micelles having monolayer structures.

I. INTRODUCTION

Vesicles are defined as a structure composed of lipid bilayers that are enclosed an aqueous compartment [1]. Phospholipids as well as synthetic double-chained surfactants are used for preparation of vesicles. The structures of micelles and emulsions which are composed of lipids are quite different from vesicles [2]. Specifically, the surface of micelles and emulsions is a lipid monolayer, while the surface of liposomes is a lipid bilayer, and the inner core of micelles and emulsions is composed of hydrocarbon chains or oily material, while the inner core of liposomes is an aqueous phase. Therefore, the ability to solubilize various compounds could be affected by the differences in these structures; micelles and emulsions can solubilize amphiphiles and organic compounds, while vesicles can solubilize (or encapsulate) organic compounds and amphiphiles found in the lipid bilayer, and inorganic compounds and amphiles found in the aqueous core. Moreover, in the case of vesicles, lipid composition and concentration, vesicle size, and number of lamellae, and so on, can also affect the ability to solubilize. In this chapter, we review the classification, preparation, and characterization of vesicles and provide examples of recent applications of solubilization of organic compounds such as lipophilic compounds, amphoteric compounds, and an alcoholic compound by vesicles.

II. PROPERTIES AND PREPARATION METHODS OF VESICLES

A. Classification of Vesicles

Vesicles are classified as shown in Table 1 in terms of number of lamellae and vesicle size. Multimembrane vesicles are divided into three groups: multilamellar vesicles, oligolamellar vesicles, and multivesicular vesicles. Multilamellar vesicles (so called onion-shaped vesicles or MLVs) were first discovered by Bangham et al. [1] Oligolamellar vesicles (OLVs) are composed of several lamellae. Unimembrane vesicles are conveniently divided into three groups in terms of vesicle size: small unilamellar vesicles (SUVs), large unilamellar vesicles (LUVs), and giant unilamellar vesicles (GUVs). Vesicles under 100 nm are usually considered SUVs, whereas those greater than 100 nm are LUVs [3]. On the other hand, reverse-phase evaporation vesicles (REVs), which are OLVs [4], are defined separately.

B. Basic Precautions for Vesicle Preparation

1. Lipids

Lipids are classified by three molecular shapes: inverted cone, cylindrical, and cone [5]. Cylindrically shaped lipids, e.g., phosphatidylcholine (PC) (>C14), sphingomyelin (SPM), phosphotidylserine (PS), phosphatidylglycerol (PG), and synthetic double-chained surfactant, are considered to form the bilayer spontaneously, while inverted-cone-shaped lipids, e.g.,

TABLE 1 Classification of Vesicles

Multimembrane vesicle	Multilamellar vesicle (MLV)
	Oligolamellar vesicle (OLV)
	Multivesicular vesicle
Unimembrane vesicle	Small unilamellar vesicle (SUV) (0.02 ~ 0.1 μm)
	Large unilamellar vesicle (LUV) (0.1 ~ 1.0 μm)
	Giant unilamellar vesicle (GUV)

Source: Ref. 2.

PC (<C8), lysoPC, ganglioside, cholic acid, and synthetic monoalkyl surfactant, are considered to form the micellar phase. Cone-shaped lipids, e.g., caldiolipin, phosphatidic acid (PA), and phosphatidylethanolamine (PE), are considered to form the hexagonal II phase. As mentioned, vesicles are easily formed with cylindrically shaped lipids alone, whereas they can be formed by the combination of cone- and inverted-cone-shaped lipids or cone, inverted cone and cylindrically shaped lipids [6].

Lipid purity is of importance in order to prepare well-defined vesicles. Even small amounts of impurities would affect such characteristics as surface charge, permeability, or stability of vesicles; it should be examined before use by thin-layer chromatography (chloroform: methanol: water = 70:30:5) [2] or high-performance liquid chromatography [7]. The peroxidation of unsaturated acyl chains of lipids would have an undesirable effect on the preparation and stability of vesicles. If necessary, an antioxidant agent (e.g., tocopherol) should be used at the same time and vesicles should be prepared in an inert atmosphere by replacing the air with nitrogen.

2. Phase Transition Temperature (*Tc*) of Lipids

In the case of lipids which have diacyl chains, the gel-liquid crystalline phase transition is observed. The phase transition temperature (*Tc*) depends on the hydrocarbon chain length and the hydrophilic region of lipids. Above *Tc*, this phase is a liquid crystal phase, and the hydrocarbon chain of the lipids is considered to be in a fluid state. Below *Tc*, the lipids form a gel phase which is considered to be in a solid state [8]. In preparing vesicles for efficient production, it is necessary to hydrate the lipids at a temperature higher than *Tc* of the lipids. Table 2 shows *Tc* of typical phospholipids used for vesicles. The permeability of incorporated compounds from vesicles is also affected by *Tc*. For the study of vesicles, *Tc* is very important.

3. Cholesterol Effect

Cholesterol is an important component of the cell membrane. One of the important effects of cholesterol on the properties of vesicles is decreasing the permeability of the incorporated compounds from the membrane because of its fluidizing effect below *Tc* (reducing the hydrophobic interaction of lipids) and its condensing effect above *Tc* (increasing the perpendicular orientation of lipids). However, in the case of vesicles containing certain organic compounds, their encapsulation efficiency tends to be reduced by the addition of cholesterol [9].

4. Osmotic Pressure

As pointed out, vesicles contain an enclosed aqueous compartment. Therefore, the difference of the osmotic pressure between the inside and

TABLE 2 Phase Transition Temperature (*Tc*) of Phospholipid

Phospholipids	*Tc* (°C)
Egg phosphatidylcholine (egg PC)	−15 ~ −7
Dimyristoylphosphatidylcholine (DMPC)	23
Dipalmitoylphosphatidylcholine (DPPC)	41
Distearoylphosphatidylcholine (DSPC)	55
Dipalmitoylphosphatidylglycerol (DPPG)	41
Dipalmitoylphosphatidic acid (DPPA)	67 (pH 6.5)
Dipalmitoylphosphatidylserine (DPPS)	51
Dipalmitoylphosphatidyletanolamine (DPPE)	60
Brain sphingomyelin (SPM)	32

Source: Ref. 3.

outside of the vesicles could affect vesicle stability. In general, MLVs have a tendency to not be affected by the osmotic pressure, as opposed to unilamellar vesicles such as LUVs and SUVs.

5. Critical Micelle Concentration of Lipids

Although lipid dispersion usually has a critical micelle concentration (CMC) in water as do artificial surfactants, the CMCs of the lipid are extremely low; for example, that of dipalmitoylphosphatidylcholine is 4.6×10^{-10} M [10]. Therefore, it is thought that vesicles easily maintain their form even if they are diluted with large amounts of water. However, care should be taken not to dilute lipid solutions below their CMC levels during experiments especially when vesicle solutions are diluted with water.

C. Typical Preparation Methods for Vesicles

Many reviews have discussed the preparation methods for vesicles [1–3, 11–16]. In this section, typical preparation methods for MLVs, SUVs, LUVs, and REVs are reviewed. In order to incorporate organic compounds into vesicles and maximize their encapsulation efficiency by any of these methods, the most general approach is one in which organic compounds are solubilized in organic solutions in which lipids are solubilized. Users of these methods interested in more details should consult the original articles.

1. Multilamellar Vesicles

MLVs are the most widely used vesicles because of their simple preparation method. Initially, appropriate lipids are dissolved in chloroform-methanol (e.g., 9:1 by volume) in a small round-bottom flask. To produce a thin lipid film on the wall of the flask, the solvent is then evaporated off under vacuum while the flask is warmed in a water bath. To remove the residual solvents completely, the flask should be allowed to stand in a desiccator under reduced pressure for several hours. Water or an aqueous buffer is added and the lipid film is hydrated at a temperature above the *Tc* of the lipids. The flask is then agitated on a Vortex mixer for more than 5 min above *Tc*. Thus the resulting milky dispersion contains MLVs. MLVs prepared by this method are said to be produced by the vortexing method, the hydration method, or Bangham's method. To attain a more homogeneous size distribution of the vesicles, they are extruded once or several times above *Tc* as described by Olson [17] through a polycarbonate membrane filter. The advantages of MLVs are their simple preparation method and a wide variety of lipid compositions, while their disadvantage is the heterogeneous size distribution of the vesicles.

2. Small Unilamellar Vesicles

Typical preparation methods of SUVs are the sonication method [18], ethanol injection method [19], cholic acid removal method [20], Triton X-100 (polyoxyethylene octylphenol ether) batch method [21], French press extrusion method [22], and prevesicle method [23]. In this section, each preparation method is reviewed briefly. The advantage of SUVs is the homogeneous size distribution of vesicles, while the disadvantage is the low encapsulation efficiency of the aqueous phase.

(a) Sonication Method [18]. After MLVs are prepared by Bangham's method, they are then ultrasonically irradiated with a bath-type sonicator or a probe-type sonicator and SUVs are prepared. However, the contamination of metal ions from the tip of the probe in the vesicle dispersion when a probe type sonicator is used should be kept in mind [24].

(b) Ethanol Injection Method [19]. SUVs are prepared by injecting an ethanol solution of lipids into water through a syringe. The dilute SUV dispersion can be concentrated by ultrafiltration or ultracentrifugation, if necessary. The ethanol should be removed by dialysis or washing on the ultrafilter.

(c) Cholic Acid Removal Method [20]. SUVs are prepared by solubilizing lipids with cholic acid and removing the cholic acid from the mixed micelles of lipids and cholic acid by dialysis or gel filtration.

(d) Triton X-100 Batch Method [21]. The principle of this method is essentially the same as the cholic acid removal method except that Triton X-100 is used instead of cholic acid. To remove Triton X-100 from mixed micelles of lipids and Triton X-100, biobeads SM-2 is used to absorb Triton X-100 selectively [25].

(e) French Press Extrusion Method [22]. SUVs are prepared by extrusion of an MLV dispersion in a French press under high pressure. Multiple extrusion of MLV dispersion results in an increase in size homogeneity.

(f) Prevesicle Method [23]. Vesicles with asymmetric lipid bilayer membranes are prepared by this method. First, W/O emulsions (prevesicle) are prepared with lipids by sonication. Then other lipids which are dissolved in organic solvents and W/O emulsions are added to water in a tube and centrifugation is carried out to prepare SUVs.

3. Large Unilamellar Vesicles

Typical preparation methods of LUVs are the Ca^{2+}-induced method [26], the ether infusion method [27], the annealing method [28], the freeze-thaw method [29], and W/O/W emulsion method [30]. The advantage of LUVs is that macromolecules can be encapsulated and the encapsulation efficiency is relatively high. On the other hand, the disadvantage is a heterogeneous size distribution of vesicles. The extrusion of vesicles is necessary to get more a homogeneous size distribution of vesicles.

(a) Ca^{2+}-Induced Method [26]. After the addition of Ca^{2+} to SUVs and incubation, LUVs are prepared by the addition of EDTA to remove Ca^{2+}. This method applies only to the vesicles obtained with acidic lipids in the presence of divalent metals.

(b) Ether Infusion Method [27]. LUVs are prepared when an ether solution of lipids is injected into a warm aqueous solution (55–65°C) with a syringe. Lipids which are solubilized by ether are successfully used.

(c) Annealing Method [28]. Sonication of MLVs below *Tc* produces SUVs with structural defects within a bilayer, and then LUVs are prepared by annealing these SUVs above *Tc* owing to the fusion of SUVs.

(d) Freeze-Thaw Method [29]. LUVs are reconstituted by sonication after a freeze-thaw step of SUVs. SUVs should be frozen quickly (e.g., in a dry ice/ethanol bath) and thawed at room temperature.

(e) W/O/W Emulsion Method [30]. This method is an application of the preparation method of W/O/W emulsions. After W/O emulsions are prepared with lipids, Span 80 (sorbitan monooleate) and *n*-hexane, *n*-hexane is removed and then a water-in-lipid mixture system is obtained. Mixing the system with an aqueous solution of hydrophilic emulsifying agent (e.g., Tween 80 (polyoxyethylene sorbitan monooleate)) results in an aqueous dispersion of lipid vesicles. Finally, dialysis of the lipid vesicle dispersion is carried out to remove the hydrophilic agent.

4. Reverse-Phase Evaporation Vesicles

Reverse-phase evaporation vesicles are large unilamellar and oligolamellar vesicles first introduced by Szoka and Papahadjopoulos [4]. Figure 1

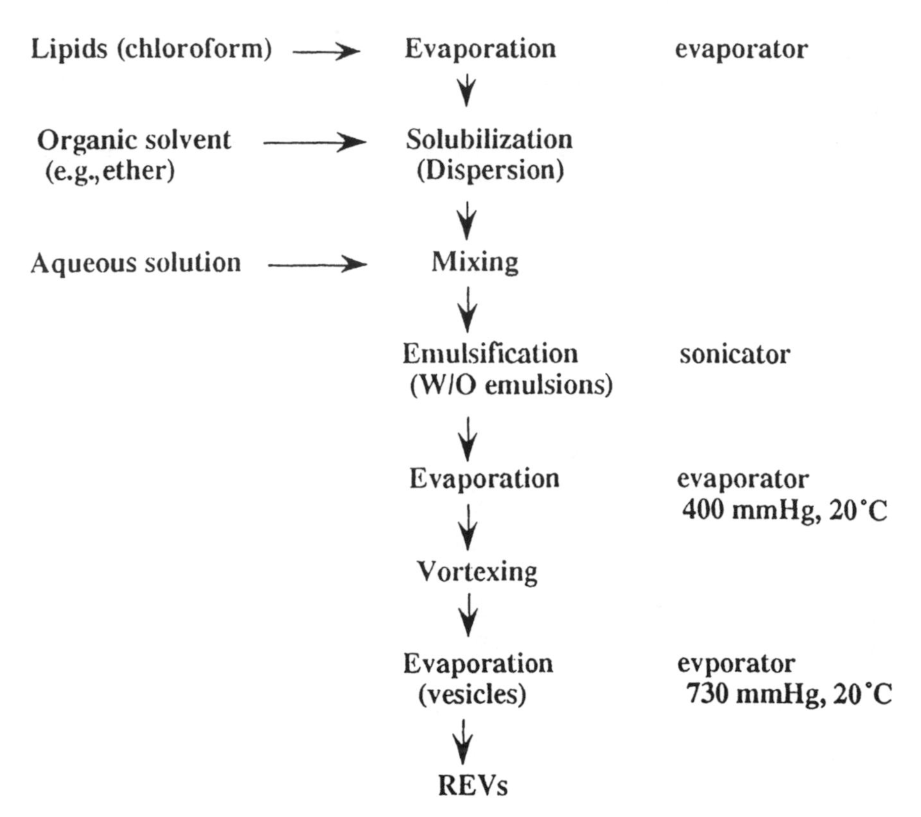

FIG. 1 Schematic preparation flow of REVs. (From Ref. 11.)

shows the schematic preparation flowchart of REVs. A W/O emulsion of lipids and an aqueous solution in an excess organic solvent such as diethylether are prepared by sonication, and the organic solvents subsequently removed by evaporation under reduced pressure. After vortexing, REVs can be formed. The advantages of this method are the high-efficiency encapsulation of the aqueous phase and the ability to encapsulate macromolecules. REVs can be reduced by extrusion through polycarbonate membranes.

5. A Spray-Drying Method for Mass Production

This method is a novel preparation method of vesicles for mass production [31]. Briefly, lipids are dissolved in a volatile organic solvent such as chloroform; in some cases, a core material such as mannitol is additionally suspended, and the organic solvent is then spray-dried. The spray-dried material can be easily hydrated with an aqueous solution because of its amorphous state and the spontaneity with which vesicles are formed by agitating.

D. Characterization of Vesicles

The characterization of vesicles prepared by any of these methods is important to confirm the results of the experiment and their application.

1. Encapsulation Efficiency

In order to evaluate the encapsulation efficiency of vesicles, the nonentrapped substances should be removed from the vesicle dispersion by gel filtration, dialysis, or ultracentrifugation [11], and then the nature and concentration of encapsulated substances should be determined. Gel filtration can be done with Sepharose CL-4B or Sephadex G-50 columns. Dialysis can be done with a cellophane tube and ultracentrifugation can be done for several times at 10^5 G for the required time.

2. Size Distribution

In order to examine the size distribution of the resulting vesicles, the quasi-elastic laser light-scattering (QELS) method, the gel filtration method, the electron microscopic method, and the NMR method are used. Other important characteristics of vesicles are

1. Permeability of bilayer membrane
2. Number of lamellae
3. Electrical surface potential
4. Fluidity of bilayer membrane
5. Distribution of lipids in the bilayer
6. Stability of vesicles

III. SOLUBILIZATION OF AMPHOTERIC COMPOUNDS BY LIPOSOMES [32]

In order to enhance the function of liposomes (for example, targetability for a specific organ and/or cell) a recognition sensor such as some amphoteric compounds having good biodegradability should be incorporated into the lipid bilayers of liposomes. In our laboratory we have solubilized some nonionic surfactants having different polyoxyethylene chain lengths into the bilayers: the nonionic surfactants used are hexadecyl polyoxyethylene ethers ($C_{16}POE_n$; n = 10, 20, and 40) supplied by Nihon Surfactant Industries (Nikko Chemicals) Co. Ltd.

A. Preparation of Liposomes Containing Nonionic Surfactants

After L-α-dipalmitoyl phosphatidylcholine (DPPC, Nihon Oil & Fats Co. Ltd.) had been dissolved in a chloroform solution, an MLV containing 2×10^{-3} mol/L of DPPC is prepared by the method of Bangham et al. [1]. Then the MLV were irradiated with a ultrasonic waves for 5 h at 50°C to form SUVs, followed by addition of a given concentration of nonionic surfactants. These mixtures are stirred for the appropriate time (12 h) with a thermostat at 30°C in order to establish their equilibrium.

B. Incorporation of a Nonionic Surfactant into a Bilayer

Figure 2 shows the effects of the concentration of various nonionic surfactants on the particle sizes of liposomes. The particle size of liposomes

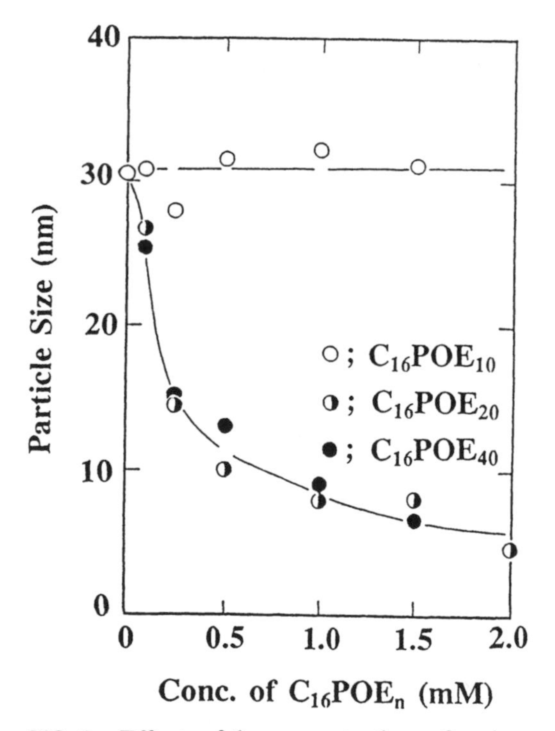

FIG. 2 Effects of the concentrations of various nonionic surfactants on the particle sizes of liposomes at 30°C. (From Ref. 32.)

containing nonionic surfactant having short polyoxyethylene chains ($C_{16}POE_{10}$) is nearly independent (about 30 nm) of surfactant concentration, whereas the particles of liposomes containing nonionic surfactant having long polyoxyethylene chains ($C_{16}POE_{20}$ and $C_{16}POE_{40}$) decrease to less than 10 nm as surfactant concentration increases. As the minimum value of liposome size is generally known to be 20 nm[33]), above 0.5 mol/L of nonionic surfactant having long polyoxyethylene chains, the DPPC liposome is not able to form; a molecular aggregate such as micelles would be obtained instead. To ascertain this, the leakage of calcein (3,6-dihydroxy-2,3-bis[*N*,*N*′-di(carboxymethyl)-aminomethyl]fluoran), which is the aqueous model material and is encapsulated into an inner water phase of liposomes, from the liposomes was determined.

Figure 3 depicts the relationship between encapsulation efficiency of liposomes and concentration of nonionic surfactants. The encapsulation

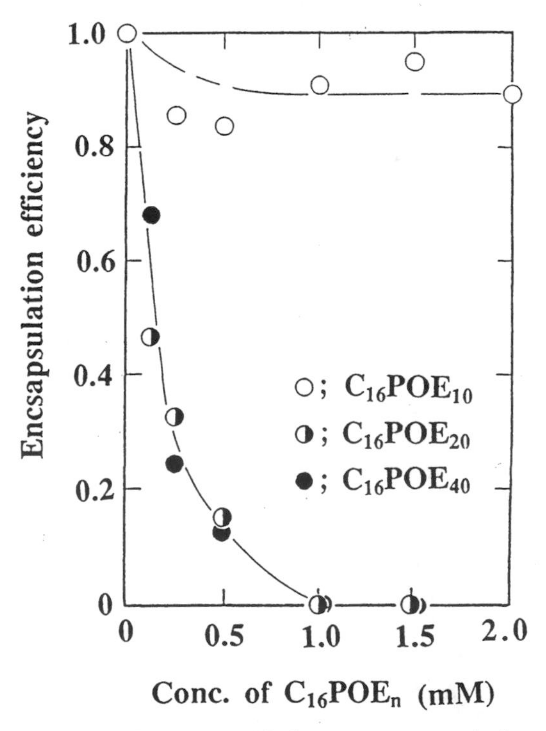

FIG. 3 The relationship between encapsulation efficiency of liposomes and concentration of nonionic surfactants at 30°C. (From Ref. 32.)

efficiency for the $C_{16}POE_{10}$ system is little changed (only 10% reduced) by the addition of the surfactant. On the other hand, for the $C_{16}POE_{20}$ or $C_{16}POE_{40}$ systems, the efficiency decreases rapidly as the surfactant concentration increases and then becomes nearly zero above 1 mmol/L of the surfactant. This means the molecular aggregate forms without an inner water phase. Specifically, above this concentration of surfactant, the liposome with nonionic surfactants would not form; the addition of nonionic surfactant may result in the destruction of the bilayer structure formed by DPPC.

To ascertain whether $C_{16}POE_{10}$ molecules become incorporated into liposomes or not, after ultracentrifuging (10^5G, 5 h, 40°C), the chemical analysis of precipitates was performed according to the methods of Greff et al. [34] and Takayama et al. [35]. As a result, the amount of DPPC is independent of surfactant concentration, and all DPPC molecules added exist in the precipitates. At the same time, all $C_{16}POE_{10}$ molecules also exist in the precipitate.

Next, we investigated molecular interactions between $C_{16}POE_{10}$ and DPPC molecules in bilayers of liposomes by DSC measurement; the cooperative effect of the phase transition of DPPC molecules decreases because of the penetration of the nonionic surfactant into the DPPC molecules.

C. Surface Charge Density of Liposomes with Nonionic Surfactants

Liposomes containing only phospholipid are unstable because of a lack of charged materials on their surface. Liposomes usually contain dicetyl phosphate (DCP), which is a charged material, to prevent the coagulation of liposome [36]. When liposomes are administered in the human body, those with negatively charged materials have little toxicity [37]. Therefore, we prepared the liposomes containing DCP and then investigated the effects of the concentration of nonionic surfactant on the ζ-potential of liposomes. As a result, every absolute value of ζ-potential decreased considerably with increasing nonionic surfactant concentrations. Moreover, the degree of depression increased with increasing polyoxyethylene chain lengths. We calculated the surface charge density of liposomes from the ζ-potential obtained on the basis of the following equations [32, 38, 39]:

$$\sigma = \left(\frac{\epsilon_r \epsilon_0 \kappa kT}{e}\right) 2 \sinh\left(\frac{e\zeta}{2kT}\right) \left[1 + \frac{2}{\kappa a \cosh^2(e\zeta/4kT)} + \frac{8 \ln\{\cosh(e\zeta/4kT)\}}{(\kappa a)^2 \sinh^2(e\zeta/2kT)}\right]^{1/2} \quad (1)$$

$$\kappa = \left(\frac{2000 N_A I e^2}{\epsilon_r \epsilon_0 k T}\right)^{1/2} \tag{2}$$

where σ is the surface charge density, ϵ_r the dielectric constant of the continuous phase, ϵ_0 the dielectric constant of the vacuum, κ the Debye-Hückel parameter, N_A the Avogadro number, k the Boltzmann constant, I the ionic strength, T the absolute temperature, e the electric charge of a single carrier, ζ the zeta potential, and a the vesicle radius. Here, the effect of a charged material in an inner water phase of liposomes on the inferred surface charge density must be neglected because the dielectric constant of DPPC forming a bilayer is considerably lower than that of water molecules, the water being a continuous phase.

The effect of concentration of nonionic surfactant on the surface charge density of liposomes is shown in Fig. 4. As can be seen in Fig. 4, the absolute value of surface charge density of the liposomes decreases with

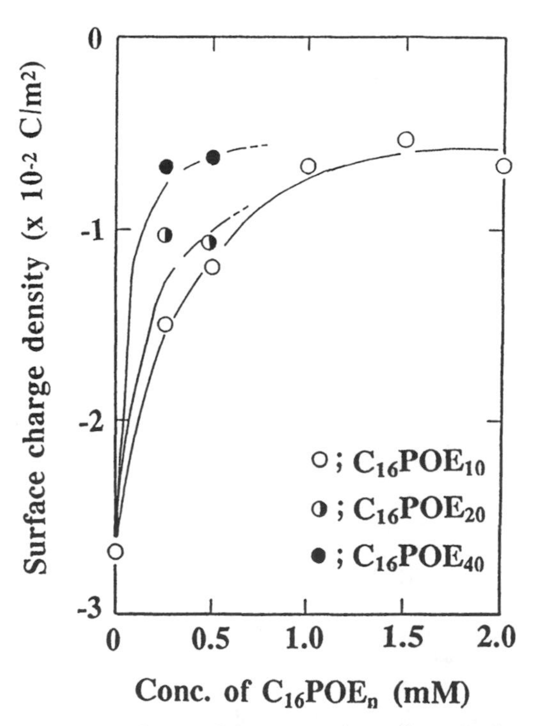

FIG. 4 Effects of concentration of nonionic surfactants on the surface charge density of liposomes at 30°C. Ionic strength is 1.65×10^{-2}. The pH of the solution is 7.5. (From Ref. 32.)

an increase in the concentration and polyoxyethylene chain lengths of nonionic surfactant involved. It can be postulated that the alkyl chains of the nonionic surfactant penetrate into DPPC bilayers. The polyoxyethylene chains hydrophilic moieties cover up the liposome surface.

In conclusion, particle size and encapsulation efficiency of the liposomes decrease rapidly with increasing concentration of nonionic surfactant possessing a long oxyethylene chain ($C_{16}POE_{20}$ and $C_{16}POE_{40}$), but show virtually no change in the case of a short chain $C_{16}POE_{10}$. The cooperative effect on the physical properties of a bilayer caused by interaction among DPPC molecules decreases with increasing concentration of the nonionic surfactant. Surface charge density of the liposomes also decreases following the addition of nonionic surfactant. The addition of a long polyethoxylated nonionic surfactant to a solution containing DPPC molecules caused structural change of molecular aggregation and in particular, in the presence of large amounts of $C_{16}POE_{20}$ and $C_{16}POE_{40}$, the DPPC liposome with nonionic surfactant does not form. A molecular aggregate such as a micelle is obtained instead, as shown in Fig. 5.

IV. SOLUBILIZATION OF LIPOPHILIC COMPOUNDS BY VESICLES [40]

Liposomes are spherical molecular aggregates consisting of a lipid bilayer membrane and an inner water phase. Needless to say, water-soluble (hy-

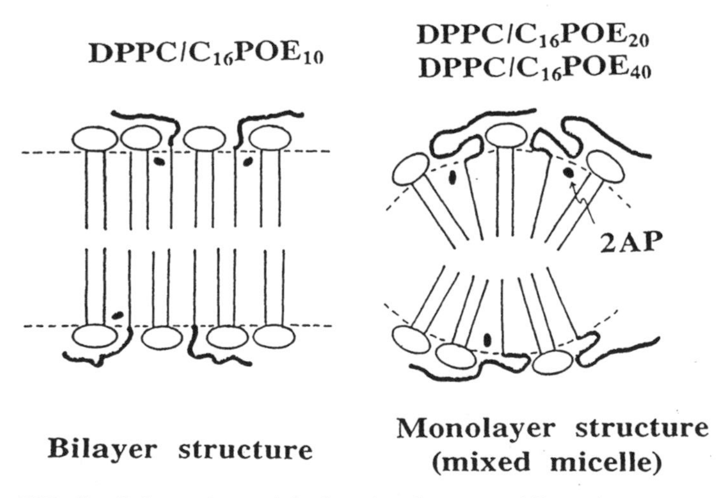

FIG. 5 Schematic model of molecular assemblies with long polyoxylated nonionic surfactant molecules. (From Ref. 32.)

drophilic) molecules can be incorporated into the inner water phase of liposomes and at the same time a hydrophobic molecule can also be incorporated into the lipid bilayer membranes being a hydrophobic moiety. In particular, numerous recent drugs gives rise to an enhanced utilization of lipid-soluble molecules.

In this section, we describe incorporation of lipophilic compounds having different hydrophilic groups such as octane, 1-octanol, and octanoic acid into the lipid bilayer membrane of liposomes.

Preparation of liposomes containing lipophilic compounds is almost the same as the method mentioned above (Sec. II). The only differences are the composition of liposomes (DPPC:cholesterol:DCP = 7:4:0.7) and the total concentration of lipid (1×10^{-2} mol/L).

A. Incorporation of a Lipophilic Compound into a Bilayer Membrane

First, we investigated the variation of absorbance of the liposome solutions containing lipophilic compounds with incubation time. The absorbance of the liposome solutions containing lipophilic compounds increased with an increase in their concentration. The absorbance of the solution with octane showed a maximum with time and then became constant. Systems containing 1-octanol, showed a small maximum, but for octanoic acid systems such maximum were not present. In this case, as a forced energy caused by shaking is given to the liposome solutions containing octane, once the absorbance increased due to the incorporation of excess octane into the bilayer membrane, after a few minutes released octane from the liposomes results in decreasing absorbance; then the absorbance becomes constant as an equilibrium distribution of the solute is attained between the outside and inside of the liposomes.

Figure 6 shows the relationship between absorbance and incubation time of liposomes incorporating various lipophilic compounds (3×10^{-2} mol/L) at 30°C. Here, arrows indicate the time required to attain a constant absorbance. As can be seen in Fig. 6, the time required to achieve constant absorbance decreases with increasing hydrophilicity of lipophilic compounds (octanoic acid < 1-octanol < octane).

We previously reported [41, 42] on the solubilization of lipophilic compounds by surfactant micelles: octane, being a nonpolar compound, is solubilized into the hydrocarbon core of the surfactant micelles, whereas 1-octanol and octanoic acid, being polar compounds, thrust the polar group out of the surface of the surfactant micelles and their alkyl chains penetrate the palisade layers of the surfactant micelles. Moreover, 1-octanol, having a weaker hydrophilicity than octanoic acid, is deeply solubilized into the palisade layers of the micelles. Furthermore, compounds

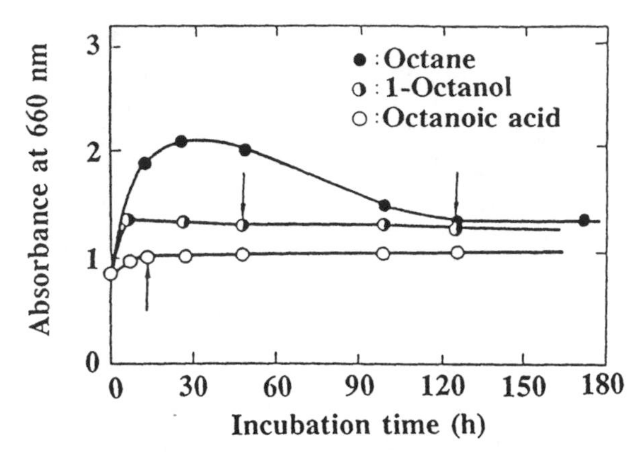

FIG. 6 Relationship between absorbance and incubation time of liposomes incorporating various lipophilic compounds at 30°C. (From Ref. 40.)

having strong hydrophilic groups offer little solubilization in micelles [43] because of their strong interaction with water molecules. The addition of suitable amounts of alcohols and/or fatty acids into the solutions containing surfactant molecules promotes to the formation of a molecular aggregate such as a microemulsion [44]. Thus, for lipophilic compounds indicating cooperative effect for forming an aggregate with DPPC molecules (in other words, strong hydrophilic compounds), the time to attain the distribution equilibrium of lipophilic compounds between the inside and outside of liposomes is short, whereas for compounds without a cooperative effect with DPPC (e.g., octane), the time allowed for the establishment of equilibrium becomes long.

Figure 7 depicts the relationship between the encapsulation efficiency of glucose, which is an aqueous model material, in liposomes and the concentration of lipophilic compounds. The encapsulation efficiency of glucose in liposomes decreases with increasing concentration of lipophilic compounds. Note that the efficiency for the system with added octane without hydrophilic groups becomes a constant (0.6) above 0.1 mol/L of octane, whereas that for 1-octanol and octanoic acid with hydrophilic groups become nearly zero above 0.1 mol/L of lipophilic compounds. The encapsulation efficiency being zero means an absence of an inner water phase in a molecular aggregate as mentioned previously. The addition of suitable amounts of alcohols and/or fatty acids to surfactant solutions results in the formation of molecular aggregates with a monolayer structure (microemulsion) [44]. That the 1-octanol and octanoic acid molecules

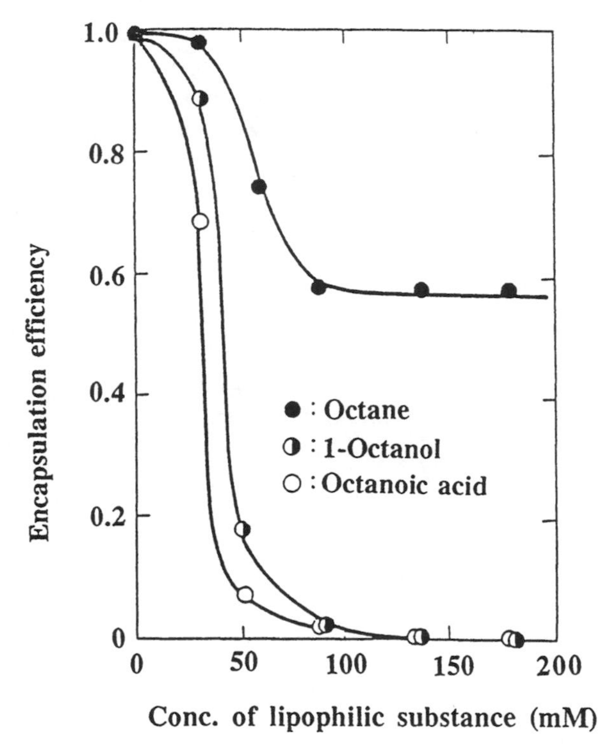

FIG. 7 Effects of concentration of lipophilic compounds on encapsulation efficiency in liposomes at 30°C. (From Ref. 40.)

have hydrophilic groups is well known to be a reversed-cone type; the addition of these compounds into the bilayer membrane of liposomes can bring about the changes of the curvature of the bilayer membrane. Thus, at higher concentrations of 1-octanol and octanoic acid, the aggregates will change from a bilayer structure to a monomolecular structure without an inner water phase.

B. Solubilization Sites of Lipophilic Compounds in a Bilayer Membrane

The microenvironment, such as micropolarity and microfluidity, determined by the fluorescence probe method is important in investigations of the solubilization sites of lipophilic compounds in the bilayer membrane of liposomes.

Pyrene is a well-known typical fluorescence probe, which, like aromatic hydrocarbons, exists in the hydrocarbon region of a bilayer membrane. Pyrene monomer fluorescence shows five predominant peaks. In the present case of pyrene, peak 3, which is strong and allowed, shows minimal intensity variation with polarity. Peak 1 shows significant intensity enhancements in polar solvents. Thus the intensity ratio of peak 1/peak 3 serves as a measure of polarity [45]. Moreover, pyrene monomer forms an excimer by face-to-face collision. Thus the excimer/monomer ratio of emission intensities (I_E/I_M) is indicative of the fluidity of the bilayer membrane [46]. The larger the I_E/I_M ratio of pyrene, the larger the fluidity becomes, while larger I_1/I_3 ratios indicate greater polarity [45, 47, 48].

Figure 8 exhibits the relationship between the I_1/I_3 ratio of pyrene in liposomes and the concentration of lipophilic compounds. Here, the I_1/I_3

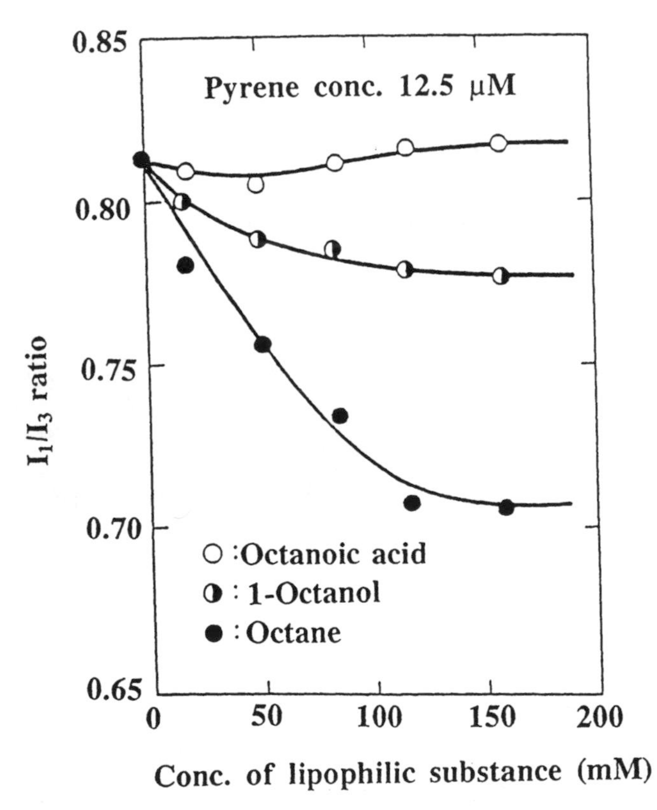

FIG. 8 Relationship between I_1/I_3 ratio of pyrene in liposomes and concentration of lipophilic compounds at 30°C. (From Ref. 40.)

ratio of pyrene in the bilayer membrane of liposomes was considerably larger than that in pure decane. As shown in Fig. 8, the micropolarity (I_1/I_3) for the added octanoic acid system is almost independent of concentration of octanoic acid. On the other hand, the micropolarity for the added 1-octanol system is slightly decreased, and that for the added octane system is considerably decreased with an increasing concentration of the lipophilic compound. It can be postulated that the I_1/I_3 ratio of pyrene solubilized in the bilayer membrane (micropolarity) is dependent on the change of environment of pyrene solubilized and influence of water molecules existing both in continuous and inner phases. In other words, when the lipophilic compounds are incorporated into the hydrocarbon core (center) of the bilayers or are prevented from interacting with water molecules, the I_1/I_3 of pyrene (in other words, micropolarity) would decrease. Namely, for the added octane system, as the I_1/I_3 decreases considerably with increasing concentration of octane, which means the octane exists in surroundings with the pyrene molecules solubilized in the hydrocarbon core of liposomes where the pyrene are protected from interaction with water molecules. For the 1-octanol system, the I_1/I_3 decreases slightly with an increase in the concentration of 1-octanol, which means the pyrene solubilized in the hydrocarbon core is slightly affected by water molecules because of an extension of lipid bilayers caused by a deep penetration of 1-octanol molecules. For the added octanoic acid system, I_1/I_3 is nearly independent of the concentration of octanoic acid, which means that the octanoic acid molecules penetrate the bulky palisade layer of the liposomes and avoid the pyrene solubilized in the hydrocarbon core of the liposomes.

To ascertain the solubilization sites in the bilayer membrane, we measured the I_E/I_M ratio of liposomes (Fig. 9). The microfluidity of the lipid bilayer membrane containing lipophilic compounds with hydrophilic groups increased with increasing concentration of lipophilic compounds; the extent of the increase is 1-octanol $>$ octanoic acid $>>$ octane. An increase of I_E/I_M for added lipophilic compound with hydrophilic groups (for example, 1-octanol) may result from the extension of the lipid bilayer membrane caused by the incorporation of its alkyl chain moieties in lipid molecules. On the other hand, octane without hydrophilic groups, deeply penetrated in the center of lipid membranes, perhaps resulting in a small increase in I_E/I_M. The glucose leakage is larger for lipophilic compounds having strong hydrophilic groups than for lipophilic compounds without such groups (Fig. 10).

From the results mentioned above, the solubilization sizes of lipophilic compounds in the liposomes are the lipophilic compound having higher hydrophobicity, which exists inside the bilayers, and that having higher

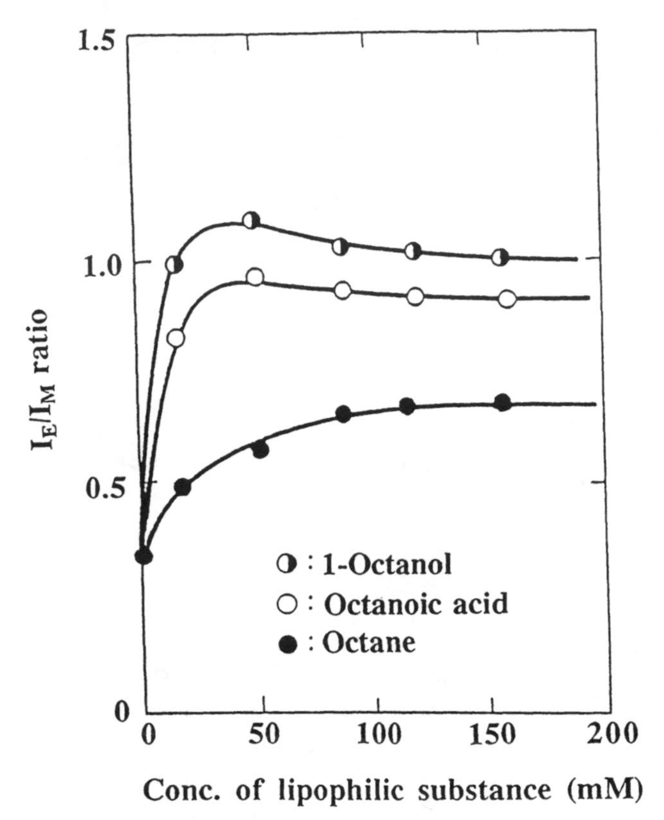

FIG. 9 Effects of lipophilic compounds on I_E/I_M ratio of pyrene in liposomes at 30°C. (From Ref. 40.)

hydrophilicity, which exists near liposome surfaces. Moreover, the excess addition of compounds having hydrophilicity results in a change of aggregate structure as shown in Fig. 11.

In conclusion, solubilization of lipophilic compounds with hydrophilic groups will bring about increases both in micropolarity and microfluidity of liposomes and will result in an increase in glucose leakage.

V. SOLUBILIZATION OF AN ALCOHOLIC COMPOUND BY VESICLES [49]

As mentioned, vesicles consisting of phospholipids, in other words liposomes, have been studied biochemically as biomembranes, drug carriers,

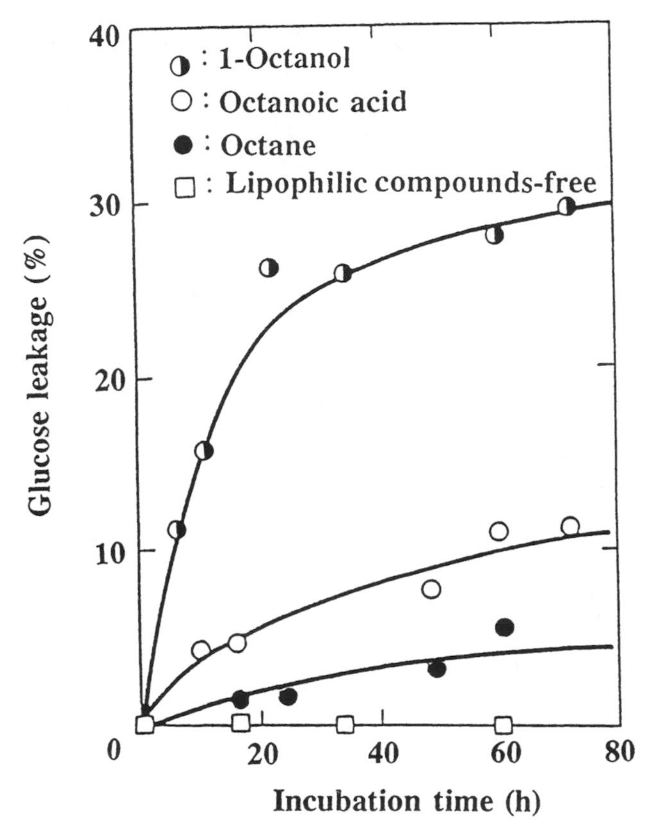

FIG. 10 The time course of glucose leakage from liposomes at 30°C. Concentration of lipophilic compound is 3×10^{-2} mol/l. (From Ref. 40.)

and so on [50–53]. Since Kunitake and Okahata [54] found that synthetic double-chained surfactants can also form vesicles, a great number of physicochemical investigations (e.g., studies of fusion [55] of vesicles, microscopic polarity [56], or fluidity [57] of vesicle bilayers) have been performed. However, there have been few studies related to the solubilization of organic compounds by vesicles and the removal of these compounds from water.

The utilization of vesicles as substitutes for micelles should lead to the development of a modification of micellar-enhanced ultrafiltration (MEUF). Therefore, in the present work, the solubilization of an alcoholic organic compound, 2-phenylethanol (PEA), in vesicles formed from dido-

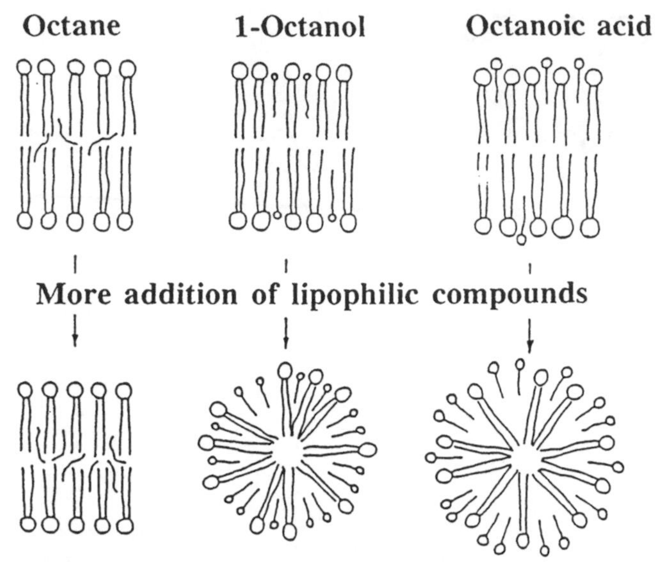

FIG. 11 Schematic model of molecular assemblies with lipophilic compounds. (From Ref. 40.)

decyldimethylammonium bromide (DDAB), a typical double-chained surfactant, is investigated using the semiequilibrium dialysis (SED) method as a prerequisite to applying vesicles in the removal of organic pollutants from water. In this section, we report mainly the solubilization equilibrium constants of PEA for DDAB vesicles together with those for dodecyltrimethylammonium bromide (DTAB) micelles, the activity coefficient of solubilized solute, and the diameter of vesicles solubilizing PEA.

A. Semiequilibrium Dialysis (SED) Method for Vesicle Solubilization

In this work, the solubilization equilibrium constant of an organic compound in an aqueous vesicular or micellar solution (K) is defined as [58–60]

$$K = \frac{X_p}{C_p} \tag{3}$$

where

$$[\mathrm{P}]_{\mathrm{tot}} = \gamma_{\mathrm{P}} X_{\mathrm{p}} C_{\mathrm{p}}^{\circ} + \left(\frac{X_{\mathrm{p}}}{1 - X_{\mathrm{p}}}\right)[\mathrm{S}]_{\mathrm{A}} \tag{4}$$

$$[\mathrm{S}]_{\mathrm{tot}} = \gamma_{\mathrm{s}}(1 - X_{\mathrm{p}})C_{\mathrm{s}}^{\circ} + [\mathrm{S}]_{\mathrm{A}} \tag{5}$$

where $[\mathrm{S}]_{\mathrm{A}}$ is the surfactant concentration in the surfactant aggregate phase, namely in the vesicular or micellar phase, C_{p}° represents a limiting concentration of PEA consistent with the pure-component standard state, C_{s}° is the concentration of either monomeric DDAB or monomeric DTAB in water, and γ_{p} and γ_{s} are activity coefficients of PEA and the surfactant, based on the pure PEA and either vesicle or micelle standard state, respectively. If the values of γ_{p} and γ_{s} in Eqs. (4) and (5) are known, it is possible to solve these equations to obtain X_{p} and $[\mathrm{S}]_{\mathrm{A}}$.

Although several forms have been used to express the relationship between the solubilization equilibrium constant and the mole fraction of solubilizate in the micellar phase, it has been recently reported that an excellent correlation is provided by the equation [59]

$$K = K_0(1 - BX_{\mathrm{p}})^2 \tag{6}$$

where K_0 is the value of the solubilization constant in the limit as X_{p} approaches zero and B is an empirical constant. In this study, therefore, Eq. (6) is used to correlate the γ_{p} value. The activity coefficient is determined by the equation

$$\gamma_{\mathrm{P}} = \frac{1}{KC_{\mathrm{p}}^{\circ}} = \frac{a}{(1 - BX_{\mathrm{p}})^2} \tag{7}$$

where $a = 1/K_0C_{\mathrm{p}}^{\circ}$.

In addition, by assuming the vesicle and micelle to be pseudophases and applying Eq. (7) to the Gibbs-Duhem equation, the activity coefficient of the surfactant, γ_{s}, can be expressed by

$$\ln \gamma_{\mathrm{s}} = \frac{2}{1 - B}\{B \ln(1 - X_{\mathrm{p}}) - \ln(1 - BX_{\mathrm{p}})\} \tag{8}$$

A nonlinear least-squares method [61, 62] is utilized to analyze SED data for each system by Eqs. (4)–(8). First, Eqs. (4) and (5) are solved simultaneously for approximate X_{p} and $[\mathrm{S}]_{\mathrm{A}}$, using analytical values of these quantities on the retentate side, values of the parameters a and B, C_{p}°, C_{s}°, and Eqs. (7) and (8). Next, $[\mathrm{S}]_{\mathrm{A}}$ on the permeate side is estimated by utilizing a, B, γ_{P}, γ_{s}, as used above and Eq. (5), assuming that X_{p} in the permeate solution is the same in the retentate solution. Further, Eq. (4) is employed to calculate the PEA concentration on the permeate side.

Finally, B and a are varied to minimize the root-mean-square deviation between the calculated and the experimental values of $[P]_{tot}$ to obtain the optimum values of the parameters.

Individual values of K are calculated from the equation [58]

$$K = (1 - X_p)\left(\frac{[P]_{tot}^{ret} - [P]_{tot}^{per}}{[P]_{tot}^{per}\,[S]_A^{ret} - [P]_{tot}^{ret}\,[S]_A^{per}}\right) \tag{9}$$

where the superscripts ret and per indicate the retentate and permeate side, respectively. This equation accounts for the presence of aggregates in the permeate. $[P]_{tot}^{ret}$ and $[P]_{tot}^{per}$ are known values, and X_p and $[S]_A^{ret}$ can be calculated by solving Eqs. (4) and (5), using Eq. (7) and the least-squares value of B. Then $[S]_A^{tot}$ can be inferred by the measurable $[S]_{tot}^{per}$, γ_s, and B as used to obtain $[S]_A^{per}$.

B. Solubilization Equilibrium Constants

It has been reported that DDAB vesicles can be formed by sonicating surfactants in aqueous solution [63]. Since DDAB vesicles with a diameter of ca. 17 nm can be prepared reproducibly [63], solutions containing these vesicles were studied by the SED method to obtain values of the solubilization equilibrium constant of PEA. The dependence of the solubilization equilibrium constant on the mole fraction of PEA (X_p) in different solutions is shown in Fig. 12. If Eq. (6) is applicable for fitting the SED data, a plot of the square root of K against X_p for each system will be linear, with a slope equal to $-\sqrt{K_0}B$. As can be seen in Fig. 12, this relationship does provide a good correlation of data for each system. Note that the solubilization constant for the DDAB-PEA system is appreciably larger than that for the DTAB-PEA system over the entire range of X_p; for DDAB, $K_0 = 35.4\ M^{-1}$ and for DTAB, $K_0 = 18.1\ M^{-1}$. Generally, the polar compounds that have a strong tendency to solubilize within the aggregates like surfactant micelles are characterized by large values of K and (originally) small values of γ_p. Namely, PEA will be solubilized more strongly by DDAB vesicles than by DTAB micelles. For the micellar solubilization of PEA by cationic dodecyldimethylamine oxide [64] (DDAO), K_0 is 17.0 M^{-1}. Thus K_0 for DDAB vesicles seems to be a relatively large value compared to that for other surfactant aggregates.

C. Activity Coefficients of the Surfactant and PEA

Information about the activity coefficient of solubilized components and the dependence of γ_p on X_p may be useful in determining the environment of these compounds within the aggregates such as vesicles and micelles.

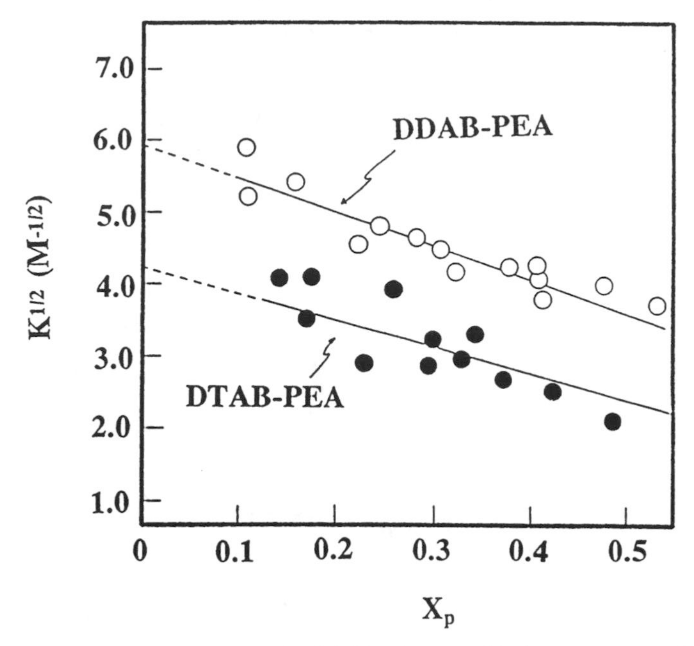

FIG. 12 Dependence of solubilization equilibrium constants for the DDAB-PEA system and the DTAB-PEA system on the mole fraction of solute at 30°C. (From Ref. 49.)

Values of γ_p near unity indicate that the environment of a solubilized species is energetically similar to that of the pure component (liquid or solid). Large values of γ_p are characteristic of compounds that solubilize in environments less favorable than the pure solute state, and values of γ_p considerably less than unity indicate that the aggregate strongly attracts the solubilizate.

The activity coefficients of DDAB and DTAB in aggregates (vesicles or micelles) plotted against X_p, are depicted in Fig. 13. Here, the values of γ_s are calculated using Eq. (7). When the mole fraction of intravesicular or intramicellar PEA, X_p, increases, these activity coefficients decrease. No significant difference between the activity coefficient of DDAB and that of DTAB is observed.

The activity coefficient of PEA solubilized in DDAB vesicles or DTAB micelles is shown in Fig. 14. As can be seen in Fig. 14, both PEA activity coefficient curves increase toward unity as X_p increases. It is generally known that the hydroxyl group of an alcohol strongly interacts with the

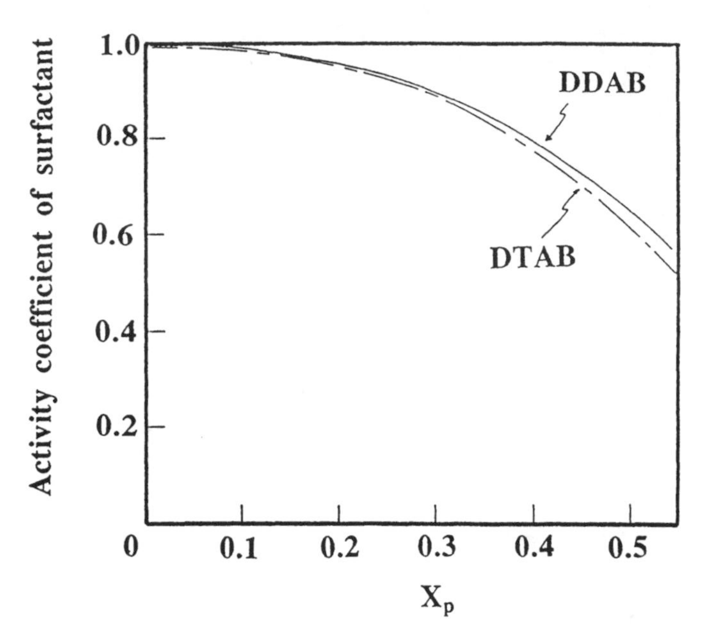

FIG. 13 Dependence of the activity coefficients (γ_s) for DDAB and DTAB on the mole fraction of PEA. (From Ref. 49.)

charged head group of the ionic surfactant [64, 66] and at the same time the activity coefficient of the solubilized alcohol is less than unity [66]. Therefore, both activity coefficient values at low X_p imply that the environments of PEA in DDAB vesicles and in DTAB micelles are highly favored energetically (compared with pure component PEA [66]), and that PEA is solubilized in the hydrophilic region of vesicle bilayers, with the hydroxyl group of PEA attracted to the charged headgroup of DDAB.

D. Competition for Sites at the Aggregate Surface

The relationship between the diameter of DDAB vesicles in the retentate solutions and the PEA mole fraction, X_p (inferred from SED experiments), is exhibited in Fig. 15. Here, the diameter of DDAB vesicles was determined by the dynamic light-scattering method using a submicron particle analyzer (Marvern Instrument, Model 4700) with an argon ion laser operating at 488 nm. The vesicular diameter increases with X_p and remains approximately constant beyond an X_p of 0.3. Generally, as organic compounds like alcohol are solubilized in the hydrophilic region of ionic micelles, the micellar diameter also increases [67].

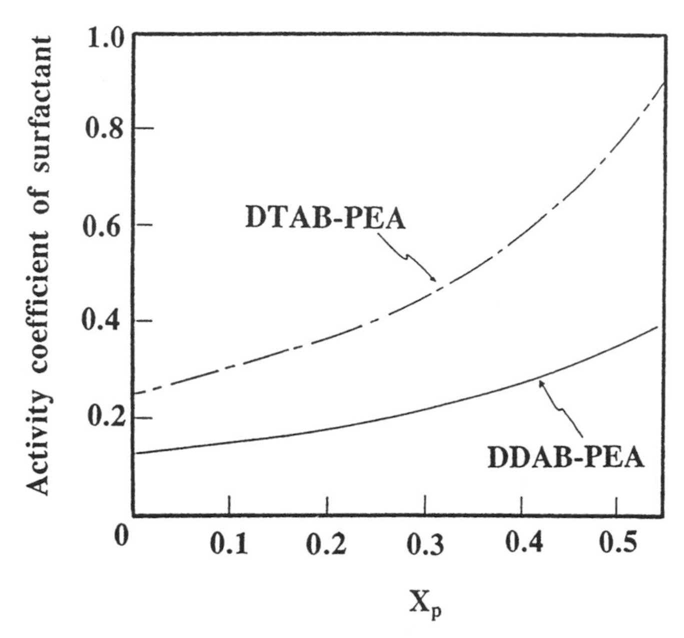

FIG. 14 Relationship between the activity coefficient (γ_p) of PEA for DDAB-PEA system and the DTAB-PEA system and the mole fraction of PEA. (From Ref. 49.)

Although equilibrium solubilization data are intrinsically valuable, these results are also suitable in indicating the locus of solubilization of molecules within aggregates like surfactant micelles. Typically, polar organic compounds such as the aliphatic alcohols, carboxylic acids, phenols, and cresol have very small values of the activity coefficient, and these values gradually increase as X_p increases. The solubilization results are consistent with other physical evidence indicating that these molecules have their headgroups anchored in the polar/ionic outer region of typical ionic surfactants. The aliphatic or aromatic moieties of these polar compounds tend to solubilize at least partly within the hydrocarbon core of the aggregate (micelle), although steric and substituent group effects may play an important role in modifying the structure and thermodynamic properties of the "intramicellar solutions" of the organic compound in the micelle.

Aliphatic hydrocarbons undoubtedly solubilize primarily within the hydrocarbon core region of the surfactant micelles. Solubilization isotherms for these very hydrophobic compounds typically exhibit activity coeffi-

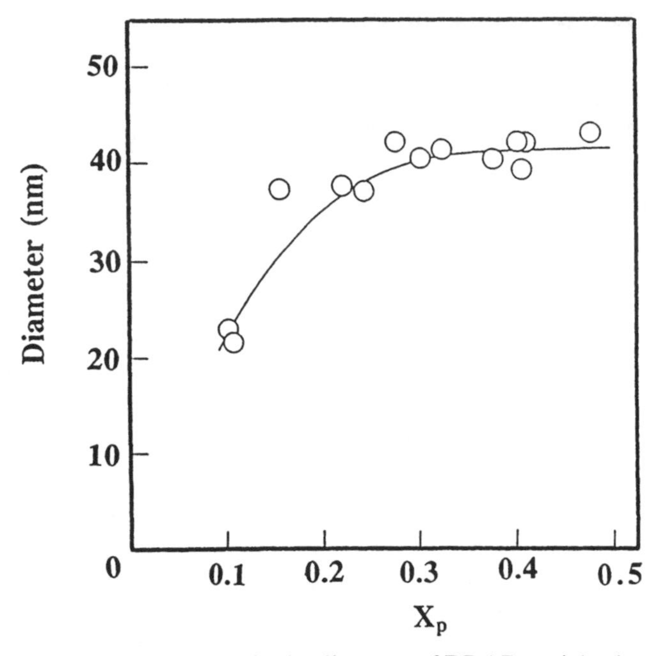

FIG. 15 The change in the diameter of DDAB vesicles in retentate solution after SED experiments. (From Ref. 49.)

cient versus X_p curves that decrease from relatively large values at infinite dilution to lower values when X_p increases toward unity. Although the activity coefficient may be as large as 5 or 10 for alkanes in surfactant micelles at small values of X_p, it should be emphasized that activity coefficients for these compounds in pure water are on the order of 10^5 (the reciprocal of the mole fraction solubility of the hydrocarbon in water).

The aromatic hydrocarbons are intermediate in behavior between highly polar compounds, which are clearly anchored in the aggregate surface region, and aliphatic hydrocarbons, which preferentially solubilize in the hydrocarbon core region.

We have reported that the empirical parameter B is related to the number of sites occupied by one PEA molecule in the headgroup region, considering that the solubilization behavior is consistent with Langmuir's adsorption [59]. If the adsorption of the polar compounds occurs with the phenolic OH group bound in the headgroup region of the aggregate, an increase in the amount of adsorption of the compound when C_p increases may be considered to deplete surface sites. The value of $2BX_p$ is equal to the fraction of the surface sites that are occupied by bound solute molecules, which indicates that the ratio of the number of surfactant molecules to the number of solubilizate molecules per site is equal to $2B$. The value $2B$ of PEA for DDAB is 1.58, and for DTAB it is 1.73. One interpretation of this observation is that each solubilizate molecule may interact strongly with one or two surfactant molecules, so these molecules can no longer act as a primary solubilization site for additional organic compound molecules. As the sites are occupied, additional PEA molecules are less attracted to the headgroup region.

All of the effects described above should occur during the solubilization of an amphiphile (PEA) in the DDAB vesicles. The observation that the size of vesicles increases as X_p varies from 0 to 0.3 (Fig. 15) may indicate that the adsorption of OH groups of PEA in the headgroup region results in a swelling of the vesicle size, analogous to the increase in micelle diameter caused by such solubilizates. Beyond $X_p = 0.3$, however, the near-constancy of the vesicle size may reflect the fact the PEA dissolves more extensively in the hydrophobic core region of the vesicle bilayer, because the surface sites have become nearly saturated with adsorbed PEA molecules. Such an effect might well leave the inner and outer surfaces of the vesicle nearly the same, causing the bilayer to increase in thickness, but not significantly increasing the diameter of the vesicles.

The smaller values of γ_P for the DDAB-PEA system, compared to γ_P values for the DTAB-PEA system in Fig. 15, indicate that PEA is solubilized more strongly by DDAB vesicles than by DTAB micelles. Previously, it has been reported that the compactness of the headgroup region in micelles makes it difficult for alcohol to solubilize into the micelle [64], and that the DDAB headgroups are also closely packed [56]. Considering these reports simultaneously, it may be somewhat difficult to account for the smaller PEA activity coefficients for the DDAB-PEA systems. However, the branching of the aliphatic moiety of DDAB may cause the packing in the head region to be less dense than in micelles of DTAB, causing solubilized PEA molecules to be more strongly bound in the headgroup region. The less rapid increase in γ_P with increasing X_p may also arise from the enhanced ability of PEA to fit into the looser (bilayer) structure of the vesicular interface.

REFERENCES

1. A. D. Bangham, M. M. Standish, and J. C. Watkins, *J. Mol. Biol. 13*: 238 (1965).
2. H. Yamauchi and H. Kikuchi, *Fragrance J. 87*: 68 (1987).
3. F. C. Szoka and D. Papahadjopoulos, *Ann. Rev. Biophys. Bioeng. 9*: 467 (1980).
4. F. C. Szoka and D. Papahadjopoulos, *Proc. Nat. Acad. Sci. U.S.A. 75*: 4194 (1978).

5. P. R. Cullis, and B. de Kruijff, *Biochim. Biophys. Acta. 559*: 399 (1979).
6. T. Kitagawa, K. Inoue, and S. Nojima, *J. Biochem. 79*: 1123 (1976).
7. M. Smith and F. B. Jungalwala, *J. Lipid Res. 22*: 697 (1981).
8. D. Chapman, in *Liposome Technology* (G. Gregoriadis, ed.), CRC Press, Florida, 1984, Vol. 1, pp. 1–18.
9. M. Arrowsmith, J. Hadgraft, and I. W. Kellaway, *Int. J. Pharm. 16*: 305 (1983).
10. R. Smith and C. Tanford, *J. Mol. Biol. 67*: 75 (1972).
11. H. Kikuchi and K. Inoue, *Cell Tech. 2*: 1136 (1983).
12. M. J. Ostro (ed.), *Liposomes* Marcel Dekker, New York, 1983.
13. G. Gregoriadis (ed.), *Liposome Technology* Vol. I, CRC Press, Florida, 1984.
14. H. Kikuchi and K. Inoue, *J. Jpn. Oil Chem. Soc. 34*: 784 (1985).
15. D. Lichtenberg and Y. Barenholz, *Methods Biol. Chem. 33*: 337 (1988).
16. N. Weiner, F. Martin, and M. Riaz, *Drug Dev. Ind. Pharm. 15*: 1523 (1989).
17. F. Olson, C. A. Hunt, F. C. Szoka, W. J. Vail, and D. Papahadjopoulos, *Biochim. Biophys. Acta. 557*: 9 (1979).
18. C. Huang, *Biochemistry. 8*: 344 (1969).
19. S. Batzri and E. D. Korn, *Biochim. Biophys. Acta. 298*: 1015 (1973).
20. Y. Kagawa and E. Racker, *J. Biol. Chem. 246*: 5477 (1971).
21. W. J. Gerritsen, A. J. Verkley, R. F. A. Zwaal, and L. L. M. Van Deenen, *Eur. J. Biochem. 85*: 255 (1978).
22. Y. Barenholz, S. Amselem, and D. Lichtenberg, *FEBS Lett. 99*: 210 (1979).
23. H. Trauble and E. Grell. Neurosci, *Res. Prog. Bull. 9*: 373 (1971).
24. H. O. Hauser, *Biochem. Biophys. Res. Commun. 45*: 1049 (1971).
25. P. W. Holloway, *Anal. Biochem. 53*: 304 (1973).
26. D. Papahadjopoulos, W. J. Vail, K. Jacobson, and G. Poste, *Biochim. Biophys. Acta. 394*: 483 (1975).
27. D. Deamer and A. D. Bangham, *Biochim. Biophys. Acta. 443*: 629 (1976).
28. R. Lawaczeck, M. Kainosho, and S. I. Chan, *Biochim. Biophys. Acta. 443*: 313 (1976).
29. M. Kasahara and P. C. Hinkel, *J. Biol. Chem. 252*: 7384 (1977).
30. S. Matsumoto, M. Kohda, and S. Murata, *J. Colloid Interface Sci. 62*: 149 (1977).
31. H. Kikuchi, H. Yamauchi, and S. Hirota, *Chem. Pharm. Bull. 39*: 1522 (1991).
32. M. Abe, T. Hiramatsu, H. Uchiyama, H. Yamauchi, and K. Ogino, *J. Jpn. Oil Chem. Soc. 41*: 136 (1992).
33. C. Huang and J. T. Mason, *Proc. Nat. Acad. Sci. U.S.A. 75*: 308 (1978).
34. R. A. Greff, E. A. Setzkorn, and W. D. Lestie, *J. Am. Oil Chem. Soc. 42*: 180 (1965).
35. M. Takayama, S. Itoh, T. Nagasaki, and I. Tanimizu, *Clin. Chem. Acta. 79*: 93 (1977).
36. P. R. Strom-Jensen, R. L. Magin, and F. Dunn, *Biochim. Biophys. Acta. 769*: 179 (1984).
37. D. A. Tyrrel, T. D. Heath, C. M. Colley, and B. E. Ryman, *Biochim. Biophys. Acta. 457*: 259 (1987).

38. H. Ohshima, T. W. Healy, and L. R. White, *J. Colloid Interface Sci. 90*: 17 (1978).
39. N. Tsubaki, Y. Nakano, Y. Goto, K. Ogino, and M. Abe, *J. Jpn. Oil Chem. Soc. 41*: 551 (1992).
40. M. Abe, Y. Takao, T. Yamamoto, K. O. Kwon, H. Yamauchi, and K. Ogino, *J. Jpn. Oil Chem. Soc. 41*: 404 (1992).
41. M. Abe, A. Shimizu, and K. Ogino, *J. Colloid Interface Sci. 88*: 319 (1982).
42. M. Abe and K. Ogino, *Yukagaku, 31*: 569 (1982).
43. M. Abe, Y. Tokuoka, H. Uchiyama, and K. Ogino, *J. Jpn. Oil Chem. Soc. 39*: 565 (1990).
44. M. Nakamae, M. Abe, and K. Ogino, *Sekiyu Gakkaishi 31*: 466 (1988).
45. K. Kalyanasundaram and J. K. Thomas, *J. Am. Chem. Soc. 30*: 2039 (1977).
46. K. Kano, *Surface 26*: 243 (1988).
47. N. J. Turro, P. L. Kuo, P. Somasundaran, and K. Wong, *J. Phys. Chem. 90*: 288 (1986).
48. G. P. L'Heureux and M. Fragata, *Biophys. Chem. 30*: 293 (1988).
49. Y. Kondo, M. Abe, K. Ogino, H. Uchiyama, J. F. Scamehorn, E. E. Tucker, and S. D. Christian, *Colloid Surf. B*: *Biointerface 1*: 51 (1993).
50. S. J. Singer and G. L. Nicolson, *Science 175*: 720 (1972).
51. M. J. Ostro and P. R. Cullis, *Am. J. Hospital Pharm. 46*: 1576 (1989).
52. K. Shinoda, I. Adachi, M. Ueno, and I. Horikoshi, *Yakugaku 110*: 186 (1990).
53. T. Hiff and L. Kevan, *Colliod Surf. 45*: 185 (1990).
54. T. Kunitake and Y. Okahata, *J. Am. Chem. Soc. 99*: 3860 (1977).
55. L. A. M. Rupert, J. B. F. N. Engberts, and D. Hoekstra, *J. Am. Chem. Soc. 108*: 3920 (1986).
56. R. McNeil and J. K. Thomas, *J. Colloid Interface Sci. 73*: 522 (1980).
57. A. S. Domazou and A. E. Mantaka-Marketou, *Ber. Bunsenges. Phys. Chem. 94*: 428 (1990).
58. S. D. Christian, G. A. Smith, E. E. Tucker, and J. F. Scamehorn, *Langmuir 1*: 564 (1985).
59. B. H. Lee, S. D. Christian, E. E. Tucker, and J. F. Scamehorn, *Langmuir 6*: 230 (1990).
60. G. A. Smith, S. D. Christian, E. E. Tucker, and J. F. Scamehorn, *J. Solution Chem. 15*: 519 (1986).
61. S. D. Christian and E. E. Tucker, *Am. Lab. 14*: 36 (1982).
62. S. D. Christian and E. E. Tucker, *Am. Lab. 14*: 31 (1982).
63. Y. Kondo, M. Abe, K. Ogino, H. Uchiyama, J. F. Scamehorn, E. E. Tucker, and S. D. Christian, *Langmuir 9*: 899 (1993).
64. H. Uchiyama, S. D. Christian, J. F. Scamehorn, M. Abe, and K. Ogino, *Langmuir 7*: 95 (1991).
65. C. M. Nguyen, J. F. Scamehorn, and S. D. Christian, *Colloid Surf. 30*: 335 (1988).
66. S. D. Christian and J. F. Scamehorn, in *Surfactant-Based Separation Processes* (J. F. Scamehorn and J. H. Harwell, eds.), Marcel Dekker, New York, 1989, Chapter 1.
67. M. Abe and K. Ogino, *J. Jpn. Oil Chem. Soc. 31*: 569 (1982).

IV
Methods of Measuring Solubilization

11

Solubilization, as Studied by Nuclear Spin Relaxation and NMR-Based Self-Diffusion Techniques

PETER STILBS Department of Physical Chemistry, The Royal Institute of Technology, Stockholm, Sweden

SYNOPSIS

Methods for quantifying solubilization into micellar systems by NMR techniques, as based on spin relaxation and pulsed field gradient (FT-PGSE) self-diffusion measurements are reviewed, together with selected results from such studies. The FT-PGSE approach is a powerful, selective, and quantitative tool capable of providing detailed information on partitioning in complex surfactant and solubilizate systems. Complementary spin relaxation methods covered aim at a characterization of solubilizate ordering and local dynamics in the micellar framework.

I. INTRODUCTION

In the past decades, NMR studies based on nuclear spin relaxation and self-diffusion have proven to be very informative with regard to the phenomenon of solubilization. When a small solubilizate molecule becomes incorporated into a surfactant aggregate, its motion becomes restricted in space, and in *most* aspects slowed down, as compared to conditions in a continuous solvent phase. NMR methods, based on nuclear spin relaxation and molecular self-diffusion are sensitive to these effects, and can actually quantify them in great detail.

The primary use of measured *spin relaxation* rates in such a situation is to provide a detailed tool for studying the *local reorientation dynamics* of the solubilized molecule. This may be informative with regard to its *average location* within the aggregate framework and, of course, to the *local chain dynamics* in itself.

Self-diffusion rates, on the other hand, provide an almost direct tool for monitoring *the partitioning* of solubilizate molecules between aqueous and micellar pseudophases, as based on the large difference in self-diffusion rates for the aggregate and the free solubilizate molecule in solution. The time-averaged self-diffusion rate between these two environments provides a direct measure of the partitioning phenomenon.

It is apparent that the parameters mentioned provide a rather detailed and also quantitative description the situation experienced by a solubilized molecule. In the following, the methods and some typical applications will be described more fully.

Early NMR work on solubilization and other aspects of surfactant aggregation in solution has been summarized in an excellent review by Chachaty [1]. Some older reviews may also be of interest to the reader [2–4].

II. MOLECULAR DYNAMICS AND ORDER IN MICELLES, AS DERIVED FROM NUCLEAR SPIN RELAXATION

NMR studies of surfactant systems in solution during the last decade have very much been influenced by a number of papers, promoting the idea of a two-step motional and spin relaxation model for molecules within a micellar framework [5–11], the theory of which was developed to its present form by Halle and Wennerström [9].

In this model, the reorientational motion of the aggregate molecules is separated into two regimes; a fast regime corresponding to the local motion of the molecule within the aggregate, and a slow regime corresponding to the global motions of the aggregate and/or molecular lateral diffusion over the curved aggregate surface. In the normal application of the method, one studies spin relaxation as a function of the NMR magnetic field strength; by an overall analysis of the data one can then deduce values of τ_c^f and τ_c^s and S, corresponding to "fast and slow" correlation times, respectively, and the so-called order parameter. The "fast motions" correspond to molecular rotation, torsional oscillations, and *trans-gauche*-isomerizations, and the "slow motions" to aggregate tumbling and internal diffusion over the curved surface. The order parameter S may assume values between 0 and 1, ranging from total internal freedom for the local motions to complete ordering within the tumbling aggregate. Typical experimental values from this approach for the methylene segments of a surfactant chain in a micelle are a few tens of picoseconds for the fast motion, a few tens of nanoseconds for the slow, and 0.1–0.2 for the S-value. For a surfactant chain one also observes a continuously increased motional freedom, and decreased molecular order from the headgroup region to the methyl end, thus also providing information on the location and order of a molecular fragment of a flexible molecule.

Most early studies were made through ^{13}C NMR studies at typical NMR fields of standard NMR spectrometers, corresponding to 15–60 MHz carbon frequency, and are basically correct. Later work has shown, however, that the quality of the experiments and the evaluation is very much improved, through selective deuterium (^2H) labeling of the molecule in question, and subsequent studies of the field-dependence of deuterium spin relaxation down to preferably 2 MHz or less (see, e.g., Ref. 8, 10, 11). This improvement is inherently related to the fact that the low-field NMR spectral window of the low-frequency nucleus ^2H more closely corresponds to the actual $1/\tau_c^s$ for the motional processes in question, which results in a larger effect on the nuclear spin relaxation processes.

A. Solubilization, as Studied by Nuclear Spin Relaxation

Similar methods can be applied to study of solubilization—the S-values of different parts of a solubilizate molecule will depend on its preferential location in the micelle, and one can also deduce reorientational rates for the solubilizates. However, only a few papers in this field have been presented. The most detailed study so far is that of Wasylishen et al. [12], which also contains a rather complete review of related work in the past. Wasylishen and co-workers concluded from their study on hydrocarbon solubilization in cationic and anionic micelles, that aromatic hydrocarbons reside near the headgroup in cationic micelles, but are evenly distributed throughout anionic SDS micelles. On the other hand, saturated hydrocarbons like cyclohexane were found to be located in the interior of micelles.

An earlier, borderline, case is worth mentioning at this point; Jansson and co-workers studied organic counterion binding (acetate and butyrate) to cationic micelles of the decylammoniate type [13]. They were able to deduce order parameter profiles for the surfactant, as well as the counterion. In the butyrate case one found clear evidence that the alkyl chain of the counterion actually "solubilizes" into the micellar interior.

B. Determining the Location of Solubilizates, from 2D Cross-Relaxation (NOESY) Experiments

Methods for proton-proton internuclear distance determination by 2D and NMR techniques were introduced by Ernst and co-workers more than a decade ago [14, 15], and have since been extensively used in the context of 3D protein and polynucleotide structural determination in solution.

They key experimental approach is based on the approximately $1/r^6$-dependent nuclear spin cross-relaxation between nearby protons. Although the experiment does not rely on alterations in spin state populations (as in the traditional one-dimensional nuclear Overhauser experiment (NOE)), the term NOESY for the cross-relaxation effects in the 2D experiment in question has been firmly established. In analogy with observed signal intensity changes in the NOE experiment, cross-peaks in a 2D NOESY map reveal which protons are close in space, and by a peak volume integration, one can deduce the approximate average distance between selected protons in a molecule or aggregate.

Undoubtedly, this could also provide a path to very valuable information concerning chain packing in micelles. However, a large complication arises from time-averaging of proton-proton distances in micelles, due to the dynamic nature of micellar systems. In relatively rigid proteins and peptides this complication will not occur at all to the same extent.

Nusselder et al. made the first attempts to use this technique in the context of micellar systems [16], following earlier work on more rigid vesicular systems [17]. Preliminary work on micellized and monomeric 1-methyl-4-dodecylpyridinium iodide was presented, but it does not appear to have been extended. Kolehmainen presented a study along similar lines at about the same time [18] on micellar solubilization in sodium cholate systems.

C. "Micellar Viscosity" from Spin Relaxation of a Solubilized Probe

It was actually first suggested by Hartley in the mid-1930s that the micellar interior is liquid-like [19]. Indeed, it is generally agreed from all kinds of experimental information that that is the case. Numerous studies have been made by several techniques to quantify this further; already covered in this review is the field-dependent spin relaxation approach (Sec. II), which provides a detailed picture of the local mobility along, e.g., a surfactant chain. However, an *overall* quantification of the "macroscopic" liquid properties of a micelle "droplet" requires some way to measure its viscosity, or related parameters. This has to be done through the "probe approach," which unfortunately has two intrinsic drawbacks:

A probe molecule will necessarily perturb the system to some extent; this effect can be minimized by selecting a probe molecule that is maximally similar to its environment.

By reducing the probe concentration, the system perturbation will be further reduced. However, in micellar systems—unlike the situation in "continuous" nonassociated liquids—a foreign probe necessarily reports on conditions in a micelle that contains at least one such probe molecule, regardless of probe concentration.

Several studies based on fluorescence depolarization of solubilized aromatic molecules have appeared in the literature, determining the "microviscosity" of the micellar environment. Such studies have very little value, since the normally quite huge and planar fluorescence probe cannot possibly report properly on a micellar system, because it will undoubtedly cause a large chemical perturbation of the probed micellar system. Also, the relation between the single reorientation rate normally evaluated from fluorescence depolarization and the viscosity of the environment is rather unclear.

By an NMR approach one can at least proceed one or two steps further. Stark et al. [20] and Stilbs et al. [21] made independent studies along these lines, utilizing spin relaxation in much smaller, and perdeuterated hydrocarbon-like molecules like aniline and *trans*-decalin. At least *trans*-decalin can be assumed to be solubilized in the micellar interior, and will be expected to report properly on this environment. The three components of the reorientational diffusion tensor can be evaluated from the NMR experiment and then further related through hydrodynamic theory to the solution viscosity. It was concluded in that study [21] that the micellar core has nearly the same viscosity as a liquid made up of a hydrocarbon with the same chain length as the surfactant monomer.

Later Wasylishen et al. [21] studied the reorientation of several types of solubilized hydrocarbons, and Canet et al. [22] the reorientation of deuterochloroform in rod-shaped alkylammonium chloride micelles. Both studies arrived at similar conclusions as those just given for *trans*-decalin in a micellar environment.

III. DETERMINATION OF MICELLE-WATER PARTITION EQUILIBRIA BY NMR METHODS

A. Partition Equilibria from Electron-Nuclear Spin Relaxation

Gao, Kwak, and Wasylishen have recently developed a very useful spin relaxation method, as an alternative to the self-diffusion method discussed below, for the study of solubilizate partitioning between micellar and aqueous pseudophases [23, 24]. It can be applied on standard spectrometer systems, and is based on electron-nuclear spin relaxation on the unsolubilized fraction of the solubilizate molecules, induced by a small fraction of added paramagnetic ions in the aqueous phase. That method and many of its applications are already covered in another chapter in this book and will therefore not be further discussed here, except to conclude that paramagnetic ion method generates results that are in very good agreement with the self-diffusion approach described as follows.

B. Partition Equilibria, as Studied by Self-Diffusion Measurements

The inherently large difference in self-diffusion coefficients for a small-to-medium solubilizate molecule ($D \approx 10^{-9}\ m^2\ s^{-1}$) and a surfactant micelle ($D \approx 0.01–0.05 \times 10^{-9}\ m^2\ s^{-1}$) provides a key to a direct determination of the fraction of solubilizate molecules that become incorporated into the micelle, since the time-averaged self-diffusion coefficient will be given by

$$D_{obs} = pD_{micelle} + (1 - p)D_{free} \tag{1}$$

where the degree of solubilization (p) assumes values in the range ($0 < p < 1$), D_{obs} denotes the observed time average, and D_{free} the self-diffusion coefficient of the solubilizate in water, in the absence of micelles. A small correction for so-called obstruction effects from the micelles should be considered, but is normally negligible.*

This equation can be rewritten in the form

$$p = \frac{D_{free} - D_{obs}}{D_{free} - D_{micelle}} \tag{2}$$

All the parameters of Eq. (2) are normally experimentally accessible. If this is not the case (e.g., a hydrocarbon may be too insoluble in water to allow an experimental determination of D_{free}), any reasonable estimation of D_{free}, based on analogy or extrapolation will not cause serious systematic errors in the measurements. Also, D_{mic} is essentially only a small correction term in Eq. 2, and any uncertainty in its actual value does not propagate very much into the p-value of Eq. 2. At concentrations well above the CMC, the surfactant diffusion rate is nearly the same as the self-diffusion rate of the micelles. For low-CMC surfactants the time-averaged contribution from surfactant monomer diffusion may contribute significantly. In such a situation one commonly determines D_{mic} by solubilizing a small amount of a highly hydrophobic molecule and monitoring its self-diffusion coefficient. Tetramethylsilane, the common NMR reference molecule "TMS," is quite useful in this context, having a strong NMR signal, in a nonoverlapping region. A disadvantage is its relatively high volatility (the boiling point is near 27°), since it may evaporate to the gas phase above the solution during the course of the experiment, causing systematic errors in the determination of D_{mic}. HMDS (hexamethyldisiloxane) is better in this respect, but is also a larger molecule; potentially it is too large to be considered nonperturbing for micelles made up by relatively short-chain surfactants, like octanoate, or similar surfactants. Desando et al. have discussed related problems in some detail [26].

These two problems are minor; however, the self-diffusion approach to micellar partition coefficients becomes increasingly inaccurate for p-values approaching 0 or 1 (p results from the difference between two numbers, that become increasingly similar (cf. Eq. 2)), but that one can

* The updated and corrected theory for obstruction effects on self-diffusion was worked out some years ago by Jönsson et al. [25].

still modify the experimental conditions, in order to move p into a suitable range, i.e., by diluting a micellar system having a solubilizate that "binds too strongly."

Further interpretation of the parameter p requires a comment; strictly speaking, p in Eq. (2) would correspond to the solubilizate fraction "diffusing with the micelle," surface bound, solubilized in the core or whatever. However, in the normal micelle-water pseudophase approach this distinction lacks any meaning (cf. also the comments in Sec. III.E.).

From the known composition of the solution, one can deduce the "volume" or "weight fraction" of the micelle and the aqueous phase. The experimental p-value can thus be translated into the partition coefficient sought.*

C. NMR Self-Diffusion Measurements through the Pulsed-Gradient Spin-Echo Experiment

In the pulsed-gradient spin-echo (PGSE) experiment on molecules carrying nuclei with gyromagnetic ratio γ, and undergoing Gaussian diffusion, characterized by the self-diffusion coefficient D, at a constant rf pulse interval (τ), the echo amplitude ($I_{2\tau}$) is described by the Stejskal-Tanner equation

$$I_{2\tau} = e^{-\gamma^2 G^2 \delta^2 (\Delta - \delta/3) D} \tag{3}$$

where δ denotes the durations of a pair of magnetic field gradient pulses separated by a time interval Δ and G denotes their amplitude [30, 31]. The normal experimental approach is to vary δ, and to monitor the echo decay for say 10–20 different δ-values. The experiment is normally done on protons, the most sensitive and abundant nucleus. It is also the one with the largest γ-value, so proton echo attenuation is particularly sensitive to diffusion effects, as compared to other nuclei.

1. The Fourier Transform Pulsed-Gradient Spin-Echo Experiment, FT-PGSE

A pulsed-gradient NMR self-diffusion approach to solubilization equilibria, for example, is impractical in the original form of the NMR experiment in question. It has been well known for decades that a direct separation

* Although the first experiments along these lines on solubilization were made through NMR self-diffusion measurements by the present author [27, 28], it should be stressed that the self-diffusion approach to aggregation phenomena had been used much earlier in the field of surfactant aggregation and counterion binding, by Lindman et al., using cumbersome radioactive tracer methods. See, e.g., Ref. 29.

of a sum of experimental exponentials requires very high signal/noise, and that the individual time constants should differ by at least a factor of 2 or 3, in order to achieve a reasonably accurate numerical separation. These conditions do not normally apply to PGSE experiment on solubilization, and thus a unresolved time-domain experiment would require several experiments in turn, with all except one component perdeuteriated, to mask the individual contributions to the echo.

Fourier transformation of the time-domain spin-echo, however, separates all spectral components of the spectrum in the frequency domain (cf. Fig. 1), where there are not normally any serious problems with signal overlap. Peak heights or integrals can then independently be fitted to the Stejskal-Tanner equation in the normal way. This modification, named FT-PGSE, was first conceived by Vold et al. [32] and later developed into a practical tool by several contributors [33–39].

In a single FT-PGSE experiment, often requiring less than 20 min measurement and evaluation time, one can now simultaneously determine self-diffusion coefficients of several constituents in a solution. The precision is often better than a few percent. The procedures have been described in detail in a 1987 review article [39]. Since that time, some trends in instrumental development make the implementation of the FT-PGSE method much easier, and extend its scope of application: First, several custom NMR probe manufacturers now do provide suitable probes for virtually any modern spectrometer. Second, developments in multidimensional NMR spectroscopy for structure determination have led to innovative and significant time-saving measurement procedures ,named "gradient enhanced spectroscopy," a high incentive for spectrometer manufacturers to provide field gradient probes and drivers as a standard option; the same equipment is normally suitable for self-diffusion measurements. Third, the newer generation of field gradient probes provide what is called active shielding of the gradients, removing much of measurement artifacts and disturbances related to overall spectrometer perturbation by magnetic field gradient pulses and eddy currents.

2. FT-Diffusion Measurements, Utilizing Gradients in the B_1 RF Field, Rather than the Static B_0-Field

Canet and co-workers [40], developed a valuable alternative to normal FT-PGSE experiments, in the form of an experiment where the magnetic field gradient is generated at the NMR frequency, by a pair of existent rf coils in a normal NMR probe. They also used this technique in an investigation of hydroxyquinoline solubilization in SDS micelles, as a function of pH.

D. Studies of Solubilization Equilibria by FT-PGSE NMR Self-Diffusion Measurements

A typical FT-PGSE experiment on solubilization is illustrated in Fig. 1.

The great advantages of FT-PGSE-based determination of partition coefficients over alternative techniques are immediately apparent:

The ease of the determination, once one has a functioning instrumental setup.

The selectivity of the measurement approach; due to the resolved multicomponent nature of the measurements, it is applicable to very complex samples. Systems with several solubilizates can in most cases be successfully investigated, and each individual partition coefficient can be

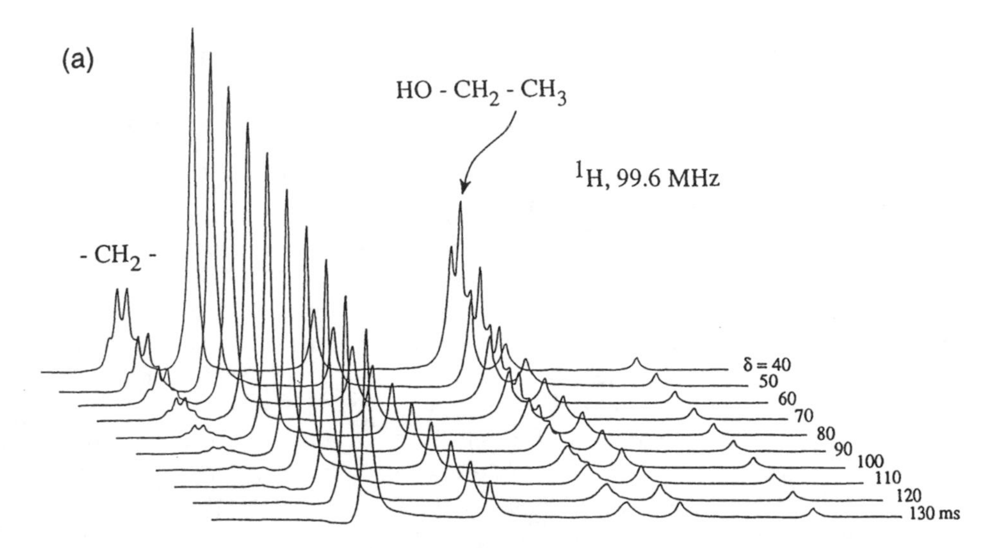

FIG. 1 A typical FT-PGSE experiment on micellar solubilization (measured at 25°C and rf pulse τ and intergradient Δ intervals equal to 140 ms). The 5-mm o.d. sample consists of micellar decyltrimethylammonium chloride in D_2O, with added solubilizates ethanol and toluene. Trace amounts of tetramethyl silane (TMS) have also been added. Spectra series (a) were recorded at a lower magnetic field gradient than those in series (b). The signal decay with δ is related to the self-diffusion coefficient in an exponential manner (cf. Eq. (3). Please note the high information content in these series of spectra: (i) ethanol is the fastest diffusing—and hence the least solubilized—component in the system. (ii) TMS is the most solubilized (100% assumed; $p = 1$), and thus diffuses at the same rate as the micelles. (iii) Toluene is quite strongly solubilized, but it is evident in the spectra that p is less than 1. (iv) The surfactant time-averaged diffusion coefficient is significantly

determined separately. For the same reason, impurities do not really perturb the measurements, at all in the same way as in alternative methods.

The quantity measured, p, being a concentration-related quantity, is well defined and easily conceived.

The very first studies by this technique [27, 28, 41–45] were focused on systematic studies of the solubilization behavior of homologous series of alcohols, methyl ketones, aromatics, and crown ethers into common surfactant micelles. Major conclusions from these studies were

In a series of n-alkyl compounds (like methanol to n-hexanol), the degree

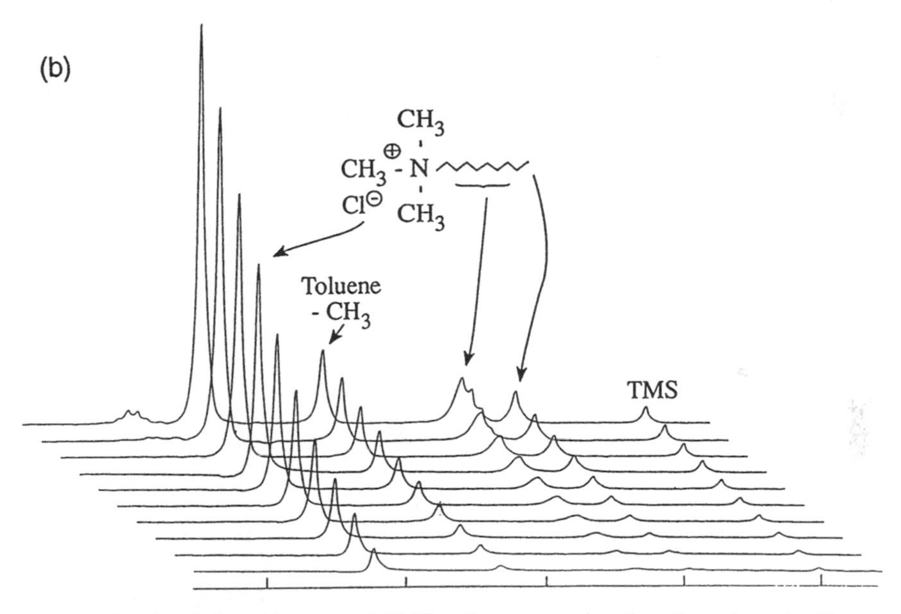

greater than both the toluene and TMS values, meaning that there is a significant part (a fraction corresponding to ≈ the CMC/(total surfactant concentration)) of free surfactant in this sample. By varying the surfactant concentration and repeating the FT-PGSE-based determination of D and p, one can actually quantify the micellization itself in great detail, and with good precision (see, e.g., Refs. cited in the section "Applications of FT-PGSE" in [39]) The data in Fig. 1 have not actually been quantified here, since they only serve as an illustration. One should note, however, that the quality of the measurements the figure allow all self-diffusion coefficients and both partition coefficients to be determined with a precision to better than a few percent, in less than 20 min experimental measurement time.

of solubilization increases strongly with alkyl chain length, quantitatively corresponding to an increment in the standard free energy for transfer of each —CH_2-group from the aqueous to the hydrocarbon pseudophase of -2.6 kJ mol^{-1} [28].

This value also applies to systems with either the aqueous or micellar phase deuterated [45]. No significant difference in solubilization behavior was actually found for any combination of perdeuterated or normal solubilizates or surfactants in normal or heavy water. This result is important, when discussing data from neutron-scattering experiments on surfactant systems or water-soluble polymers, where deuteriation is common practice for inducing contrast in the experiments.

Chain branching decreases the degree of solubilization, as compared to *n*-alkyl compounds, probably as a result of packing constraints in the micellar environment [28, 42].

At low concentrations, the degree of solubilization is essentially independent of solubilizate concentration. There is, of course, a limit to this behavior. It was clearly observed that SDS micelles actually break down upon addition of increased amounts of short-chain alcohols, to what is normally named bicontinuous microemulsion systems [41].

Crown ether "solubilization" was investigated, and it was found that larger crown ethers bind strongly to SDS micelles, while smaller crown ethers or dioxane do not [42]. This trend does not occur for cationic or nonionic micelles and is evidently related to the sodium complexation capability of the larger crown ethers. The binding force is thus not really solubilization, but rather electrostatic attraction to the SDS micellar surface.

For nonionic surfactants of the $C_{12}E_6$ type, it was found that saturated hydrocarbons, like cyclohexane, solubilize in the hydrocarbon micellar interior only, while benzene appears to be equally distributed throughout the hydrocarbon center, as well as the ethylene oxide surface region [42].

Solubilization in perfluorinated surfactants or solubilization of perfluorinated molecules is also possible by the techniques described, provided one has access to ^{19}F spin echo NMR instrumentation, for measuring these diffusion coefficients. A study of solubilization of *n*-alcohols and aromatics in sodium perfluorooctanoate micelles [44] indicated that the degree of solubilization is lower than in sodium octanoate micelles, and correspondingly, that also the increment in free energy for transfer from the aqueous to the fluorocarbon pseudophase corresponds to -2.3 kJ mol^{-1} per —CH_2-group, a somewhat lower value than for the hydrocarbon pseudophase (-2.6, vide supra). This effect was ascribed to the less nearly ideal fluorocarbon/hydrocarbon interaction.

The degree of solubilization in vesicles can also be investigated by the same methods. Benzene and tetracaine (a local anesthetic) solubilization in phospatidylcholine systems was investigated [43]. It was found that NMR transverse (T_2) spin relaxation times are considerably shorter in vesicular systems, also for the solubilizate molecules, making the application somewhat less powerful than for micellar systems, but still the sought partition coefficients could be determined to good precision.

E. Final Comments

In Sec. III.B. the significance of the concept of the degree of solubilization (p) was discussed, concluding that in the self-diffusion approach it will relate to the fraction of solubilizate molecules that diffuse with the micelle. The self-diffusion approach to solubilization has occasionally been criticized on these grounds, arguing that "solubilizate molecules diffusing with the micelle are not all necessarily solubilized in the normal sense—a significant fraction may be associated with the surface, or similar." While, in principle, this argument is correct, this is, to my own judgment, unlikely to be any significant source of error. It is physically unreasonable that any significant fraction of typical solubilizate molecules would be distributed that way.

However, it is interesting to note in this context that the paramagnetic ion-induced spin relaxation approach of Gao, Kwak, and Wasylishen (Sec. II. A.) and the self-diffusion approach evidently lead to p-values in very good agreement in essentially all cases tested. Therefore, if the paramagnetic ion method (affecting molecules in the aqueous phase) and the self-diffusion approach (monitoring molecules that diffuse with the micelle) lead to similar results, one can conclude that the above-mentioned argument is experimentally disproved and that the average solubilization site must predominantly be the "micellar core" for typical solubilizate molecules, and only rarely the interfacial region. Similar indications are found from nuclear spin relaxation data on solubilized molecules, as discussed in Sec. II. A.

It should finally be mentioned that very recent methodological progress in the field of FT-PGSE may be potentially relevant to future studies of solubilization by FT-PGSE NMR:

Morris and Johnson have implemented [46] and applied [47] a 2D-like approach named DOSY, dispersing FT-PGSE information the normal spectral dimension as well as in "the self-diffusion coefficient domain." The techniques may extend the field of application of FT-PGSE to more complex systems, since the component NMR band shapes do become separated in the diffusion dimension. The application of DOSY to multicomponent micellar solubilization has been described [48].

Similarly, Schulze and Stilbs [49] have suggested an alternative experimental approach, involving a deconvolution of severely overlapping component band shapes in FT-PGSE, by multivariate statistical methods. This technique may find similar use in future studies on solubilization in more complex systems.

ACKNOWLEDGMENTS

The author wishes to express thanks to all co-workers in the past, especially Björn Lindman, for stimulating collaboration in the field of FT-PGSE and its applications in surfactant science. This work has been supported by the Swedish Natural Sciences Research Council (NFR).

REFERENCES

1. C. Chachaty, *Prog. Nucl. Magn. Reson. 19*: 183 (1987).
2. B. Lindman and H. Wennerström, in *Solution Behaviour of Surfactants. Theoretical and Applied Aspects* (K. L. Mittal and E. J. Fendler, eds.), Plenum Press, New York, (1982), Vol. 1, pp. 3–25.
3. B. Lindman and P. Stilbs, in *Physics of Amphiphiles, Micelles, Vesicles and Microemulsions* (V. Degiorgio and M. Corti, eds.), XC Corso, Soc. Italiana di Fisica, Bologna, 1985, pp. 94–121.
4. P. Stilbs, in *Surfactants in Solution* (K. L. Mittal and B. Lindman, eds.), Plenum Press, New York, (1984), Vol. 2, pp. 917–22.
5. U. Henriksson, L. Ödberg, J. C. Eriksson, and L. Westman, *J. Phys. Chem. 81*: 76 (1977).
6. H. Wennerström, B. Lindman, O. Söderman, T. Drakenberg, and J. B. Rosenholm, *J. Am. Chem. Soc. 101*: 6860 (1979).
7. H. Walderhaug, O. Söderman, and P. Stilbs, *J. Phys. Chem. 88*: 1655 (1984).
8. O. Söderman, U. Henriksson, H. Walderhaug, and P. Stilbs, *J. Phys. Chem. 89*: 3693 (1985).
9. B. Halle and H. Wennerström, *J. Chem. Phys. 75*: 1928 (1981).
10. O. Söderman, D. Canet, J. Carnali, U. Henriksson, H. Nery, H. Walderhaug, and T. Wärnheim, in *Microemulsion Systems* (H. L. Rosano and M. Clausse, eds.), Marcel Dekker, New York and Basel, (1987), pp. 145–161.
11. O. Söderman, U. Henriksson, and U. Olsson, *J. Phys. Chem. 91*: 116 (1987).
12. R. E. Wasylishen, J. T. Kwak, Z. Gao, E. Verpoorte, J. B. MacDonald, and R. M. Dickson, *Can. J. Chem. 69*: 822 (1991).
13. M. Jansson, P. Li, U. Henriksson, and P. Stilbs, *J. Chem. Phys. 93*: 1448 (1989).
14. R. R. Ernst, G. Bodenhausen, and A. Wokaun, *Principles of NMR in One and Two Dimensions* Oxford, 1987.
15. A. Kumar, R. R. Ernst, and K. Wüthrich, *Biochem. Biophys. Res. Comm. 95*: 1 (1980).

16. J. J. Nusselder, J. B. F. N. Engberts, R. Boelens, and R. Kaptein, *Recl. Trav. Chim. Pays-Bas. 107*: 105 (1988).
17. J. F. Ellena, W. C. Hutton, and D. S. Cafiso, *J. Am. Chem. Soc. 107*: 1530 (1985).
18. E. Kolehmainen, *Magn. Reson. Chem. 26*: 760 (1988).
19. G. S. Hartley, *Aqueous Solutions of Paraffin Salts*, Hermann, Paris, 1936.
20. R. E. Stark, M. L. Kasakevich, and J. W. Granger, *J. Phys. Chem. 86*: 334 (1982).
21. P. Stilbs, H. Walderhaug, and B. Lindman, *J. Phys. Chem. 87*: 4762 (1983).
22. D. Canet, T. Turki, A. Belmajdoub, and B. Diter, *J. Phys. Chem. 92*: 1219 (1988).
23. Z. Gao, R. Wasylishen, and J. C. T. Kwak, *J. Phys. Chem. 93*: 2190 (1989).
24. Z. Gao, J. C. T. Kwak, R. Lebonte, D. G. Marangoni, and R. Wasylishen, *Colloids Surf. 45*: 269 (1990).
25. B. Jönsson, H. Wennerström, P.-G. Nilsson, and P. Linse, *Colloid Polym. Sci. 264*: 77 (1986).
26. M. A. Desando, G. Lahajnar, I. Zupancic, and L. W. Reeves, *J. Mol. Liq. 47*: 171 (1990).
27. P. Stilbs, *J. Colloid Interface Sci. 80*: 608 (1981).
28. P. Stilbs, *J. Colloid Interface Sci. 87*: 385 (1982).
29. B. Lindman and B. Brun, *J. Colloid Interface Sci. 42*: 388 (1973).
30. E. O. Stejskal and J. E. Tanner, *J. Chem. Phys. 42*: 288 (1965).
31. J. Kärger, H. Pfeifer, and W. Heink, in *Advances in Magnetic Resonance*, (W. S. Warren, ed.) Academic Press, New York, 1988, Vol. 12, p. 1.
32. R. L. Vold, J. S. Waugh, M. P. Klein, and D. E. Phelps, *J. Chem. Phys. 48*: 3831 (1968).
33. T. L. James and G. G. McDonald, *J. Magn. Reson. 11*: 58 (1973).
34. J. Kida and H. Uedaira, *J. Magn. Reson. 27*: 253 (1977).
35. P. Stilbs and M. E. Moseley, *Chem. Scr. 13*: 26 (1978–79).
36. P. Stilbs and M. E. Moseley, *Chem. Scr. 15*: 176 (1980).
37. P. T. Callaghan, C. M. Trotter, and K. W. Jolley, *J. Magn. Reson. 37*: 247 (1980).
38. P. T. Callaghan, *Aust. J. Phys. 37*: 359 (1984).
39. P. Stilbs, *Prog. NMR Spectrosc. 19*: 1 (1987).
40. A. Belmajdoub, D. Boudot, C. Tondre, and D. Canet, *Chem. Phys. Lett. 150*: 194 (1988).
41. P. Stilbs, *J. Colloid Interface Sci. 89*: 547 (1982).
42. P. Stilbs, *J. Colloid Interface Sci. 94*: 463 (1982).
43. P. Stilbs, G. Arvidson, and G. Lindblom, *Chem. Phys. Lipids 35*: 5048 (1984).
44. J. Carlfors and P. Stilbs, *J. Colloid Interface Sci. 103*: 332 (1985).
45. J. Carlfors and P. Stilbs, *J. Colloid Interface Sci. 104*: 489 (1985).
46. K. F. Morris and C. S. Johnson, Jr., *J. Am. Chem. Soc. 115*: 4291 (1993).
47. K. F. Morris, Thesis, University of North Carolina, 1993.
48. K. F. Morris, P. Stilbs, and C. S. Johnson, Jr., *Anal. Chem. 66*: 211 (1994).
49. D. Schulze and P. Stilbs, *J. Magn. Resonance, A., 105*: 54 (1993).

12

The Partitioning of Neutral Solutes Between Micelles and Water as Deduced from Critical Micelle Concentration Determinations

C. TREINER Laboratoire d'Electrochimie, Université Pierre et Marie Curie, URA CNRS 430, Paris, France

SYNOPSIS

The partition coefficient P of neutral solutes between micelles and water can be deduced from the analysis of the decrease of the initial rate of change of the critical micelle concentration (CMC) of a surfactant with the addition of the solute, provided an adequate model of surfactant/solute interaction in dilute solutions can be defined. The various models available in the literature are critically reviewed, and the results deduced from various experimental approaches are compared. P values are calculated for over 200 surfactant/solute systems using a suggested best procedure based upon experimental data. A tentative group-contribution approach is next proposed in terms of log P values for the partitioning of a number of neutral molecules between cationic micelles and water. It is shown that the log P group-contribution for various polar moieties follows the same order than that obtained in the two-phase 1-octanol/water binary. The increase of CMC sometimes observed upon addition of small or large neutral solute concentration is interpreted using preferential solvation models.

I. INTRODUCTION

Surfactants are amphiphilic ions or molecules which are essentially characterized by their ability to adsorb at interfaces and to form spontaneously organized aggregates in hydrogen-bonded solvent systems and particularly in water above a critical micelle concentration (CMC). These aggregates, acting either as micro oil droplets, as adsorption sites or as any combination of both of these idealized models may solubilize hydrophobic solutes. Any additive to a surfactant solution will change the value of the CMC. Thus, it was long ago observed that upon addition of *neutral* solutes the CMC may increase or decrease depending upon the solute properties. The CMC decreases for so-called hydrophobic solutes and increases often with hydrophilic ones. Under specific conditions, a CMC decrease may be followed at higher solute concentrations by a CMC increase. The former observation is attributed to a micellar solubilization and the latter one to a classical solvent effect, i.e., to the preferential interaction of the monomer surfactant with the solute molecules, the consequence being that the CMC is lowered because the micelles have been destabilized by solute addition. However only relatively recently have several attempts been made to relate the rate of change of the CMC upon solute addition, to the partition coefficient P, defined as the solute concentration ratio between the micelles and the solvent (usually, water).

The need for such determinations stems from the ill-defined notion of solubility in micellar systems. Although numerous solubility determinations have been made in micellar systems (particularly in the pharmaceutical field [1]), the very definition of solute solubilization is at best obscure in a microsystem whose physicochemical properties are continuously modified by the solute addition. Moreover, the information needed to introduce activity coefficients is seldom available [2, 3]; only apparent thermodynamic quantities can be determined, even at a dilute solute concentration.

From a practical point of view, solubility experiments whenever feasible are essential as they define a phase boundary; however in order to understand the fundamental aspects of the solubilization phenomenon, the solute activity within the micelle has to be controlled. Indeed, a number of physicochemical methods have been devised to overcome these difficulties such as vapor pressure [2–4], head-space gas-chromatography (GC) [5–8], calorimetry [9–11], partial molar volume [12, 13] and heat capacity [14], semiequilibrium dialysis [15], NMR [16–18], fluorescence [19] gel filtration [20], ultrafiltration [21], Krafft point changes upon solute addition [22, 23], and UV spectroscopy chemical shifts [24, 25].

The determination of partition coefficients by CMC measurements is a simple alternative to these various methods provided that a correct theoretical analysis of the phenomenon has been performed. A number of different theories have been recently proposed for such an evaluation, none of which being completely satisfactory; thus it seems appropriate to consider the validity and the shortcomings of these theories in order to suggest improved evaluation procedures for inferring CMC-based partition coefficients. The interpretation and significance of such parameters are discussed in other chapters in this volume.

We shall in the following first recall the derivation of an equation which takes into account both CMC effects outlined above, namely, the initial increase or decrease of the CMC upon neutral solute addition [26, 27]. Then various alternatives to this approach will be examined and the derived results will be compared with those obtained from other physicochemical methods. Literature CMC changes with variation in neutral solute concentrations will be computed using the best available theories in order to offer a self-consistent set of data. Some generalizations will be attempted. Finally, the effect of high solute concentrations inducing CMC minima will be briefly considered. Only the behavior of neutral compounds will be discussed.

II. CALCULATIONS OF PARTITION COEFFICIENTS ACCORDING TO VARIOUS CHEMICAL MODELS

The available theories may be divided, somewhat formally, into two groups whether they represent the surfactant chemical potential in ionic micelles by a conventional chemical potential (group "a") or as an electrochemical potential (group "b").

The first theoretical derivation to be analyzed presents some aspects which are common to other theoretical approaches. Thus calculation details will be described below which will not be repeated thereafter. The pseudophase model will be adopted. It is used by most (but not all [12]) authors who have been concerned with the determination of partition coefficients from CMC data. [Note: A uniform glossary of symbols could not be adopted for the various relevant theories without confusion with the notation of the orginal publications; therefore each new symbol will be redefined whenever necessary.]

1a. Consider the case of a 1-1 ionic surfactant completely dissociated in water. According to the pseudophase model, the CMC is the monomer saturation solubility in the aqueous phase. Neglecting activity coefficients contributions its chemical potential may be written as

$$\mu_{s,w} = \mu^{\circ}_{s,w} + 2RT \ln X^{c}_{w} \tag{1}$$

where w stands for water and c for CMC. The surfactant chemical potential in the pseudo micellar phase may be tentatively described by the relation [12]

$$\mu_{s,M} = \mu^{\circ}_{s,M} + (1 + \beta)RT \ln X_{M} \tag{2}$$

in which the effect of the electrical potential created by the micelle is implicitly taken into account by assuming a micellar degree of counterion dissociation β. X_M is the surfactant mole fraction in micelles.

Noting that $X_M = 1$ for the pure micelle and writing equations similar to (1) and (2) in the presence of an additive N, one finally obtains

$$\Delta G_{t(w,wN)} = \mu_{w,N} - \mu^{\circ}_{w} = 2RT \ln \frac{X^{c}_{w}}{X^{c}_{w,N}} + (1 + \beta)RT \ln X_{M} \tag{3}$$

where the left side represents the standard free energy of transfer of one mole of *surfactant* from the aqueous phase to the micellar pseudophase in presence of the solute N.

The solute mole fraction Y_M within the micelle is defined by

$$X_M = 1 - Y_M \tag{4}$$

If the solubilized solute mole fraction is sufficiently small, $\ln(1 - Y_M) = -Y_M$. Assuming a distribution process, $Y_M = PY_N$ where Y_N is the total solute mole fraction. P is the mole fractional partition coefficient between micelles and water. Then Eq. (3) becomes

$$\Delta G_{t(w,wN)} = 2RT \ln \frac{X^{c}_{w}}{X^{c}_{w,N}} - (1 + \beta)RTPY_N \tag{5}$$

It has been shown before [28] that in ternary solutions where two components are dilute with respect to the third one, any standard thermodynamic quantity may be expanded as a function of both solute concentrations in terms of pairwise, triplet, and higher-order terms. This is equivalent to the McMillan-Mayer procedure for excess thermodynamic quantities.

Using the properties of cross-differential derivatives in the dilute aqueous solutions below the CMC, i.e., with respect both to solute, N, and surfactant, one can write

$$\left[\frac{\delta \Delta G_{t(w,ws)}}{\delta m_s}\right]_{m_N} = \left[\frac{\delta \Delta G_{t(w,wN)}}{\delta m_N}\right]_{m_s} = 2\nu g_{Ns} + 6\nu g_{NNs} + \cdots \tag{6}$$

where $\Delta G_{t(w,ws)}$ is the standard free energy of transfer of the neutral *solute* from the aqueous phase to the micellar phase, g_{Ns} and g_{NNs} are pairwise and triplet interaction order terms, and ν is the number of species; g_{Ns} is equivalent to a second virial coefficient specialized to a standard thermodynamic function. It can be shown that

$$g_{Ns} = 2.303RTk_N m_N \tag{7}$$

where k_N is the familiar Setchenow constant expressed in the mole fraction scale and m_N is the solute molarity. The right-hand side of Eq. (7) is multiplied by 2.303 because the Setchenow constant is traditionally expressed in the decimal logarithm scale.

Introducing Eq. (7) into Eq. (5) leads, at the limit of small solute concentration, to the equation

$$\log \frac{\mathrm{CMC_w}}{\mathrm{CMC_{w,N}}} = \frac{1}{2}\left[k_N + \frac{P(1 + \beta)M}{2303}\right] m_N \tag{8}$$

where M is the molecular weight of water. $\mathrm{CMC_w}$ and $\mathrm{CMC_{w,N}}$ are the molar surfactant CMC in absence and in presence of solute at concentration m_N. For a nonionic surfactant, $\beta = 0$ and the coefficient before the bracket is 1. Equation (8) is a generalized form of a Setchenow Eq. (26):

$$\log \frac{\mathrm{CMC_w}}{\mathrm{CMC_{w,N}}} = K_M m_N \tag{8'}$$

where K_M is a constant. The Setchenow constant k_N may be either positive or negative. In the case of surfactants in the presence of organic solutes in water, k_N is known to be negative (salting-in effect) [29–31]. This salting constant can be evaluated from the following empirical relation [27]:

$$k_N = 0.637 - 0.014n\{\mathrm{CH_2}\} - 0.1464R \tag{9}$$

where n is the number of methylene groups in the surfactant aliphatic chain and R is the hard-sphere diameter of the solutes, which can be evaluated from van der Waals volumes [32].

Equation (9) holds for tetraalkylammonium bromides and most certainly for akylsulfates as well. In the case of chloride derivatives for example, a value of 0.074 should be added to Eq. (9) in order to account for the increased salting-out effect produced on neutral solutes in water by chloride over bromide ions [30, 33]. Such a correction has little practical effect on the calculated partition coefficients. Also, it was considered that the introduction of an ethoxy group on a solute molecule did not produce a significant effect on the Setchenow constant because of the compensation effects of the methylene and ether moieties.

It is clear then that if the solute is not solubilized in the micelles, which is the case with predominantly hydrophilic molecules, P is equal or close to zero, and the CMC must increase with increasing m_N. Hence, by measuring a CMC change in the presence of such solutes, one determines a Setchenow constant, i.e., a classical medium effect. If the solute is solubilized in the micelles, then the CMC should decrease. Thus, Eq. (8) seems to be able to interpret most observed CMC variations in the limit of low additive concentrations. As will be shown on the last part of this review, Eq. (5) can be likewise applied to the more general case of large additive concentrations.

Note that the various approximations leading to Eq. (8) (constant partition coefficient, limiting expansion of the logarithm argument) prohibit its expansion to solute concentrations higher than the limiting slope.

De Lisi et al. [12], using the same approach but with solute + micelle interactions described by a mass-action law, obtain a slightly different equation which, on the molar basis, is

$$\log \frac{\mathrm{CMC_w}}{\mathrm{CMC_{w,N}}} = \frac{1}{2}\left[k_N + \frac{(1+\beta)P}{2.303(1+Pm_N)}\right]m_N \tag{10}$$

Equation (10) reduces to Eq. (8) as m_N goes to zero.

Table 1 presents some experimental K_M values taken from the literature and compares them with Setchenow constants determined experimentally or calculated by Eq. (9). K_M is equal or close to $(1/2)k_N$, depending on the relative hydrophilicity of the compounds which raise the CMC. For hydrophobic molecules, K_M is increasingly positive as the molecules become more hydrophobic because of the appearance of the partition coefficient P in Eq. (8).

If, for very large P values, the numerical influence of the Setchenow term is negligible, it does have an effect in intermediate cases. For example [27], the addition of ethyl methyl ketone (EMK) to $C_{12}Br$ increases the

TABLE 1 Characteristic Parameters of Micellar Solubilization for Polar Solutes in Ionic Surfactant Solutions at 25°C[a]

	$C_{12}Br$		NaDS		
	K_M	$(1/2)k_N$	K_M	$(1/2)k_N$	Ref.
Methanol	0.0	−0.037	0.03	−0.037	
Ethanol	0.05	−0.112	0.20	−0.112	45
1-Propanol	1.19	−0.125	0.51	−0.125	45
2-Propanol	0.13	−0.125	0.32	−0.125	45
1-Butanol			1.5	−0.169	5
1-Pentanol			4.5	−0.212	5
1-Hexanol			14.7	−0.257	5
1-Heptanol			34.9	−0.301	5
Urea	−0.086	−0.088	−0.042	−0.088	34
Thiourea	−0.054	−0.098			34
Acetamide	−0.128	−0.101	−0.050	−0.101	34
Acetone	−0.06	−0.102	+0.015	−0.102	34
N,N-Dimethylurea	−0.212	−0.164	−0.14	−0.164	34
Tetramethylurea	−0.261	−0.218	−0.14	−0.218	34
p-dioxane	−0.06	−0.168	−0.15	−0.168	34
N,N-Dimethyl formamide	−0.165	−0.130	−0.13	−0.13	34

[a] K_M is the constant of Eq. (8′); k_N is the Setchenow constant calculated from Eq. (9) for both surfactants. Results from Ref. 34 were recalculated from the raw experimental data.

CMC by twice as much as hexamethylphosphotriamide (HMPT). This result could be interpreted as an indication of a larger solubilization in the former case than in the latter. However, as the HMPT molecule is larger than that of EMK, so the Setchenow term overshadows the solubilization effect. As a consequence the complete Eq. (8) shows that the P value for HMPT, albeit small, is larger than that of EMK.

For those systems which raise the CMC, the generalized Setchenow Eq. (8) may be valid in a large concentration domain. A typical example is presented in Fig. 1, which shows the variation of the CMC of $C_{12}Br$ with acetamide molality [34] where the linear variation extends to 6 molar. For ethylene glycol [35] with the nonionic surfactant OPE_{9-10} (Triton X-100) in water, the linear portion of CMC variation is smaller but still close to 3 molar at 25°C. For hydrophobic solutes, the second term of Eq. (8) more than compensates for the salting effect; then the CMC decreases with m_N with a positive K_M. The range of validity of Eq. (8) with respect

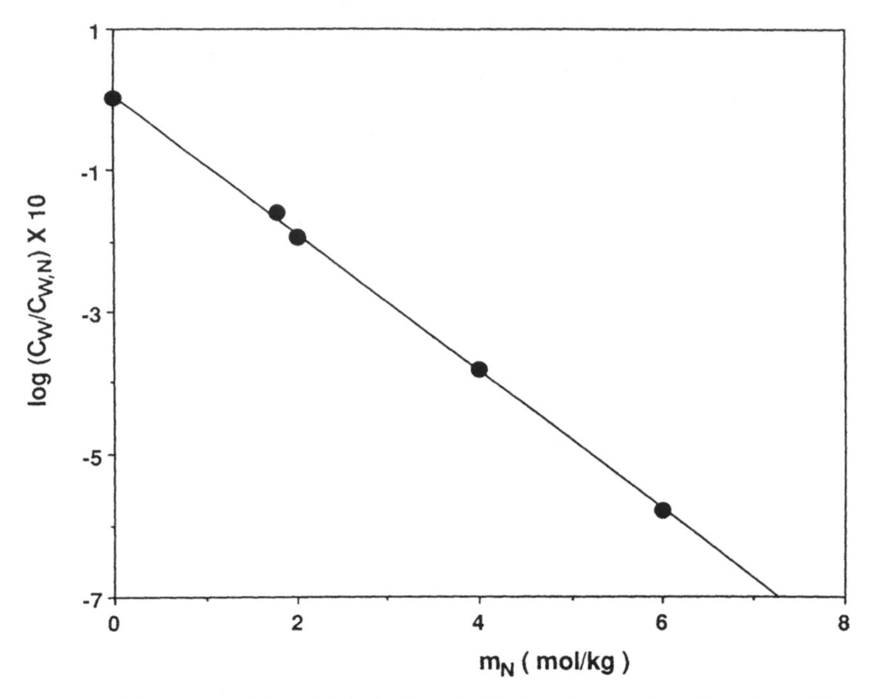

FIG. 1 Variation of the CMC of trimethyldodecylammonium bromide with acetamide concentration at 25°C. (From Ref. 34.)

to solute concentration is smaller for those solutes which decrease the CMC. Hence in the case of 1-butanol with NaDS, the careful measurements of Perez-Villar et al. [36] show that the departure from Eq. (8) begins around 0.2 molar (Fig. 2). This linear domain further decreases as the solutes become more hydrophobic.

The questions raised by the CMC minimum observed in the presence of some neutral solutes will be addressed in the last portion of this chapter.

Several approximations or assumptions are implicit in the above derivation. For example, it is assumed that the degree of dissociation of the micelles is independent of solute addition. This assumption is in apparent disagreement [37–39] with experimental evidence, at least in the case of polar solutes; the question may be different for nonpolar ones [40]. This point will be discussed further. The pseudophase model is used and all

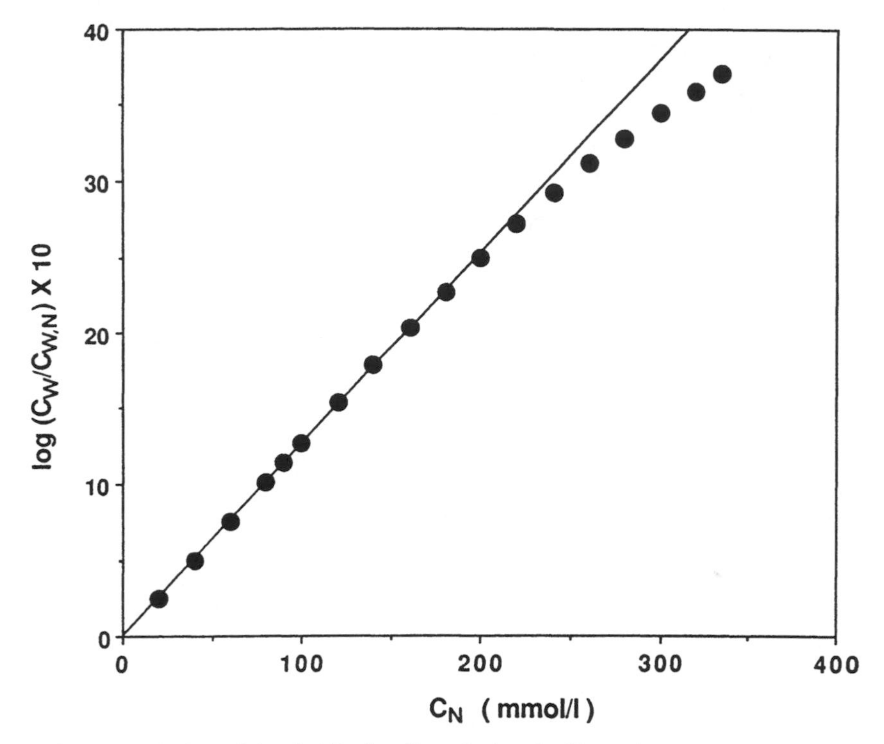

FIG. 2 Variation of the CMC of sodium dodecylsulfate with 1-butanol concentration. (From Ref. 36.)

activity coefficients are assumed equal to 1. Moreover an explicit electrical potential term is ignored in Eq. (2). These and other approximations will be discussed below along with other thermodynamic derivations which have been recently proposed to extract partition coefficients from CMC determinations. Only the main features of each calculation will be reported.

2a. Hall's treatment of the critical micelle concentration [41, 42] makes use of the Kirkwood-Buff theory of solutions. It emphasizes the relation between this theory and the Gibbs adsorption formalism by using quantities similar to the excess quantities of surface thermodynamics to describe the behavior of the chemical potentials. One of the main feature of this approach is that the micelle can be considered as a neutral aggregate as suggested by Mijnlieff [43].

In the case of solute solubilization, the relevant equation which expresses the effect of a solute on the CMC in the absence of added salt is

$$\left[\frac{\delta \ln \mathrm{CMC}}{\delta m_\mathrm{N}}\right]_{T,p,\mathrm{CMC}} = \frac{\overline{N}_\mathrm{N}/\overline{N}_\mathrm{s} m_\mathrm{N}}{1 + \left[(1-\alpha)/z_2 + (1 + (1-\alpha)/z_2)\dfrac{\delta \ln \gamma}{\delta \ln m_\mathrm{N}}\right]} \tag{11}$$

$\overline{N}_\mathrm{N}$ and $\overline{N}_\mathrm{s}$ are respectively the average number of solute molecules and the number of surfactant molecules per micelle and γ is the surfactant activity coefficient.

$$\alpha = (1 + z_2)\beta \tag{12}$$

β is the negative adsorption of the co-ions and z_2 is the valency of the counterion. It must be stressed that Eq. (11) contains no derivative of α.

The numerator of the right-hand side of Eq. (11) may be considered as a solute partition coefficient. If the solution is assumed to be ideal, this equation becomes, in the case of a 1-1 surfactant,

$$\left[\frac{\delta \ln \mathrm{CMC}}{\delta m_\mathrm{N}}\right]_{T,p,\mathrm{CMC}} = \frac{1}{2}\frac{P}{1-\beta} \tag{13}$$

P is expressed here on the molar basis. As with Eq. (8), the partition coefficient obtained is about half as large as it would be take in the case of a real two-phase system.

3a. The thermodynamic approach of Perez-Villar et al. [36] is based on a treatment of micellar solutions put forward by Motomura and collaborators [44] where the micelle is considered as a bulk phase having the thermodynamic properties of an adsorbed surfactant film. An attractive feature of this model is that the micellar phase is not explicitly taken into account, so that the troublesome question of the degree of dissociation of ionic micelles is avoided.

As applied to the case of a micellar phase above the CMC containing both surfactant and solute, the two basic equations are written starting with the Gibbs-Duhem equation at constant T and P. These equations apply to the aqueous phase whether below or above the CMC. In the premicellar region one has

$$n_\mathrm{w}\, d\mu_\mathrm{w}^\mathrm{m} + \delta_\mathrm{a} n_1^\circ\, d\mu_\mathrm{a}^\mathrm{m} + \delta_\mathrm{c} n_1^\circ\, d\mu_\mathrm{c}^\mathrm{m} + n_2^\circ\, d\mu_2^\mathrm{m} = 0 \tag{14}$$

In the postmicellar region one can write

$$n_w \, d\mu_w^{aq} + n_a \, d\mu_a^{aq} + n_c \, d\mu_c^{aq} + n_2 \, d\mu_2^{aq} = 0 \tag{15}$$

δ is the number of anions, a, or cations, c: $\delta = \delta_c + \delta_a$; n_1° and n_2° are respectively the number of moles of surfactant and of solute.

Equation (15) expresses the solute concentration which *remains* in the aqueous phase while solubilization occurs in the postmicellar region. From Eq. (14) and (15), after some straightforward mathematical manipulation, one obtains the mole fraction of solute in the micellar phase as

$$x_2^m = \hat{x}_2\delta \frac{\text{CMCM}_w - \hat{x}_1 D}{\hat{x}_1(1 + \hat{x}_1\hat{x}_2 D(1 - \delta) + \delta\text{CMCM}_w} \tag{16}$$

with

$$D = \frac{d \ln \text{CMC}}{d\hat{x}_2} \tag{17}$$

and

$$\hat{x}_2 = \frac{x_2(1 + M_w\text{CMC})}{x_2 + M_w\text{CMC}} \tag{18}$$

and

$$x_2 = \frac{n_2^\circ}{n_w + n_1^\circ + n_2^\circ} \tag{19}$$

where $\hat{x}_1 + \hat{x}_2 = 1$; M_w and n_w are the molecular weight and the number of moles of water.

The partition coefficient is expressed classically in the mole fraction scale as

$$P = \frac{x_2^m}{x_2} \tag{20}$$

In order to make direct use of CMC versus solute concentration data, which are often presented under a mathematical form equivalent to Eq. (8′), Eq. (17) may be replaced by

$$D = 2.303K_M \left[\frac{1000\text{CMC}}{1 - (m_N/(m_N + 1000\text{CMC})^2}\right] \tag{21}$$

The essential assumption of this approach is the ideal two-phase model. The only experimental parameter to be introduced in Eq. (16) is the variation of the CMC with solute concentration. This theory differs from most other calculations in that it allows P to change with solute concentration even though the system is assumed ideal. This point will be examined below.

1b. Shirahama and Kashiwabara's approach [45] is the first of that group of theories dealing with micellar solubilization which considers explicitly the electrical potential created by the micelles in the surfactant chemical potential expression.

The solubilization phenomenon in ionic micelles is formally described by most authors by two basic effects: an increase of the micellar entropy of mixing and a decrease in micelle surface charge density. The first effect is implicitly taken into account in the above procedures; the second effect may be described through the introduction of an electrochemical potential [46].

Thus, instead of Eq. (2), Eq. (22) is assumed to be

$$\mu_{\mathrm{M}}^{\mathrm{s}} = \mu_{\mathrm{M.N}}^{\circ} + kT \ln X_{\mathrm{M}} f_{\mathrm{M}} + e\Phi^{\circ} \tag{22}$$

where e is the charge of the electron and Φ° is the electrical potential created by the micelle. In this model, the electrical potential created by the free ions is assumed equal to zero.

Shirahama and Kashiwabara [45] estimate the potential Φ° of a charged sphere by solving numerically the Poisson-Boltzmann equation using tabulated values with the proper boundary conditions. They also introduce values of the activity coefficient f_{M} for aliphatic alcohols, utilizing hydrophobic standard free-energy changes taken from the literature. Their final result (on the mole fraction scale), for aliphatic alcohols in NaDS solutions is presented in differential form, assuming that the standard potential terms are independent of solute concentration.

$$\frac{d \ln \mathrm{CMC}}{dY_{\mathrm{N}}} = \vartheta P \tag{23}$$

ϑ was found constant and equal to -2.15 for NaDS with 1-alkanols from 1-butanol to 1-heptanol.

Later, Hayase and Hayano [5, 47] determined P by a gas-chromatographic (GC) method and using CMC measurements as a function of solute addition showed that for NaDS + 1-alkanols at 25°C, the constant in Eq. (23) should be equal to $\vartheta = -0.82$. This coefficient was coined ISA (Interaction of Surfactant and Additive) by these authors.

2b. In the above derivation the change of the surface charge density with solute concentration was calculated from the Poisson-Boltzmann equation. Some authors [48] chose to make this parameter appear explicitly.

TABLE 2 The ISA Coefficient of Eq. (23) as Deduced from Various Experimental Procedures for Polar Additives

Surfactant	ϑ	t (°C)	Ref.
$C_{10}Na$	0.59	43.8	51
$C_{12}Na$	0.71	43.8	51
$C_{14}Na$	1.00	43.8	51
$C_{16}Na$	1.33	43.8	51
$C_{12}Na$	0.67	25	50
$C_{12}Na$	0.69	25	48
$C_{12}Na$	0.82	25	47
$C_{12}Na$	0.95	40	100
$C_{12}Na$	0.56	55	100
$C_{12}Br$	0.32	25	68
$C_{14}Br$	0.48	25	68
$C_{16}Br$	0.58	25	68
$C_{16}Br$	0.78	25	52
$C_{16}Br$	0.81	25	102
NaDec[a]	2.0	25	25(b)

[a] Sodium deoxycholate.

The surfactant electrochemical potential in the micellar phase is written in a manner which combines Eqs. (2) and (23):

$$\mu_{M,N} = \mu^{\circ}_{M,N} + (1 + \beta)RT \ln X_M + F\Phi \tag{24}$$

Shinoda's result [46] is used for the evaluation of Φ at the CMC:

$$\ln \Phi = \ln A + 2 \ln \sigma - \ln \text{CMC} \tag{25}$$

where A is a constant and σ the surface charge density; their final result takes the form of Eq. (23) with the slope ϑ equal to

$$\vartheta = \frac{1}{2}\left[\beta + 3 - 2\left(\frac{\delta \ln \beta}{\delta Y_M}\right)\right] \tag{26}$$

This calculation provides a physical meaning to the coefficient ϑ: it is proportional to the degree of dissociation of the micelles and to the rate of change of the degree of dissociation with the addition of a neutral solute. Using Hayase and Hayano's P values (as deduced from from GC experiments), and their CMC determinations, the constant ϑ was found equal to 0.69 for NaDS with linear alkanols.

However, if the change of electrical potential Φ° with neutral solute addition is small, for example as the result of compensation effects, it was shown that from Eq. (26), one obtains

$$\vartheta = \tfrac{1}{2}(1 + \beta) \tag{27}$$

With this condition, Eq. (23) is similar to Eq. (8) when derived with respect to m_N. Furthermore there is no dependence of β upon micellar composition, in agreement with Hall's Eq. (13). Manabe et al. deduced from their result that ϑ should be considered as the ionic degree of dissociation of the pure micelles: $\vartheta = 2$ corresponds to complete dissociation.

The compensation effects suggested above may be understood by consideration of the change of the apparent charge σ of ionic micelles upon addition of a nonionic surfactant [49]. Mixed micelles are then formed just as in the solubilization case. As a first approximation one can write

$$\sigma = \beta(1 - X_M) \tag{28}$$

Using a potentiometric method with a sodium ion-selective electrode, it was shown that for the NaDS + dodecylpolyoxyethlene [23] (POE) system, σ remains constant over a large micelle composition domain. Thus, $\sigma = 0.29 \pm 0.03$ from pure NaDS micelles to mixed NaDS + POE aggregates up to a mole fraction $x = 0.6$. When the nonionic component contains only four oxyethylene groups, nearly the same result is obtained: $\sigma = 0.25 \pm 0.03$ in the same micellar composition range. Thus, the increase of β as a result of the release of counterions when the nonionic surfactant is incorporated in the micelles is exactly compensated by the decrease of ionic charges through the change of the mixed micelle composition. This implies that the derivative $d\Phi/dX_M$ is equal to zero, a result which justifies the approximation of Eq. (27).

3b. Abu-Hamdiyyah and collaborators chose a more empirical viewpoint [50–53]. They showed empirically that upon solute addition, the rate of change of the degree of dissociation of ionic micelles may be linearly related to the rate of change of the CMC according to the relation

$$\ln\left(-\frac{d \ln X_w^c}{dY_N}\right) = \ln\left(\frac{d \ln \beta}{dY_N}\right) + b \tag{29}$$

where b is a constant.

Then starting also from Eq. (25), which they derive with respect to solute mole fraction within the micelles, they obtain, assuming $d\Phi/dY_N = 0$ and a linear variation of the charge density with micellar composition, the following relationship:

$$-\frac{1}{2}\left(\frac{d \ln \mathrm{CMC}}{dX_{\mathrm{N}}}\right) + \left(\frac{d \ln \beta}{dX_{\mathrm{N}}}\right) = P \tag{30}$$

Once Eq. (29) has been established experimentally, P may be calculated from Eq. (30). Furthermore, using Eqs. (25) and (30) they derive for the ϑ coefficient of Eq. (23) the relation

$$\vartheta = 2 - 2\left(\frac{d \ln \beta}{dY_{\mathrm{M}}}\right) \tag{31}$$

As the coefficient ϑ has been shown to be a constant, at least for surfactant + neutral solute systems, Eq. (31) implies that all additives induce the same effect on the micelle degree of condensation at the same solute mole fraction within the micelle. The same conclusion had been obtained for a series of alcohols in NaDS solutions [47].

The final equation is presented as [53]

$$\ln P = \left\{\ln\left[-\frac{d\,\mathrm{CMC}}{dX_{\mathrm{N}}}\left(\frac{55.5}{\mathrm{CMC_w}}\right)\right]\right\} - \ln \vartheta \tag{32}$$

Note that as the result of using a negative variation of CMC with solute addition, the ϑ coefficient is here counted as positive as in Eq. (27) contrary to Hayase and Hayano's formulation.

1c. Finally, a different approach to the determinations of P, which is not based on CMC variation but is closely related to it, may be mentioned here. It relies on the concept of a partial molal conductivity. This quantity is deduced from differential conductivity versus surfactant concentration representations [54, 55]. The fundamental equation is

$$\kappa = \bar{\kappa}_{\mathrm{f}} C_{\mathrm{f}} + \bar{\kappa}_{\mathrm{M}} C_{\mathrm{M}} \tag{33}$$

where κ, $\bar{\kappa}_{\mathrm{f}}$, and $\bar{\kappa}_{\mathrm{M}}$ are respectively the measured conductivity and the conductivity change with surfactant concentration below and above the CMC. Deriving this equation with respect to solute concentration above the CMC and rearranging leads to the final equation:

$$\frac{d\kappa}{dC_{\mathrm{N}}} = (\bar{\kappa}_{\mathrm{f}} - \bar{\kappa}_{\mathrm{M}})k + \left[\bar{\kappa}_{\mathrm{cmc}}\left(\frac{d\beta}{dX_{\mathrm{N}}}\right) - (\bar{\kappa}_{\mathrm{f}} - \bar{\kappa}_{\mathrm{M}})k\right]j \tag{34}$$

k is equal to the familiar derivative $d\mathrm{CMC}/dC_{\mathrm{N}}$.

$$j = \frac{P}{1 + P(C - \mathrm{CMC})} \tag{35}$$

The relevant experimental quantity here is the conductivity change with solute addition at constant surfactant concentration above the CMC. This variation is linear, at least in the lower concentration range. The corresponding slopes are obtained at several surfactant concentrations and plotted as a function of j for various P values. The ordinate intercept of the best straight line enables the calculation of the partition coefficient.

Another version of essentially the same approach has been presented by the same research group using as the experimental parameter the variation of the unbound counterion concentration in the presence of various additives. A potentiometric technique is then employed [56].

III. RESULTS

A. Experimental Techniques

Two experimental quantities are needed for the determination of P by most of the above procedures: the CMC and the degree of counterion condensation. The CMC in the presence of neutral solutes can be determined using a number of experimental techniques: the most popular one for ionic surfactant solutions is the electrical conductivity method. It enables one to extract both experimental quantities needed from a single experimental run. Automatic devices enable fast and accurate CMC determinations [57]. In most cases, this method cannot be used for CMC determinations in the presence of added salt or obviously not for nonionic surfactants, which explains the paucity of CMC data in the presence of neutral solutes for such systems.

The degree of counterion dissociation is most often determined as the ratio of the slope of the conductivity versus solute concentration plot below and above the CMC. Potentiometric (emf) measurements with membrane or ion-selective electrodes have also been used [38, 39], although some electrodes may not work properly (non-Nernstian behavior) in the absence of added salt. Differences between results obtained from conductivity and emf results have been noted by Zana and co-workers [39]. The emf method seems to produce larger β values (more dissociated micelles) than the conductivity method when β is calculated from the slope ratio. Large discrepancies may even be found when β is calculated from conductivity experiments, depending upon the model used [58]. This occurs for example in comparing the slope ratio mentioned above and that inferred from Evans [59] procedure, using the equation

$$(1000S_1 - \lambda_c)\left(\frac{n - m}{n^{4/3}}\right)^2 + \lambda_c\left(\frac{n - m}{n}\right) - 1000S_2 = 0 \tag{36}$$

where S_1 and S_2 are respectively the slopes of the conductivity versus surfactant concentration curve, below and above the CMC, n and m are the number of monomers per micelle and the number of counterions of conductance λ_c adsorbed on the micelle; β is then equal to $(n - m)/n$.

The influence of the chosen value of n on the calculated β is small: thus, as n is varied from 60 to 80, the change of β is 0.01. In addition, for most surfactants in water, λ_c can be approximated by the limiting value at infinite dilution. These data may be found in all textbooks of electrochemistry. Typically, the difference between the results obtained through the procedure with $\beta = S_1/S_2$ and that obtained through Eq. (36) is of the order of 0.2 unit of degree of counterion condensation, the latter equation providing higher degree of association than the former. In our opinion Eq. (36) should be preferred. In any case the choice of the proper technique is even more crucial when the *derivative* of β with respect to the solute addition has to be included in the calculation.

The general picture which emerges from the available data is a decrease of counterion association upon addition of polar solutes [37, 39, 58]. A single case of the opposite behavior has been noted for an anesthetic molecule, halothane, in NaDS micelles [60]. Nonpolar molecules do not greatly affect the value of β, at least for low additive concentration [40].

There are relatively few systematic disagreements between authors about CMC determinations. However it is worthwhile to point out some atypical results. For example, acetone decreases the CMC of ionic surfactants at low solute concentrations [58, 60]. Some authors have missed this decrease by working at higher concentrations [34, 37]. Interpretation of ^{19}F NMR measurements in the case of a partially fluorinated anionic surfactant [61] leads to the conclusion that even highly hydrophilic molecules such as acetone may be considered as solubilized, confirming the reason for the initial CMC decrease observed. Some curious results of a CMC decrease for C_{16}Br [62] upon D-fructose addition have been reported. Some noted discrepancies may be found among some of the results for 1-alkanols in C_{16}Br solutions (see Appendixes A and B). The "best" results for these and perhaps a few other systems might be identified by comparison between data sets as suggested below. Finally, two isolated cases have been published in the literature showing a small initial increase of CMC upon solute addition, followed by a CMC decrease. It concerns aromatic alcohols in benzyldimethyltetradecylammonium chloride [63] and in polyoxyethylenemonohexadecylether solutions [64].

B. Comparison of Partition Coefficients as Deduced from Various Experimental Techniques

It is important to compare the results obtained from the various theories outlined above and also to compare these results to those derived by different physicochemical approaches. Table 3 presents such a comparison in the much-documented case of 1-butanol in NaDS solutions at 25°C. Results from Eq. (8), (16), and (34) are used as they may be considered as independent relations. β is taken as a weighed average equal to 0.30 [65–67] in Eq. (8). The other data are taken directly from the literature. They are all expressed on the mole fraction basis. [Note: Recall that the partition coefficient may be expressed in molality (P_m), molarity (P_c), and mole fraction (P_x) scales. These are related by the expressions: $P_x = 55.5P_m$, $P_c = P_m/\overline{V}$, $P_x = 55.5P_m\overline{V}$, where $\overline{V}$ is the surfactant partial molar volume at high dilution. Table 4 presents some representative $\overline{V}$ values.]

It is clear from inspection of the 11 independent results of Table 3 that most of these data agree well. They agree among the CMC-based procedures and they also agree with most of the other data available.

There are two exceptions to this general observation: the somewhat smaller value deduced from the NMR measurements (whether deduced

TABLE 3 Comparison of Partition Coefficients P for 1-Butanol in Sodium Dodecylsulfate Micellar. Solutions at 25°C according to Various Experimental Techniques

P	Method	Reference
305[a]	Conductance (Eq. (16))	36
327	Conductance (Eq. (34))	55
276	Conductance (Eq. (8))	80
322	Density	13
305	Density	12
346	Potentiometry	56
300	Gas chromatography	5
180	NMR	16(a)
134	NMR	16(b)
513	Calorimetry	9
250	Krafft point[b]	22

[a] Value corresponding to infinite solute dilution.
[b] Value obtained at 15°C.

TABLE 4 Partial Molar Volumes at Infinite Dilution in Water at 25°C for Some Surfactants of Appendix A

Surfactant	$\overline{V}°$ (cm^3mol^{-1})	Ref.	Surfactant	$\overline{V}°$ (cm^3mol^{-1})	Ref.
$C_{10}Br$	205.45	14	$C_{10}COONa$	164.17	98
$C_{12}Br$	288.2	14	$C_{10}COOK$	175.12	98
$C_{14}Br$	319.8	14	$C_{12}COOK$	205.49	98
$C_{16}Br$	351.4	14	C_8Na	173.40	99
			$C_{10}Na$	206.55	99
			$C_{12}Na$	238.0	99

from Fourier transform pulse-gradient spin echo [16a] or from paramagnetic relaxation experiments [16b]) and the somewhat high result obtained from calorimetric experiments.

It must be stressed that in most *P* determinations of Table 3, the solute activity is controlled under comparable experimental conditions. For example in NMR experiments, diffusion coefficients are determined at a solute/surfactant molar ratio of about 2 [16–18]. In gas-chromatographic measurements, this ratio is usually around 0.5. [5, 6] In the case of calorimetric measurements, the alcohol + surfactant ratio is usually varied from about 0.5 to 5. This ratio is lower for partial volume measurements. In the case of CMC experiments, the surfactant concentration is varied up to about five times the CMC with the solute concentration kept at a minimum. It is of the order of 0.02 molar or less for alkanols, which means that in the case of NaDS, the molar ratio of solute to surfactant is on the order of 1 or lower. In terms of the more significant solute per micelle ratio, these numbers are of course much lower and the lower solute concentration limit is only defined by the minimum observable CMC change. In the case of a high *P* value such as with 1-decanol the solute mole fraction in the micelle may be kept down to 0.004 [68], which ensures that real limiting partition coefficients are measured.

Comparison with direct *solubility* determinations have been made in the cases of alkanols. An increased solute solubility is observed with increasing surfactant concentration followed by a sudden solubility decrease [39, 69, 70]. This phenomenon is believed to be the consequence of a micellar structural change upon solute addition. The saturated partition coefficient obtained under these conditions is lower than the "real" *P* value obtained under controlled (dilute) solute activity. Such comparisons are usually not possible with solid substances. In any case, a meaningful comparison between apparent (solubility) and real P values could only be performed using measured activity coefficients. These are usually not known except for a few cases [2, 3, 21].

C. Comparison of CMC-Based Partition Coefficients

Comparison of experiment with theory has been performed with Ruckenstein and Rao's work on micellization and solubilization [71]. One important feature of that theory is the prediction of a dramatic decrease of P with increasing solute incorporation in the micelles at small solute concentration. For example, calculations show for 1-butanol in NaDS micelles, an increase of P from 100 at a mole fractional solute occupancy Y_M of 0.50 to 1070 for $Y_M = 0.050$. These results do not fully agree with experimental data. GC [72], vapor pressure [3] as well as fluorescence [19] experiments show that P may be constant for NaDS + 1-alkanol systems up to a mole fraction in micelles of 0.25, and then decrease more or less rapidly toward the solubility value with increasing solute addition. Note, however, that according to some NMR experiments, P does not vary with solute concentration at least in the case of alcohols [18]. The example is provided of 1-pentanol in $C_{12}Br$ solutions where the P value is constant over the whole solute concentration range investigated, i.e. up to 0.148 mol/kg. This concentration is about 5 times higher than that employed in GC experiments.

P calculations using Eq. (16) show a much smaller decrease of P with solute concentration than that predicted by Rao and Ruckenstein' theory at low solute concentration. Nevertheless, the agreement between calculated and theoretical values at higher concentration values for 1-butanol in NaDS micelles is remarkable (Fig. 3).

Figure 4 presents the same comparison in the case of GC-based experiments for 1-pentanol in NaDS micellar solutions [72] where the variation of P is presented as a function of solute mole fraction within the micellar pseudophase. The same conclusion can be made: there is an excellent agreement between experiment and theory at higher solute concentration corresponding to a solute mole fractional occupancy above 0.25.

Derzanski and co-workers [73] propose an analysis of CMC changes which is essentially based upon the Nagarajan and Ruckenstein thermodynamic approach. Application of their final equation necessitates a number of experimental parameters which are usually not available such as the variation of micelle polydispersity and aggregation number with solute addition. A curious aspect of this theory is that the increase of CMC with solute addition is considered to reflect of either artefacts (impurity of the solute, finite solute concentration used, ill-defined instrumental CMC) or

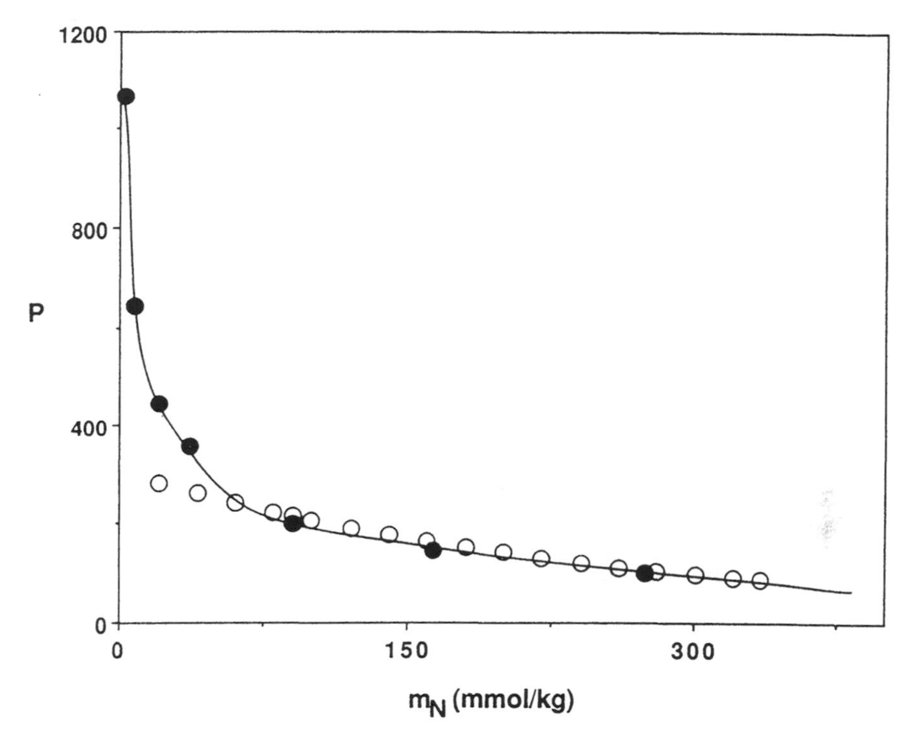

FIG. 3 Comparison of experimental CMC-based [36] (○) and theoretical [71] (●) partition coefficients as a function of solute concentration for 1-butanol in sodium dodecylsulfate micelles.

theoretical difficulties (constant surfactant chemical potential) but not a physical phenomenon.

1. Some Discrepancies

In order to further inquire about the reliability of the various calculations discussed above the variation of a solute partition coefficient with surfactant chain length is important. In effect, if all theories used lead to the expected result that for a given surfactant, *P* increases with the hydrophobicity of the solute, the effect of surfactant chain length show some conflicting results.

Briefly summarized, contrary to all other derivations, Eq. (32) produces *P* values for polar solutes which decrease with increasing surfactant aliphatic chain length in a homologous surfactant series. As this equation is

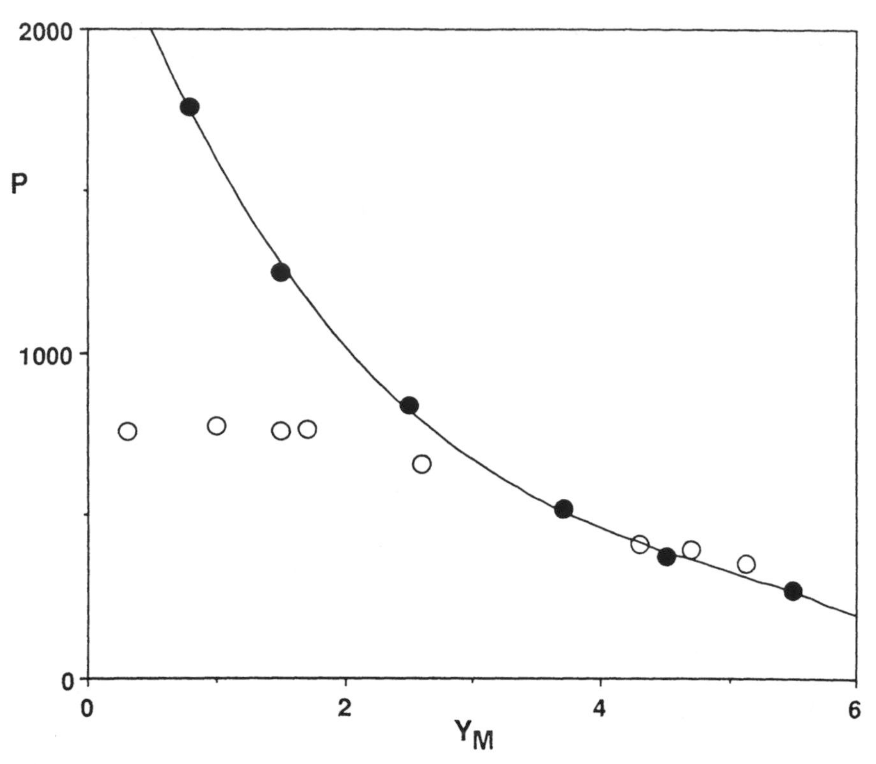

FIG. 4 Comparison of experimental GC-based [72] (○) and theoretical [71] (●) partition coefficients as a function of solute mole fraction within the micelles for 1-pentanol in sodium dodecylsulfate micelles.

the only one which includes a contribution from the derivative of counterion association with solute addition and because a large number of systems have been analyzed with Eq. (32), some discussion of these results is necessary in order to be able to recommend some reference equations for CMC-based *P* values.

Abu-Hamdiyyah's calculations, using the numerous CMC data in tetraalkylsulfate solutions with alkylchain length from C_{10} to C_{16} at 43.8°C, lead to a decrease of *P* with increasing surfactant chain length for polar solutes [51] and a *P* increase for nonpolar ones [53]. This result arises as the consequence of the use of Eq. (32) for the former solutes and of Eq. (8) (with $\beta = 0$) for the latter ones. This procedure is justified by the experimental observation that the solubilization of nonpolar solutes apparently does not modify the micellar counterion degree of dissociation and therefore that the second term of Eq. (30) should be zero. As a typical case amongst a variety of polar compounds, according to Eq. (32), the P value of benzocain decreases from 8560 to 3660 as the alkylsulfate chain length increases from C_{10} to C_{16} [51].

It was of interest to recalculate P from Abu-Hamdiyyah and co-workers raw experimental results, applying Eq. (8). The original data are presented in a reduced form as CMC = $CMC_w - aC_N$, where CMC_w is the CMC without additive. In dilute solutions the following approximate relation holds:

$$\frac{a}{CMC_w} = 2.303K_M \tag{37}$$

One finds then that for the case of benzocain mentioned above, P increases from 7320 to 10200. The β values were taken from the literature [74] and k_N calculated from Eq. (9) at 25°C assuming a negligible temperature effect. The same reverse behavior applies to all the data obtained at 43.8°C. The same comparison was performed on a partition study at 25°C with the cationic surfactant series from $C_{12}Br$ to $C_{16}Br$ [68]. Results are presented in Appendix B. Little or no variation of P with alkylchain length is observed according to Eq. (8) or its analogues, a result which is again at variance with the decreasing P values deduced when using Eq. (32). As will be shown below, Eq. (16) leads to the same general conclusion as that deduced from Eq. (8).

Examination of the literature is instructive at this point. Using Eq. (34), Manabe et al. [56] show that in the alkylsulfate surfactant series from C_9 to C_{12}, P is constant from 1-butanol to 1-pentanol but increases with surfactant chain length for 1-hexanol and 1-heptanol. Shinoda's classical data [75] on the variation of CMC for 1-alkanols in the presence of potassium octanoate to tetradecanoate solutions also shows a slight increase of P values for most alcohols when calculated using Eqs. (8) or (16); taking the case of 1-pentanol and Eq. (8), P varies smoothly from 200 to 770 (with K_N varying from -0.28 to -0.37 and the β values taken as equal to 0.23, 0.27, and 0.26 respectively for potassium dodecanoate, tetradecanoate, and hexadecanoate [74]). P values for 1-pentanol as derived from partial molar volume measurements also increase with surfactant chain length in the trimethylalkylammonium bromide series [76]. The 1-alkanol solubility data of Harkins and Oppenheimer [77] in the potassium carboxylate series show an increase with surfactant chain length for the smaller surfactant alkylchain length and a near constancy at larger values.

Thus, it appears that in most documented cases, P slightly increases with alkylchain length or remains independent of alkylchain length for polar as well as for nonpolar solutes in a homologue surfactant series. This result is in opposition with the decrease of P obtained from the use of a model based upon the change of β with solute addition. It seems that such a model overemphasizes the contribution of the variation of the surface charge density by solute incorporation to the overall solubilization phenomenon. This conclusion is in agreement with the calculations leading to Eqs. (8), (13), and (28). It must also be pointed out that Eq. (32) depends heavily on the experimental determination of the variation of β with added solute. These data are deduced from an analysis of a conductivity versus concentration plot, the validity of which may be questioned, as suggested above.

2. Recommended Procedure

In view of the above discussion it is possible to suggest a "best available" procedure for the calculation of CMC-based P values. Relations which rely upon the determination of empirical ϑ values necessitate some additional technique for the calculation of P. Thus, it seems that the independent Eqs. (8) and (16) are the most useful ones for the calculation of micellar P values from CMC measurements.

Figure 5 presents a direct comparison on a log/log scale of P values at 25°C as deduced from Eqs. (8) and (16). A large number of data have been collected from different sources. All data were not used for lack of space. They were recalculated using the original reduced K_M values [27, 78–80] as described above or Abu-Hamdiyyah and Kumari [68] and Shinoda's data from the raw experimental results (75) using the approximate correcting factor [37]. Equation (16) was used at the limit of low (m_N = 0.001 mol/kg) concentration.

The two calculations agree remarkably well above a partition coefficient of about 250 up to a value of about 40,000, the highest literature value available. In all cases only CMC-based data are used. The larger values obtained from Eq. (8) at low P values are most certainly due to the salting constant term whose numerical effect becomes negligible at higher P values.

The results of Figure 5 suggest that Eqs. (8) and (16) should be favored for the determination of P values for neutral solutes, the former equation being restricted to the limit of low concentrations. Equation (8) should be favored over Eq. (16) for low P values (below about 250 on the mole fraction basis, which corresponds to the case of 1-butanol in most surfactant solutions). Equation (34) seems an interesting alternative, but too few results are available in the literature for a valuable comparison with the other two equations.

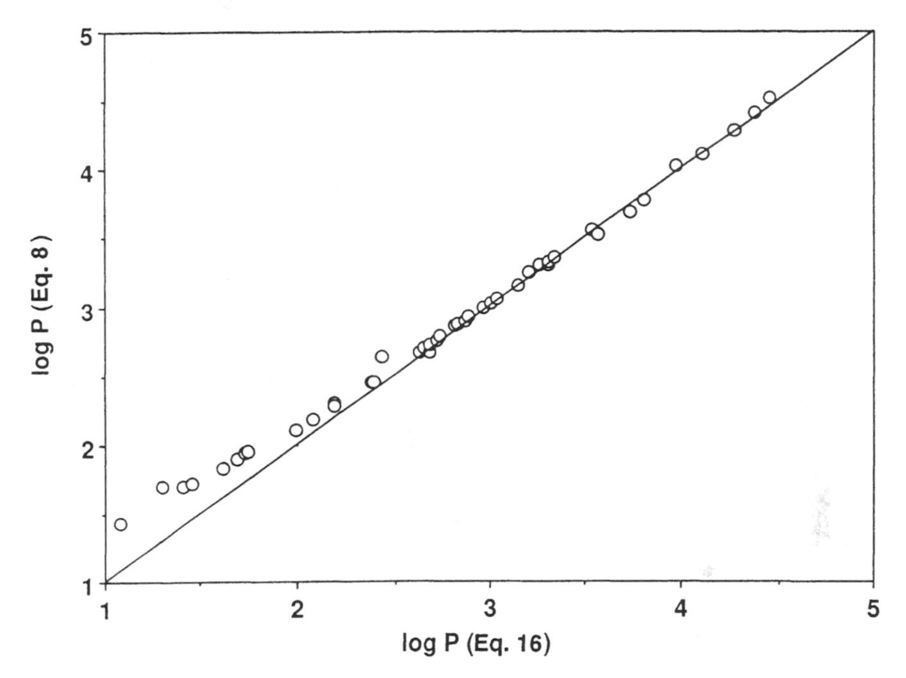

FIG. 5 Comparison of CMC-based P values (on a log/log scale) as calculated from Eqs. (8) and (16).

Finally, some authors have focussed their attention on the determination and the significance of the ϑ coefficient of Eq. (23). Table 2 collects most ϑ values available in the literature according to various sources and at different temperatures (see page 395).

In the following discussion Eq. (8) will be used, as it will be concerned essentially with limiting P values.

IV. DISCUSSION

A. Computed Literature CMC-Based Partition Coefficients

1. Ionic Surfactants

The literature on the variation of CMC with solute concentration which is amenable to mathematical analytical is not very abundant and concerns almost exclusively ionic surfactants. Furthermore, some authors publish their data, unfortunately, only in graphical form. It is the case among many other examples of the CMC decreasing effect of crown ethers on sodium decanoate [81], parabens on NaDS [20], or phenylalkanols in sodium deoxycholate [25b] solutions.

As stated in the Introduction section and further developed in the last section of this Chapter, many CMC data concern solutes which are soluble enough in water to be considered as cosolvents. These solutes have been studied in a concentration domain well above the region of the limiting law as defined by Eq. (8) and the other analogue functions. This concentration domain depends upon the properties of the solutes and the surfactants as indicated by the typical examples displayed on Fig. 1 to 3. Thus, depending on the relative amplitude of the standard chemical potentials and that of the partition coefficient terms (Eq. (3)), the deviations may start at lower or higher solute concentrations.

In order to present a self-consistent ensemble of data it was considered necessary to reanalyze all available CMC determinations in the presence of neutral solutes. For that purpose, K_M values of Eq. (8′) were recalculated from the published CMC results only when sufficient data points in the low-solute-concentration region were available. This coefficient seems the most adequate reduced parameter for testing all equations besides Eq. (8) or (16). Note again that the CMC data presented in the literature in the reduced form of a linear variation with solute concentration were used to evaluate K_M by applying the approximate correction factor [37].

The results of the calculations at 25°C are presented on Appendixes A and B. β values were taken from the literature [74, 82] and k_N was calculated from Eq. (9) for all solute + surfactant binaries. It must be stressed that no discrimination was exercised against any set of CMC-based log P values presented in these tables.

The results of Abu Hamdiyyah and Rahman in NaDS solutions at 43.8°C [51, 53] were not included in the tables as neither k_N nor β are known at that temperature.

Some sort of discrimination procedure is necessary so that the data ensemble presented are not only self-consistent but offer also some degree of confidence.

Two general rules can be suggested. The first one stems from many analyzed data in the case of alkanols and other aliphatic series of solutes. One observes that, on average, the methylene group contribution on a linear aliphatic chain, is equal to 0.50 log P unit, provided that the polar group is remote from it by at least two methylene groups. In the case of the alkanol series, this rule applies starting with 1-propanol. This rule applies also well for log P values in the two-phase system 1-octanol + water [83].

The first criterion is fulfilled by most polar aliphatic data regardless of ionic charge or type of headgroup as can be deduced from inspection of the data of Appendix A. The methylene group contribution appears closer to log $P = 0.40$, when aromatic molecules are considered.

The second rule may be discussed in conjunction with Figs. 6 and 7. In Fig. 6, data obtained from the same research group have been plotted on a log P scale for solutes which have been studied in NaDS and $C_{12}Br$ solutions at 25°C [80]. A linear plot is obtained for all data available in the two surfactant solutions, except for phenol and, to a lesser extent, perfluorobutanol. The other solutes concern aliphatic as well as aromatic molecules, polar and nonpolar, with the most important and diverse polar moieties. This plot shows that the driving force for micellar solubilization is, as is well known, predominantly hydrophobic and consequently essentially unspecific. The behavior of two nonorthodox solutes can be rationalized by noting that for phenol, which is somewhat dissociated into the anionic phenoxide group, its solubilization in a cationic micelle is stabi-

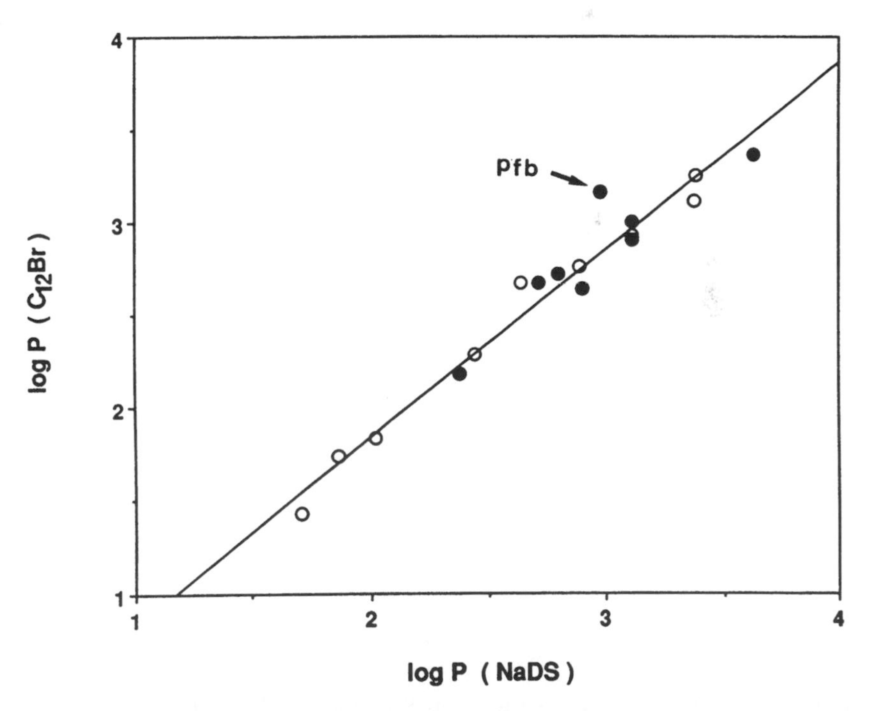

FIG. 6 Comparison (on a log/log scale) of CMC-based partition coefficients for polar solutes in trimethyldodecylammonium bromide and in sodium dodecylsulfate micelles: alcohol [○] and nonalcohol [●] molecules from Appendix A: Pfb = perfluorobutanol.

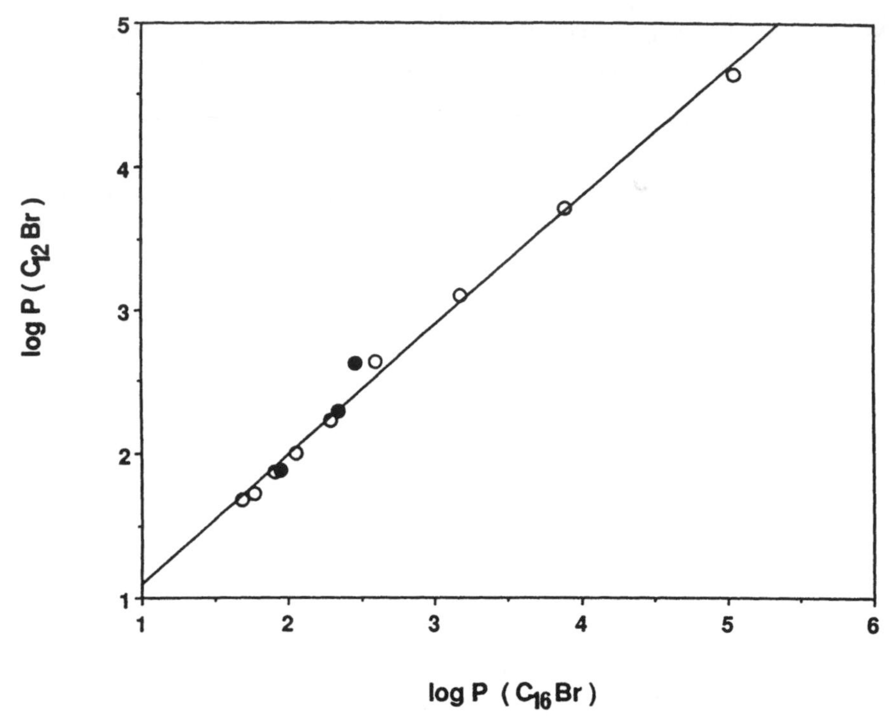

FIG. 7 Comparison (on a log/log) scale of CMC-based partition coefficients in trimethyldodecyl and trimethylhexadecylammonium bromide micelles: alcohol [○] and nonalcohol [●] molecules from Appendix B.

lized by coulomb forces, so that its P value is larger for the $C_{12}Br$ than for NaDS micellar solutions.

The same type of plot is presented using the results for $C_{12}Br$ and $C_{16}Br$ of Appendix B (Fig. 7). The slight scatter provides an estimation of the reliability of such data as no specific effect is expected.

Thus, it seems that at least for common ionic surfactants, plots such as those of Fig. 6 and 7 should help to discriminate among the solutes those whose behavior points out an experimental problem, or in a few cases, solutes which present a specific interaction with the surfactant. This is apparently the case when results for hydrogenated and perfluorinated compounds are compared. P values for perfluoroalkanols do not meet criterion 2 with the 1-alkanols in LiDS and LiFOS micelles.

2. Nonionic Surfactants

As noted above, CMC values for hydrophobic solutes in nonionic surfactants solutions are lacking. Most certainly, the question of the polydispersity of most nonionic surfactants and that the greater difficulty in obtaining CMC values for these surfactants explains the paucity of data for such systems. One may point out the atypical initial CMC caused by increase of aromatic solutes (phenol, benzylalcohol, and 2-phenylethanol) in the presence of polyoxyethylene monohexadecylether [64], which is followed by a CMC decrease prevent the calculation of a partition coefficient. Nishikido et al. [84] derive P values for lower alkanol derivatives in polyoxyethylene dodecylethers solutions with various oxyethlene chain length $\bar{n}$. The actual CMC values were not published. P was calculated using the Shinoda et al. semiempirical approach [46]. According to this analysis, the partition coefficient of 1-pentanol increases with oxyethylene chain length from 155 for $\bar{n} = 6$ to 450 for $\bar{n} = 41$.

In the case of Triton X-114, Derzhanski et al. [73] find a positive CMC change for 1-butanol and fructose, whereas the CMC change is negative for 1-octanol and glucose. P values cannot be readily calculated from their data. Other CMC data are available for Triton X-100 with predominantly hydrophilic solutes such as urea and dextrose. The P values obtained applying Eq. (8) are accordingly very small [85].

3. The log P_{oct} Scale

The log P scale for neutral molecules in binary systems is of great value for the evaluation of aqueous solubilities using experimental or calculated values with one of the group contribution approaches available (84). Particularly useful is the octanol + water binary. This system has been used for the correlation of solubility data in micellar systems for compounds belonging to the same group series, e.g., steroids (1).

The following expressions have been proposed for NaDS and $C_{12}Br$ micellar solutions [80]:

$$\text{NaDS: } \log P(x)_{mic} = 1.88 + 0.80 \log P(c)_{oct} \quad (38)$$

$$C_{12}\text{Br: } \log P(x)_{mic} = 1.51 + 0.91 \log P(c)_{oct} \quad (39)$$

These relations were obtained using measured P_{oct} values in the original molarity scale (c) and collected CMC-based P data (mole fraction scale) in the relevant micellar systems as deduced from Eq. (8) These correlations were obtained using aliphatic polar molecules (alcohols, aldehydes, ketones, nitriles, ethers, esters). Aromatic molecules, whether polar or nonpolar, did not fit as well in the correlation as the aliphatic molecules and were discarded. Nonpolar molecules (alkanes) and halogenated derivatives were definitely out of the line and therefore were not considered when eq. (38) and (39) were established. Other correlations of the same type may be constructed from the data of the Appendixes.

The usefulness of such correlations for the evaluation of unknown P data even for a class of molecules which were discarded in the above correlations may be shown by the example of the partition coefficient of ester derivatives of parahydroxybenzoic acids (parabens) in NaDS solutions. These data had been determined by Goto and Endo [20] using a gel filtration method and by Iwatsuru and Shimizu [24] from UV spectra chemical shifts (both experiments performed under controlled solute activity). Table 5 compares these experimental results to those obtained using relation [38]. The log P_{oct} values were taken from the Leo and Hansch data bank [86]. The agreement is rather satisfactory. Other examples of that sort may be found using literature values. It may be pointed out that the methylene group contribution in the paraben series is only 0.40 log P units as compared to the 0.50 for aliphatic molecules. It is noteworthy that the methylene group is the only one for which no difference is found in the 1-octanol + water binary system whether it is attached to an aromatic ring or to an aliphatic group [83].

Using the numerous partition coefficients values available in the 1-octanol + water binary system [86], the log P_{oct} scale may also be employed on a log/log basis as a discrimination procedure by plotting micellar results versus two-phase binary data. However, the scatter is then larger than when only micellar solutions are considered as in Fig. 7 and 8.

B. Tentative Group-Contribution Approach to Micellar Solubilization

Additive schemes based on group-contribution approaches have been found extremely useful for the evaluation of thermodynamic data. Such an approach for the evaluation of log P values was never attempted for

TABLE 5 Experimental Gel-Filtration [20] and UV Spectra [24] and Calculated (Eq. (38)) Partition Coefficients for *para*-Hydroxybenzoate Derivatives (parabens) in NaDS Solutions at 25°C (mole fraction basis)

Solute	log P [20]	log P [24]	log P (calc.) Eq. 38
Methylparaben	3.51	3.58	3.45
Ethylparaben	3.93	3.94	3.88
Propyparaben		4.18	4.31
Butylparaben	4.86	4.66	4.74

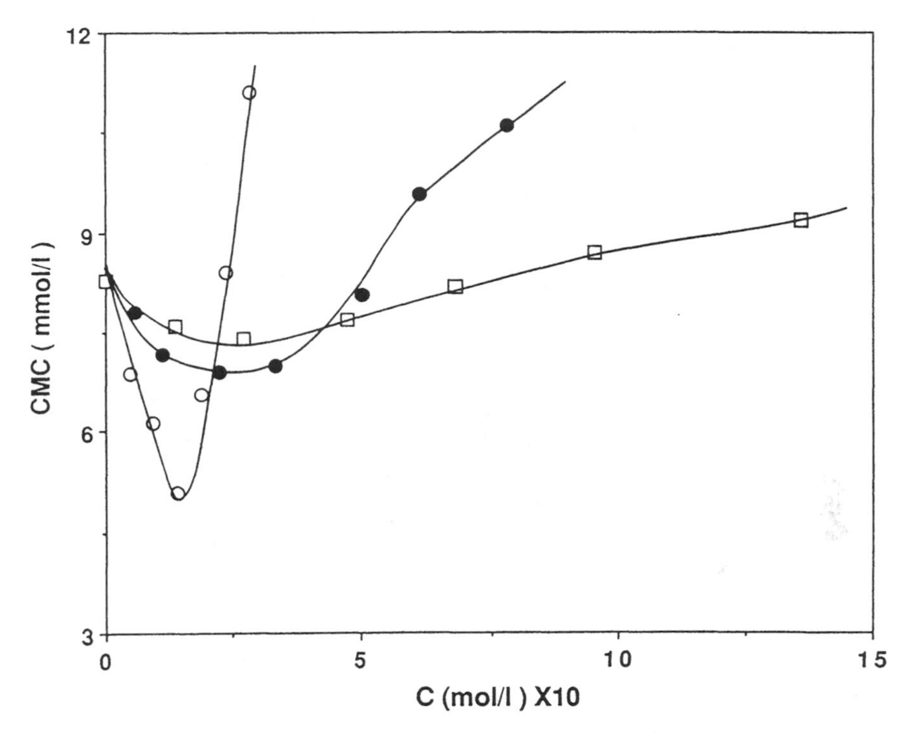

FIG. 8 Variation of CMC with aliphatic ketone concentration (redrawn from Ref. 92): 2-propanone (□), 2-butanone (●), and 2-pentanone (○).

micellar solutions. It was thought that a group-contribution approach was reasonable in the present case because of the lack of specific effects displayed by many neutral solutes in ionic micellar solutions. Plots such as that presented in Fig. 7 indicates that specific effects could be less important with micelles than with two-phase systems such as octanol + water. It has been shown for example that halogenated molecules may form complexes with alcohols [87] with evident complications in terms of group-additivity schemes. This type of specificity has not been observed with ionic surfactants and neutral molecules. Also, the present data have shown that neither the alkylchain length nor the substitution of headgroups greatly influence the partitioning of neutral molecules.

An additional evidence of the relative lack of specificity of neutral solutes in solubilization is provided by the case of 1-pentanol using the data of Appendix A where *P* values have been obtained in 13 different surfactant solutions. Putting aside the low *P* value obtained for potassium octanoate with its high CMC (CMC = 0.4 molar), an average log P = 2.83 ± 0.10 is obtained for 1-pentanol and the 12 other ionic surfactants. The same conclusion was obtained from the comparison of the P values of a barbituric acid such as butobarbital in a series of surfactants solutions. P was found constant for surfactants of chain length larger than 12, irrespective of charge, or type of head-group [88] (see Table 6).

Table 7 presents the tentative group contribution for polar aliphatic molecules in micellar solution. The cationic surfactant $C_{12}Br$ was chosen because the largest variety of polar moieties may be found with this surfactant using the data of Appendix A. The log P contribution of a —CH_2— group was assigned the round value 0.50 and the =CH— and —CH_3 groups were estimated as 0.5 log $P(CH_2)$ and 1.5 log $P(CH_2)$ respectively. No proximity, branching or specific effects were introduced. Thus, such a list of partial contributions should be refined in the future.

The main observations which may be drawn from inspection of the log P polar group contributions for aliphatic molecules in the case of a cationic surfactant relate to the classification of the polar moieties into three different classes according to their preferential behavior toward micelles or water:

Ether, secondary, and tertiary amines contributions oppose micellar solubilization.

Hydroxyl, ester, and ketone contributions are close to zero.

Nitrile and aldehyde groups as well as halogenated atoms contribute favorably to micellar solubilization with log P contributions of the same order of magnitude as for the methyl or the methylene groups.

As the contribution of a methylene group is assigned a single value regardless of the surfactant, differences of log P from one surfactant to

TABLE 6 Partition Coefficient of Butobarbital in Various Micellar Solutions as Deduced from Solubility Measurements at 25°C

Surfactant	log P
$C_{14}BzCl$	3.32
$C_{14}Cl$	3.35
$C_{14}Br$	3.26
$C_{12}Na$	3.26
$C_{12}E_{23}$	3.32

Source: Ref. 88.

TABLE 7 Group-Contribution Estimation of log *P* Values (mole fraction basis) for Polar Aliphatic Molecules in $C_{12}Br$ Micellar Solutions at 25°C[a]

Group		log *P*
tertiary amine	=N—	−0.95
secondary amine	>NH	−0.55
ether	>O	−0.6
hydroxyl	—OH	−0.05
ketone	>C=O	−0.05
ester	—C(=O)—O—	−0.05
nitrile	—CN	+0.4
aldehyde	—CHO	+0.8
bromine	—Br	+0.8
chlorine	—Cl	+0.9

[a] Contributions based upon log $P(CH_2)$ = 0.5, log $P(CH)$ = 0.5 log $P(CH_2)$, log $P(CH_3)$ = 1.5, log $P(CH_2)$.

another must be caused by the molecule polar group. Particularly interesting, if confirmed, is the behavior of the second class of polar moieties. Hydroxyl, ester, and ketone show no preferential interaction between micelles and water. From inspection of Appendixes A and B, it may be proposed that for the classical surfactant NaDS, an average value of 0.20 log *P* unit should be added to all group contributions of Table 7. This means that the hydroxyl moiety, for example, is more closely bound to the anionic than to the cationic ammonium headgroup. This behavior was expected. It is curious that the aldehyde group, which displays a positive contribution to log *P* should be more strongly bound than the hydroxyl moiety to the ionic headgroups. An alternative suggestion would be that the former moiety is, as an average, somewhat buried inside the micellar core. Such information could be useful for the qualitative prediction of cosurfactants for microemulsion formation. Further speculation should await the availability of more data on related surfactant systems.

The case of aromatic molecules seems peculiar. For example, in the case of $C_{12}Br$, the average log *P* value for those aromatic molecules with only one polar moiety such as benzylalcohol, benzylamine, benzonitrile, acetophenone, benzaldehyde, and phenylacetamide is 2.83 ± 0.15. The value for benzene is close to this average, being equal to log $P = 3.0$. Thus a group-contribution approach such as that proposed for aliphatic molecules may not be valid for aromatic ones because of a strong proximity effect due to the aromatic moiety. A parallel situation is encountered for the two-phase 1-octanol + water system. Proximity effects induced by the benzene ring reduces dramatically the favorable interaction between polar moieties and water [84].

V. SOLUTES AS COSOLVENTS IN AQUEOUS SURFACTANT SOLUTIONS

As we have shown, theoretical difficulties restrict the calculation of partition coefficients as deduced from CMC experiments to the relatively low solute concentration range, in the decreasing portion of the CMC versus solute concentration curve. In these experiments, the surfactant and the neutral molecule may both be considered as solutes in water, i.e., in most cases the properties of the solvent are not greatly modified by the presence of both compounds. This is no longer true at higher solute concentration. This situation is worth mentioning in the context of this discussion.

It has long been known that after an initial decrease, the CMC of some surfactants may increase again upon further addition of a neutral solute; then, above a critical solute concentration, no CMC is observed. This behavior is well documented, especially for alcohols and a number of moderately hydrophilic compounds [89–92]. It has also been discussed in review papers [93]. In some cases, the CMC may be observed in the whole composition domain from pure water to the pure organic component; it is the case, for example, for polyoxyethylene glycol derivatives such as PEG 400 with the nonionic surfactant Triton X-100 [94] or with the cationic surfactant $C_{16}Br$ [95].

The general equation (5) may be used as a starting point for the discussion of such systems. In this situation the solute may be considered as a cosolvent to water. Equation (5) may be rewritten as

$$2RT \ln \frac{X_w^c}{X_N^c} = \Delta G_{t(w,wN)} + (1 + \beta)RTPY_N \tag{40}$$

ΔG_t is the standard free energy of transfer of the *surfactant* ions or molecules from water to the mixed solvents. This quantity can be experimentally obtained either by solubility measurements [96] when such experiments are feasible or by vapor pressure experiments [97]. It is well known that ΔG_t for tetraalkylammonium salts (whether they form micelles in water or not) decreases (becomes more negative) with the addition of a cosolvent [31, 58] as a result of favorable interactions between most organic solvents and hydrophobic salts in water.

It may be therefore concluded that in the most general case, the CMC change upon the addition of neutral molecules measures both the solute partitioning and the Gibbs transfer free energies of the surfactant ions or molecules from pure water to the mixed solvents. Solute partitioning initially decreases the CMC as demonstrated, for example, by Eq. (8). When the micelles are destabilized by the addition of large solute quantities (or, more to the point, the water structure which favors micelle formation has been destroyed), the micellar solubilization phenomenon ultimately disappears, P goes to zero, and only the standard transfer free energy term from cosolvent to water remains. This term being negative in the case of predominantly hydrophobic ions or molecules in water, the CMC must increase, hence the CMC minimum. Stated differently, as the solute concentration is increased, the micelle aggregation decreases and consequently the partition coefficient must also decrease until the second term of Eq. (40) vanishes. At the same time β increases toward unity.

The value Y_N at the *CMC minimum* can be obtained by differentiating Eq. (40) with respect to Y_N one obtains

$$Y_N = -\frac{\Delta G_t}{(1 + \beta)RTP} \tag{41}$$

For a given surfactant, the variation of the logarithm-dependent ΔG_t term varies much less with solute structure than P. Furthermore, the variation of β with solvent composition is not large. Therefore, the value of Y_N for which the CMC is at a minimum, depends essentially on the value of P. Thus, as P increases with aliphatic chain length in a homogenous solute series, Y_N should get closer to pure water. This result is in agreement with experimental evidence as shown in the case of aliphatic ketones in NaDS [92] (Fig. 8). The value of Y_N depends also, although to a lesser extent, on the effect of the solutes on the physicochemical properties of water through ΔG_t. Hence, the CMC minimum is attained at lower solute concentrations with ketones than with the alcohols, which may accommodate in water through hydrogen bonds. It is also for the latter type of solutes that the CMC increase are detected in the largest cosolvent concentration domain. This was shown on the examples of Figs. 1 and 2. Here P may be taken as equal to zero in the whole cosolvent concentration domain (see Table 1).

In conclusion, the reversed CMC trend with respect to solute concentration depends, on one hand, upon the ΔG_t value, i.e., the preferential solvation (with respect to water) of the surfactant by the cosolvent and, on the other hand, by the interaction of the cosolvent with the micelles; the latter effect has two consequences: micellar solubilization at low concentration and micellar destabilization at higher cosolvent concentration.

To the best of our knowledge, the only other theoretical treatment of the effect of large cosolvent concentrations on the CMC is that of Hall

[42]. The approach leading to Eq. (10) at the limit of m_N going to zero was expanded using the same formalism to large concentration values. As mentioned above, this approach is based on the thermodynamics of adsorption. Only the case of nonionic surfactants was considered. Here the relevant equation reads

$$\left[\frac{\delta m_s}{\delta m_N}\right]_{T,p,\mathrm{CMC}} = \frac{[\overline{N}_N^+/\overline{N}_s - \overline{N}_{1N}^+]\Delta}{kT/m_s - \overline{N}_{1s}^+[\overline{N}_N^+/\overline{N}_s - \overline{N}_{1N}^+]\Delta} \tag{42}$$

with $\Delta = (\delta\mu_N/\delta m_N)_{m_s}$,

$\overline{N}_N^+$ is the average amount of solute molecules adsorbed by a micelle, and $\overline{N}_{sN}^+$ is the average amount of adsorbed solute by the monomeric surfactant.

According to Eq. (42), for the CMC to increase, the quantity between brackets should change sign, which means that the solute should be more adsorbed by the monomeric molecules than by the micelles.

This physical conclusion is similar to that deduced from Eq. (8) although a numerical application would not be straightforward; there is a competition between solute + surfactant and solute + micelle interactions, both effects being numerically of opposite sign.

VI. CONCLUSIONS

The determination of partition coefficients of neutral solutes in micellar solutions can now be determined with relative confidence from CMC determinations. Equations have been proposed which seem adequate for that purpose. These equations are by no means free of criticisms. The assumed degree of counterion condensation change upon solute addition is not adequately taken into account. The surfactant + solute mixing is assumed ideal in all theories. The pseudophase model has to be employed because the change of micellar aggregation number, which would enable us to treat the data with the mass-action model, is not known for most systems. However, the agreement observed on a vast range of P values between two completely different theoretical approaches as shown on Fig. 6 is extremely encouraging.

The simplicity and reliability of the conductivity technique makes the CMC-based P values a very attractive method of investigation for the case of ionic surfactants. When compared to other experimental approaches this procedure has the great advantage that it can be applied to solutes of very low aqueous solubility. This is not the case with most of the methods used for the establishment of the list in Table 3. Thermodynamic techniques enable us to determine valuable physicochemical parameters which are essential for the interpretation of the micellar solubilization phenomenon but they are too sensitive to small micellar structural changes upon solute addition for the evaluation of precise partition coefficients. This is clearly demonstrated in the case of partial molar enthalpies or heat capacities [10]. They are also limited to the study of a range of solutes of medium partition coefficient values (typically, in the case of NaDS, from 1-butanol to 1-heptanol). The profile of the plot of the experimental thermodynamic quantities versus surfactant concentration cannot be analyzed using standard distribution or chemical models for lower and larger *P* values. Spectroscopic methods, such as NMR (using diffusion coefficients) or the study of chemical shifts, suffer from essentially the same shortcomings, although the reasons are somewhat different.

Thus, the CMC-based partition coefficient procedure has a wide application domain which can only be compared in scope to the classical *solubility* technique with all the thermodynamic advantages of the use of nonsaturated solutions and without the need of finding the proper analytical chemical method for solute concentration determination.

Future work in this domain could focus on the following directions.

1. The dependence of the partition coefficient upon solute concentration should be better understood in relation to both theoretical and experimental aspects. In particular, the reason for the contradiction between the nonsolute dependency of the NMR and calorimetric results, on the one hand, and the dependency of the vapor pressure, head-space gas-chromatographic, gel-filtration, and fluorescence results, on the other hand, should be investigated.

2. The study of the micellar solubilization of solutes other than alcohols in ionic surfactant solutions should be pursued so that the proposed additivity scheme could be further refined. Authors should publish their actual CMC data so that this information can be used in further theoretical or compilation studies.

3. Information on the micellar solubilization behavior of neutral solutes in nonionic surfactant solutions is greatly needed in view of the increasingly widespread uses of these systems in modern applications. It may be expected that here also *P* values would not be too dependent upon specific effects and that some reference surfactant could be chosen whose micellar solubilization capabilities could serve as model system for the understanding of the behavior of this class of solute + micelle systems.

4. Salt effects on the CMC of surfactants in the presence of solubilized molecules have not been studied to any great extent. Given the influence of salts on the structure of micellar media and the importance of salt effects on ionic surfactant solution properties, the CMC-based method could of great value in the evaluation of the effect of structure on micellar solubilization.

APPENDIX A

Partition Coefficients of Polar Solutes in Ionic Micellar Solutions as Deduced from CMC Determinations (Eq. (8)) at 25°C

No.	Solute	Surfactant	K_M	$\log_{10} P(x)$	Reference
1	methanol	NaDS	0.03	1.27	[80]
2	ethanol	NaDS	0.20	1.71	[80]
3	1-propanol	NaDS	0.51	2.02	[80]
4	2-propanol	NaDS	0.32	1.86	[80]
5	1-butanol	NaDS	1.5	2.44	[80]
6	*tert*-butanol	NaDS	1.4	2.41	[80]
7	1-pentanol	NaDS	4.5	2.89	[80]
8	1-hexanol	NaDS	14.7	3.39	[80]
9	1-heptanol	NaDS	34.9	3.64	[80]
10	chloroform	NaDS	3.0	2.72	[80]
11	Toluene	NaDS	3.63	2.58	[80]
12	phenol	NaDS	5.30	2.96	[80]
13	benzylalcohol	NaDS	3.40	2.64	[80]
14	phenylacetamide	NaDS	7.6	3.11	[80]
15	perfluorobutanol	NaDS	5.97	2.98	[80]
16	butyronitrile	NaDS	1.66	2.38	[80]
17	3-phenoxy,1-propanol	NaDS	14.3	3.38	[80]
18	2-phenoxyethanol	NaDS	7.5	3.11	[80]
19	carbon tetrachloride	NaDS	4.6	2.90	[80]
20	propane-1,3-diol	NaDS	0.04	1.45	[50]
21	butane-1,4-diol	NaDS	0.11	1.62	[50]
22	hexane-2,5-diol	NaDS	0.67	2.21	[50]
23	hexane-1,6-diol	NaDS	1.06	2.38	[50]
24	decane-1,10-diol	NaDS	45.6	3.96	[50]
25	allylthiourea	NaDS	0.50	2.11	[50]
26	butyramide	NaDS	0.45	2.07	[50]
27	δ-valerolactam	NaDS	1.26	2.45	[50]
28	1-butylurea	NaDS	1.48	2.51	[50]
29	caprolactam	NaDS	1.42	2.50	[50]
30	phenol	NaDS	2.36	2.69	[50]
31	benzylalcohol	NaDS	2.66	2.74	[50]
32	2-phenylethanol	NaDS	4.36	2.95	[50]
33	3-phenylpropanol	NaDS	9.2	3.27	[50]
34	4-ethylphenol	NaDS	12.3	3.39	[50]
35	benzene	NaDS	1.74	2.60	[50]
36	cyclohexane	NaDS	21.1	3.62	[50]
37	hexane	NaDS	163	4.51	[50]
38	phenol	NaDS	2.61	2.73	[104]

(continued)

No.	Solute	Surfactant	K_M	$\log_{10} P(x)$	Reference
39	phenylmethanol	NaDS	2.70	2.75	[104]
40	phenylethanol	NaDS	5.17	3.02	[104]
41	phenylpropanol	NaDS	9.54	3.28	[104]
42	phenylbutanol	NaDS	21.7	3.63	[104]
43	phenylpentanol	NaDS	53.4	4.02	[104]
44	phenylhexanol	NaDS	123.8	4.39	[104]
45	ethanol	LiDS	0.10	1.62	[101]
46	1-propanol	LiDS	0.41	2.04	[101]
47	1-butanol	LiDS	1.31	2.48	[101]
48	1-pentanol	LiDS	3.64	2.90	[101]
49	1-hexanol	LiDS	8.68	3.26	[101]
50	trifluoroethanol	LiDS	0.23	1.84	[101]
51	pentafluoropropanol	LiDS	1.10	2.40	[101]
52	heptafluorobutanol	LiDS	4.30	2.96	[101]
53	nonafluoropentanol	LiDS	9.20	3.28	[101]
54	ethanol	LiFOS	0.27	1.88	[101]
55	1-propanol	LiFOS	0.69	2.20	[101]
56	1-butanol	LiFOS	1.74	2.56	[101]
57	1-pentanol	LiFOS	4.46	2.96	[101]
58	1-hexanol	LiFOS	7.40	3.18	[101]
59	trifluoroethanol	LiFOS	0.42	2.02	[101]
60	pentafluoropropanol	LiFOS	2.88	2.77	[101]
61	heptafluorobutanol	LiFOS	8.89	3.25	[101]
62	nonafluoropentanol	LiFOS	23.9	3.68	[101]
63	1-butanol	NaPFO	1.72	2.55	[79]
64	1-pentanol	NaPFO	3.43	2.84	[79]
65	1-hexanol	NaPFO	7.63	3.19	[79]
66	1-heptanol	NaPFO	16.1	3.51	[79]
67	perfluorobutanol	NaPFO	5.97	3.28	[79]
68	1-butanol	KPFO	1.02	2.34	[79]
69	1-pentanol	KPFO	2.24	2.67	[79]
70	1-hexanol	KPFO	5.02	3.01	[79]
71	1-heptanol	KPFO	10.6	3.33	[79]
72	phenol	KPFO	0.53	2.13	[79]
73	benzylalcohol	KPFO	1.04	2.37	[79]
74	perfluorobutanol	KPFO	6.80	3.13	[79]
75	1-butanol	C_7COOK	0.41	2.05	[75]
76	1-pentanol	C_7COOK	0.85	2.30	[75]
77	1-hexanol	C_7COOK	3.9	2.92	[75]
78	1-heptanol	C_7COOK	9.0	3.28	[75]
79	1-octanol	C_7COOK	25	3.71	[75]
80	1-nonanol	C_7COOK	62	4.10	[75]
81	1-decanol	C_7COOK	122	4.40	[75]

(continued)

No.	Solute	Surfactant	K_M	$\log_{10} P(x)$	Reference
82	1-butanol	C_9COOK	0.70	2.24	[75]
83	1-pentanol	C_9COOK	1.58	2.55	[75]
84	1-hexanol	C_9COOK	4.8	3.01	[75]
85	1-heptanol	C_9COOK	16.2	3.53	[75]
86	1-octanol	C_9COOK	30.6	3.80	[75]
87	1-nonanol	C_9COOK	107	4.34	[75]
88	1-decanol	C_9COOK	203	4.62	[75]
89	1-butanol	$C_{11}COOK$	0.57	2.18	[75]
90	1-pentanol	$C_{11}COOK$	3.14	2.83	[75]
91	1-hexanol	$C_{11}COOK$	5.55	3.02	[75]
92	1-heptanol	$C_{11}COOK$	15	3.49	[75]
93	1-octanol	$C_{11}COOK$	52	4.03	[75]
94	1-nonaol	$C_{11}COOK$	121	4.40	[75]
95	1-decanol	$C_{11}COOK$	270	4.74	[75]
96	1-butanol	$C_{13}COOK$	0.61	2.20	[75]
97	1-pentanol	$C_{13}COOK$	3.57	2.89	[75]
98	1-hexanol	$C_{13}COOK$	6.09	3.11	[75]
99	1-heptanol	$C_{13}COOK$	19.9	3.61	[75]
100	1-octanol	$C_{13}COOK$	62	4.11	[75]
101	1-nonanol	$C_{13}COOK$	161	4.52	[75]
102	1-decanol	$C_{13}COOK$	503	5.01	[75]
103	ethanol	$C_{12}Br$	0.05	1.43	[80]
104	1-propanol	$C_{12}Br$	0.19	1.83	[80]
105	2-propanol	$C_{12}Br$	0.13	1.73	[80]
106	1-butanol	$C_{12}Br$	0.72	2.28	[80]
107	*tert*-butanol	$C_{12}Br$	0.26	1.95	[80]
108	1-pentanol	$C_{12}Br$	2.50	2.76	[80]
109	1-hexanol	$C_{12}Br$	7.90	3.25	[80]
110	dioxane	$C_{12}Br$	−0.10	1.6	[80]
111	ethylmethylketone	$C_{12}Br$	0.088	1.70	[80]
112	diethylamine	$C_{12}Br$	0.25	1.94	[80]
113	triethylamine	$C_{12}Br$	2.64	2.79	[80]
114	propylacetate	$C_{12}Br$	1.13	2.45	[80]
115	chloroform	$C_{12}Br$	2.05	2.67	[80]
116	diethylether	$C_{12}Br$	0.22	1.90	[80]
117	tetrahydrofuran	$C_{12}Br$	0.114	1.70	[80]
118	cyclohexanone	$C_{12}Br$	0.72	2.30	[80]
119	ethylbutyrate	$C_{12}Br$	3.14	2.86	[80]
120	*N*-methylpyrrolidone	$C_{12}Br$	−0.21	1.62	[80]
121	phenol	$C_{12}Br$	9.97	3.33	[80]
122	benzylalcohol	$C_{12}Br$	2.30	2.67	[80]
123	benzylamine	$C_{12}Br$	2.12	2.70	[80]
124	benzonitrile	$C_{12}Br$	8.05	2.93	[80]

(continued)

No.	Solute	Surfactant	K_M	$\log_{10} P(x)$	Reference
125	benzaldehyde	$C_{12}Br$	3.27	2.87	[80]
126	acetophenone	$C_{12}Br$	4.92	3.03	[80]
127	phenylacetamide	$C_{12}Br$	3.56	2.90	[80]
128	butyronitrile	$C_{12}Br$	0.57	2.18	[80]
129	3-phenoxy, 1-pro-	$C_{12}Br$	9.60	3.31	[80]
130	panol	$C_{12}Br$	3.74	2.93	[80]
131	2-phenoxyethanol	$C_{12}Br$	4.53	3.00	[80]
	benzene				
132	toluene	$C_{12}Br$	10.6	3.36	[80]
133	chlorobenzene	$C_{12}Br$	17.0	3.56	[80]
134	halothane	$C_{12}Br$	2.3	2.72	[80]
135	dichloromethane	$C_{12}Br$	1.17	2.45	[80]
136	carbon tetrachloride	$C_{12}Br$	1.27	2.64	[80]
137	butylbromide	$C_{12}Br$	5.33	3.06	[80]
138	butoxyethanol	$C_{12}Br$	0.70	2.28	[80]
139	perfluorobutanol	$C_{12}Br$	6.85	3.16	[79]
140	butyraldehyde	$C_{12}Br$	1.73	2.57	[103]
141	1-butanol	$C_{12}Br$	0.65	2.24	[16b]
142	EG-butylether	$C_{12}Br$	0.87	2.35	[16b]
143	diEG-butylether	$C_{12}Br$	0.87	2.34	[16b]
144	triEG-butylether	$C_{12}Br$	0.80	2.32	[16b]
145	1-butanol	$C_{14}Cl$	0.95	2.33	[103]
146	1-pentanol	$C_{14}Cl$	3.35	2.84	[103]
147	1-hexanol	$C_{14}Cl$	10.7	3.33	[103]
148	benzylalcohol	$C_{14}Cl$	4.1	2.92	[103]
149	acetophenone	$C_{14}Cl$	6.20	3.10	[103]
150	butyraldehyde	$C_{14}Cl$	2.30	2.68	[103]
151	1-pentanol	$C_{14}BzCl$	2.5	2.72	[103]
152	benzylalcohol	$C_{14}BzCl$	4.0	2.92	[103]
153	phenol	$C_{16}Br$	5.2	3.08	[52]
154	benzylalcohol	$C_{16}Br$	6.6	3.18	[52]
155	2-phenylethanol	$C_{16}Br$	9.0	3.31	[52]
156	3-phenylpropanol	$C_{16}Br$	21.2	3.68	[52]
157	1-butanol	$C_{16}Br$	0.94	2.38	[52]
158	1-hexanol	$C_{16}Br$	9.85	3.35	[52]
159	1-octanol	$C_{16}Br$	88	4.29	[52]
160	1-decanol	$C_{16}Br$	375	4.92	[52]
161	butyramide	$C_{16}Br$	0.47	2.18	[52]
162	caprolactam	$C_{16}Br$	0.84	2.36	[52]
163	butylurea	$C_{16}Br$	1.83	2.65	[52]
164	benzocaine	$C_{16}Br$	51.5	4.06	[52]
165	ethylparaben	$C_{16}Br$	75	4.28	[52]
166	perfluorobutanol	$C_{16}Br$	9.45	3.33	[52]

(continued)

No.	Solute	Surfactant	K_M	$\log_{10} P(x)$	Reference
167	1-butanol	$C_{16}Br$	1.6	2.58	[102]
168	1-pentanol	$C_{16}Br$	4.4	3.00	[102]
169	1-hexanol	$C_{16}Br$	18.3	3.61	[102]
170	1-heptanol	$C_{16}Br$	39.2	3.94	[102]

NaDS: sodium dodecylsulfate; LiDS: lithium dodecylsulfate; LiFOS: lithium perfluorooctanoate; NaPFO: sodium perfluorooctanoate; KPFO: potassium perfluorooctanoate; C_7COOK: potassium octanoate; C_9COOK: potassium decanoate; $C_{11}COOK$: potassium dodecanoate; $C_{13}COOK$: potassium tetradecanoate; $C_{12}Br$: trimethyldodecylammonium bromide; $C_{14}Br$: trimethyltetradecylammonium bromide; $C_{16}Br$: trimethylhexadecylammonium bromide; $C_{14}Cl$: trimethyltetradecylammonium chloride; $C_{14}BzCl$: benzyldimethyltetradecylammonium chloride; EG: ethyleneglycol.

APPENDIX B

Effect of Surfactant Alkylchain Length on Neutral Solutes Partition Coefficients in Cationic Micellar Solutions as Deduced from CMC Determinations at 25°C (Eq. (8)): mole fraction basis[a]

Solute	$C_{12}Br$	$C_{14}Br$	$C_{16}Br$
1-propanol	1.72	1.75	1.77
2-propanol	1.68	1.69	1.68
1-butanol	2.23	2.26	2.29
2-butanol	2.00	2.03	2.05
2-methyl,2-propanol	1.87	1.91	1.90
butylurea	2.62	2.50	2.46
tert-butylurea	2.29	2.38	2.34
1-hexanol	3.10	3.18	3.18
cyclohexanol	2.63	2.65	2.60
caprolactam	1.88	1.92	1.95
1-octanol	3.71	3.87	3.89
1-decanol	4.64	4.80	5.04

[a] Partition coefficients are presented as $\log_{10} P$ as in Table 3. The β values for the C_{12}, C_{14}, and C_{16} derivatives are respectively equal to [14] 0.22, 0.19, 0.16. k_N was calculated using Eq. (9).
Source: Ref. 68

REFERENCES

1. D. Attwood and A. T. Florence, *Surfactant Systems*, Chapman and Hall, London, 1983.
2. E. E. Tucker and S. D. Christian, Faraday Symposium No. 17, The *Hydrophobic Interaction*, The Royal Society of Chemistry, 1982.

3a. S. D. Christian, E. E. Tucker, G. A. Smith, and D. S. Bushong, *J. Colloid Interface Sci. 113*: 439 (1986).
3b. G. A. Smith, S. D. Christian, E. E. Tucker, and J. F. Scamehorn, *J. Colloid Interface Sci. 113*: 254 (1988).
4. M. Fromon, A. K. Chattopadhyay, and C. Treiner, *J. Colloid Interface Sci. 102*: 14 (1984).
5. K. Hayase and S. Hayano, *Bull. Chem. Soc. Jpn. 50*: 83 (1977).
6. C. Treiner, J. F. Bocquet, and C. Pommier, *J. Phys. Chem. 90*: 3052 (1986).
7. C. Treiner, A. A. Khodja, and M. Fromon, *Langmuir 3*: 729 (1987).
8. C. M. Nguyen, J. F. Scamehorn, and S. D. Christian, *Colloids Surf. 30*: 355 (1988).
9. R. de Lisi, C. Genova, and V. Turco Liveri, *J. Colloid Interface Sci. 95*: 428 (1983).
10. S. Milioto, D. Romancino, and R. de Lisi, *J. Solution Chem. 16*: 943 (1987).
11. R. Bury, E. Souhalia, and C. Treiner, *J. Phys. Chem. 95*: 3824 (1991).
12. R. de Lisi, V. Turco Liveri, M. Castagnolo, and A. Inglese, *J. Solution Chem. 15*: 23 (1986).
13. M. Manabe, K. Shirahama, and M. Koda, *Bull. Chem. Soc. Jpn. 49*: 2904 (1976).
14. R. de Lisi, S. Milioto, and R. Triolo, *J. Solution Chem. 17*: 673 (1988).
15. S. D. Christian, G. A. Smith, E. E. Tucker, and J. F. Scamehorn, *Langmuir 1*: 564 (1985).
16a. P. Stilbs, and J. Colloid *Interface Sci. 87*: 385 (1982); (b) D. Marangoni and J. C. T. Kwak, *Langmuir 7*: 2083 (1991).
17. J. Carlfors and P. Stilbs, *J. Colloid Interface Sci. 104*: 489 (1985).
18. Z. Gao, R. E. Washlishen, and J. C. T. Kwak, *J. Phys. Chem., 93*: 2190 (1989).
19. E. B. Abuin and E. A. Lissi, *J. Colloid Interface Sci. 95*: 198 (1983).
20. A. Goto and F. Endo, *J. Colloid Interface Sci. 66*: 26 (1978).
21. S. J. Dougherty and J. C. Berg, *J. Colloid Interface Sci. 48*: 110 (1974).
22. S. Kaneshina, H. Kamaya, and I. Ueda, *J. Colloid Interface Sci. 83*: 589 (1981).
23. C. Treiner and A. K. Chattopadhyay, *J. Colloid Interface Sci. 98*: 447 (1984).
24. M. Iwatsuru and K. Shimizu, *Chem. Pharm. Bull. 34*: 3348 (1986).
25a. S. G. Bertolotti, N. A. Garcia, and H. E. Gsponer, *J. Colloid Interface Sci. 129*: 406 (1989).
25b. H. Kawamura, M. Manabe, T. Tokunoh, H. Saiki, and S. Tokunaga, *J. Solution Chem. 20*: 817 (1991).
26. C. Treiner, *J. Colloid Interface Sci. 90*: 444 (1982).
27. C. Treiner, *J. Colloid Interface Sci. 93*: 33 (1983).
28. J. E. Desnoyers, M. Billon, S. Leger, G. Perron, and J. P. Morel, *J. Solution Chem. 5*: 681 (1976).
29. C. Treiner, *Canad. J. Chem. 59*: 2518 (1981).
30. R. Aveyard and R. Heselden, *J. Chem. Soc. Faraday Trans. I 70*: 312 (1975).

31. R. Aveyard and R. Heselden, *J. Chem. Soc. Faraday Trans. I 69*: 1953 (1974).
32. A. Bondi, *J. Phys. Chem. 68*: 441 (1964).
33. M. Fromon and C. Treiner, *J. Chem. Soc. Faraday Trans I, 75*: 1837 (1979).
34. M. F. Emerson and A. Holtzer, *J. Phys. Chem. 71*: 3320 (1967).
35. A. Kay and G. Nemethy, *J. Phys. Chem. 75*: 809 (1971).
36. V. Perez-Villar, V. Mosquera, M. Garcia, C. Rey, and D. Attwood, *Colloid Polym. Sci. 268*: 965 (1990).
37a. S. Miyagishi, *Bull. Chem. Soc. Jpn. 47*: 2972 (1974).
37b. S. Miyagishi, *Bull. Chem. Soc. Jpn. 48*: 2349 (1975).
38. A. J. Jain and R. P. B. Singh, *J. Colloid Interface Sci. 81*: 536 (1981).
39. R. Zana, S. Yiv, C. Strazielle, and P. Lianos, *J. Colloid Interface Sci. 80*: 208 (1981).
40. W. K. Mathews, J. W. Larsen, and M. Pikal, *Tetrahedron Lett. 6*: 513 (1972).
41. D. G. Hall, *J. Chem. Soc. Faraday Trans II 68*: 1439 (1972).
42. D. G. Hall, in *Aggregation Processes in Solution*, Studies in Physical and Theoretical Chemistry No. 26 (E. Wyn Jones and J. Gormelly, eds.), Elsevier, 1983.
43. P. F. Mijnlieff, *J. Colloid Interface Sci. 33*: 255 (1970).
44. K. Motomura, M. Yamanaka, and M. Aratono, *J. Colloid Polym. Sci. 262*: 948 (1984).
45. K. Shirahama and T. Kashiwabara, *J. Colloid Interface Sci. 36*: 65 (1971).
46. K. Shinoda, T. Nagakawa, B. Tamamushi, and T. Isemura, in *Colloidal Surfactants* Academic Press, 1963.
47. K. Hayase and S. Hayano, *J. Colloid Interface Sci. 63*: 446 (1978).
48. M. Manabe, M. Koda, and K. Shirahama, *J. Colloid Interface Sci. 77*: 189 (1980).
49. C. Treiner, A. A. Khodja, and M. Fromon, *J. Colloid Interface Sci. 128*: 416 (1989).
50. M. Abu-Hamdiyyah and C. El-Danab, *J. Phys. Chem. 87*: 5443 (1983).
51. M. Abu-Hamdiyyah and A. I. Rahman, *J. Phys. Chem. 89*: 23377 (1985).
52. M. Abu-Hamdiyyah, *J. Phys. Chem. 90*, 1345 (1986).
53. M. Abu-Hamdiyyah and I. A. Rahman, *J. Phys. Chem. 91*: 1530 (1987).
54. M. Manabe, H. Kawamura, A. Yamashita, and S. Kokunaga, *J. Colloid Interface Sci. 115*: 147 (1987).
55. M. Manabe, H. Kawamura, A. Yamashita, and S. Tokunaga in *The Structure, Dynamics and Equilibrium Properties of Colloidal Systems* (D. M. Bloor and E. Wyn-Jones, eds.), Academic Press, New York, 1990.
56. M. Manabe, H. Kawamura, S. Kondo, M. Kojima, and S. Tokunaga, *Langmuir 6*: 1596 (1990).
57. G. I. Mukhayer, S. S. Davis, and E. Tomlinson, *J. Pharm. Sci. 64*: 147 (1975).
58. C. Treiner, A. Lebesnerais, and C. Micheletti, in *Thermodynamic Behavior of Electrolytes in Mixed Solvent Systems* II Adv. Chem. Ser. No. 177, (W. F. Furter, ed.), ACS, Washington DC, 1979.

59. Evans H. C., *J. Chem. Soc.* 579 (1956).
60. T. Yoshida, S. Kaneshina, H. Kamaya, and I. Ueda, *J. Colloid Interface Sci. 116*: 458 (1987).
61. N. Muller and T. W. Johnson, *J. Phys. Chem. 73*: 2042 (1969).
62. S. K. Kanungo and B. K. Sintra, *J. Ind. Chem. Soc. 61*: 964 (1984).
63. E. Tomlinson, D. Guveli, S. S. Davies, and J. B. Kayes, in *Solution Chemistry of Surfactants* (K. L. Mittal, ed.), Vol. 1, Plenum Press, New York, 1979 p. 355.
64. D. E. Guveli, S. S. Davies, and J. B. Kayes, *J. Colloid Interface Sci. 86*: 213 (1982).
65. G. Charbit, F. Dorion, and R. Gaboriaud, *J. Colloid Interface Sci. 106*: 265 (1965).
66. J. Georges and J. W. Chen, *J. Colloid Interface Sci. 113*: 143 (1986).
67. T. Sasaki, M. Hattori, J. Sasaki, and K. Nukina, *Bull. Chem. Soc. Jpn. 48*: 1397 (1975).
68. M. Abu-Hamdiyyah and K. Kumari, *J. Phys. Chem. 94*: 2518 (1990).
69. R. de Lisi, S. Milioto, and R. Triolo, *J. Solution Chem. 18*: 905 (1989).
70. P. Lianos and R. Zana, *J. Colloid Interface Sci. 101*, 587 (1984).
71. I. V. Rao and E. Ruckenstein, *J. Colloid Interface Sci. 113*: 375 (1986).
72. C. Treiner, A. A. Khodja, M. Fromon, and J. Chevalet, *J. Solution Chem. 18*: 217 (1989).
73. A. Derzhanski, S. Panayotova, G. Popov, and I. Bivas, in *Surfactants in Solution* (K. L. Mittal and P. Bothorel, eds.), Vol. IV, Plenum Press, New York, 1986, p. 333.
74. G. C. Kresheck in *Water, A Comprehensive Treatise* Vol. IV, Plenum Press, New York, 1975, p. 238.
75. K. Shinoda, *J. Phys. Chem. 58*, 1136 (1954).
76. R. de Lisi, S. Milioto, and R. E. Verrall, *J. Solution Chem. 19*: 665 (1990).
77. W. J. Harkins and H. Oppenheimer, *J. Am. Chem. Soc. 71*, 808 (1049).
78. C. Treiner and A. K. Chattopadhyay, *J. Colloid Interface Sci. 98*: 447 (1984).
79. C. Treiner and A. K. Chattopadhyay, *J. Colloid Interface Sci. 109*: 101 (1986).
80. C. Treiner and M. H. Mannebach, *J. Colloid Interface Sci. 118*: 244 (1987).
81. E. Vikingstad and J. Bakken, *J. Colloid Interface Sci. 74*: 8 (1980).
82. D. F. Evans, M. Allen, B. W. Ninham, and A. Fouda, *J. Solution Chem. 13*, 87 (1984).
83. J. T. Chou and P. C. Jurs, in *Physical Chemical Properties of Drugs* Med. Research Series, Vol. 10 (S. H. Yalkowski, A. A. Sinkula and S. C. Valvani, eds.), Marcel Dekker, New York, 1980.
84. N. Nishikido, Y. Moroi, H. Uehara, and R. Matuura, *Bull. Chem. Soc. Jpn. 47*: 2634 (1974).
85. P. Bhattacharya and I. N. Basumallick, *Ind. J. Chem. 26A*: 25 (1987).
86. C. Hansch and A. Leo, *Substituent constants for Correlation Analysis in Chemistry and Biology*, Pomona College, CA, 1979.
87. T. Fujita, T. Nishioka, and M. Nakajima, *J. Med. Chem. 20*: 1071 (1977).

88. C. Treiner, M. Nortz, and C. Vaution, *Langmuir 6*: 1211 (1990).
89. M. L. Corrin and W. D. Harkins, *J. Chem. Phys. 14*: 640 (1946).
90. H. N. Singh and S. Swarup, *Bull. Chem. Soc. Jpn. 51*: 1534 (1978).
91. D. E. Guveli, J. B. Kayes, and S. S. Davies, *J. Colloid Interface Sci. 72*: 130 (1979).
92. M. Milliaris and W. Binana-Limbele, *Progr. Coll. Polym. Sci. 84*: 83 (1991).
93. L. Magid, in *Surfactants in Solution*, Vol. 1, K. L. Mittal, ed., Plenum Press, New York 1977.
94. B. G. Sharma and A. K. Rakshit, in *Surfactants in Solution* (K. L. Mittal, ed.), Vol. VII, Plenum Press, New York, 1986.
95. I. Ionescu, L. S. Romanesco, and F. Nome, in *Surfactants in Solution* Vol. II, (K. L. Mittal and B. Lindman, ed.), Plenum Press, New York, 1982.
96. K. Shirahama and R. Matuura, *Bull. Chem. Soc. Jpn. 38*: 373 (1965).
97. C. Treiner and A. Lebesnerais, *J. Chem. Soc. Faraday Trans. I 73*: 44 (1977).
98. E. M. Wooley and T. E. Burchfield, *J. Phys. Chem. 89*: 714 (1985).
99. S. Vass, T. Torok, G. Jakli, and E. Berecz, *J. Phys. Chem. 93*: 6553 (1989).
100. K. Kayase, S. Hayano, and H. Tsubota, *J. Colloid Interface Sci. 101*: 336 (1984).
101. Y. Muto, K. Yoda, N. Yoshida, K. Esumi, K. Meguro, W. Binana-Limbele and R. Zana, *J. Colloid Interface Sci. 130*: 165 (1989).
102. Y. Miyashita and S. Hayano, *J. Colloid Interface Sci. 86*: 344 (1982).
103. C. Treiner, unpublished results.
104. H. Kawamura, M. Manabe, Y. Miyamoto, Y. Fujita, and S. Tokinawaga, *J. Phys. Chem. 93*: 5536 (1989).

13

Vapor Pressure Studies of Solubilization

EDWIN E. TUCKER Department of Chemistry and Biochemistry, University of Oklahoma, Norman, Oklahoma

SYNOPSIS

Classical types of physical measurements involving pressure and volume have provided substantial contributions to our understanding of solubilization. King (Chapter 2 in this volume) has provided a very good historical summary of pioneering efforts which have involved pressure or related activity measurements to determine the solubilization of volatile compounds in surfactant micelles. The focus of this chapter is on the use of precise vapor pressure measurements to determine the extent of solubilization of volatile nonpolar liquids in several types of surfactants, including mixed surfactants. An area of particular interest is the significant dependence of the solubilization constant upon change in the activity of the volatile solute in the surfactant solution.

I. SOLUTE VAPOR PRESSURE APPARATUS

A vast number of experiments have been directed toward determination of solubilization equilibrium constants for a variety of substances in surfactant micelles. Most investigations have been concerned with solubilization in situations where a pure substance is in contact with solutions of varying surfactant concentration. This particular technique is known as a maximum additive concentration (MAC) or a molar solubilization ratio (MSR) determination [1]. The MAC/MSR method results in a solubilization constant determined at unit solute activity. Much less emphasis has been placed on determination of solubilization constants at reduced solute activity. The assumption that solubilization constants determined at unit solute activity or at reduced solute activity (in the Henry's law region) are identical has been made frequently but rarely tested. For solutes which are relatively volatile, activity determination via vapor pressure measurements provides a means of precisely and accurately following the dependence of the solubilization constant upon solute activity.

Figure 1 gives a general outline of the major components of the vapor pressure apparatus developed at the University of Oklahoma [2, 3]. A central sample flask, of accurately known volume, is contained in a constant temperature water bath. The flask is composed of borosilicate glass with a glass to stainless steel transition joint (not shown) at the top for an access port and fitting attachments. Exterior lines to the pressure transducer and to the HPLC valve consist of small-diameter stainless steel tubing wrapped with a heating tape to keep these areas at a temperature (ca. 50°C) above the highest temperature at which experiments are usually performed to prevent vapor condensation. The low-sensor-volume-pressure transducer is also kept at constant temperature (50°C) above the sample bath temperature. Pressure gauges or transducers used in these experiments have been either of the quartz Bourdon tube kind (Texas Instruments, Inc. or Mensor Corp. products) or vibrating quartz crystal type (Paroscientific, Inc.).

The pressure gauge, the digital thermometer, the system vacuum valve, and the solenoid valves controlling operation of the HPLC valve are all interfaced to a small computer. The water bath temperature is controlled by a two-part system to within ±0.003°C over periods of 5–24 h. The

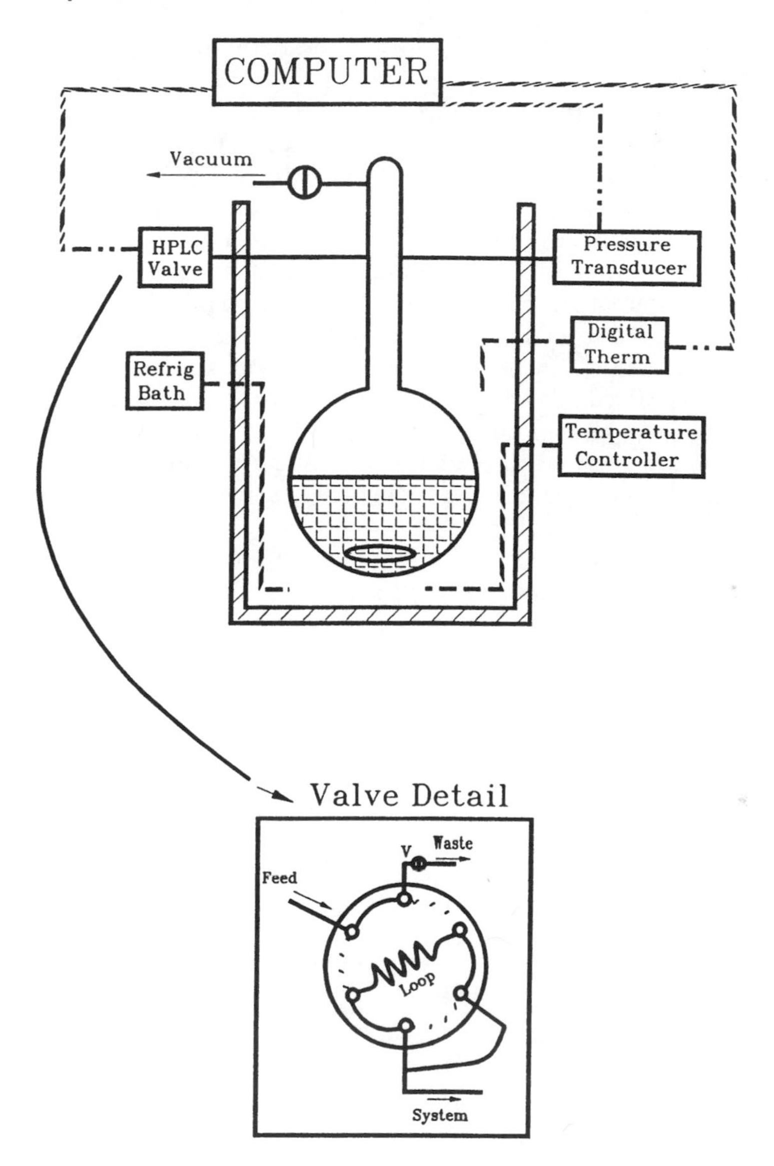

FIG. 1 Schematic of solute vapor pressure apparatus.

solution in the sample chamber is agitated by a stirring bar to hasten attainment of equilibrium. Pressure within the apparatus is measured with a resolution of ca. 0.001 torr in the range 0 to 1 atm.

The solute vapor pressure (SVP) apparatus relies upon the flash evaporation of pure solute liquid, in a constant temperature (50°C) external sample loop of an HPLC valve, into the evacuated sample chamber which contains water or an aqueous surfactant solution under its own vapor pressure. The boxed portion of Fig. 1 (labeled Valve Detail) gives a schematic of a six-port external sample loop HPLC valve in the inject position. The sample loop is filled from a reservoir of degassed pure solute which is at the valve temperature. In the fill position the loop is connected to the solute reservoir (Feed) and (through a stop valve) to an evacuated waste reservoir at room temperature. When the stop valve (V) is opened by remote control, liquid flows through the loop aided by a driving force which is due to the reservoir being elevated in both height and temperature with respect to the waste vessel.

The HPLC valve sample loop is calibrated by doing a blank experiment, i.e., delivering successive loop volumes of solute into the totally evacuated apparatus sample chamber. Previous experiments have shown that successive increments of liquid benzene (ca. 22 μL) can be delivered with a precision and reproducibility of about 1 part in 6000 [4]. Table 1 gives sample data for a calibration with three replicate injection sequences.

A typical experiment is begun by delivering a weighed quantity (volume of 50 to 200 mL) of pure water or surfactant solution into the dry sample

TABLE 1 Total Pressure of Benzene Vapor Recorded in Three Series of Liquid Benzene Additions from HPLC Valve

	Pressure (torr)		
Increment #	Series 1	Series 2	Series 3
1	6.500	6.499	6.497
2	12.990	12.990	12.991
3	19.480	19.479	19.482
4	25.954	25.956	25.957
5	32.417	32.419	32.418
6	38.898	38.901	38.903
7	45.361	45.365	45.365
8	51.800	51.808	51.808
9	58.230	58.239	58.240
10	64.648	64.664	64.661

chamber through an access port. The port is then sealed and the solution is degassed while stirring until the pressure in the apparatus is essentially that of the solution only. Water vaporized during this procedure is caught in a preweighed trap at liquid nitrogen temperature and a correction is made to the total solution quantity. A small computer monitors the system pressure and temperature. When these are at equilibrium (within 0.003 torr and 0.002°C, respectively), both ends of the sample loop are opened to the system and the liquid vaporizes into the system. The computer monitors the total system pressure over time as part of the added vapor dissolves in the liquid. At equilibrium, which is reached within 20–45 min, pressure and temperature values are recorded and another sample is subsequently injected until the total system pressure is approximately that of the pure solute liquid being added.

The overall result of the SVP experiment is the delivery of a known (reproducible) quantity of volatile liquid into a sample vessel of known volume containing a known quantity of liquid (either pure water or aqueous surfactant solution). At equilibrium, the vapor pressure above the sample liquid is composed of the partial pressures of the volatile solute and the solvent water. In most experiments described here, the partial pressure of the solvent water is essentially fixed because very low concentrations of added solute are not sufficient to have a significant Raoult's law effect, i.e., the mole fraction of solvent water changes only very slightly. The partial pressure of the volatile solute is readily determined at any point by subtracting the initial pressure (for either water or surfactant solution) from the measured total equilibrium pressure. Since the required volumes and masses of components are known, the SVP experiment can be used to provide very accurate data for the activity of a volatile component above a solution of known composition.

II. VAPOR PRESSURE DATA TREATMENT

In order to utilize the SVP approach to accurately determine the extent of solubilization of a volatile solute by surfactant solutions, it is first necessary to obtain reliable data on the concentration of the solute in pure water as a function of solute activity. We define a dimensionless partition coefficient (K_d) for the transfer of solute from the vapor phase to aqueous solution as

$$K_d = \frac{C_S}{C_V}$$

where the subscripts S and V represent solution and vapor phases, respectively, and C is in molarity units. In the absence of measurable deviations from ideality of the solute in either vapor or solution phase, the solute vapor phase concentration is simply P/RT, where P is the partial pressure of the solute, R is the ideal gas constant in appropriate units, T is temperature in K, and the solution concentration is in moles of solute per liter. The activity of the solute is P/P^0, where P^0 is the vapor pressure of pure solute at the given temperature. Although the present discussion assumes, for simplicity, that the volatile solute is ideal in both vapor and aqueous solution, deviations from ideality on the order of 2–3% have previously been accurately measured and accounted for in solute vapor pressure work [4]. Such corrections have been employed where necessary in all the experimental data to be discussed in this chapter.

Figure 2 gives a comparison of the activity of benzene vapor above solutions of benzene in pure water and benzene in aqueous (0.1 M) sodium dodecylsulfate solution, respectively, as functions of benzene concentration in the two solutions at 25°C. The data plotted for the benzene-water solution nearly follow a straight line, showing that the distribution coefficient changes very little as a function of the benzene activity in water. Data for the benzene-aqueous SDS solution show continuous curvature, implying that the partition coefficient for benzene from vapor into aqueous SDS solution is increasing substantially with increasing benzene activity. If it is assumed that the benzene (in the bulk aqueous solution) which is not associated with the micellar pseudophase behaves ideally, then the horizontal distance between the two curves is a direct measure of the solubilized benzene. The total benzene concentration in the aqueous surfactant solution can then be represented as the sum of benzene in bulk aqueous solution and benzene solubilized in the micellar pseudophase:

$$[C_6H_6](\text{tot}) = [C_6H_6](H_2O) + [C_6H_6](\text{mic})$$

An equilibrium partition or solubilization constant for the transfer of solute from bulk water to micellar pseudophase might be represented in a number of ways which are discussed in Chapter 1 of this volume. A convenient expression used in this laboratory is

$$K_m = \frac{X_{mic}}{C_S}$$

where X_{mic} is the mole fraction of organic solute in the micelle and C_S is the molar concentration of organic solute in the bulk aqueous solution. The mole fraction of the solute in the micelle can be calculated from the moles of solute solubilized and the moles of surfactant existing in micellar form. This latter quantity requires a knowledge of the critical micelle concentration of the surfactant, which is usually small in relation to the total surfactant concentration. In the case of ionic surfactants, a minor

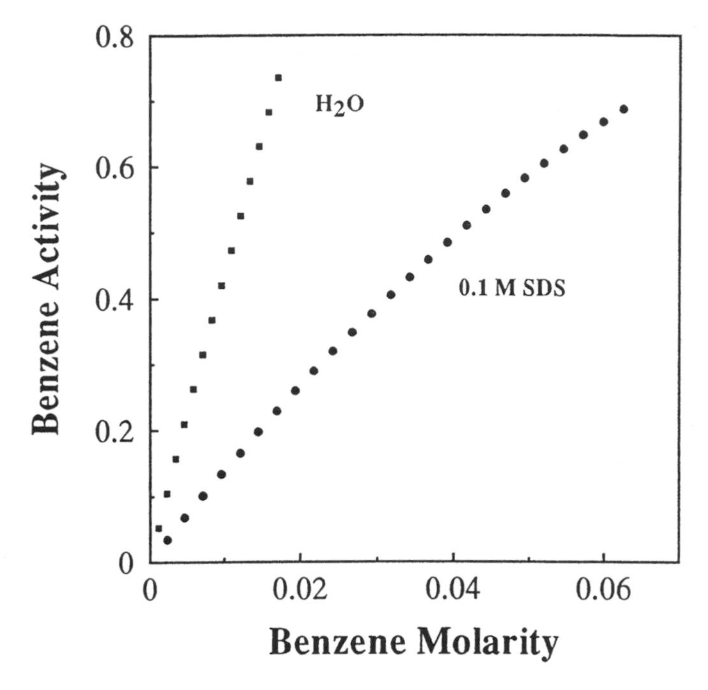

FIG. 2 Benzene activity vs. concentration for solutions in pure water and in 0.1 M sodium dodecylsulfate.

correction to C_S may be required due to the "salting out" of the organic solute. That is, the effective Henry's law constant will increase for the volatile solute in the presence of aqueous ions relative to the situation for pure water.

Since the SVP technique provides a direct measure of the activity of the volatile solute, it is a simple matter to calculate the activity coefficient of the solute in the micellar environment. The activity coefficient of the solute is expressed by

$$\gamma_{\text{mic}} = \frac{f_A}{[f_A^0 X_{\text{mic}}]}$$

where f_A is the fugacity (very nearly equal to the partial pressure) of the solute and f_A^0 is the fugacity of the pure solute in its standard state. The solute activity coefficient in the micellar environment provides a direct characterization of the energetics of transfer of a solute molecule from its standard state into the micelle. A solute activity coefficient less than unity in the micelle reflects a negative free energy of transfer from the pure solute standard state, and, conversely, an activity coefficient greater than unity reflects a positive free energy of transfer of solute from its standard state into the micelle. In the limit as the activity of the volatile solute approaches unity, it can be seen that there is a reciprocal relationship between the solubilization constant and the activity coefficient of the solute in the micelle:

$$K_m = \frac{1}{[\gamma_{mic} C_S^0]}$$

where C_S^0 is the solubility of the solute in pure water at saturation or unit solute activity.

III. SOLUBILIZATION OF VOLATILE AROMATICS AND HALOCARBONS

Several SVP studies in this laboratory have focused on the solubilization of benzene and related volatile aromatics in a variety of aqueous micellar systems. One early and detailed study was that of the system benzene-sodium octylsulfate [5]. Solubilization data were obtained at several temperatures as well as at surfactant concentrations both below and above the critical micelle concentration (CMC). Since the CMC for octylsulfate is rather high (ca. 0.13 M), this system offered an opportunity to investigate the interaction of benzene with the surfactant monomer and premicellar aggregates as well as the solubilization of benzene by the micelle. Each individual experiment at a given surfactant concentration in water provides a curve similar to that in Fig. 2. It is possible to fit such data to an arbitrary power series to very high precision and obtain empirical parameters to be used for interpolating purposes. Figure 3 shows how many such sets of data may be utilized to obtain unique information about the progress of solubilization as a function of solute activity in solutions of varying surfactant concentration.

Each smooth curve in Fig. 3 is drawn through interpolated points at a constant aqueous benzene monomer concentration (or activity). The smooth curves represent the increase in total benzene concentration in solution, at fixed benzene activity, with increase in SOS molarity. The CMC would be represented by the intersection of the straight-line segments of a particular curve. Small amounts of benzene are solubilized or complexed below the CMC and a much larger effect is evident above the CMC. The CMC of sodium octylsulfate appears to decrease with increasing benzene activity in solution.

Several hundred data points for benzene in sodium octylsulfate solutions of varying concentration and at temperatures in the range 15–45°C

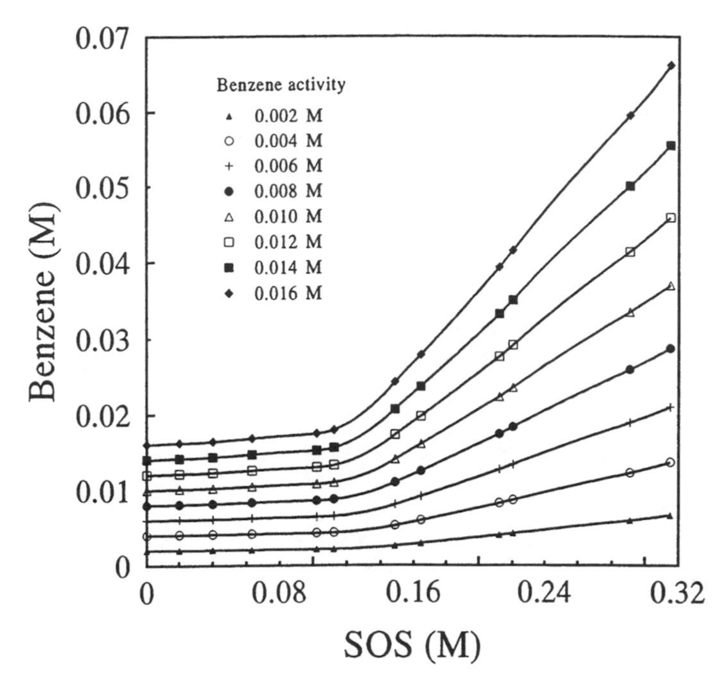

FIG. 3 Benzene aqueous solution concentrations at fixed benzene activities as a function of sodium octylsulfate concentrations at 25°C.

were obtained. Data from 25°C solubilization measurements will be compared with comparable data for other surfactant solutions. Solubilization constants from SVP measurements for benzene in several different surfactant solutions at 25°C are plotted in Fig. 4 as a function of benzene mole fraction in the micelle. The surfactants include two anionics (sodium octylsulfate and sodium dodecylsulfate [6, 7]), an amphoteric (dimethyldodecylamine oxide or DDAO) [8], a cationic (cetylpyridinium chloride or CPC), and a nonionic nonylphenol with an average of 10 ethylene oxide groups attached (NPEO10). The nonionic surfactant was a GAF product with the trade name Igepal CO660 [6]. The benzene solubilization constants vary by about a factor of 4 throughout the range of surfactants studied. At least for the ionic surfactants, this variation is primarily a function of the length of the hydrocarbon tail of the surfactant. As might be expected, *K* is smallest for the 8-carbon SOS and almost doubles when the 12-carbon SDS or DDAO are used.

The solubilization constant further increases by almost a factor of 2 (relative to that in SDS) when the cationic surfactant CPC with a 16-

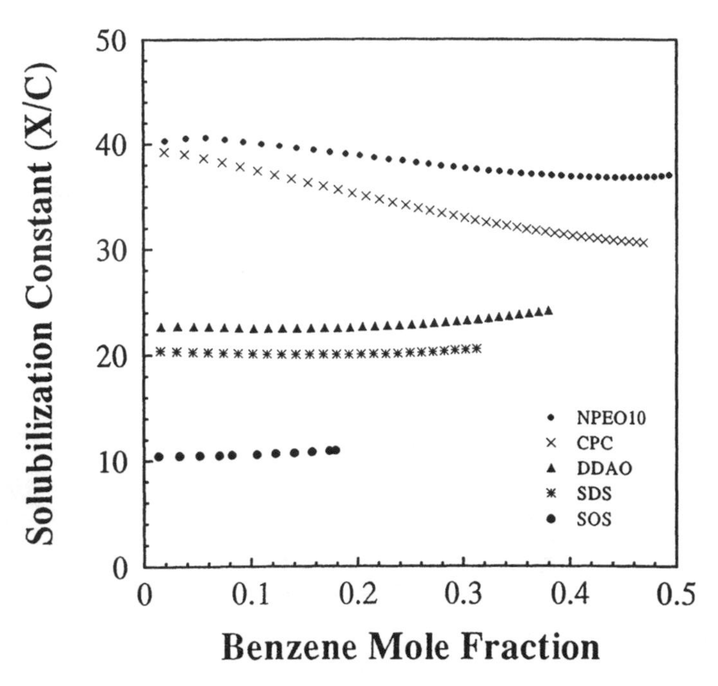

FIG. 4 Benzene solubilization constant vs. micellar mole fraction of benzene in five different surfactant solutions at 25°C.

carbon chain is employed. Previous literature reports, summarized in an earlier review [9], have suggested that there may be a strong enough interaction between aromatic solubilizates and cationic headgroups to localize the aromatic molecule in the headgroup vicinity of the micelle. Figure 5 gives activity coefficients for benzene as a function of its mole fraction in the individual surfactant micelles mentioned above. All of the limiting activity coefficients for benzene solubilization are larger than unity, which suggests that benzene is in a less energetically favorable environment in any of the micelles relative to the environment of a benzene molecule in pure benzene. For comparison, utilizing known properties of the benzene-water system [4] and a limiting distribution coefficient for benzene between water and *n*-hexadecane [10], the limiting activity coefficient of benzene in *n*-hexadecane can be calculated as 0.95. Thus, the activity coefficient information provided by the SVP method does not suggest the presence of a strong specific interaction of benzene with any of the five micellar systems since the limiting activity coefficient values for benzene are larger than unity in all the micellar systems in Fig. 5.

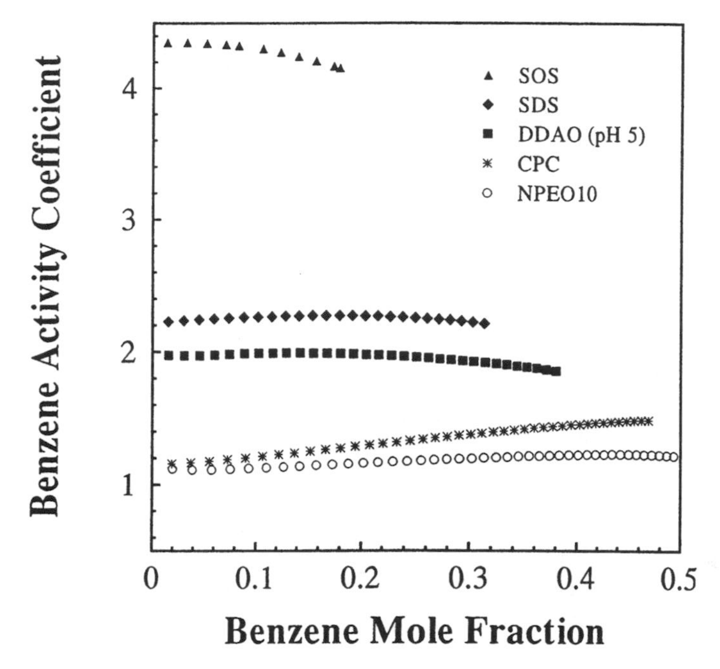

FIG. 5 Benzene micellar activity coefficient vs. micellar mole fraction in five different surfactant solutions at 25°C.

SVP measurements have provided information concerning the effect of substituent groups on solubilization of volatile aromatics in CPC micelles. Figures 6 and 7 show the solubilization constant and activity dependence on the mole fraction of benzene, fluorobenzene, toluene, and chlorobenzene in 0.1 M aqueous CPC at 25°C [6, 7]. The benzene and fluorobenzene data were obtained with the automated SVP apparatus, while the toluene and chlorobenzene data were obtained through use of a manually operated vapor pressure apparatus. The solubilization constants for these four aromatics in CPC increase as the water solubility of the aromatic molecule decreases in the series from benzene to chlorobenzene. There is a slight dependence of the solubilization constant upon solute mole fraction in the micelle with the K_m decreasing (and the activity coefficient increasing) throughout most of the range.

The limiting activity coefficients shown in Fig. 7 (at low aromatic solute mole fraction in the micelle) are all larger than unity, showing that the excess free energy of transfer from the pure liquid reference state into

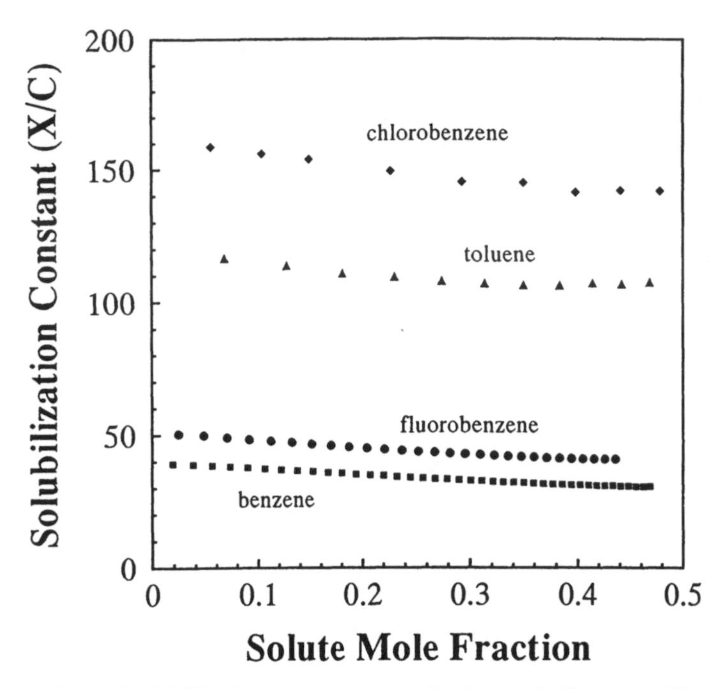

FIG. 6 Solubilization constant vs. micellar mole fraction of benzene and substituted benzenes in 0.1 M *n*-hexadecylpyridinium chloride solution at 25°C.

dilute solution in the micelle is positive. The limiting activity coefficient for toluene in CPC micelles from SVP data appears to be ca. 1.5 from Fig. 7. It is of interest to compare this value with activity coefficients derived from gas chromatography measurements for toluene in several pure hydrocarbon solvents. The limiting activity coefficients for toluene in *n*-heptane, *n*-hexadecane, and squalane are 1.51, 0.96, and 0.69, respectively [11, 12].

There is substantial information in curves of solubilization behavior versus solute mole fraction in the micelle as typified by Figs. 6 and 7. However, it is difficult to attach much quantitative meaning to such curves because of the lack of detailed information concerning the structural transformations occurring in the micellar solution. As the solute activity increases, the number of monomer units in an average micelle is changing and the micelle may even be changing from a roughly spherical structure to a rodlike one. In the limit of unit solute activity, we know that the product of the activity coefficient and the solute mole fraction will become unity. Since the solute mole fraction in the micelle will always be less

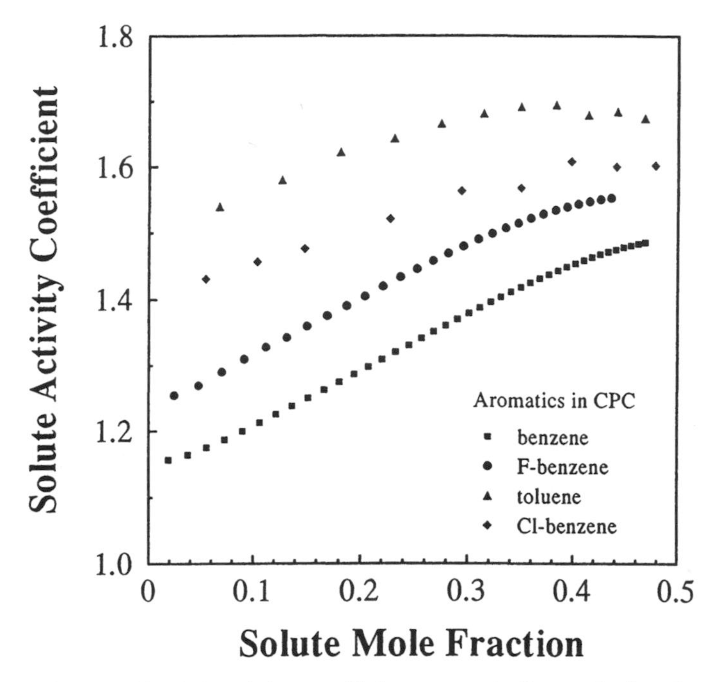

FIG. 7 Micellar activity coefficient vs. micellar mole fraction of benzene and substituted benzenes in 0.1 M *n*-hexadecylpyridinium chloride solution at 25°C.

than unity at unit solute activity, the activity coefficient of the solute will be larger than one at that point.

Because chlorinated hydrocarbons are common pollutants of soil and groundwater, some interest has arisen in the possibilities of enhanced removal of halocarbons by surfactant solubilization [13]. Limited data are available in the literature concerning the solubilization behavior of common volatile halocarbon contaminants [14, 15]. It has generally been assumed that the solubilization constant is independent of solute activity. That is, that the same solubilization constant would be obtained at either low solute activity or at unit solute activity. Figure 8 shows the dependence of the solubilization constant upon mole fraction in the micelle for trichloroethylene (TCE) solubilized by cationic (CPC) and nonionic (DNPEO18) surfactants at 30°C [16, 17]. The nonionic is a dinonylphenol-based surfactant with an average of 18 ethylene oxide units attached. Over the mole fraction range studied, the solubilization constants for TCE decrease more than 25% from the limiting values at low solute activity. Dependence of the activity coefficients upon micellar mole fraction for

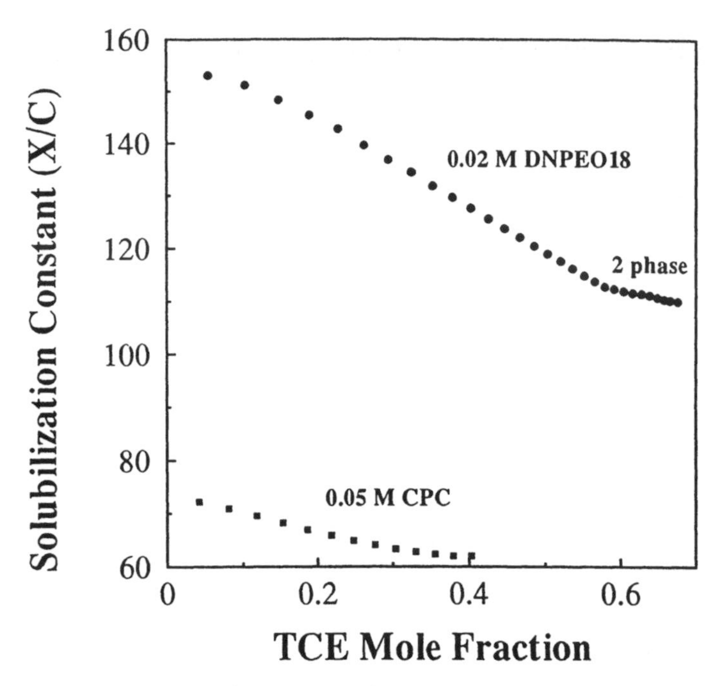

FIG. 8 Solubilization constant for trichloroethylene vs. its micellar mole fraction in two surfactants at 30°C.

TCE in the two solutions is shown in Fig. 9. The limiting activity coefficient for TCE in CPC at low TCE activity is ca. 1.4 at 30°C while $\gamma_{TCE}(\text{mic})$ is ca. 0.65 in the limit for the large nonionic DNPEO18. An interesting type of behavior is observed for the TCE-DNPEO18 curve in Fig. 9 (and in Fig. 8 also). The break in the curve at ca. 0.6 mole fraction TCE in the nonionic micelle coincides with phase separation in the aqueous surfactant solution. The aqueous solution becomes milky at this composition. Since the activity of TCE over the solution is substantially less than unity, it is assumed that the separating phase is not a pure solute phase but is a phase enriched in both TCE and nonionic surfactant.

IV. SOLUBILIZATION OF CYCLOHEXANE AND HEXANE

Figures 10 and 11 show the solubilization constants and activity coefficients, respectively, for cyclohexane in three different 0.1 M surfactant

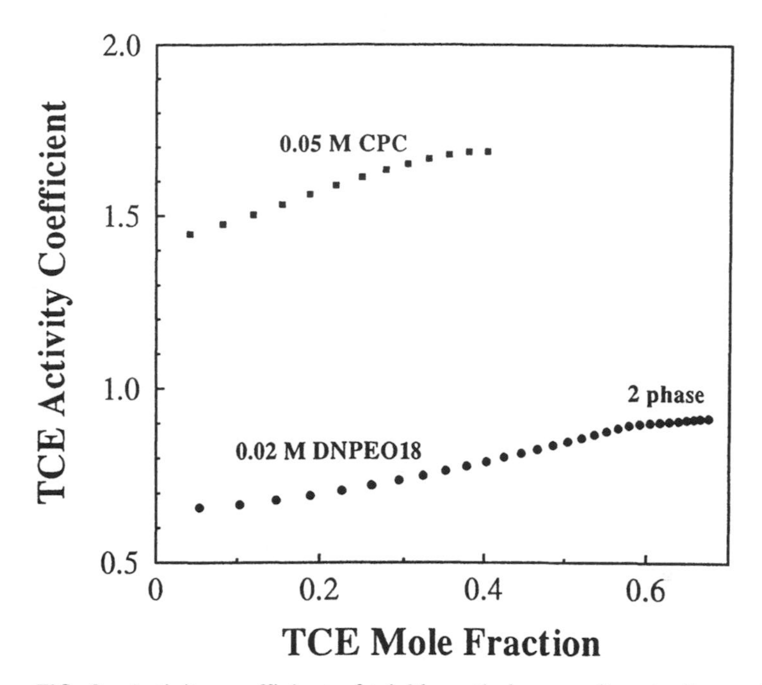

FIG. 9 Activity coefficient of trichloroethylene vs. its micellar mole fraction in two surfactants at 30°C.

solutions at 25°C from SVP measurements [6, 18]. In contrast with data for benzene solubilization, these data demonstrate a substantial increase in K_m (and decrease in γ_{mic}) with increasing cyclohexane mole fraction in the micelle. The limiting activity coefficients for cyclohexane in all surfactants are appreciably larger than unity. The Laplace pressure effect is probably most prominent among the various reasons which have been advanced for these large activity coefficients (Chapter 2, this volume). One possible approach toward modifying the Laplace effect is to repeat the experiment with a substantially larger surfactant concentration in an attempt to shift the micelle configuration toward a rodlike shape. Figure 12 shows the activity coefficient of cyclohexane at 25°C as a function of mole fraction in CPC micellar solutions of 0.1 and 0.5 M concentrations, respectively. The activity coefficient does decrease by ca. 10% in the 0.5 M CPC solution. However, it may be noted that the limiting activity coefficient of cyclohexane in a pure hydrocarbon solvent like *n*-hexadecane is approximately 0.95 [19]. Thus the transfer of cyclohexane from

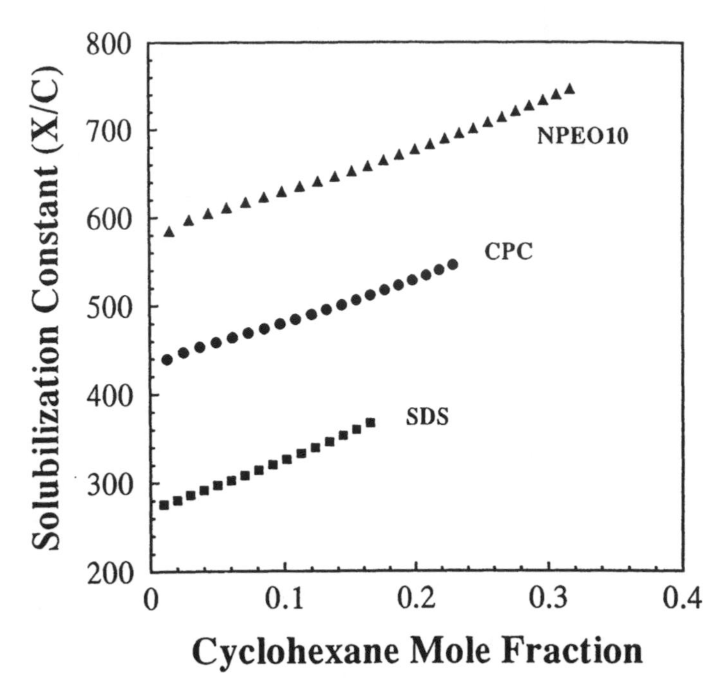

FIG. 10 Solubilization constant for cyclohexane as a function of its micellar mole fraction in three different surfactants (0.1 M solution) at 25°C.

its pure state into the 0.5 M CPC micellar solution still occurs with a positive excess free energy relative to the transfer of cyclohexane into dilute solution in the hydrocarbon solvent.

SVP solubilization data have also been obtained for hexane in several different surfactant solutions over a range of temperatures. Figure 13 displays the concentration dependence of solubilization constants for hexane in three surfactant solutions, while Fig. 14 shows the activity coefficient dependence for hexane in the surfactant solutions as a function of hexane mole fraction in the micelles. Both the solubilization constants and the activity coefficients for hexane in these surfactants are substantially larger than those for cyclohexane, which might be anticipated from the smaller water solubility of hexane relative to that of cyclohexane. Possible changes in Laplace pressure effects upon the solubilization of hexane have been investigated by obtaining data for a high concentration (ca. 0.6 M) of CPC at 25°C. At this concentration, the configuration of the micelles would be expected to be more nearly rodlike. Figure 15 compares the

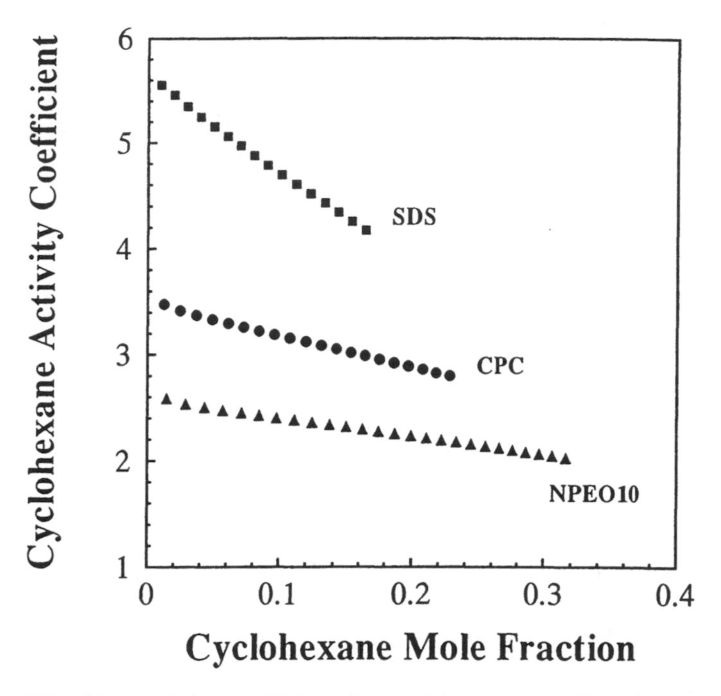

FIG. 11 Activity coefficient for cyclohexane as a function of its micellar mole fraction in three different surfactants (0.1 M solution) at 25°C.

solubilization data for hexane in 0.1 and 0.6 M aqueous solutions of CPC. The limiting activity coefficient of hexane in the surfactant solution is reduced by 10–15% in the 0.6 M solution of CPC. The limiting activity coefficient of hexane in 0.6 M aqueous CPC is still ca. 4.1, while the activity coefficient of hexane in dilute solution in n-hexadecane is ca. 0.92 [20].

V. SOLUBILIZATION IN MIXED MICELLES

Application of the SVP method to study of solubilization of volatile compounds in mixed micelles is straightforward and leads to new types of information. One interesting example is the solubilization of benzene in solutions of the amphoteric surfactant dimethyldodecylamine oxide (DDAO) [8]. By modifying the pH of aqueous DDAO solutions from slightly basic (pH = 8) to acidic (pH = 2), DDAO changes from a nonionic surfactant to a cationic surfactant. At an intermediate pH of 5, DDAO

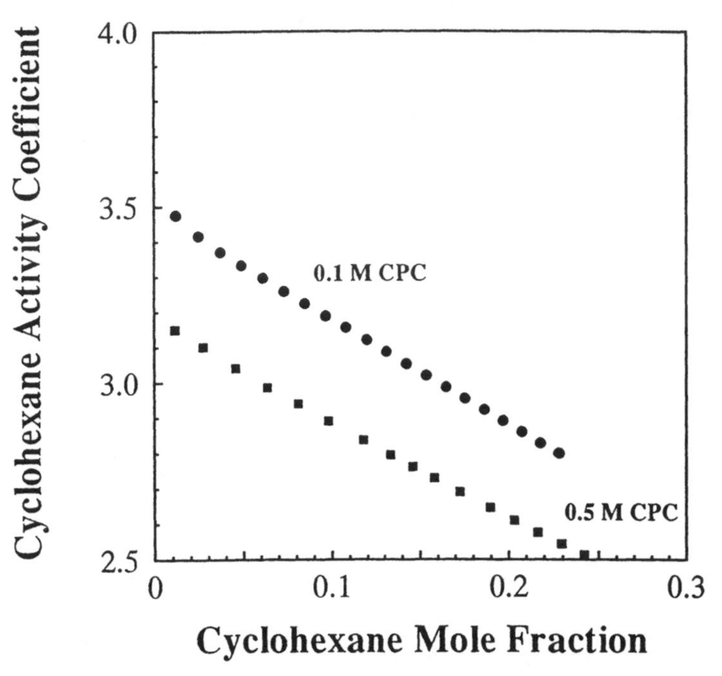

FIG. 12 Cyclohexane activity coefficient vs. its micellar mole fraction at two different *n*-hexadecylpyridinium chloride solutions at 25°C.

micelles are expected to be composed of nearly a 50:50 mixture of nonionic and cationic monomers. Figure 16 shows the solubilization constant of benzene in DDAO micelles at 35°C at three pH values as a function of benzene mole fraction. Although the three curves approach the same solubilization constant at higher benzene mole fraction, there are interesting differences in the curves at low benzene mole fraction. The solubilization constant is somewhat larger for the totally cationic micelle at pH = 2 than for the micellar system at the other pH values. One replicate data set for the pH = 2 condition is shown, which illustrates the good reproducibility of the SVP measurement. The solubilization constant for the mixed cationic/nonionic micelle at pH = 5 is slightly smaller than that for the nonionic micelle at pH = 8. Extensive results for the benzene-DDAO system at other temperatures (and in the presence of 0.06 M NaCl) are quite similar to these curves.

The solubilization constant for benzene in DDAO micelles shows subtle differences as the micellar structure is affected by pH. There is some evidence that the mixed DDAO micelle (cationic-nonionic) is larger in

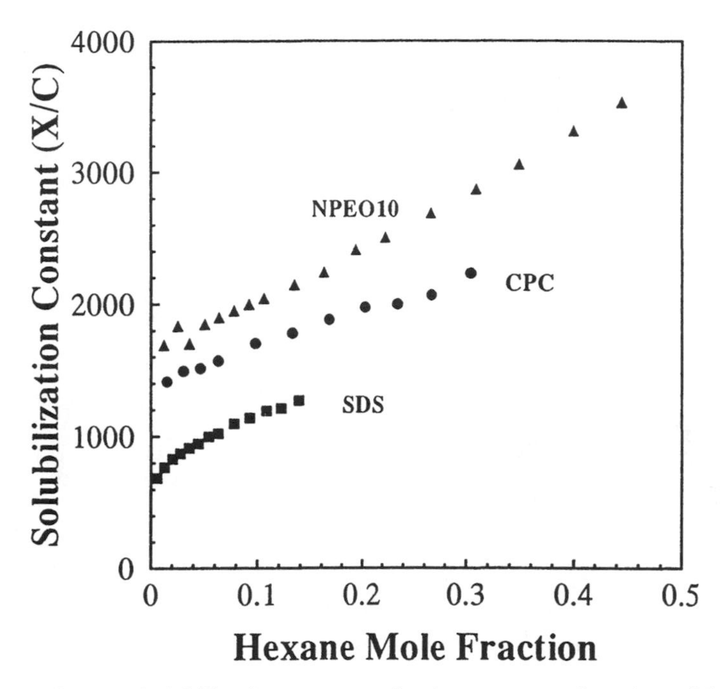

FIG. 13 Solubilization constant for hexane as a function of its micellar mole fraction in three different surfactants (0.1 M solution) at 25°C.

size than either the pure cationic micelle at pH 2 or the pure nonionic at pH 8 [21, 22]. Even though the mixed micelle is larger, the solubilization constant for benzene is smaller for this micellar system. Another example of the solubilization of benzene in mixed cationic-nonionic micelles is given in Fig. 17. The limiting solubilization constant of benzene in the mixed micelle is less than that in either the pure cationic CPC or the nonionic NPEO10 solutions [6]. This is similar to the behavior of benzene in the DDAO micelles.

Figure 18 shows the dependence of the solubilization constant for cyclohexane in the pure components and the mixture of cationic CPC and nonionic NPEO10. The cyclohexane solubilization constant in the mixed micelle has a concentration dependence which is somewhat different from that for benzene in the cationic/nonionic mixed micelle. The cyclohexane K_m shows more nearly ideal behavior in that it has values between those for the pure cationic or nonionic micelles, while the benzene solubilization constant is less in the mixed micelle than in either pure micelle over most of the available mole fraction range.

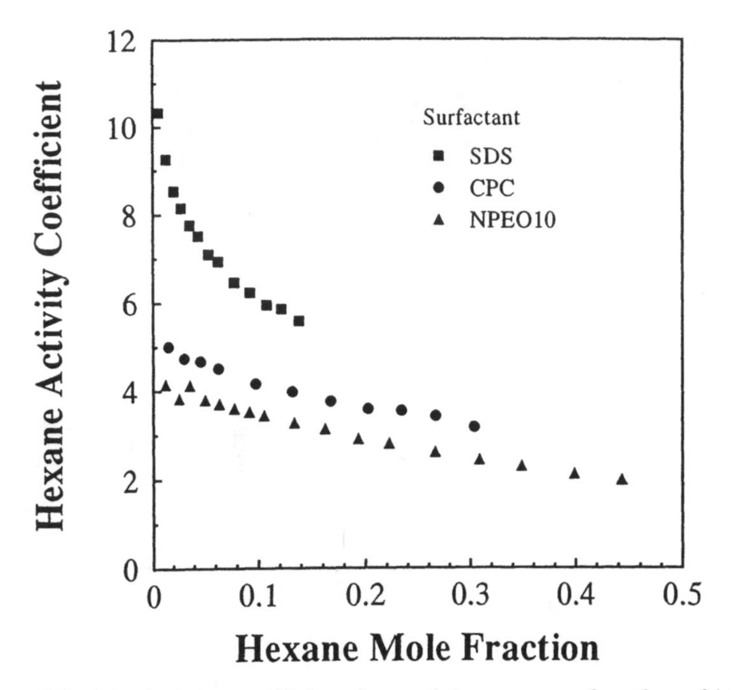

FIG. 14 Activity coefficient for cyclohexane as a function of its micellar mole fraction in three different surfactants (0.1 M solution) at 25°C.

Additional information is available for other mixed micelle systems [6]. For example, the solubilization of benzene in 50:50 mixed anionic-nonionic micelles such as SDS-NPEO10 shows intermediate concentration dependence comparable to the cyclohexane-CPC-NPEO10 system above. Hexane solubilized in the 50:50 SDS-NPEO10 mixture has a concentration dependence of the solubilization constant little different from that in the NPEO10 alone.

The SVP method is capable of providing unique and detailed information concerning the effect of structural changes in mixed micelles upon the solubilization constant. However, this information cannot be fully utilized without some specific assessment of the actual structural characteristics of the mixed micelles.

VI. THERMODYNAMICS OF SOLUBILIZATION

Solubilization of hydrocarbons in aqueous micelles has generally been viewed qualitatively as engendering a partial reversal of the unusual ef-

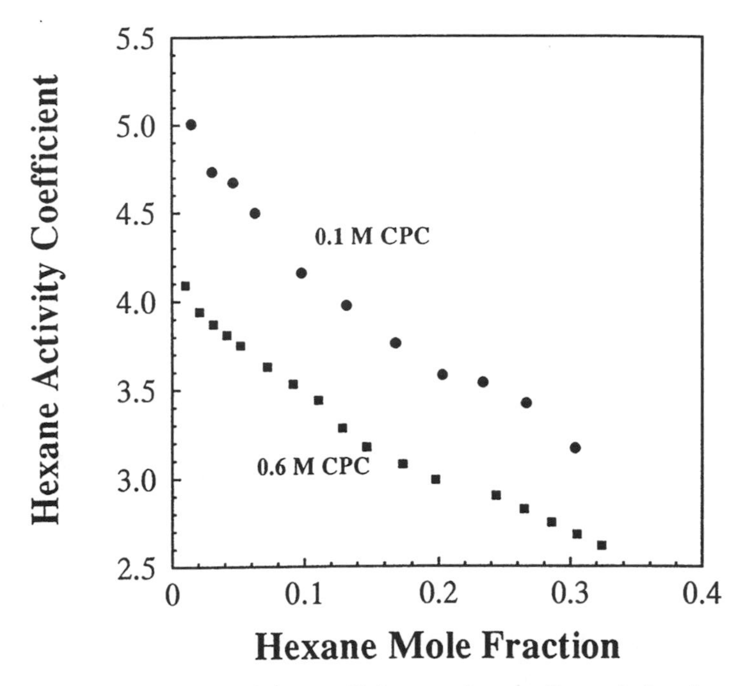

FIG. 15 Hexane activity coefficient vs. its micellar mole fraction at two different *n*-hexadecylpyridinium chloride solutions at 25°C.

fects evident when a hydrocarbon is transferred into an aqueous environment. Infinite dilution activity coefficients for hydrocarbons in water are quite large. For example, the limiting activity coefficient for benzene in water is approximately 2500 at 25°C [4, 23]. The limiting activity coefficient for the chlorinated hydrocarbon trichloroethylene in water [17] is ca. 5600 at 30°C while those for cyclohexane or hexane in water are substantially larger. These extremely large activity coefficients for hydrocarbons in water may be contrasted with the much smaller activity coefficients determined by the SVP method for saturated and unsaturated hydrocarbons in aqueous micelles.

The solubility of hydrocarbons and substituted hydrocarbons in water usually shows a minimum with temperature because of the large changes in the heat capacities of these molecules when dissolved in water. The heat of solution of a liquid hydrocarbon in liquid water is zero at the temperature of minimum solubility. Examples of the temperature of minimum solubility in water are 16, 24, and 31°C for benzene, cyclohexane, and trichloroethylene, respectively [4, 17].

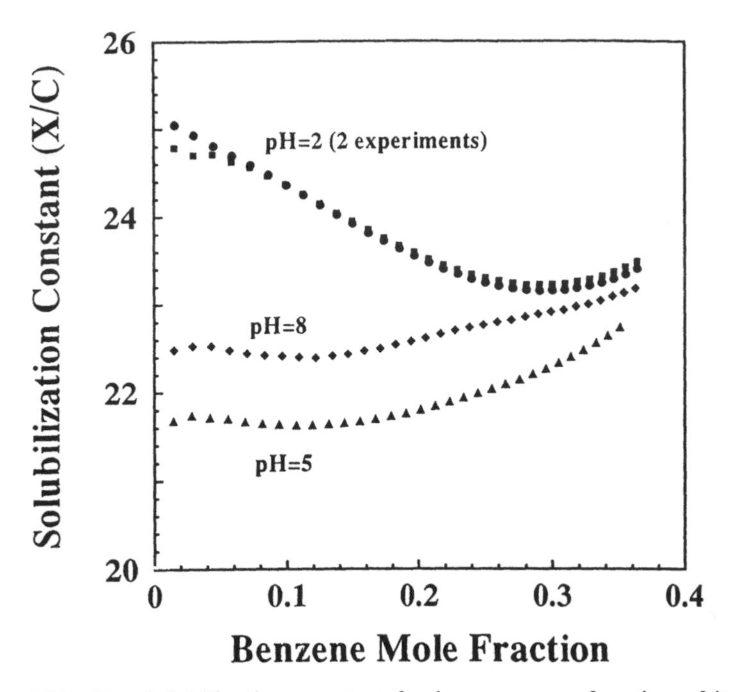

FIG. 16 Solubilization constant for benzene as a function of its micellar mole fraction in 0.1 M *N*,*N*-dimethyldodecylamineoxide solutions at different pH values at 35°C.

The behavior of micelles and the solubilization of hydrocarbon solutes in aqueous micellar solution also reflect temperature-dependent phenomena related to what might be called hydrophobic solvation. For example, the CMC of sodium octylsulfate reaches a minimum in water at 27°C [24]. Careful studies of the temperature dependence of solubilization constants for solutes such as benzene and trichloroethylene in surfactant micelles reveal small maxima in solubilization constants versus temperature. The solubilization constant for benzene in sodium octylsulfate micelles reaches a maximum at very nearly the temperature at which the CMC of sodium octylsulfate is at a minimum. Because of the large heat capacity changes in the solubilization reaction, the enthalpy of solubilization of benzene in SOS is slightly endothermic (+1.2 kJ/mole) at 25°C, near zero at 27°C, and then becomes exothermic above 27°C [5]. It is important to note that this enthalpy is based on the standard state of benzene being that of the infinitely dilute solution in water. Substantially less temperature dependence will be evident in the solubilization enthalpy when it is based on a

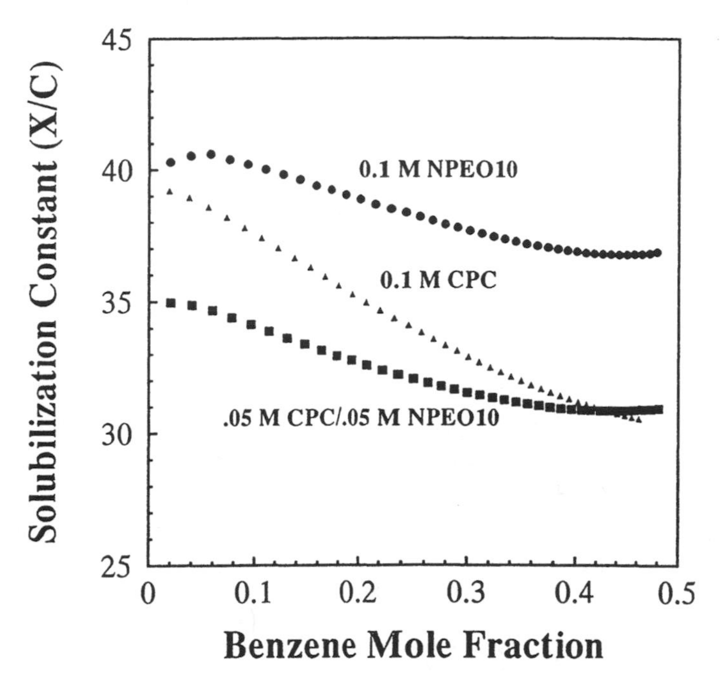

FIG. 17 Solubilization constant for benzene as a function of its micellar mole fraction in individual and mixed cationic-nonionic surfactant solutions at 25°C.

standard state of benzene in its pure liquid. For example, the enthalpy of transfer of benzene from liquid benzene into water at infinite dilution is +2.3 kJ/mole at 25°C [4]. With pure liquid benzene as the reference state, the solubilization enthalpy for benzene in aqueous sodium octylsulfate micelles is +3.5 kJ/mole at 25°C. The temperature dependence of the solubilization enthalpy based on the pure liquid solute as the reference state will be substantially reduced due to the cancellation of a large portion of the water-based heat capacity effects.

All of the detailed vapor pressure studies of the solubilization of benzene, cyclohexane, and trichloroethylene in anionic, cationic, and nonionic micelles support the result that the enthalpy for solubilization of these molecules is endothermic at 25°C [5, 6, 16, 18]. These slight increases in enthalpy and excess free energy (from the activity coefficient measurements) which occur in the solubilization process suggest the extent to which surfactant micelles in dilute solution are somewhat less effective as solvents compared with the pure solute liquids. Additionally, the fact

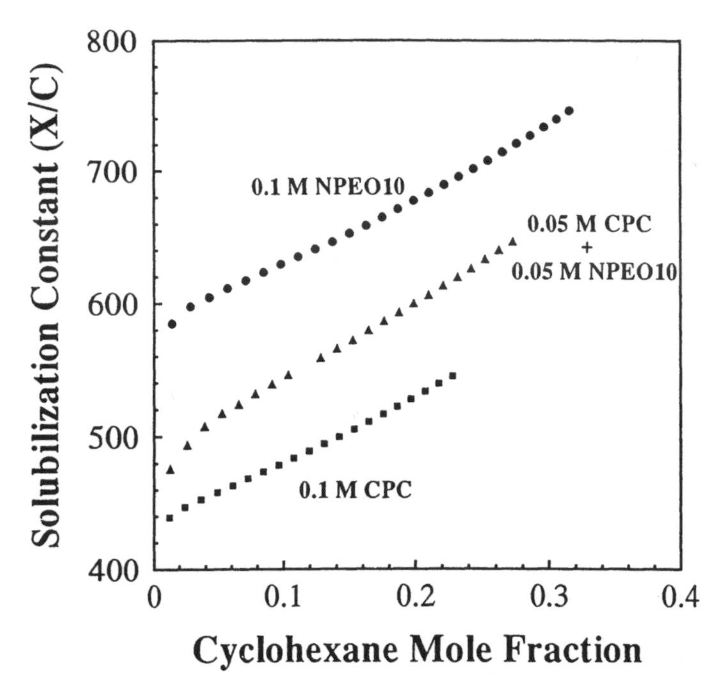

FIG. 18 Solubilization constant for cyclohexane as a function of its micellar mole fraction in individual and mixed cationic-nonionic surfactant solutions at 25°C.

that solubilization of benzene in cationic or anionic micelles is endothermic compared with pure benzene as a reference state places a definite upper limit on the possible interaction of the polarizable aromatic with an ionic micelle.

VII. CONCLUSION

The solute vapor pressure method provides the most direct and accurate method available for determining solubilization equilibrium constants of volatile compounds in aqueous micelles. The SVP method can provide detailed information concerning the activity coefficient of a solute as a function of its concentration in micellar systems. No other experimental method can provide comparable information on the dependence of solubilization on solution composition. However, the SVP method does not provide any direct structural information. For a more complete understanding of changes in solubilization with composition, it will be necessary to combine structural information from other techniques concerning the size and shape of micelles with the very precise equilibrium data from the vapor pressure method.

REFERENCES

1. P. H. Elworthy, A. T. Florence, and C. B. MacFarlane, *Solubilization by Surface Active Agents*, Chapman and Hall, London, 1968.
2. E. E. Tucker and S. D. Christian, *J. Chem. Thermodynamics 11*: 1137 (1979).
3. E. E. Tucker and S. D. Christian, *J. Phys. Chem. 83*: 426 (1979).
4. E. E. Tucker, E. H. Lane, and S. D. Christian, *J. Solution Chem. 10*: 1 (1981).
5. E. E. Tucker and S. D. Christian, *Faraday Symp. Chem. Soc. 17*: 11 (1982).
6. G. A. Smith, Ph.D. dissertation, University of Oklahoma, Norman, OK, 1986.
7. G. A. Smith, S. D. Christian, E. E. Tucker, and J. F. Scamehorn, *ACS Symp. Ser. No. 342*, 184 (1987).
8. M. Parande, S. D. Christian, E. E. Tucker, and H. Uchiyama, unpublished work, University of Oklahoma.
9. L. Sepulveda, E. Lissi, and F. Quina, *Adv. Colloid Interface Sci. 25*: 1 (1986).
10. J. Li and P. W. Carr, *Anal. Chem. 65*: 1443 (1993).
11. J. H. Park and P. W. Carr, *Anal. Chem. 59*: 2596 (1987).
12. J. H. Park, A. Hussam, P. Couasnon, D. Fritz, and P. W. Carr, *Anal. Chem. 59*: 1970 (1987).
13. C. C. West and J. H. Harwell, *Environ. Sci. Technol. 26*: 2324 (1992).
14. K. T. Valsaraj, A. Gupta, L. J. Thibodeaux, and D. P. Harrison, *Water Res. 22*: 1173 (1988).
15. C. C. West, *ACS Symp Ser. No. 491*: 149 (1992).
16. H. Uichyama, E. E. Tucker, S. D. Christian, and J. F. Scamehorn, *J. Phys. Chem. 98*: 1714 (1994).
17. E. E. Tucker, H. Uchiyama, S. D. Christian, and J. F. Scamehorn, unpublished work, University of Oklahoma.
18. G. A. Smith, S. D. Christian, E. E. Tucker, and J. F. Scamehorn, *J. Colloid Interface Sci. 130*: 254 (1989).
19. M. Barbe and D. Patterson, *J. Phys. Chem. 82*: 40 (1978).
20. S. Weiguo, A. X. Qin, P. J. McElroy, and A. G. Williamson, *J. Chem. Thermodynamics 22*: 903 (1990).
21. S. Ikeda, M. Tsunoda, and H. Maeda, *J. Colloid Interface Sci. 70*: 448 (1979).
22. H. Uchiyama, S. D. Christian, J. F. Scamehorn, M. Abe, and K. Ogino, *Langmuir 7*: 95 (1991).
23. J. Li, A. J. Dallas, D. I. Eikens, P. W. Carr, D. L. Bergmann, M. J. Hait, and C. A. Eckert, *Anal. Chem. 65*: 3212 (1993).
24. E. D. Goddard and G. C. Benson, *Can. J. Chem. 35*: 986 (1957).

14

Comparison of Experimental Methods for the Determination of the Partition Coefficients of *n*-Alcohols in SDS and DTAB Micelles

D. GERRARD MARANGONI Department of Chemistry, Saint Francis Xavier University, Antigonish, Nova Scotia, Canada

JAN C. T. KWAK Department of Chemistry, Dalhousie University, Halifax, Nova Scotia, Canada

SYNOPSIS

Micelles are often used in applications because of their ability to solubilize nonaqueous solutes that are originally insoluble in an aqueous solution. Micellar solubilization is an important facet of current studies in micellar systems, and is an integral part of a number of processes including detergency, drug delivery, tertiary oil recovery, and micellar catalysis. Many experiments have been used to determine the degree of partitioning of nonaqueous solutes between aqueous and micellar phases. In this chapter we will compare the distribution coefficients (K_x values) obtained from many different techniques, and we will examine closely the partitioning of some typical solubilizates (n-alcohols, benzene, and benzyl alcohol) between aqueous and micellar phases, as determined by a number of different methods, in two well-studied surfactant micelles (SDS and DTAB). Finally, we will discuss some of the assumptions of an NMR experiment which may be used to examine solubilization equilibria in a number of aggregated systems, including surfactant micelles.

I. INTRODUCTION

A. General Background and Overview

Micellar solubilization and mixed micelle formation have been studied intensively for many years, and remain an important facet of current studies in micellar systems. Solubilization is probably best defined in a fashion similar to that used by Sepulveda, Lissi and Quina in their exhaustive review article [1], as the incorporation of neutral molecules into a micellar aggregate. Solubilization is an integral part of a number of processes including detergency, drug delivery, tertiary oil recovery, and micellar catalysis to name but a few.

A number of techniques have been developed to interpret the phenomenon of solubilization including vapor pressure techniques, conductance experiments, NMR self-diffusion experiments, techniques based on luminescence probing, and the thermodynamic methods. In this chapter we will compare the distribution coefficients (K_x values) obtained from many different techniques, and we will describe a new method to determine K_x, developed in our laboratory, based on the measurement of NMR spin-lattice relaxation rates of solubilizate protons. Solubilization of neutral species is a complicated process and it will be shown that a phenomenological interpretation in terms of simplified models such as the pseudophase model or the mass action model can often lead to K_x values that are widely scattered.

In this review, we will concentrate on the partition of some typical solubilizates (*n*-alcohols, benzene, and benzylalcohol) between aqueous and micellar phases, as determined by a number of different methods. In order to arrive at a meaningful comparison, we limit ourselves mainly to two of the more popular choices of surfactants, sodium dodecylsulfate (SDS) and dodecyltrimethylammonium bromide (DTAB). We have examined the values of K_x for the *n*-alcohols, with a particular emphasis on *n*-butanol and *n*-pentanol, benzene, and benzyl alcohol. We will begin with a brief discussion of the mass-action model and the pseudophase model for micelle formation, and their application to the interpretation of solubilization. We will briefly review those experimental techniques not covered in other chapters, in particular the NMR paramagnetic relaxation experiment, and a few other less common methods. Finally, we will present a collection of data for *n*-alcohols, benzene, and benzyl alcohol in SDS and DTAB micelles, and discuss the variations in the K_x values reported from different measurements.

B. The Mass Action and Pseudophase Models of Micelle Formation: Interpretation of Solubilization

The simplest thermodynamic model for micelle formation, and by extension also for micellar solubilization, is the pseudophase or phase separation model. In this model, the micelles are described as a separate phase, similar to an oil phase where the oil molecules have a limited solubility in water. Another analogy is that of an oil-in-water emulsion, where the oil is simply treated as a separate phase. Thus, when the surfactant concentration exceeds the CMC, a micellar phase separates off, with uniform properties typical for a given surfactant. The phase separation model applied to uni-univalent ionic surfactants in the absence of other added electrolyte gives the relation between the standard Gibbs energy of micellization, $\Delta_{\mathrm{m,ps}}G^\circ$, and the critical micelle concentration, CMC, as

$$\Delta_{\mathrm{m,ps}}G^\circ = 2RT \ln \gamma_\pm \mathrm{CMC} \quad (1)$$

In other words, the micelles are treated as a separate, neutral phase, and the maximum free surfactant activity is given by $\gamma_{\pm}$CMC. The phase separation model therefore does not take into account effects due to partial ionization of the micelles, and the aggregation number of the micelle is not a variable in the thermodynamic description. Solubilization is treated as a simple distribution equilibrium of additive A between aqueous and micellar phases

$$\mu_{A,aq} = \mu_{A,mic} \tag{2}$$

When the mole fraction scale is used, the equilibrium constant K is given by

$$K_x = \frac{\gamma_{A,mic} X_{A,mic}}{\gamma_{A,aq} X_{a,aq}} \tag{3}$$

and when for neutral additives the activity coefficients γ are close to unity

$$K_x = \frac{X_{A,mic}}{X_{A,aq}} \tag{4}$$

$$X_{A,mic} = \frac{n_{A,mic}}{n_{A,mic} + n_{S,mic}} \tag{5}$$

$$X_{A,aq} = \frac{n_{A,aq}}{n_{S,aq} + n_{A,aq} + n_w} \tag{6}$$

Nonideal mixing in the micellar phase can be accounted for in $\gamma_{A,mic}$ via a number of model approximations.

The mass-action model treats micelle formation as an equilibrium process leading to micelles with distinct aggregation numbers and chemical potentials. If N is the surfactant aggregation number for an ionic surfactant, and z the number of associated counterions, the aggregation equilibrium can be written as

$$NS + zX \rightleftarrows (S_N X_z)_{mic}$$

The micellar aggregation number N is, in principle, variable, and micelles with different aggregation numbers may coexist; however, it is easiest to assume that a single value for N dominates. In this case, the standard Gibbs energy of micellization, again for a uni-univalent ionic surfactant, is given by

$$\Delta_{m,ma} G^\circ = \left(2 - \frac{z}{N}\right) RT \ln(\mathrm{CMC}) \tag{7}$$

z/N is the degree of counterion binding, an important variable in solubilization of neutral additives. We can now include the additive A in a stepwise solubilization equilibrium

$$(S_N X_z)_{mic} + A_{aq} \rightleftarrows (S_N X_z A_1)_{mic} \qquad K_1$$

$$(S_N X_z A_1)_{mic} + A_{aq} \rightleftarrows (S_N X_z A_2)_{mic} \qquad K_2$$

etc. It is clear that even if we do not allow N and z to vary, a large number of parameters is required for such a description. In practice the techniques described later in this chapter only determine the additive concentration in the micellar phase and in the aqueous phase. In this situation the use of the phase separation model is adequate to describe the solubilization equilibrium, and it is the model used by most investigators, with K_x (Eq. (3) or (4) as the main experimental parameter. At the same time, the measurement of N as a function of additive concentration can be achieved, e.g., by fluorescence techniques, and together with the determination of K_x as a function of additive concentration, it is, at least in principle, possible to obtain a more complete description of these mixed systems based on the mass-action approach.

C. The Gibbs Energy of Transfer of Solubilizates from Water to Micellar Phases

The emphasis in a number of recent articles on solubilization has been on the interpretation of the partition coefficient or distribution coefficient in terms of thermodynamic transfer functions, most notably the Gibbs energy of transfer, $\Delta_t G^\circ$ [1–3]. In order to calculate $\Delta_t G^\circ$ suitable reference states must be chosen. For example, in terms of the pseudophase model of micelle formation, the Gibbs transfer energy, $\Delta_t G^\circ$, can be represented as follows

$$\Delta_t G^\circ = \mu^\circ_{A,mic} - \mu^\circ_A = -RT \ln K_x \qquad (unitary\ scale) \tag{8}$$

$$\Delta_t G^\circ = \mu^\circ_{A,mic} - \mu^\circ_{A,aq} = -RT \ln K_c \qquad (molarity\ scale) \tag{9}$$

In these equations, $\mu^\circ_{A,mic}$ is the standard Gibbs energy of the solubilizate in the micellar phase, in the mole fraction or molarity scale, and μ°_A and $\mu^\circ_{A,aq}$ are the standard molar Gibbs energies in the solution phase, again in the appropriate scale. Discussions on the use of either reference state can be found in articles by Tanford [4] and Ben-Naim [5]. For our purposes we have chosen, for purely arbitrary reasons, to express all the partition coefficients in terms of the mole-fraction-based K_x, and we have recalculated all the solubilization data to be consistent with the definition of K_x. Note that in defining mole fractions the solvent is specifically excluded as a component of the micellar phase.

Solubilization can be viewed in a manner similar to micelle formation for pure surfactants, i.e., as a compromise between the tendency for hydrophobic solute to avoid unfavorable contacts with water and to disperse in a nonpolar medium (i.e., the surfactant micelle). In the case of polar neutral solubilizates, there is an added tendency for the polar groups to be near the polar micellar surface. A rigorous thermodynamic description of the process of solubilization includes a description of all possible electrostatic and hydrophobic contributions to the overall decrease in the Gibbs transfer energy [6]; we will mainly concentrate on the hydrophobic contributions to the standard Gibbs energy of transfer. For a homologous series of solubilizates (e.g., the *n*-alcohols or the *n*-alkanes), the value of $\Delta_t G°$, as given by Eq. (8), generally increases in a linear fashion:

$$\Delta_t G° = (a - bn_c)RT \tag{10}$$

where a and b are constants for the particular homologous series, and n_c is the number of carbon atoms in the solubilizate chain. Thus, $\Delta_t G°$ can be factored into individual group contributions (i.e., $\Delta_t G°$ [hydrophobic] + $\Delta_t G°$ [hydrophilic]), and for a homologous series, the hydrophilic contribution to the Gibbs energy is essentially constant. Therefore, the decrease reflects the incremental contribution of a specific group to $\Delta_t G°$, e.g., the CH_2 group, to the hydrophobic part of the transfer Gibbs energy. Note that similar trends in $\Delta_t G°$ for the partitioning of neutral hydrophobic solutes between water and hydrocarbon solvents are well known [4]. The transfer Gibbs energy, $\Delta_t G°$, can be divided into entropic and enthalpic contributions

$$\Delta_t G° = \Delta_t H° - T\Delta_t S° \tag{11}$$

where $\Delta_t H°$ and $\Delta_t S°$ are the enthalpy and entropy of transfer, respectively, obtained by determining K_x as a function of temperature, or from a combination of calorimetric measurements with the Gibbs energies. Spink and Colgan have determined calorimetrically the transfer enthalpies and transfer Gibbs energies of *n*-alcohols to sodium deoxycholate micelles [7], and have found that the partition coefficient, K_x, is not particularly sensitive to changes in the absolute temperature (i.e., $\Delta_t G°$ is essentially constant). However, $\Delta_t H°$ and $\Delta_t S°$ change dramatically, indicating that the driving forces for solubilization of hydrophobic solutes, like the driving forces for micellization, are dependent on the temperature at which the process occurs.

D. Solubilization and Mixed Micelle Formation

Since the physical and chemical properties of the micellar interior are generally believed to resemble those of a liquid hydrocarbon, it is not surprising that a sparingly soluble solute would preferentially solubilize in the micellar interior. If this compound is an alcohol, or another hydrophobic molecule, this facilitates micellization due to mixed micelle formation. Mixed micelles composed of surfactant, alcohol, and water have been studied extensively over the past few years [8]. The importance of medium-chain-length alcohols as cosurfactants in microemulsion systems has added to the interest in their micellar partition coefficients [9].

The addition of alcohols to ionic surfactant solutions has been found to decrease the CMC [10–12]. This decrease is due to the incorporation (solubilization) of the alcohol in the micellar interior, and the relative value of the decrease of CMC for different alcohols as solubilizates must be related to their different degrees of incorporation into the micelle.

The critical micelle concentration is not the only micellar parameter affected by alcohol solubilization. Mean activities of the surfactant [13], counterion binding [14], aggregation numbers [15], and the micellar surface charge density [16] are also affected when alcohol is incorporated in the micelles. Therefore, a reasonable estimate of the alcohol partition coefficient is essential for interpreting the effects that added solutes have on the properties of micellar systems.

There has been considerable discussion about the location of the solubilized alcohol in the micelle [17]. Some authors have suggested "alcohol-swollen" micelles [18] with an alcohol core to explain the apparent large degrees of solubilization at moderate to high alcohol mole fractions. This would be consistent with the observed high solubility of alcohols in hydrocarbon solvents. However, the commonly held view is that at least in the low alcohol concentration range the alcohol is solubilized with the hydrocarbon tail in the micellar interior and the hydroxyl group in the headgroup [19].

We may express the fraction of the total concentration of alcohol that is distributed in the mixed micelles as the p-value of the solubilizate

$$p = \frac{n_{A,\mathrm{mic}}}{n_{A,\mathrm{t}}} \tag{12}$$

where $n_{A,\mathrm{mic}}$ is the number of moles of additive solubilized in the micelles, and $n_{A,\mathrm{t}}$ is the total number of moles of additive. A large number of techniques have been used to determine the micellar bound fraction (or p-value) of the solubilizate in the micellar phase, as is clear from our tabulated survey. A number of these techniques are described in other chapters of this volume, or in the exhaustive review by Sepulveda, Lissi, and Quina [1]. In Sec. II we will describe in some detail a new NMR method for determining p-values, a method particularly suitable to the case of water-soluble additives such as alcohols, ethoxylates, amines, etc. This method has the ^{1}H spin-lattice relaxation time, T_1, of solubilizate protons as the observable quantity, and uses the difference in T_1-value in the presence or absence of paramagnetic relaxation agents in the aqueous phase to determine p. The method is generally applicable to many of the same systems as the NMR diffusion (FT-PGSE NMR) method, but does not need field-gradient attachments and techniques.

It is important to distinguish between the case of sparingly soluble additives, where the total solubility of the additive is greatly enhanced by the presence of micelles, and soluble additives, where it is probably more appropriate to speak of a partitioning of the additive between the solvent phase and the micellar phase, as expressed by the value of p in Eq. (12). In the case of partitioning, most methods are based on the detection of a change in a solute parameter X which is the average of the values of X for solute in the aqueous phase and in the micelle [6]

$$X_{obs} = pX_m + (1 - p)X_s \tag{13}$$

where X_{obs} is the observed (average) value of X, and X_m and X_s are the values of X in micellar and solvent phases respectively. Methods based on thermodynamic quantities (H_{mix}, V_{mix}), diffusion (FT-PGSE NMR, Taylor dispersion, or tracer), changes in extinction coefficient, etc., are all based on this assumption.

II. NMR SPECTROSCOPY OF MICELLAR AND MIXED MICELLAR SOLUTIONS: THE NMR PARAMAGNETIC RELAXATION ENHANCEMENT METHOD FOR DETERMINING *p*-VALUES

NMR spectroscopy is widely recognized as a powerful tool for the investigation of many physicochemical phenomena. Since the advent of the pulsed Fourier transform NMR spectrometer, studies can be made on a wide number of nuclei with good resolution. NMR is an extremely versatile tool for working in micellar systems, since the transfer of a monomer from the bulk solution to the micellar phase is generally accompanied by changes in a large number of NMR parameters (e.g., chemical shifts and relaxation rates). NMR experiments have provided important insight into the nature of the solubilization process, and, hence, mixed micelle formation, from the determination of the p-value of an additive, the location of the solubilizates in the micellar phase, and the dynamics of the hydrocarbon chains comprising the micellar interior. A brief discussion of some NMR basics and the paramagnetic relaxation experiment will follow [20–22].

When a sample is placed in a static magnetic field, nuclei with the same value of the magnetogyric ratio, γ, in different chemical environments experience different magnetic shieldings produced by the electronic environment surrounding the nucleus. In the presence of a strong applied magnetic field, these different shieldings give rise to differences in the resonance frequencies of the nuclei, given by

$$\nu_i = \left(\frac{\gamma}{2\pi}\right) B_0(1 - \sigma_i) \tag{14}$$

The chemical shift scale is a dimensionless scale which expresses the resonance frequencies as chemical shifts, δ's, with respect to the resonance frequency of the reference

$$\delta = 10^6 \frac{\nu_s - \nu_r}{\nu_r} \approx 10^6(\sigma_r - \sigma_s) \tag{15}$$

where σ_r and σ_s are the shielding constants for the reference and sample, respectively. For ^1H and ^{13}C, the reference generally used is tetramethylsilane, or TMS. Hence, when surfactant amphiphiles are transferred from the bulk solution, differences in the environment surrounding the nuclei of interest give rise to slight changes in its resonance frequency of the nuclei of interest. For solubilizates dissolved in a micellar solution, the observed chemical shift is a weighted average of the chemical shift of the solubilizate (additive) in the bulk solution and the micellar phase (assuming a two-site solubilization process)

$$\delta_{A,obs} = p\delta_{A,mic} + (1 - p)\delta_{A,aq} \tag{16}$$

where $\delta_{A,obs}$ is the observed (measured) chemical shift, $\delta_{A,mic}$ is the chemical shift of the additive in the micellar phase, and $\delta_{A,mic}$ is the chemical shift of the solubilizate in the bulk aqueous phase. In most cases, ^1H chemical shifts show only very minor changes upon solubilization in a micelle. An exception is the case where the surfactant contains a phenyl group, in which case large ^1H chemical shift changes are observed upon micellization and solubilization, and these shifts can in principle be used to determine the p-value. In the absence of this ring-current effect ^1H chemical shifts are too small to be suitable for p-value determinations. ^{13}C chemical shifts could be used, but the lack of sensitivity makes this method unsuitable for most solubilization studies.

When a collection of spins is placed in a magnetic sample, a net magnetization in the z direction, M_0 develops. If the spin system is perturbed from its equilibrium M_0 value by a small field, B_l, which rotates at the Larmor frequency in the x-y plane, the theory of Bloch [20–22] predicts that the rate at which the spin system will proceed toward the equilibrium magnetization M_0 is given by

$$\frac{dM_z}{dt} = \frac{M_z - M_z^\circ}{T_1} \tag{17}$$

where M_z is the magnetization along the z axis at time t and T_1 is the time constant that fully characterizes the first-order process. Integrating the above equation yields

$$\ln(M_z^\circ - M_z) = -\frac{\tau}{T_1} + \ln C \tag{18}$$

The rate at which the spin system proceeds toward the equilibrium distribution, $R_I = 1/T_I$, depends on the ease with which the spin system can transfer energy to the surroundings (i.e., the "lattice"). Hence, R_I is referred to as the spin-lattice relaxation rate. A number of mechanisms are available to the spin system for transferring the spin energy to the lattice, including dipole-dipole, spin rotation, scalar relaxation, interaction with unpaired electrons, and for nuclei with $I > 1/2$, an additional mechanism, quadrupolar relaxation, is available. The enhancement of the rate of relaxation of the spin system due to its interaction with unpaired electrons will be discussed in some detail below.

Paramagnetic ions or molecules can contribute to the relaxation of a nuclear spin system in one of two ways: by dipole-dipole relaxation by the electron magnetic moment, or by the transfer of some unpaired electron spin density to the relaxing nucleus itself. The basic equations describing the relaxation of nuclei involved in complexation with paramagnetic complexes were derived by Solomon, Bloembergen, and Morgan [23, 24]. In the event that the nuclei and paramagnetic ions do not form stable complexes, paramagnetic relaxation enhancement is due to the dipole-dipole relaxation between the nucleus of interest and the electron magnetic moment. In this case, the paramagnetic contribution to the spin lattice relaxation rate of the nuclei is given by [23–26]

$$\begin{aligned}\Delta R_I &= R_{I,\mathrm{aq}}^{\mathrm{p}}(\mathrm{A}) - R_{I,\mathrm{aq}}(\mathrm{A}) \\ &= \left(\frac{8\pi}{15}\right)\left(\frac{\mu_0}{4\pi}\right)^2 (\gamma_\mathrm{I}\gamma_\mathrm{S}\hbar)^2\, S(S+1)\left(\frac{N_\mathrm{el}\,\tau}{b^3}\right)[3\,J(\omega_\mathrm{I}) + 7J(\omega_\mathrm{S})]\end{aligned} \tag{19}$$

where $R_{I,\mathrm{aq}}^{\mathrm{p}}(\mathrm{A})$ and $R_{I,\mathrm{aq}}(\mathrm{A})$ are the spin-lattice relaxation rates in the presence and absence of paramagnetic ions, respectively, S is the total electron spin, γ_I and γ_S are the magnetogyric ratios of the nuclei and the electron, respectively, and ω_I and ω_S are the Larmor frequencies of the nucleus and electron, respectively. N_el is the number density of the electrons, b is the distance of closest approach of the paramagnetic agent to the nucleus, τ is the translational correlation time (defined as $b^2/(D_I + D_S)$, where D_I and D_S are the translational diffusion coefficients of spin I and spin S), and $J(\omega)$ is the spectral density function. The nature of the paramagnetic enhancement of the spin-lattice relaxation rate can be deduced by measuring the relaxation rate as a function of the nuclei containing solute, in the presence of a fixed quantity of the paramagnetic species. If the rate enhancement is intramolecular, the relaxation rate will be dependent on the solute concentration, and the rate enhancement will be described by the Solomon-Bloembergen-Morgan relationships [23–26]; if there is no dependence of the relaxation rate on the solute concentration, the paramagnetic relaxation enhancement is intermolecular, and is described by Eq. (19). The utility of paramagnetic relaxation enhancement for determining the concentration of micellar bound solubilizates will be further explored in the next section.

Two NMR experiments for determining the p-value are the NMR self-diffusion experiment [27], and the recently developed NMR paramagnetic relaxation experiment [28]. The NMR self-diffusion experiment has been discussed previously; it is a well-established method for measuring solubilization equilibria [27], which has contributed greatly to our understanding of many simple and complex surfactant systems. The NMR paramagnetic relaxation experiment estimates the p-value of the solubilizate based on the difference in the relaxation rates of the solubilizate, in aqueous and micellar solutions, in the presence and absence of a very small concentration of paramagnetic ion. If the paramagnetic probe (ion) and the surfactant headgroup have the same charge, the paramagnetic ion should reside exclusively in the aqueous phase, away from the micellar headgroup. Assuming that the paramagnetic ion does not influence the relaxation of micellar bound species, a solubilizate (e.g., an alcohol) which is distributed between the aqueous and micellar phases, will be affected by the paramagnetic ion in one of two ways. If the solubilizate distribution favors the aqueous phase, its relaxation rate will be enhanced significantly by the paramagnetic ion. Conversely, if the distribution favors the micellar phase, relaxation enhancement will not be significant.

Under conditions of fast exchange, the observed relaxation rate for the solubilizate is a weighted average of the rates for the solubilizate in the aqueous and the micellar phase. When there are no paramagnetic species in the solution, the observed rate of spin-lattice relaxation is described by the usual equation

$$R_{I,\text{obs}}(\text{A}) = R_{I,\text{aq}}(\text{A}) + p(R_{I,\text{mic}}(\text{A}) - R_{I,\text{aq}}(\text{A})) \tag{20}$$

where $R_{I,\text{obs}}(\text{A})$ is the observed spin-lattice relaxation rate of the solubilized species, $R_{I,\text{mic}}(\text{A})$ is its relaxation rate in the micellar phase, and

$R_{I,aq}(A)$ is its rate of relaxation in the water phase. When paramagnetic ions are present, the observed rate is written

$$R^{P}_{I,obs}(A) = R^{P}_{I,aq}(A) + p(R_{I,mic}(A) - R^{P}_{I,aq}(A)) \tag{21}$$

Since the paramagnetic ion is assumed not to have an influence on the micellar bound solubilizate (i.e., $R_{I,mic}(A)$ is unaltered), subtracting Eq. (20) from Eq. (21) and rearranging allows the fraction of the solubilizate in the micellar phase, $p = n_{A,mic}/n_{A,t}$, to be calculated

$$p = 1 - \frac{R^{P}_{I,obs}(A) - R_{I,obs}(A)}{R^{P}_{I,aq}(A) - R_{I,aq}(A)} \tag{22}$$

The degree of solubilization, p, can be used to obtain the mole-fraction-based partition coefficient, K_x, using Eqs. (4)–(6) [28]. Gao et al. have determined the p-values of a number of solubilizates in anionic SDS and cationic DTAB micellar systems, using the paramagnetic relaxation experiment [28]. These authors compared their results for the p-values of a number of organic solubilizates, to those determined by Stilbs, using the NMR self-diffusion experiment for the same solubilizates [28]. Generally, it has been found that the results obtained with the paramagnetic relaxation method are in good agreement with those determined by means of the NMR-diffusion method [29, 30].

The NMR paramagnetic relaxation experiment has also been applied to a determination of the p-value of polymers in surfactant micelles. Polymer diffusion is generally slow and difficult to measure, and hence, problems may arise in using the FTPGSE method to measure the p-values of polymers in polymer-surfactant systems. The NMR paramagnetic relaxation experiment also has the advantage that it can be used routinely on any FT spectrometer, whereas special probes and hardware modifications are necessary for the FT-PGSE experiment [27]. The precision of the distribution coefficients measured by the NMR paramagnetic relaxation experiment is comparable to what can be achieved with the FTPGSE experiment. An error analysis of the method is given later in this chapter. The FTPGSE experiment has been applied successfully in microemulsion systems [27]; the application of the NMR paramagnetic relaxation experiment in microemulsion systems is currently under investigation [31].

III. OTHER METHODS TO DETERMINE PARTITION COEFFICIENTS

A. Overview

The data collected in Sec. IV (Refs. 32–94) have been obtained by a large number of different methods. Perhaps the most direct of these methods, in a thermodynamic sense, is the vapor pressure method. When a solute is dissolved in a micellar solution, its vapor pressure will be lowered relative to the vapor pressure of a solution of the same total additive concentration, which does not contain micelles

$$\mathrm{vp_m} = (1 - p)\mathrm{vp_0} \tag{23}$$

where $\mathrm{vp_m}$ is the solute vapor pressure above the micellar solution and $\mathrm{vp_0}$ the solute vapor pressure above the solution without surfactant micelles. Equation (23) assumes ideal behavior of the solute; if not, appropriate activity coefficients need to be introduced. This method has been applied widely to the case of soluble solubilizates, including many of the alcohols reported here. The vapor pressure may be determined directly, or if necessary by head-space chromatography. Calorimetric and volumetric methods (microcalorimetry, H_{mix}, H_{sol}, Φ_{C_p}, V_{mix}, Φ_V) are based on the assumption that the respective calorimetric or volumetric properties follow Eq. (13), but in many cases the results of these methods would also be influenced by changes in CMC or counterion binding. Extrapolations to zero additive concentration and a variety of activity corrections for surfactant and additive may be needed; a recent article by Nguyen and Bertrand [75] presents a detailed description of the data treatment for measurements of heats of dilution in alcohol–sodium dodecylsulfate solutions. A few less well-known methods are described briefly in the following sections.

B. Total Solubility Method

The relative simplicity of the total solubility method has made it attractive to many authors. In the total solubility method, excess solute (i.e., solute above its saturation molality in water) is added to a certain concentration of a micellar surfactant solution. Under the assumption that the solubility of the neutral species is not affected by the presence of micelles, the concentration-based partition coefficient, K_c, is written

$$K_c = \frac{C_{\mathrm{A,mic}}}{C_{\mathrm{A,aq}}} = \frac{C_{\mathrm{A,t}} - C_{\mathrm{A,0}}}{C_{\mathrm{A,0}}} \tag{24}$$

where $C_{\mathrm{A,0}}$ represents the saturation molality for the additive in the solvent, and $C_{\mathrm{A,t}}$ is the total concentration of added solute. An alternative concentration-based partition coefficient, K'_c can be obtained

$$K'_c = \frac{C_{\mathrm{A,t}} - C_{\mathrm{A,0}}}{C_{\mathrm{A,0}}(C_{\mathrm{S,t}} - \mathrm{CMC} + C_{\mathrm{A,mic}})} \tag{25}$$

where $C_{\mathrm{S,t}}$ is the total surfactant concentration. A plot of $C_{\mathrm{A,t}}/C_{\mathrm{A,0}}$ versus $C_{\mathrm{S,t}}$ should be linear with a slope of K'_c. Total solubility methods are generally the easiest to employ experimentally, with the total solute saturation concentration in the micellar solution being determined by gas chromatography [32], turbidity [33], and density measurements [34]. The total solubility method yields a value for the partition coefficient, K_x or K_c, at one specific solute concentration, i.e., the saturation concentration of the solute in the aqueous phase. On the other hand, the NMR methods determine p and K_x at low solute concentrations. For solutes with high partition coefficients, the micellar composition may change significantly at high solute concentrations, leading generally to a decrease in K_x [35–37]. All total solubility determinations of the solute partition coefficient suffer from the disadvantage that the micellar solution is necessarily saturated with solute. Thus, the micellar structure may be perturbed significantly, and these perturbations may not be reproducible among different families of solubilizates. Note, however, that $\Delta_t G°$ (CH_2), obtained from the decrease in the transfer Gibbs energy in a homologous series (from $\Delta_t G° = -RT \ln K_x$) versus n_c, is in very good agreement with the accepted value for that property [34].

C. Krafft Point Method [38]

In terms of the pseudophase model for micelle formation, the Krafft point is interpreted as the temperature at which the hydrated solid surfactant melts. Thus, the chemical potential of the surfactant in the micellar phase can be obtained easily from an application of the colligative properties (in this case, the depression of the Krafft point, or the "melting point"). The mole fraction of the surfactant in the micellar phase, $X_{S,mic}$, is related to the Krafft point temperature, T_0, in the absence of additives, according to the thermodynamic relationship

$$\ln X_{S,mic} = \frac{-\Delta_{fus}H°}{R}\left(\frac{1}{T} - \frac{1}{T_0}\right) \tag{26}$$

where T is the Krafft point in the presence of additives and $\Delta_{fus}H°$ is the enthalpy change when the solid, hydrated surfactant "melts" to form micelles. For a surfactant micelle with added alcohol, $X_{S,mic} = 1 - X_{A,mic}$ where $X_{A,mic}$ is the mole fraction of the additive in the micellar phase. Hence, the fraction of alcohol in the micelles can be obtained from the variation in the Krafft point with an increase in the micellar mole fraction of added solubilizate.

$$\ln(1 - X_{A,mic}) = \frac{-\Delta_{fus}H°}{R}\left(\frac{1}{T} - \frac{1}{T_0}\right) \tag{27}$$

$X_{A,mic}$ can be written as $K_x X_{A,aq}$. In the limit as $X_{A,mic}$ approaches 0,

$X_{A,aq} \approx n_{A,aq}/n_{H_2O} \approx n_{A,t}/55.5$; substituting in the molality of the alcohol, $m_{A,t}$, we obtain

$$\ln\left(1 - K_X \frac{m_{A,t}}{55.5}\right) = \frac{-\Delta_{fus}H^\circ}{R}\left(\frac{1}{T} - \frac{1}{T_0}\right) \tag{28}$$

and the partition coefficient is obtained from Krafft point measurements as a function of the molality of the additive in the mixed solvent.

D. Raman Spectroscopic Method

A Raman spectroscopic method for determining partition coefficients has been described by Shih and Williams [39]. Their method is based on determining the C-D stretching region of a deuterated micellar solubilized additive, and comparing the micellar spectrum to the spectrum of the solubilizate in aqueous and hydrocarbon media, respectively (two-site solubilization model). Under the assumption that the Raman spectrum of the solubilizate in the micellar phase is adequately represented by the solubilizate spectrum in the hydrocarbon medium (e.g., heptane), the solubilizate/surfactant spectrum is fitted to a linear combination of the spectrum of the additive in the two reference states (water and hydrocarbon). The fraction of the solubilizate having a spectral profile similar to the spectrum of the solubilizate in the hydrocarbon phase (the *p*-value) is determined by least-squares fitting of the spectra.

E. Electron Spin-Echo Modulation Method

A method for determining the partition coefficients of solubilizates, based on the modulation of the dipolar interaction between the unpaired electron on a solubilized ESR probe molecule and the surrounding deuteriums of the water (D_2O) or of solubilized additives, has been described by Baglioni and Kevan [40]. The magnitude of the dipolar interaction is expressed as the deuterium modulation depth and is directly proportional to the density of the surrounding deuteriums and inversely proportional to the average distance between the spin probe and the deuterium nuclei. The experiments are conducted in rapidly frozen micellar solutions (4.2 K) in order to prevent isotropic averaging of the electron-nuclear dipolar interactions that are responsible for the modulation effects, as well as to take advantage of the increased sensitivity and slower electron relaxation rates at lowered temperatures. A plot of the normalized modulation depth versus the additive concentration results in an increase in the normalized modulation depth up to a specific concentration of additive where the micellar surface is saturated with deuterium (plateau region). The point at which the plateau concentrations are reached reflects changing degrees of perturbation of the micellar surface, which can be interpreted as representing differences in the degrees of additive solubilization in the micellar surface region. By comparing the concentration at which the plateau region of the normalized modulation depth is reached for an extremely hydrophobic solute (e.g., 1-octanol) with the plateau regions for other solubilized alcohols, the fraction of alcohol associated with the micellar surface (the *p*-value) can be obtained.

IV. COMPARISON OF K_x VALUES

The values obtained for the partition coefficients of *n*-alcohols ($C_3OH \rightarrow C_8OH$), benzyl alcohol (BzOH), and benzene (Bz) in SDS and DTAB micelles are presented in Tables 1–8; assorted K_x values for alcohol and aromatics in CTAB micelles are presented in Table 9. We will begin by discussing the data reported for the solubilization of C_4OH and C_5OH (1-butanol and 1-pentanol) in SDS and DTAB micelles (Tables 2,3). From Table 3, it is clear that the reported values of K_x for C_5OH in SDS vary widely depending on the method used to obtain the distribution coefficient, ranging from 190 (total solubility measurement [67]) to 1232 (mass action model applied to heat capacities and volumes [55]) and 1215 (phase-separation model applied to heat of solution measurements [75]). All values reported here can be assumed to be derived from proper experimental procedures, and accordingly we have no justification to delete any particular data in the calculated averages. We have calculated the average value of K_x for C_5OH in SDS from all the experimental results to be 763 $\pm$ 252 (error denotes one standard deviation within the set), with most of the values clustered in the range 700–800. The value for the partition coefficient obtained using the NMR paramagnetic relaxation experiment is con-

TABLE 1 K_x Values for C_3OH (1-propanol) in SDS and DTAB Micelles from a Number of Different Experimental Methods

K_x (SDS)	Method (ref.)	K_x (DTAB)	Method (ref.)
92 $\pm$ 24	FT-PGSE [29]	100	MAM ($\Delta_{mix}H°$) [47]
113 $\pm$ 33	NMR-PRE [28, 41]	46 $\pm$ 25	NMR-PRE [42]
70 $\pm$ 16	NMR-PRE [42]	68	Cond. data [43]
105	Cond. data [43]	33	Cond. data [47]
194	MAM ($\Delta_{dil}H^{ex}$) [44]	78	MAM (V's) [48]
100	MAM (V's) [45]	55	Cond. data [49]
115 $\pm$ 34	ESEM [40]		

TABLE 2 K_x Values for C_4OH (1-butanol) in SDS and DTAB Micelles from a Number of Different Experimental Methods

K_x (SDS)	Method (ref.)	K_x (DTAB)	Method (ref.)
151 ± 35	FT-PGSE [29]	144 ± 32	NMR-PRE [42, 53, 54]
231 ± 33	Taylor disp. [53]	191	Cond. [43]
42[a]	Total. sol. [32][a]	94	Cond. [47]
300	Vap. press. [51, 52]	222	MAM ($\Delta_{mix}H°$) [46]
146 ± 22	NMR-PRE [42, 53, 54]	166	MAM ($\Delta_t H°$) [49]
275	Cond. data [43]	144 ± 6	MAM (V's) [58]
250	Krafft pt. [38]	216	MAM ($\Delta_{dil}H°$) [66]
472 ± 30	MAM (C_p's, V's) [55]	205	MAM (C_p's) [56]
213	MAM (V's) [56]	262	Cond. data [49]
350	MAM ($\Delta_t H°$) [57]		
322 ± 105	MAM (V's) [58]		
333 ± 39	MAM (V's) [45]		
350	MAM ($\Delta_{dil}H^{ex}$) [44]		
267	Cond. data [59, 60]		
670	MAM (ϕ_V's) [61]		
346	Counter. bind. [62]		
250 ± 22	Cond. data [63, 64]		
209 ± 20	Cond. data [65]		

[a] Surfactant = lithium dodecylsulfate.

siderably below the value of K_x averaged over all techniques. It is possible that the partition coefficients are dependent on the alcohol concentration, since the highest values of K_x have been extrapolated to an infinitely dilute alcohol solution [75], whereas the lowest values, the total solubility determinations, are necessarily measured in alcohol-saturated solutions (high alcohol concentrations). This observation is in agreement with the work of Abuin and Lissi [81] and Hartland et al. [71]. When we examine the partition coefficients for C_5OH in DTAB, presented in Table 3, we note that the partition coefficients for this alcohol in the cationic micelle are less scattered (average value for all reported data 542 ± 117).

The results for the measurement of K_x of C_4OH in SDS micelles (Table 2), also show considerable scatter. The lowest value for K_x is again derived from a total solubility experiment, with the highest values being obtained from a mass-action model fit of solution thermodynamic data [55, 61]. The average value for K_x is 288 ± 135; this time, however, both NMR determinations of the partition coefficient (FT-PGSE [29, 30] and the NMR-PRE [42, 53, 54]), although in good agreement with each other, are below the statistical average for this quantity, but within one standard

TABLE 3 K_x Values for C_5OH (1-pentanol) in SDS and DTAB Micelles from a Number of Different Experimental Methods

K_x (SDS)	Method (ref.)	K_x (DTAB)	Method (ref.)
611 ± 34	FT-PGSE [29]	586 ± 78	FT-PGSE [30]
865 ± 97	FT-PGSE [30]	575	Cond. [43]
190	Total sol. [67]	610	MAM (C_p's, V's) [77]
212[a]	Total sol. [68]	611	MAM (V's, K's) [78]
722	Vap. p. [51, 52]	544	MAM ($\Delta_{\text{mix}}H^\circ$) [46]
850	Microcal. [69]	511 ± 33	MAM (V's) [70]
790	Vap. p. [70]	722	MAM ($\Delta_{\text{dil}}H^\circ$) [66]
718	Krafft pt. [38]	500	MAM (ϕ_{C_p}'s, ϕ_v) [79]
378 ± 50	NMR-PRE [42]	401 ± 53	NMR-PRE [42]
710	Fluor. (dil) [71]	536 ± 116	NMR-PRE [28, 41]
776	Cond. [43]	288	Cond. data [47]
1232 ± 50	MAM (C_p's, V's) [55]	611	MAM (C_p's) [56]
623	Part. mol. vol. [72]		
927 ± 50	MAM (V's) [45, 58]		
1166	MAM ($\Delta_{\text{dil}}H^{\text{ex}}$) [44]		
604 ± 70	ESEM [40]		
867	Pulse rad. [73]		
785	Gas. chrom. [74]		
900[a]	Gas. chrom. [74]		
650	Gas. chrom. [74]		
750	Cond. data [59, 60]		
944	MAM (ϕ_v's) [61]		
1030	Counter. bind. [62]		
696 ± 75	Cond. data [63, 64]		
661 ± 70	Cond. data [65]		
1215 ± 33	PSM ($\Delta_{\text{sol}}H^\circ$) [75]		
722	Cond. data [76]		

[a] Surfactant = lithium dodecylsulfate.

deviation reported for all values in Table 2. When we examine the experimental results for the determination of K_x for C_4OH in DTAB micelles (Table 2), we again observe a distribution of experimental results, but here also the measurements are somewhat less scattered than what we have observed for C_4OH in SDS micelles. In this case the average K_x value for all reported measurements is 183 ± 51, the value obtained with the NMR-PRE determination of K_x is in good agreement with the average value of all reported partition coefficients in Table 2.

Fewer measurements have been reported for C_3OH (Table 1), C_6OH (Table 4), C_7OH (Table 5), and C_8OH (Table 6). For the higher alcohols,

TABLE 4 K_x Values for C_6OH (1-hexanol) in SDS and DTAB Micelles from a Number of Different Experimental Methods

K_x (SDS)	Method (ref.)	K_x (DTAB)	Method (ref.)
2045 ± 276	FT-PGSE [29]	1778	Cond. data [43]
1787 ± 712	Taylor disp. [50]	870	Cond. data [47]
690	Total sol. [67]	1007	NMR-PRE [42]
760	Total sol. [80]	1554	MAM (V's) [48]
758[a]	Total. sol. [32][a]	2319	Cond. data [57]
2250	Vap. press. [51, 52]	1002	Total sol. [82]
2160	Krafft pt. [38]		
1900 ± 200	Vap. press. [40]		
2700	MAM (C_p's, V's) [55]		
820	Fluor. (sat'd) [81]		
3350	Fluor. (dilute) [81]		
2455	Cond. data [43]		
1719	Part. mol. vol. [72]		
3053	MAM ($\Delta_{dil}H^{ex}$) [44]		
2750	MAM (V's) [45]		
806 ± 106	NMR-PRE [42]		
2316	Cond. data [59, 60]		
2390	Counter. bind. [62]		
1876 ± 159	Cond. data [63, 64]		
1941 ± 170	Cond. data [65]		

[a] Surfactant = lithium dodecylsulfate.

of course, solubility in water is limited, and the data reported present conditions at or close to the aqueous solubility limit. For C_6OH, C_7OH, and C_8OH in SDS the values reported from total solubility measurements still fall well below the aggregate of the other reported values, probably reflecting the higher degree of saturation of the micelle by the solubilizate in this method.

Figures 1 (for SDS) and 2 (for DTAB) present all values reported in Tables 1–6 in the form of $+RT \ln K_x$, the negative of the standard Gibbs energy of transfer given by Eq. (8). The solid lines in these figures represent the least-squares straight lines through the aggregate averages (as marked on the graphs) for each alcohol. The excellent correlation with Eq. (10) is most notable. From this correlation, we derive a value of bRT equal to $+2.38 \pm 0.02$ kJ mol^{-1} for linear alcohols in SDS, and $+2.59 \pm 0.07$ kJ mol^{-1} for linear alcohols in DTAB. Given this strong correlation, it is probably justified to assume that these averages are close to the "real" K_x values, and the data presented in Tables 1–6 can be judged accordingly.

TABLE 5 K_x Values for C_7OH (1-heptanol) in SDS and DTAB Micelles from a Number of Different Experimental Methods

K_x (SDS)	Method (ref.)	K_x (DTAB)	Method (ref.)
4261 ± 1104	FT-PGSE [29]	3996	MAM ($\Delta_{mix}H^{\circ}$) [47]
2650	Total. sol. [34]		
1750	Total sol. [67]		
6020	Vap. press [51, 52]		
6200	Vap. press. [83]		
5380	Krafft pt. [38]		
5754	Cond. data [43]		
4965	Part. mol. vol. [72]		
8159	MAM ($\Delta_{dil}H^{ex}$) [44]		
8048	MAM (V's) [45]		
5761	Cond. data [59, 60]		
4880	Counter. bind. [62]		
4733 ± 412	Cond. data [63, 64]		
6266 ± 250	Cond. data [65]		

A number of investigators have reported measurements of K_x values for typical aromatic solubilizates, benzene (Bz) and benzyl alcohol (BzOH), in SDS and DTAB micelles (Tables 7,8). With only a few exceptions the data reported for each of these solutes are in reasonable agreement; the correspondence of the two NMR methods (FT-PGSE and PRE) is especially encouraging. The case for benzene in SDS deserves special attention, since it has been reported by Hétu et al. [84], that their value for the partition coefficient, obtained from fitting of solution thermodynamic data to a mass-action model, was about double the values of K_x from other techniques (Table 7). Hétu et al. reported similar observations for alcohols in micellar systems [55], and this is certainly in good agree-

TABLE 6 K_x Values for C_8OH (1-octanol) in SDS and DTAB Micelles from a Number of Different Experimental Methods

K_x (SDS)	Method (ref.)	K_x (DTAB)	Method (ref.)
17500 ± 10000	FT-PGSE [29]	3957 ± 2082	NMR-PRE [42]
30000 ± 15000	Taylor disp. [50]	9782	Cond. data [49]
4500	Total sol. [34]		
8743 ± 4450	NMR-PRE [42]		
33980	Part. mol. vol. [72]		

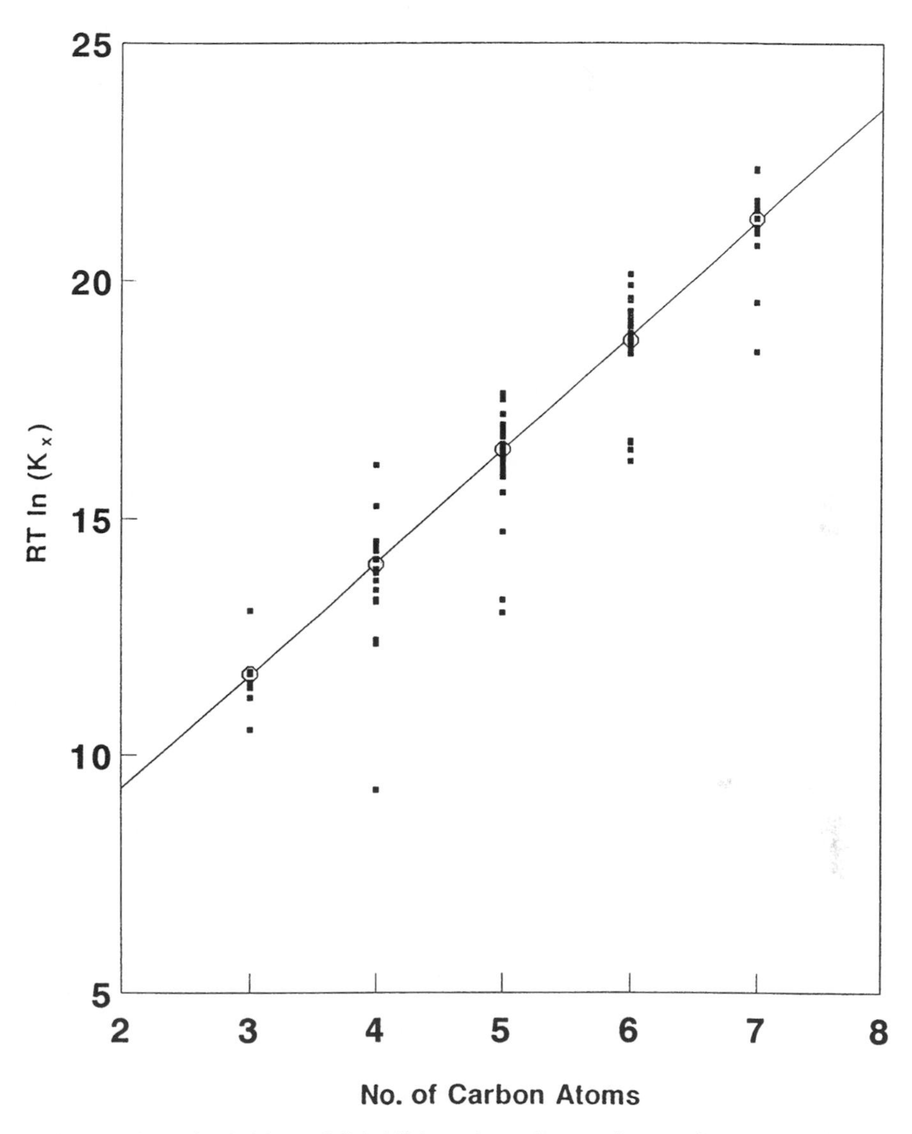

FIG. 1 Plot of $-\Delta_t G^\circ = RT \ln(K_x)$ against the number carbon atoms in the alcohol for n-alcohols in SDS micelles. The circles correspond to the negative of the Gibbs transfer energy calculated from the averaged K_x value, while the solid lines correspond to the least-squares lines for these averaged Gibbs transfer energies vs. the alcohol carbon number.

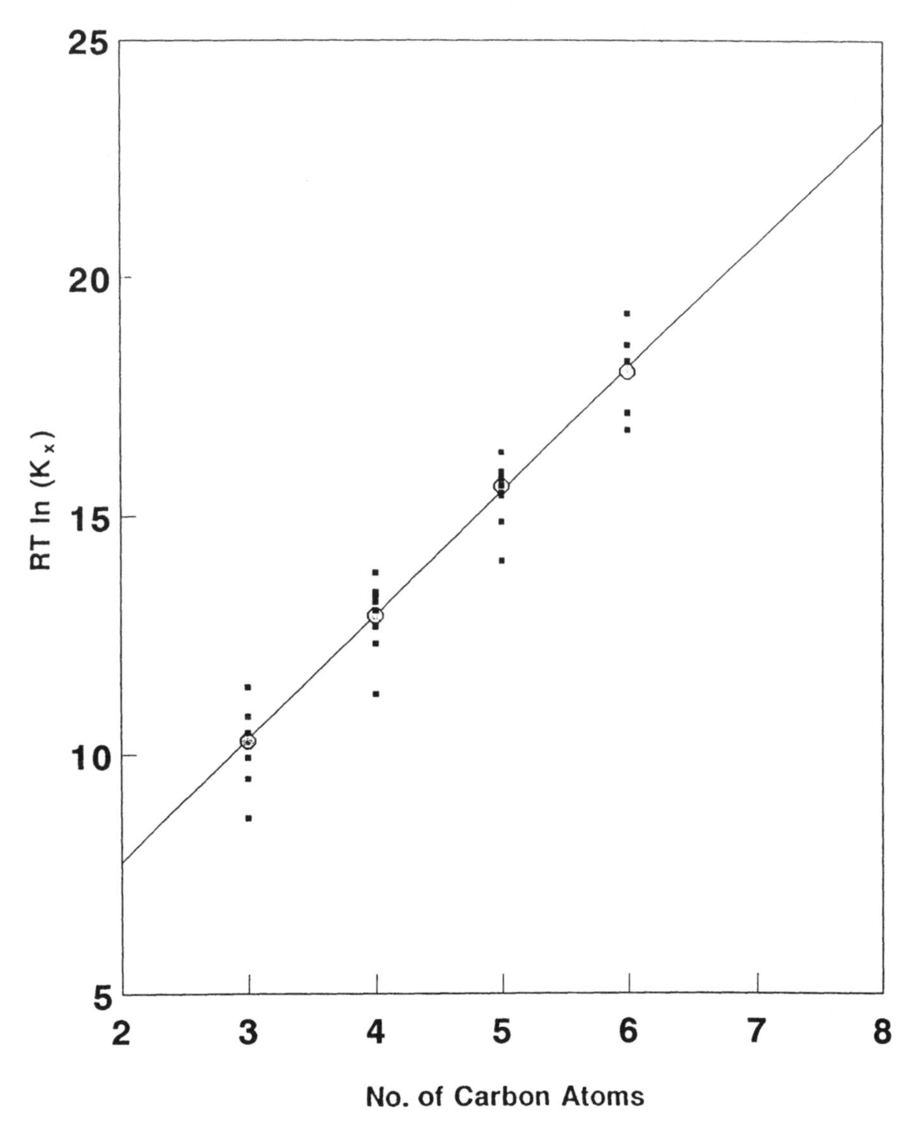

FIG. 2 Plot of $-\Delta_t G^\circ = RT \ln(K_x)$ against the number of carbon atoms in the alcohol for *n*-alcohols in DTAB micelles. The circles correspond to the negative of the Gibbs transfer energy calculated from the averaged K_x value, while the solid lines correspond to the least-squares lines for these averaged Gibbs transfer energies vs. the alcohol carbon number.

TABLE 7 K_x Values for Bz (C_6H_6) in SDS and DTAB Micelles from a Number of Different Experimental Methods

K_x (SDS)	Method (ref.)	K_x (DTAB)	Method (ref.)
2220	MAM (V's, C_p's) [84]	1248 ± 343	NMR-PRE [28, 41]
1022 ± 79	FT-PGSE [29]	1253 ± 285	FT-PGSE [88]
1200	Fluores. [85]	1000	Cond. data [43]
1100	NMR-PRE [42]	888	MAM ($\Delta_t H°$) [49]
1094	Fluor. [86]		
1288	Cond. data [43]		
1000	MAM ($\Delta_t H°$) [49]		
1100	Total sol. [87]		

ment with the data collection reported here for *n*-alcohols, Bz, and BzOH in SDS and DTAB micelles.

All measurements of partition coefficients require assumptions regarding the experimental methodology. In a number of cases, far-reaching model assumptions are necessary. The excellent reviews by Nguyen, Christian, and Scamehorn [3], and by Sepulveda, Lissi, and Quina [1] discuss a number of these assumptions, and the papers by Hétu, Roux, and Desnoyers [84] and by Nguyen, Venable, and Bertrand [75] describe the assumptions made in the thermodynamic model calculations on which the calorimetric methods are based. It is clear that the measurement of the partition coefficients is not a simple matter (judged from the scatter in the partition coefficients in this chapter), and that error limits and model assumptions of each specific method or investigation need to be examined.

TABLE 8 K_x Values for BzOH (benzyl alcohol) in SDS and DTAB Micelles from a Number of Different Experimental Methods

K_x (SDS)	Method (ref.)	K_x (DTAB)	Method (ref.)
376 ± 49	FT-PGSE [29]	517 ± 68	FT-PGSE [30]
578 ± 79	FT-PGSE [30]	473 ± 98	NMR-PRE [28, 41]
376 ± 82	NMR-PRE [28, 41]	411 ± 65	NMR-PRE [42]
360 ± 62	NMR-PRE [42]	1000	Cond. data [43]
437	Cond. data [43]	252 ± 29	Raman spect. [39]
440	Cond. data [47]		
1220	Diff. UV spec. [89]		

TABLE 9 Assorted K_x Values for Miscellaneous Alcohols in CTAB Micelles from a Number of Different Experimental Methods

Alcohol	K_x (CTAB)	Method (ref.)
C_3OH	28	Total sol. [68, 89]
	320	Microcal. [90]
	30	Cond. data [49]
C_4OH	56	Total sol. [68, 89]
	550	Microcal. [90]
	161	Cond. data [49]
	99 ± 2	Cond. data [91]
C_4E_1	500	MAM ($\Delta_{dil}H^{ex}$) [44]
	148 ± 40	Cond. data [92]
	113 ± 30	Cond. data [93]
C_5OH	205	Total sol. [68, 89]
	1500	Microcal. [87]
	833	MAM (ϕ_{C_p},ϕ_v) [79]
C_6OH	566	Total sol. [68, 89]
	4400	Microcal. [90]
	1453	Cond. data [49]
	1558 ± 48	Cond. data [91]
C_7OH	777	Total sol. [68, 89]
C_8OH	7551	Cond. data [49]
	13905 ± 740	Cond. data [91]
BzOH	405	Microcal. [90]
	210	Fluor. [94]
	1041 ± 186	Cond. data [91]

V. DISCUSSION OF THE NMR PARAMAGNETIC RELAXATION ENHANCEMENT METHOD

The NMR paramagnetic relaxation enhancement (NMR-PRE) method is a recent addition to the arsenal of methods now available to study partitioning of water-soluble solutes in micellar systems. The method does not depend on model assumptions for the micelle to extract the p-value or K_x, and in most cases it can be applied to the same systems as the FTPGSE NMR diffusion method described in detail elsewhere in this volume. In this final section we will give a critical discussion of the assumptions and conditions which govern the NMR-PRE method.

The tables of K_x values demonstrate that the results of the NMR-PRE method are generally in good agreement with those of the FT-PGSE NMR diffusion method, and with the average values for all methods for the cationic surfactants, and for nonpolar solubilizates such as benzene for SDS and DTAB. On the other hand, for *n*-alcohols in SDS the method seems to yield lower results, although even here in many cases the agreement with the average of all methods is reasonable. In principle, NMR experiments (both the FT-PGSE and the NMR-PRE) measure directly the number of nuclei in the "free" versus the "bound" state, assuming a two-site model for solubilization holds, and should be free from assumptions about activity coefficients of solubilizates in micellar systems.

The main assumption with the NMR-PRE method is that the relaxation rate of the micellar bound solubilizate ($R_{l,\mathrm{mic}}(\mathrm{A})$) is unaffected by the paramagnetic ion in the aqueous solution, and that the measurement of $R_l^{\mathrm{p}}(\mathrm{A})$ includes only the contribution to the relaxation of the aqueous solubilized species by the paramagnetic ion. Admittedly, it is difficult to exclude the possibility that either the 3-carboxylate-proxyl anion or the Mn^{2+} (aq) cation has some interaction with the micelle, thereby affecting the relaxation rate of the micellar bound solubilizate. In using the method, the possibility that the paramagnetic ions may have an effect on the additive solubilized in the micellar phase, should always be checked by monitoring the change in the relaxation rate of the surfactant headgroups in the presence and absence of paramagnetic ions; in the systems reported these checks have consistently shown that the paramagnetic ion has little effect on the relaxation rate of protons in the surfactant headgroup region. As an example, spin-lattice relaxation rates of the protons in the α-CH_2 position of SDS micelles, using the sodium salt of 3-carboxylic-proxyl as paramagnetic probe, are constant within experimental error up to 10 mmolal of 3-carboxylate-proxyl [39]. In DTAB micelles, using Mn^{2+}(aq) as the paramagnetic probe, the relaxation rate for the *N*-methyl groups on the surfactant is not affected significantly until the paramagnetic ion concentration exceeds 1.0 mmolal [42]. Such results indicate that the concentration of the paramagnetic relaxation agent should be kept as low as possible. However, it is usually necessary to maintain a high concentration of paramagnetic probe, since the error in the *p*-values from the NMR-PRE is dependent on differences in the degree of relaxation enhancement of the solubilizate remaining in aqueous solution and the degree of relaxation enhancement for the same solubilizate in an aqueous solution in the absence of surfactant.

In Table 10, we present the measured ^1H spin-lattice relaxation times of the α-CH_2 group of 0.044 *m* 1-butanol, and of the phenyl protons of 0.058 molal benzyl alcohol, and the degrees of solubilization (*p*-values) of the alcohols in a 0.162 molal DTAB micellar solution at 35°C (calculated from Eq. (20)), using different concentrations of Mn^{2+}(aq). In Table 11,

TABLE 10 ^{1}H Relaxation Times and Distribution Constants (p) of 0.044 molal 1-Butanol and 0.058 molal Benzyl Alcohol in DTAB Micelles[a] as a Function of the Concentration (mmolal) of $MnCl_2 \cdot 6H_2O$

$C(MnCl_2)$	$T_{1,aq}(A)$/s	$T^p_{1,aq}(A)$/s	$T^p_{1,obs}(A)$/s	$T^p_{1,obs}(A)$/s	p
			1-Butanol		
0.00	4.77 ± 0.19		3.72 ± 0.15		
0.20		2.12 ± 0.08		2.17 ± 0.08	0.27 ± 0.11
0.40		1.16 ± 0.05		1.51 ± 0.06	0.39 ± 0.06
0.60		0.93 ± 0.04		1.17 ± 0.05	0.32 ± 0.06
0.80		0.76 ± 0.03		0.96 ± 0.04	0.30 ± 0.05
1.00		0.63 ± 0.03		0.79 ± 0.03	0.28 ± 0.05
2.00		0.32 ± 0.02		0.45 ± 0.02	0.32 ± 0.04
			Benzyl alcohol (phenyl protons)		
0.00	11.6 ± 0.46		3.56 ± 0.14		
0.40		1.83 ± 0.08		2.23 ± 0.09	0.64 ± 0.06
0.80		1.21 ± 0.05		1.72 ± 0.07	0.59 ± 0.04
1.00		0.97 ± 0.04		1.55 ± 0.06	0.61 ± 0.03
2.00		0.50 ± 0.03		1.00 ± 0.04	0.62 ± 0.03

[a] c_{DTAB} = 0.162 molal; T = 308 K.

the measured ^{1}H spin-lattice relaxation times and the distribution constants (p) of the same concentration of 1-butanol in 0.243 molal SDS micelles at 25°C, as a function of the concentration of 3-carboxylate-proxyl, are presented. The reported error limits in the relaxation times T_1 (approximately 4%), are a reflection of the deviation of the relaxation time data in separate series of measurements. Note that these error limits

TABLE 11 ^{1}H Relaxation Times and Distribution Constants (p) of 0.044 molal 1-Butanol in SDS Micelles[a] as a Function of the Concentration (mmolal) of 3-Carboxylate-Proxyl

C(Proxyl)	$T_{1,aq}(A)$/s	$T^p_{1,aq}(A)$/s	$T^p_{1,obs}(A)$/s	$T^p_{1,obs}(A)$/s	p
0.0	3.38 ± 0.12		2.43 ± 0.10		
2.0		1.47 ± 0.06		1.54 ± 0.06	0.38 ± 0.09
5.0		0.88 ± 0.04		1.14 ± 0.05	0.45 ± 0.06
7.5		0.59 ± 0.02		0.81 ± 0.03	0.42 ± 0.05
10.0		0.43 ± 0.02		0.64 ± 0.03	0.43 ± 0.03
15.0		0.31 ± 0.01		0.50 ± 0.02	0.46 ± 0.04
20.0		0.23 ± 0.01		0.36 ± 0.02	0.42 ± 0.04

[a] c_{SDS} = 0.243 molal; T = 298 K.

are much larger than the estimated error in T_1 obtained from the least-squares fitting procedure in the NMR software, which are generally less than 1%. The estimated error in p (dp) is calculated from the following equation:

$$dp = \left(\frac{(dR^{P}_{I,\mathrm{mic}}(A))^2 + (dR_{I,\mathrm{mic}}(A))^2}{R^{P}_{I,\mathrm{aq}}(A) - R_{I,\mathrm{aq}}(A)} + \frac{(R^{P}_{I,\mathrm{mic}}(A) - R_{I,\mathrm{mic}}(A))^2}{(R^{P}_{I,\mathrm{aq}}(A) - R_{I,\mathrm{aq}}(A))^4} ((dR^{P}_{I,\mathrm{aq}}(A))^2 + (dR_{I,\mathrm{aq}}(A))^2) \right)^{1/2} \quad (29)$$

The results in Tables 10 and 11 indicate clearly that the error in p, as calculated from Eq. (29), decreases fairly rapidly as the paramagnetic ion concentration increases; optimal results are obtained for DTAB micelles when the Mn^{2+}(aq) concentration exceeds 0.60 mmolal, and for SDS micelles when the concentration of 3-carboxylate-proxyl anions exceeds 10 mmolal. As well, it can be seen that within the experimental error represented by dp, the p-values are constant.

At this point in the discussion, it is useful to reiterate the main assumptions of the NMR paramagnetic relaxation experiment stated previously in the introduction of this chapter:

1. Fast exchange of solubilizate between the micellar phase and the aqueous phase is occurring (lifetime in the micellar phase is in the order of 10^{-6}–10^{-3} s)
2. The paramagnetic ions do not form stable complexes with the solubilizate
3. The addition of paramagnetic ions to the aqueous phase has no effect on the spin-lattice relaxation rate of the solubilizate located in the micellar phase.

We will now examine how these assumptions have been verified for alcohols in SDS and DTAB micellar systems, using 3-carboxylate-proxyl and Mn^{2+}(aq) as the paramagnetic probes. Regarding the first assumption, it is known that the micellar phase/aqueous phase exchange times for alcohols are much greater than the spin-lattice relaxation times we have encountered in our measurements (exchange times of alcohols have been found to be on the order of 10^{-5} s [73], whereas the NMR spin-lattice relaxation times are usually in the range of 0.2 to 3.0 s).

Secondly, we must be certain that the paramagnetic ions do not form stable complexes with the additives. The Solomon-Bloembergen-Morgan relationships that were used in the derivation of Eq. (19) are those for the case where the enhancement of the relaxation rate of the solubilizate by the paramagnetic probe is by dipolar coupling of the electron and proton spins, and where complex formation between the additive and the paramagnetic probe is negligible [42]. However, if a stable complex is formed, the solubilization equilibrium may be perturbed (i.e., the distribution of the solubilizate may be shifted toward the aqueous phase), or the relaxation rate of the solubilizate will be dependent on the solubilizate concentration. Gao et al. have examined and verified this assumption for the solubilization of increasing amounts of C_5OH in 0.162 molal DTAB micelles, at a constant Mn^{2+}(aq) concentration [28]. Their results indicate quite clearly that the Mn^{2+}(aq) does not form complexes with the neutral alcohol (indicated by the constancy of the R_1 of increasing amounts of C_5OH at constant Mn^{2+}(aq) concentration), and that the solubilization equilibria were not perturbed (i.e., the p-values are unchanged). As well, from Tables 10 and 11 we see that increasing the amount of paramagnetic ion has no effect on the measured p-values, indicating condition 2 is satisfied. Similar results were obtained when this test was applied to the solubilization of C_4OH and 2-butoxyethanol in SDS micelles [95].

The third condition, that the relaxation rate of the solubilizate in the micelles is unchanged by the paramagnetic ion, is more difficult to verify. For example, if the paramagnetic probe and the solubilizate form a complex that solubilizes near the micellar surface, or if the noncomplexed paramagnetic probe is not completely repelled from the micellar surface due to a decrease in the electrostatic repulsions, there exists the possibility that $R_{1,\mathrm{mic}}$(A) will be altered, and would have to be replaced by $R^{p}_{1,\mathrm{mic}}$(A) in Eq. (21), rendering the NMR-PRE unusable in this situation. The possibility that the paramagnetic agent may be influencing the micelle-bound solubilizate can unfortunately not be tested with with NMR experiments; as discussed before, the constancy of the R_1 values of surfactant protons in the headgroup region may be taken as reasonable evidence for the lack of an effect. Fluorescence quenching techniques may provide good evidence that the probe molecule does not affect the micelle-bound solubilizate [96]. It is known that nitroxide radicals will quench the fluorescence of an aromatic probe molecule like pyrene [97]. In the presence of increasing amounts of 3-carboxylate-proxyl, the fluorescence emission of an aqueous solution of pyrene was observed to decrease in a regular fashion, indicating that quenching of the pyrene luminescence by the paramagnetic 3-carboxylate-proxyl was occurring. However, when the same experiment was carried out with pyrene solubilized in SDS micelles, the fluorescence of the pyrene probe was unaffected by increasing concentrations of 3-carboxylate-proxyl, indicating that the probe was repelled completely from the micellar surface, and resided exclusively in the aqueous phase. Not surprisingly, when the same concentrations of 3-carboxylate-proxyl were added to a solution of 0.0750 molal DTAB containing solubilized pyrene, the fluorescence is quenched according to the static equation of Turro and Yekta [98]. However, for a cationic quencher, the pyrene/

Cu^{2+}(aq) probe/quencher pair (similar to Mn^{2+}(aq)), the luminescence of the DTAB solubilized pyrene probe was also found to be constant as the aqueous quencher concentration was increased, again indicating that the Cu^{2+}(aq) resided away from the micellar surface [99].

In the case of benzylalcohol as solubilizate, the *p*-value reported in 0.243 m SDS using 3-carboxylate-proxyl as the paramagnetic ion, is 0.67, leading to $K_x = 360 \pm 62$ (Table 8). This is in excellent agreement with the value reported by Stilbs using the FTPGSE NMR diffusion method [29], although another value for the same quantity reported by Stilbs is higher [30].

These observations may be considered as evidence that Mn^{2+}(aq) and 3-carboxylate-proxyl are reasonable choices for paramagnetic probe molecules when using the NMR-PRE method to obtain information on the solubilization of neutral species in cationic and anionic micelles respectively.

Finally, Gao et al. presented some results for the determination of the *p*-value of a solubilizate using different paramagnetic probes [42]. By comparing the results for the ^1H T_1 values and the *p*-values of 0.058 molal benzyl alcohol in 0.243 molal SDS micelles solution measured using 3-carboxylate-proxyl and $Mn(EDTA)^{2-}$ as the paramagnetic probes, these authors demonstrate that the *p*-values obtained using 3-carboxylate-proxyl and $Mn(EDTA)^{2-}$ are comparable, although the values obtained with $Mn(EDTA)^{2-}$ do appear to be slightly higher. The average of the *p*-values obtained with 3-carboxylate-proxyl (0.67) agrees very well with the distribution constant determined by Stilbs ($p = 0.67$) [29], using the FT-PGSE self-diffusion method. In addition, the *p*-values of benzyl alcohol in SDS, measured using the NMR paramagnetic relaxation experiment with the two separate probes, are independent of the paramagnetic ion concentration; again, the error in *p* decreases as the paramagnetic ion concentration increases. Note that a much higher concentration of 3-carboxylate-proxyl is needed to obtain a similar enhancement of spin-lattice relaxation rate compared to $Mn(D_2O)_6^{2+}$ or $Mn(EDTA)^{2-}$, which is expected from the difference in the number of unpaired electrons in the manganese containing species compared with the 3-carboxylate-proxyl. In agreement with Table 11, the optimum concentration of 3-carboxylate-proxyl was found to be around 10 mmolal.

VI. CONCLUSIONS

It is obvious from the results for the partition coefficients for *n*-alcohols and some aromatic solubilizates in two well-characterized micellar systems, i.e., SDS and DTAB, that values obtained with different methods may vary widely; even for those systems where experimental errors should be smallest, e.g., C_4OH and C_5OH in SDS, the extremes of the reported values for K_x differ by a factor of 10! Some of this variation may be due to a dependence of K_x on the solubilizate concentration [35–37, 71, 81, 84], or, more precisely, on the mole fraction of the solubilizate in the micelle. In general, the concentration dependence of solubilization, in particular for the case of water-soluble additives, is in need of further, careful experimental work using appropriate methods based on well-clarified assumptions.

A comparison of the results for K_x values as reported in Tables 1–8 does not lead to a simple generalization regarding the relative merit of different methods. The expectation that direct concentration methods such as the FT-PGSE NMR diffusion and NMR-PRE techniques, and direct activity methods such as the vapor pressure method should give similar results, is only partially supported by the data. Even for these methods the systematic differences which are sometimes observed are obscured by statistical fluctuations. This problem is even worse when we include data derived from model-dependent techniques such as the calorimetric, volumetric, and conductometric methods. Some statistical weight can be placed on the fact that the K_x values averaged over all measurement techniques, as reported in Figs. 1 and 2, closely follow the prediction of Eq. (9), leading to reasonable values for the factor b, which represents the hydrophobic effect contribution to $\Delta_t G^\circ$ of one added CH_2 group in the solubilizate. It is therefore reasonable to assume that the average $\Delta_t G^\circ$ values in Figs. 1 and 2 represent the best values for each alcohol in SDS and DTAB, and the results from each technique should be compared to these values. In general we conclude that the measurement of the partition coefficients of solubilizates in micellar solutions is not a simple exercise. For each method due attention should be paid to the various experimental and theoretical considerations necessary to extract partition coefficient data for a complicated process such as the solubilization of a water-soluble neutral solubilizate in micellar systems.

The NMR paramagnetic relaxation enhancement method (NMR-PRE) has been described in some detail. We conclude that in particular that the partition coefficients in SDS micelles obtained using this technique are somewhat lower than those obtained from other techniques. The method seems to yield its best results in cationic micelles, using Mn^{2+}(aq) as the relaxation agent. The lower values in SDS may be due to a breakdown of the key assumption in the method, i.e., the exclusion of the relaxation agent from the micellar surface. In most cases however, the agreement between the two concentration-based NMR methods (FTPGSE and NMR-PRE) is satisfactory.

SYMBOLS

K_x	mole-fraction-based partition coefficient
K_c	molarity-based partition coefficient
$\Delta_{m,ps}G^\circ$	standard Gibbs energy of micellization from the phase separation model
$\Delta_{m,ma}G^\circ$	standard Gibbs energy of micellization from the mass-action model
$\gamma_\pm$	mean surfactant activity coefficient
A	additive
S	surfactant
$\mu_{A,aq}$	chemical potential of the additive in the aqueous phase
$\mu_{A,mic}$	chemical potential of the additive in the micellar phase
$X_{A,aq}$	mole fraction of additive in the aqueous phase
$X_{A,mic}$	mole fraction of additive in the micellar phase
z	number of associated counterions
$\gamma_{A,aq}$	activity coefficient of the additive in the aqueous phase
$\gamma_{A,mic}$	activity coefficient of the additive in the micellar phase
NMR	nuclear magnetic resonance
$\Delta_t G^\circ$	Gibbs energy of transfer
a	intercept of the plot of the free energy of transfer of a homologous series of hydrocarbon solutes vs. the number of carbon atoms in the solute
b	slope of the plot of the free energy of transfer of a homologous series of hydrocarbon solutes vs. the number of carbon atoms in the solute
n_c	number of carbon atoms in a hydrocarbon solute
$\Delta_t H^\circ$	enthalpy of transfer
$\Delta_t S^\circ$	entropy of transfer
CMC	critical micelle concentration
SDS	sodium dodecylsulfate
DTAB	dodecyltrimethylammonium bromide
N	surfactant aggregation number
$C_{A,mic}$	concentration of micellar solubilized additive
$C_{A,t}$	total counterion concentration
α	fraction of free counterions
p	fraction of solubilized alcohol
$n_{A,mic}$	concentration of additive in the micellar phase
$n_{A,t}$	total concentration of additive
$\hbar$	Planck's constant/2π
I	spin quantum number
μ	magnetic dipole moment

γ	magnetogyric ratio
B_0	static magnetic field strength
ω_0	Larmor frequency
σ	shielding constant
δ	chemical shift
σ_r	reference shielding constant
σ_s	shielding constant of sample
TMS	tetramethylsilane
M_z°	total macroscopic magnetic moment
N	total number of nuclei
k_B	Boltzmann's constant
T	absolute temperature
B_1	radio-frequency magnetic field strength
M_z	z component of the macroscopic magnetic moment
T_1	spin-lattice relaxation time
R_1	spin-lattice relaxation rate
$R_1^p(A)$	relaxation rate of solubilizate in the presence of paramagnetic ions
$R_{1,aq}(A)$	relaxation rate of solubilizate in the aqueous phase
γ_I	magnetogyric ratio of spin I
γ_S	magnetogyric ratio of spin S
ω_I	Lamor frequency of spin I
ω_S	Lamor frequency of spin S
N_{el}	number density of electrons
b_c	distance of closest approach of spin I and spin S
$J(\omega)$	spectral density at angular frequency ω
$R_{1,obs}(A)$	relaxation rate of an additive distributed between the aqueous and the micellar phase
$R_{1,mic}(A)$	relaxation rate of micellar solubilized additive
$R_{1,aq}(A)$	relaxation rate of aqueous solubilized additive
FT	Fourier transform
FT-PGSE	Fourier transform, pulsed-gradient spin echo
$\Delta_{fus}H^\circ$	enthalpy of fusion for the process of a hydrated solid surfactant melting to form a micelle
T	Krafft temperature of surfactant in the presence of additives
T_0	Krafft temperature of surfactant in the absence of additives
MAM (meth)	mass-action model fitting of solution thermodynamic data where meth. indicated the data chosen for the fit, e.g., V = volumes, $\Delta_{dil}H^\circ$ measurements, C_p = heat capacity measurements, and k = compressibilities

CTAB	cetyltrimethylammonium bromide
vp_0	additive vapor pressure above a solution in the absence of micelles
vp_m	additive vapor pressure above a micellar solution
C_3OH	1-propanol
C_4OH	*n*-butanol
C_4E_1	ethylene glycol mono-*n*-butyl ether (2-butoxyethanol)
C_5OH	1-pentanol
C_6OH	1-hexanol
C_7OH	1-heptanol
C_8OH	1-octanol
Bz	benzene
BzOH	benzyl alcohol

REFERENCES

1. Sepulveda, L.; Lissi, E.; Quina, F. *Adv. Colloid Interface Sci. 25*: 1 (1986).
2. Melo, E. C. C.; Costa, S. M. B. *J. Phys. Chem. 91*: 5635 (1987).
3. Nguyen, C. M.; Christian, S. D.; Scamehorn, J. F. *Tenside Surf. Det. 25*: 6 (1988).
4. Tanford, C. *The Hydrophobic Effect: The Formation of Micelles and Biological Membranes*, 2nd ed., Wiley; New York, 1980.
5. Ben-Naim, A. *J. Phys. Chem. 82*: 797 (1978).
6. Lindman, B.; Wennerström, H. *Top. Current Chem. 87*: 1 (1980).
7. Spink, C. H.; Colgan, S. *J. Phys. Chem. 87*: 888 (1983).
8. Holland, P. M.; Rubingh, D. N., (eds.), *Mixed Surfactant Systems*, ACS Symp. Ser., No. 501, American Chemical Society, Washington DC, 1992.
9. Zana, R.; Yiv, S.; Straizelle, C.; Lianos, P. *J. Colloid Interface Sci. 80*: 208 (1981).
10. Scamehorn J. F. (ed.), *Phenomena in Mixed Surfactant Systems*, ACS Symp. Ser., No. 311, American Chemical Society, Washington DC, 1986.
11. Mittal, K. L., (ed.), *Micellization, Solubilization, and Microemulsions*, Plenum, New York, 1977.
12. Harwell, J. H.; Scamehorn, J. F., (eds.), *Surfactant Based Separation Processes*, Surf. Sci. Ser., No. 33, Marcel Dekker, New York, 1989.
13. Yamashita, F.; Perron, G.; Desnoyers, J. E.; Kwak, J. C. T. *J. Colloid Interface Sci. 114*: 548 (1986).
14. Vikingstad, E.; Kvammen, O. *J. Colloid Interface Sci. 74*: 16 (1980).
15. Almgren, M.; Swarup, S. *J. Colloid Interface Sci. 91*: 256 (1983); *J. Phys. Chem. 86*: 4212 (1982); *J. Phys. Chem. 87*: 876 (1983).
16. Kibblewhite, J.; Drummond, C. J.; Greiser, F.; Healy, T. W. *J. Phys. Chem. 91*: 4658 (1987).
17. Zana, R.; Candeau, S. *J. Colloid Interface Sci. 84*: 206 (1981).
18. Lianos, P.; Zana, R.; *J. Colloid Interface Sci. 101*: 587 (1984).

19. Yiv, C.; Zana, R.; Ulbricht, W.; Hoffman, H. *J. Colloid Interface Sci. 80*: 224 (1981).
20. Harris, R. K. *Nuclear Magnetic Resonance Spectroscopy: A Physico-Chemical View*, Longman Press, Essex, UK, 1986.
21. Wasylishen, R. E. in *NMR Spectroscopy Techniques* (Lichter, R. L.; Dybowski, C., eds.), Marcel Dekker, New York, 1986, pp. 45–91.
22. Farrar, T. C. *An Introduction to Pulse NMR*, Farragut Press, Chicago, 1987.
23. Solomon, I; Bloembergen, N. *J. Chem. Phys. 25*: 261 (1956).
24. Bloembergen, N; Morgan, L. O. *J. Chem. Phys. 34*: 842 (1961).
25. Hwang, L.-P.; Freed, J. H. *J. Chem. Phys. 63*: 4017 (1975).
26. Freed, J. H. *J. Chem. Phys. 68*: 3034 (1978).
27. Stilbs, P. *Prog. NMR Spectros. 19*: 1 (1987).
28. Gao, Z.; Wasylishen, R. E.; Kwak, J. C. T. *J. Phys. Chem. 93*: 2190 (1989).
29. Stilbs, P. *J. Colloid Interface Sci. 87*: 385 (1982).
30. Stilbs, P. *J. Colloid Interface Sci. 89*: 547 (1982).
31. Gao, Z.; Wasylishen, R. E.; Kwak, J. C. T., unpublished results.
32. Muto, Y.; Yoda, K.; Yoshida, N.; Esumi, K.; Meguro, K. *J. Colloid Interface Sci. 130*: 165 (1989).
33. Gettins, J.; Hall, D.; Jobling, P. L.; Rassing, J. E.; Wyn-Jones, E. *Trans. Faraday Soc. 74*: 1957 (1978).
34. Hoiland, H.; Ljosland, E.; Backlund, S. *J. Colloid Interface Sci. 101*: 467 (1984).
35. Lee, B.-H.; Christian, S. D.; Tucker, E. E.; Scamehorn, J. F. *J. Phys. Chem. 95*: 360 (1991).
36. Christian, S. D.; Tucker, E. E.; Smith, G. A.; Bushong, D. S. *J. Colloid Interface Sci. 113*: 439 (1986).
37. Nguyen, C. M.; Scamehorn, J. F.; Christian, S. D. *Colloids and Surfaces 30*: 335 (1988).
38. Kaneshina, S.; Kamaya, H.; Ueda, I. *J. Colloid Interface Sci. 83*: 589 (1981).
39. Shih, L. B.; Williams, R. W. *J. Phys. Chem. 90*: 1615 (1986).
40. Baglioni, P.; Kevan, L. *J. Phys. Chem. 91*: 1516 (1987).
41. Gao, Z. Ph.D. thesis, Dalhousie University, 1989.
42. Gao, Z.; Labonte, R.; Kwak, J. C. T.; Marangoni, D. G.; Wasylishen, R. E. *Colloids and Surfaces 45*: 269 (1990).
43. Treiner, C.; Mannebach, M.-H. *J. Colloid Interface Sci. 118*: 243 (1987).
44. De Lisi, R.; Milioto, S. *J. Solution Chem. 17*: 245 (1988).
45. De Lisi, R.; Lizzio, A.; Milioto, S.; Turco-Liveri, V. *J. Solution Chem. 15*: 623 (1986).
46. De Lisi, R.; Milioto, S.; Tuco-Liveri, V. *J. Colloid Interface Sci. 117*: 64 (1987).
47. Treiner, C. *J. Colloid Interface Sci. 93*: 33 (1983).
48. De Lisi, R.; Milioto, S.; Castagnolo, M.; Inglese, A. *J. Solution Chem. 19*: 767 (1990).
49. Causi, S.; De Lisi, R.; Milioto, S. *J. Solution Chem. 19*: 995 (1990).
50. Leaist, D. G. *J. Solution Chem. 20*: 175 (1990).

51. Hayase, K.; Hayano, S.; Tsubota, H. *J. Colloid Interface Sci. 101*: 336 (1984).
52. Hayase, K.; Hayano, S. *Bull. Chem. Soc. Jpn. 50*: 83 (1977).
53. Marangoni, D. G.; Kwak, J. C. T. *Langmuir 7*: 2083 (1991).
54. Marangoni, D. G.; Rodenhiser, A. P.; Thomas, J. M.; Kwak, J. C. T. in *Mixed Surfactant Systems* (Holland, P. M.; Rubingh, D. M., eds.), ACS Symp. Ser., American Chemical Society, Washington, DC, 1992, Chap. 11.
55. Hétu, D.; Roux, A. H.; Desnoyers, J. E. *J. Solution Chem. 16*: 529 (1987).
56. De Lisi, R.; Milioto, S. *J. Solution Chem. 16*: 767 (1987).
57. Abu-Himidiyyah, M.; Kumari, K. *J. Phys. Chem. 94*: 2518 (1990).
58. De Lisi, R.; Turco Liveri, V.; Castagnolo, M.; Inglese, A. *J. Solution Chem. 15*: 23 (1986).
59. Abu-Himidiyyah, M. *J. Phys. Chem. 89*: 2377 (1985).
60. Abu-Himidiyyah, M.; El-Danab, C. *J. Phys. Chem. 87*: 5443 (1983).
61. De Lisi, R.; Genova, C.; Testa, R.; Turco Liveri, V. *J. Solution Chem. 13*: 121 (1984).
62. Manabe, M.; Kawamura, H.; Kondo, S.; Kojima, M.; Tokunaga, S. *Langmuir 6*: 1596 (1990).
63. Manabe, M.; Kawamura, H.; Sugihara, G.; Tanaka, M. *Bull. Chem. Soc. Jpn. 61*: 1551 (1988).
64. Manabe, M.; Koda, H. *Bull. Chem. Soc. Jpn. 51*: 1599 (1978).
65. Hayase, K.; Hayano, S. *Bull. Chem. Soc. Jpn. 51*: 933 (1978).
66. De Lisi, R.; Milioto, S.; Castagnolo, M.; Inglese, A. *J. Solution Chem. 16*: 373 (1987).
67. Hoiland, H.; Ljosland, E.; Backlund, S. *J. Colloid Interface Sci. 114*: 9 (1986).
68. Lianos, P.; Zana, R. *Chem. Phys. Lett. 76*: 62 (1980).
69. Bury, R.; Treiner, C. *J. Solution Chem. 18*: 499 (1989).
70. Treiner, C.; Khodja, A. A.; Fromom., M.; Chevalet, J. *J. Solution Chem. 18*: 217 (1989).
71. Hartland, G. V.; Grieser, F.; White, L. R. *J. Chem. Soc. Faraday Trans. 1 83*: 591 (1987).
72. Manabe, M.; Shirahama, K.; Koda, M. *Bull. Chem. Soc. Jpn. 49*: 2904 (1976).
73. Almgren, M.; Grieser, F.; Thomas, J. K. *J. Chem. Soc. Faraday Trans. 1 75*: 1674 (1979).
74. Treiner, C.; Bocquet, J.-F.; Pommier, C. *J. Phys. Chem. 90*: 3052 (1986); Treiner, C.; Khoda, A. A.; Fromon, M. *Langmuir 3*: 729 (1987).
75. Nguyen, D.; Venable, R. L.; Bertrand, G. L. *Colloids and Surfaces 65*: 231 (1992).
76. Treiner, C. *J. Colloid Interface Sci. 90*: 444 (1982).
77. De Lisi, R.; Fisicaro, E.; Milioto, S. *J. Solution Chem. 18*: 403 (1989).
78. De Lisi, R.; Milioto, S.; Verrall, R. E. *J. Solution Chem. 19*: 97 (1990).
79. De Lisi, R.; Milioto, S.; Triolo, R. *J. Solution Chem. 17*: 673 (1988).

80. Hoiland, H.; Blokhus, A. M.; Backlund, S. *J. Colloid Interface Sci. 107*: 576 (1985).
81. Abuin, E. B.; Lissi, E. A. *J. Colloid Interface Sci. 95*: 198 (1983).
82. Weers, J. G. *J. Am. Oil Chem. Soc. 67*: 340 (1990).
83. Spink, C. H.; Colgan, S. *J. Colloid Interface Sci. 97*: 41 (1984).
84. Hétu, D.; Roux, A. H.; Desnoyers, J. E. *J. Colloid Interface Sci. 122*: 418 (1988).
85. Hirose, H.; Sepulveda, L. *J. Phys. Chem. 85*: 3689 (1981).
86. Abuin, E. B.; Valenzuela, E.; Lissi, E. A. *J. Colloid Interface Sci. 101*: 401 (1984).
87. Simon, S. A.; McDaniel, R. V.; McIntosh, T. J. *J. Phys. Chem. 86*: 1449 (1982).
88. Kawamura, H.; Manabe, M.; Miyamoto, Y.; Fujita, Y.; Tokunaga, S. *J. Phys. Chem. 93*: 5536 (1989).
89. Gettins, J.; Hall, D.; Jobling, P. L.; Rassing, J. E.; Wyn-Jones, E. *Trans. Faraday Soc. 74*: 1957 (1978).
90. Treiner, C.; Chattopdyhay, A. K.; Bury, R. *J. Colloid Interface Sci. 104*: 565 (1985).
91. Abu-Himidiyyah, M. *J. Phys. Chem. 90*: 1345 (1986).
92. Quirion, F.; Drifford, M. *Langmuir 6*: 786 (1990).
93. Quirion, F.; Magid, L. J. *J. Phys. Chem. 90*: 5193 (1986).
94. Lissi, E. A.; Abuin, E. B., Rocha, A. M. *J. Phys. Chem. 84*: 2406 (1980).
95. Marangoni, D. G.; Kwak, J. C. T., unpublished results.
96. Gao, Z.; Wasylishen, R. E.; Kwak, J. C. T. *J. Chem. Soc. Faraday Trans. 87*: 947 (1991).
97. Blatt, E,; Chatelier, R. C.; Sawyer, W. H. *Photochem. Photobiol. 39*: 477 (1984).
98. Turro, N. J.; Yekta, A. *J. Am. Chem. Soc. 100*: 5951 (1978).
99. Marangoni, D. G.; Rodenhiser, A. P.; Kwak, J. C. T., unpublished results.

V
Applications of Solubilization

15

Solubilization and Detergency

GUY BROZE Advanced Technology, Colgate-Palmolive Research & Development, Inc., Milmort, Belgium

SYNOPSIS

To deliver products satisfying different consumers, the detergent industry has to master not only the formulations but also the ways they interact with the different types of soils and surfaces. The rules to optimize the solubilization of oily materials in surfactant aqueous solutions are discussed as a function of temperature, amphiphile structure, additives in water, and oil nature. The solubilization of oil or grease in nonaqueous media is also discussed in terms of solvent solubility parameters and molar volume.

I. THE CHALLENGE OF THE DETERGENT INDUSTRY

The detergent industry is faced with the challenge to deliver consumer products able to remove a great variety of stains from a great variety of substrates with a minimum effort by the user, with safe raw materials. The soils and stains to be removed are not the same in fabric care (laundry cleaning) and in hard surface care (floor, window, bathroom cleaning). Nevertheless, they can be classified into groups: greasy soils, particulate soils, protein and starchy soils, bleachable (oxidizable) soils, etc. In laundry cleaning, the substrate is a soft fiber, but its chemical nature can be hydrophilic, such as cotton, or more hydrophobic, such as polyester and even polyamid fibers. In hard surface care, both surface hardness and chemical nature can be very different: metal, plastic, glass, ceramic, wood, marble, etc.

As the products are to be used by regular, untrained consumers, the formulated products must be safe for the user. They also need to be safe to the substrate, and to the environment, as the product will eventually be disposed. At the present time, safety to user, substrate and environment are no longer enough. The current trend is to formulate consumer products from renewable raw materials. This is excellent from an ecological standpoint, but it significantly reduces the degrees of freedom of the formulator.

The kinetics of cleaning is another essential factor to take into account. The consumer expects a product that will deliver its cleaning function in a reasonable period of time, with a minimum of mechanical action.

Compared to other industrial sectors such as pharmacology and the pesticide industry, the household cleaning industry must meet additional challenges. The formulator has only limited control over some very important parameters, such as product usage concentration, temperature, quality of water (hardness), time of contact with the substrate, composition of the soils, etc. This problem can only be addressed by formulation in such a way that the impact of the lesser controllable parameters on product performance is minimized, and by providing the consumer with guidelines (recommended temperature, amount of detergent required as a function of water hardness and soil, etc.)

It is not surprising, therefore, that for a long time, detergent recipes remained based on empiricism and were obtained by trial and error. Today, the formulation of more effective and safer household cleaners requires more and more multidisciplinary scientific inputs. Physical chemistry is probably the most relevant discipline, but organic chemistry, rheology, biochemistry, and applied mathematics (statistics) are also playing an essential role in solving the detergent industry challenges.

II. GREASE REMOVAL MECHANISMS

The objective of a grease cleaner is to remove grease from a substrate. This can be achieved by essentially two mechanisms: roll-up, according to which grease is detached from the substrate but keeps its nature, and solubilization, according to which grease is "dissolved" in the washing liquid. In fact, in the later case, the grease is dispersed in tiny droplets, small enough not to scatter light significantly.

A. Roll-Up

The roll-up mechanism involves several successive steps. First, the surfactant molecules present in the liquid phase (washing liquor) adsorb on the grease as well as on the substrate. This step results in the reduction of grease-water and substrate-water interfacial tensions. This step can be referred to as the wetting step.

The surface tensions at these two interfaces are reduced to the same range as the substrate-grease interfacial tension. Increasing the water-grease surface at the expense of the grease-substrate now requires much less mechanical energy, and convection or mild mechanical agitation may be enough to detach a grease droplet from the surface.

Whether or not a thin grease layer remains on the substrate depends on the actual value of the grease-substrate interfacial tension. If the grease-substrate interfacial tension is smaller than the water-substrate interfacial tension, a thin film of grease will remain. This is the case for instance when degraded grease is chemisorbed to the surface of the substrate, for instance by calcium ion bridging.

The above description is appropriate if the grease is liquid. If the temperature is below the melting point of the grease or if the grease contains significant amounts of solids, the grease layer is solid or pasty and a drop cannot be formed. Grease removal can be possible nevertheless if the adsorbed surfactant is able to penetrate the grease and plastify it (make it more liquid). Practice shows that this can be achieved by some C12–C14 alkyl-chain-based surfactants.

The grease removed from the substrate remains suspended in the washing liquor as coarse droplets, surrounded by a layer of adsorbed surfactant molecules (milky emulsion). This emulsion is not physically stable. When grease droplets collide, coalescence can occur and a grease layer eventually forms. In fabric cleaning, the surface area of the substrate is very large. The grease droplets can deposit on the fabric surface, leading to so-called "redeposition" phenomena.

The turbidity of the washing liquor is sometimes used as an estimate of the grease cutting ability of a given surfactant system. This is valid only within a very narrow range of conditions, because turbidity is not only a function of the amount of grease emulsified but also of particle size.

B. Solubilization

Above the critical micelle concentration (CMC), surfactant molecules in aqueous solution associate in submicronic aggregates called "micelles." The hydrophobic part of the surfactant occupies the core of the micelles and the hydrophilic head groups are pointing outside toward the water.

The size of a micelle is determined to a large extend by the length of the hydrocarbon chain. Typical micelle sizes range from 5 nm to less than 100 nm. Micelles are much smaller than the wavelength of light, so a micellar solution is essentially transparent to the eye. Quasielastic light scattering nevertheless allows the characterization of micellar systems.

The shape of a micelle is determined by the ratio of the hydrodynamic volumes of the hydrophilic and hydrophobic parts of the surfactant molecules. A highly hydrophilic surfactant will usually form globular micelles. A less hydrophilic surfactant will tend to generate rodlike micelles. An even less hydrophilic surfactant may form disklike micelles, but usually it will not form a stable micellar system and separates into two immiscible phases.

The "solubilization" mechanism implies the solubilization of grease in the inner (hydrocarbon) core of the micelles rather than in the water phase. "Solubilization" corresponds to the formation of swollen micelles. Their size is comparable to that of to the original micelles, although slightly bigger, but still far below the wavelength of light, so the system appears transparent. Swollen micelle systems can be considered as oil-in-water microemulsions, since as opposed to grease droplet emulsions, they are physically stable.

III. RULES TO OPTIMIZE OILY MATERIAL SOLUBILIZATION

The amount of grease that a given system can "solubilize" or "microemulsify" depends on several parameters: temperature, surfactant structure, oil/grease nature, and presence of substances dissolved in the water. The way solubilization capacity is affected is clearly established in the case of nonionic surfactant-based systems.

A. General Phase Behavior of Ethoxylated Fatty Alcohols

The interest in ethoxylated nonionic amphiphiles is very great, both in the industry and in the academic world. Their solution properties are really astonishing and unique. The length of their hydrocarbon, lipophilic tail, as well as their water-soluble, polyethoxylated head group can be almost continuously varied, resulting in tailor-made hydrophilic lipophilic balance (HLB). Moreover, similar HLBs can be achieved with different amphiphile molar volumes. As a result, they can maintain oil and water (or brine) in a single, microemulsion phase without any cosurfactant, in contrast to ionic amphiphiles [1].

Another essential advantage of nonionic amphiphiles compared to ionic amphiphiles is that they are not easily salted out. They can accordingly be formulated in the presence of a high electrolyte concentration. Unlike anionics, they are not particularly sensitive to alkaline-earth cations.

For these reasons, they are key candidates for practical applications in multiple areas such as oil recovery, pesticide industry, household cleaning, cosmetics, fine organic synthesis, etc. [1–5].

The isothermal phase behavior of brine-oil-amphiphile mixtures can be represented in a triangular phase diagram. Theoretically, only mixtures of three pure substances can be represented in a triangle; nevertheless, pseudoternary phase diagrams can be obtained by a clever grouping of ingredients. Specifically, diluted electrolyte aqueous solutions (brines) can be considered as single ingredient. Such a practice can, in some cases, lead to errors and must be used with care.

The phase behavior of brine-oil-amphiphile systems can be reduced to three main classes, corresponding to the Winsor classification:

Winsor I, when the amphiphile is significantly more soluble in the brine than in the oil.

Winsor II, when the amphiphile is significantly more soluble in the oil than in the brine.

Winsor III, when the surfactant partition coefficient between the brine and the oil is close to the unity. The Winsor III phase topography is characterized by a triangular three phase domain near the brine-oil axis.

As widely described in the literature [6–9], the phase topography of such ternary systems strongly depends on several factors such as the amphiphile HLB, the type of the oil, the presence of additives in water (electrolytes, hydrotropes, e.g.), and thermodynamic fields such as temperature and pressure.

1. Effect of Temperature

The effect of temperature on ethoxylated nonionic amphiphiles is now understood. The presence of a lower consolute temperature (cloud point) in the water-amphiphile binary mixture phase diagram implies that the amphiphile-water compatibility decreases as temperature rises. On the other hand, the presence of a higher consolute temperature (haze point) in the amphiphile-oil binary mixture phase diagram implies that the amphiphile-oil compatibility increases as the temperature rises. The net result is a shift in solubility of the amphiphile from the water phase to the oil phase as the temperature rises. The details of the temperature evolution of these ternary systems has been very well described by Kahlweit et al. [6]. We shall only recall the most essential features (see Fig. 1).

At a sufficiently low temperature, the system is a Winsor I. The tie lines of the two-phase domain are oriented towards the oil corner. An isothermal critical point (plait point) can be observed near the oil corner if the temperature is above the haze point temperature.

When the temperature is raised, a critical tie line "opens" to give a narrow three-phase domain, with a second plait point close to the small side of the three-phase triangle. The critical temperature at which the Winsor III configuration appears is referred to as Tl.

Increasing the temperature allows the three-phase triangle to further "open." At the so-called phase inversion temperature (PIT), the amphiphile rich corner of the three-phase triangle is just located at a point characterized by equal amounts of oil and water. The amount of amphiphile at this point ($\tilde{\gamma}$) is the minimum ever needed to maintain oil and water in a single, usually microemulsion phase.

At a higher temperature, the three-phase triangle closes through a second critical tie line, oriented toward the water corner. This critical temperature at which the Winsor III configuration vanishes to give The Winsor II is referred to as Tu.

If the temperature is further increased, nothing special occurs. The tie lines are pointing to the water corner. If the cloud point temperature (CPT) is not reached yet, a plait point is observed close to the water phase.

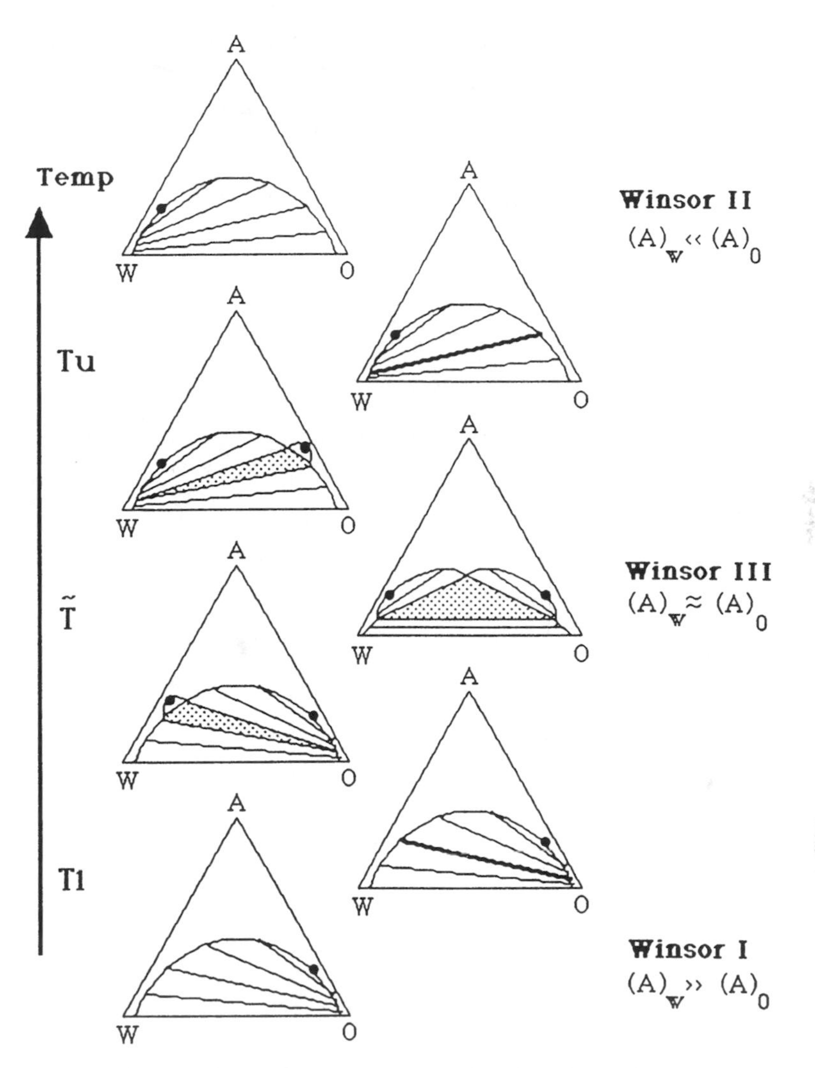

FIG. 1 Typical effect of temperature on the phase behavior of water-oil-ethoxylated nonionic amphiphile systems.

A phase triangle occurs because the hydrophilic part of the amphiphile is not compatible with the oil and the lipophilic part of the amphiphile is not compatible with water. Increasing compatibilities results in a direct transition from Winsor I to Winsor II. For a very specific set of conditions, the three-phase triangle just vanishes. Such a set of conditions defines a tricritical thermodynamic point (TCP). See Fig. 2.

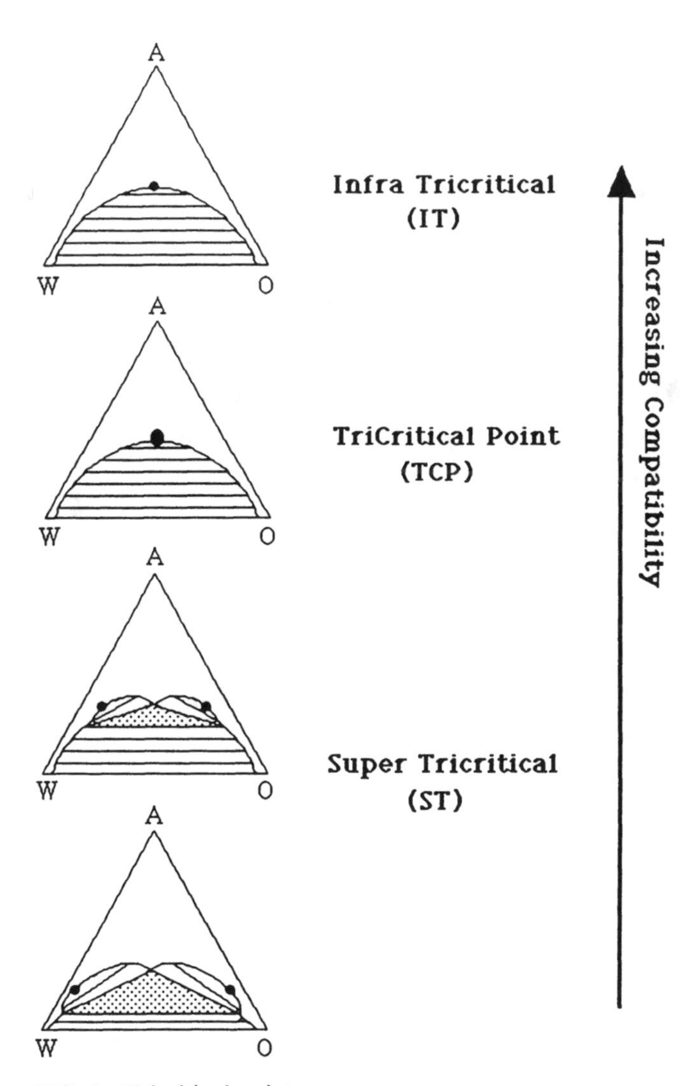

FIG. 2 Tricritical point.

If at any temperature a system is able to exhibit a three-phase domain, we call it supertricritical (ST); if at whatever temperature a system is unable to give a three-phase domain, we call it infratricritical (IT). Of course, if a ST system can be continuously modified to give an IT system, a tricritical point is necessarily encountered.

To provide a quantitative tool to predict the aspect of nonionic amphiphile surfactant-based phase diagrams, Busier and Ravey [10] proposed an equation giving the temperature at which a given phase diagram topography occurs as a function of the

Nonionic amphiphile HLB (calculated according to Griffin)
Alkane carbon number of the oil (ACN)
Salinity (S) in grams of electrolyte per liter of aqueous solution

Applied to the phase inversion temperature (in °C), the equation is

$$\mathrm{PIT} = -160 + 15.5(\mathrm{HLB}) + 1.8(\mathrm{ACN}) + bS \quad (1)$$

where b is a function of the electrolyte. For example, $b = -0.68$°C L/g for sodium carbonate and -0.25°C L/g for sodium chloride.

Similar equations are proposed for the CMC and CPT. The value of b is reported to be almost invariant with the salt concentration and is almost the same for CMC, CPT, and PIT.

Such an equation is of limited practical interest as it does not take into account for the amphiphile molar volume and the oil polarity.

2. Effect of the Amphiphile

It is well known that the PIT increases with the amphiphile hydrophilicity (the HLB). Nevertheless, the HLB concept, although powerful, appears to be very poor at predicting the minimum amount of amphiphile in the amphiphile phase. The amphiphile molar volume is a very useful complement to the HLB. The higher the amphiphile molar volume, the smaller the minimum amount of amphiphile in the amphiphile phase (the better the surfactant "coupling capacity").

Reducing the amphiphile molar volume also reduces the supertricritical state of the system. As a typical example, we found that the water-decanol acetate-diethylene glycol monobutyl ether ternary system is infratricritical, with the PIT at 29°C, for 38 wt.% amphiphile; the water-decanol acetate-triethylene glycol monohexyl ether is supertricritical, with the PIT at 24°C, for 29 wt.% amphiphile.

3. Effect of Brine

Water can be added with electrolytes and/or with highly soluble nonionic molecules. The effects of each type of additive will be considered individually, first on binary brine-nonionic amphiphile mixtures, followed by oil containing systems.

(a) Electrolytes. C13–C15 secondary alcohol ethoxylated with an average of 9 moles ethylene oxide is used as the nonionic amphiphile. The cloud point temperature is obtained by heating at 1°C/min a 1 wt.% solution of the nonionic amphiphile in the considered brine. As the cloud point is not very dependent on the nonionic amphiphile concentration, the 1% value should be an acceptable estimation of the lower consolute temperature.

It is currently believed that electrolytes always exhibit a salting-out effect (negative b value). Most electrolytes actually do cause salting out, but exceptions do exist, such as sodium perchlorate monohydrate (b = +0.35°C L/g) and potassium thiocyanate (b = +0.44°C L/g).

The effect of an added electrolyte on the solubility of an organic compound in water was discovered by Hofmeister about one century ago [11]. He found that the effect is mainly determined by the nature of the anion. The solubility of an organic compound in water is decreased by inorganic anions in the order:

sulfate > chromate > carbonate > chloride > nitrate

and perchlorate and thiocyanate increase the solubility.

The "lyotropic series" actually modifies the solvent properties of water and the effects are observed in various domains, all of them dealing with organic molecules and water.

The adsorption of a polymer from its aqueous solution onto a surface is favored by salting-out salts (sulfates, phosphates, . . .), whereas desorption is favored by salting-in salts.

The helix-random coil transition temperature of a protein is increased by salting-out salts and reduced by salting-in salts.

In fact, the salting-out salts promote the most ordered state (helix, adsorbed polymer, phase separation) and salting-in salts promote the most disordered state (random coil, solution).

The lyotropic series appears to apply in the same way to the aqueous solution properties of nonionic amphiphiles.

(b) Effect of the Electrolytes on the Cloud Point Temperature of Nonionic Amphiphiles. We measured the effect of a series of electrolytes on the CPT of an industrial C13–C15 secondary fatty alcohol ethoxylated with an average of 9 moles ethylene oxide. Fig. 3 demonstrates the linear relation between the CPT and the electrolyte concentration.

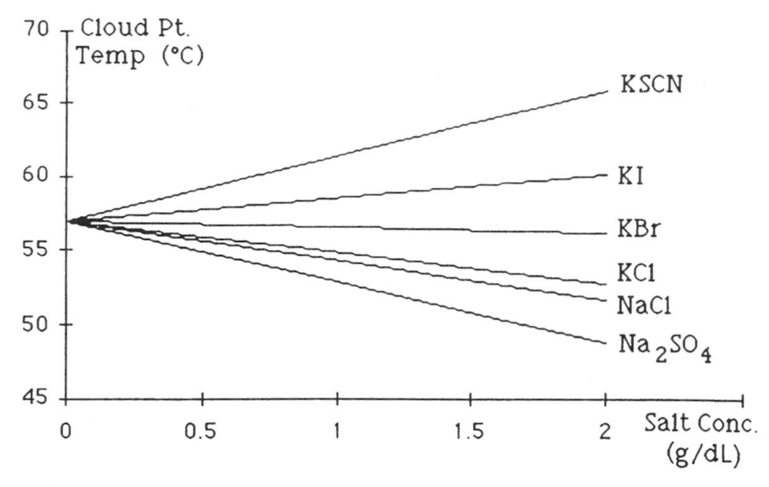

FIG. 3 Effects of electrolyte type and concentration on the cloud point temperature of *sec*C13-C15 alcohol ethoxylated with nine ethylene oxide groups.

For this particular system, the following equations are proposed:

$$CPT = 57.2 + bs \text{ (CPT in °C and } s \text{ in wt.\%)} \tag{2}$$

or

$$CPT = 57.2 + BS \text{ (CPT in °C and S in mole/L)} \tag{3}$$

Table 1 reports the salt coefficients measured with different salts.

(c) Small Water-Soluble Organic Molecules. We obtained very similar effects with small (not surface active) water-soluble organic molecules (e.g., the ''hydrotrope'' character of urea). To take account for these chaotropic effects in an equation like (1), (2), or (3), we introduce an additional term, CM. Equation (3b) can now be written:

$$CPT = 57.2 + BS + CM \tag{4}$$

where B is the molar salt coefficient, S is the electrolyte molar concentration (in mole/L), C is the ''molar chaotropy index,'' and M is the organic molecule molar concentration (in mole/L). One can also define a ''weight chaotropy index,'' c (in wt.%). The chaotopy indexes of several water soluble organic molecules are presented in Table 2.

Highly positive chaotropy indexes are obtained for organic molecules exhibiting an amphiphilic character (although the hydrophobic part is not

TABLE 1 Salt Coefficients for *sec*C13–C15 Alcohol Ethoxylated with Nine Ethylene Oxide Groups

Salt	Molecular mass	*b* (K dL/g)	*B* (K L/mole)
NaCl	58.5	−2.7	−15.8
KCl	74.5	−2.1	−15.6
KBr	119	−0.4	−4.8
KI	166	1.55	25.7
NaBr	103	−0.4	−4.1
LiCl	38.5	−2.2	−8.5
CsCl	168.5	−0.9	−15.2
Na_2SO_4	142	−4.6	−65.3
K_2SO_4	174	−3.75	−65.3
Na_2CO_3	106	−6.1	−64.7
KSCN	97	4.4	42.7
$NaClO4 \cdot H_2O$	140.5	3.5	49.2

TABLE 2 Chaotropy Indexes of Organic Molecules

Molecule	Molecular mass	*c* (K dL/g)	*C* (K L/mole)
Urea	60	0.75	4.5
Methyl urea	74	1.5	11.1
Ethyl urea	88	2.45	21.6
n-Butyl urea	116	4.9	56.8
Acetamid	59	1.05	6.2
n-Butyl amid	87	3.7	32.2
n-Propyl amine	59	5.6	33.0
n-Butyl amide	73	7.0	51.1
Methanol	32	1.65	5.3
Ethanol	46	2.55	11.7
n-Propanol	60	2.6	15.6
n-Butanol	74	−1.65	−12.2
2-Propanol	60	2.85	17.1
Acetic acid	60	1.7	10.2
Glycine	75	−2.6	−19.5
Valine	117	−0.5	−5.9
Leucine	131	0.75	9.8

long enough to lead to micellization). Moreover, as long as the organic molecule is far below its solubility limit, the higher the internal "contrast" between the hydrophilic and hydrophobic parts, the higher the chaotropy index.

So far, the highest values recorded are, on a molar basis, *n*-butyl urea ($C = 56.8$ K L/mole) and, on a weight basis, *n*-butyl amine ($C = 7.0$ K dL/g).

For substituted ureas, amines, amide, simple amino acids, methanol, and ethanol, the molar chaotropy index follows perfectly well the empirical quadratic equation:

$$C = 2.2n^2 + 4.275n + p \tag{5}$$

where n is the number of carbon atoms in the hydrophobic part of the molecule and p is a constant characteristic of the hydrophilic part of the molecule. Table 3 gives p values of several hydrophobic groups as well as the maximum number of carbon atoms that can be attached. For longer hydrocarbon chains, Eq. (5) holds no longer, as the molecule approaches its solubility limit.

Fig. 4 shows how the molar chaotropy indexes of substituted ureas fit Eq. (5).

(d) Oil Containing (pseudo) Ternary Systems. The effects of different electrolytes on the phase inversion temperature of the diethyleneglycol monobutyl ether (diethylene glycol monobutyl ether, model amphiphile), tridecyl acetate (C13Acetate, oil), brine pseudoternary system are illustrated in Fig. 5.

The PIT data are obtained, as a function of temperature, by monitoring the number and the volumes of coexisting phases in 15-mL glass tubes. The tubes contain equal weights of brine and oil; the proportion of amphiphile varies from 15 to 40 wt.%, depending on the amphiphile efficacy. With low molecular mass amphiphiles, the phase separation is fast and

TABLE 3 *P* Coefficients

Headgroup	p (K L/mole)	n_{max}
$H_2N—CO—NH—$	4.5	4
$H_2N—$	0	4
$H_2N—CO—$	0	3
$H_3N^+—CH(CO_2^-)—CH_2—$	−23	3
$HO—CH_2—$	5	1

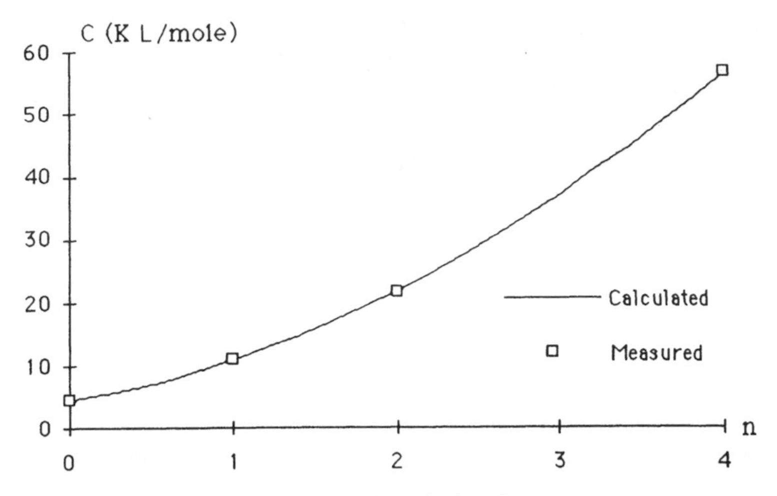

FIG. 4 Quadratic model fitting for alkylated ureas.

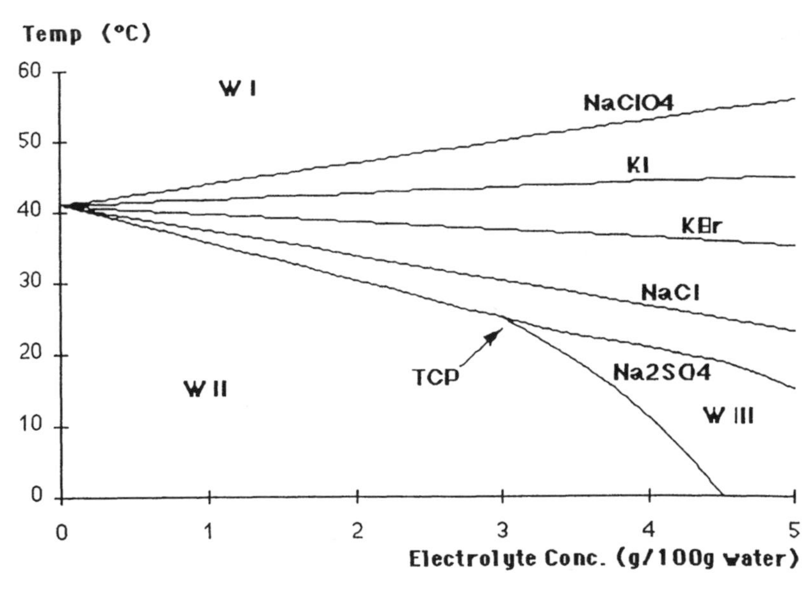

FIG. 5 Effects of electrolyte type and concentration on the phase behavior of diethylene glycol monobutyl ether-tridecanol acetate-brine system (infratricritical).

the evaluation can be done after 30 min at each temperature. The PIT corresponds to the temperature at which the weight of the water-rich phase equals the weight of the oil-rich phase. The measured phase volumes need to be corrected for the volumic mass of each phase. The accuracy of the PIT is better than 2°C.

The diethylene glycol monobutyl ether-C13Ac-water ternary system is infratricritical, as the phase topography shifts directly from Winsor I to Winsor II as the temperature is increased. At 41°C, the tie lines are parallel to the oil-water axis and the plait point is located at an equal weight fraction of oil and water (α = 0.50) and for an amphiphile concentration of 0.38 wt.% (γ = 0.38). The infratricritical state is due to the lower water-oil incompatibility and to the low molar volume of the amphiphile.

Sodium perchlorate and potassium iodide increase the PIT, as they do the CPT of C13–C15 alcohol ethoxylated with 9 ethylene oxide groups. The other electrolytes reduce the PIT, in the same order as they do for CPT.

The lyotropic series remains perfectly applicable to brine-oil-amphiphile systems.

Sodium sulfate concentrations above 3 wt.% induce the appearance of a three-phase triangle, which translates into a supertricritical state. High levels of salting-out electrolytes reduces the water availability for organic molecule solubilization. In the present case, the brine-oil and the brine-amphiphile incompatibilities are increased, leading to the formation of an amphiphilic phase. A tricritical point necessarily exists in the neighborhood of 3% sodium sulfate.

The ethyleneglycol monobutyl ether-decane-water system is supertricritical. Fig. 6 shows what happens when electrolytes are added to water and Fig. 7, when small organic molecules are added to water.

The salting-out sodium chloride induces a reduction of both Tu and Tl, along with an increase of the Tl-Tu temperature gap. The slightly salting-in potassium iodide has no effect on Tu, but increases Tl, reducing the Tu-Tl temperature gap. In the low concentration range at least, the electrolytes keep on changing the water capability to solubilize organic molecules according to their position in the Hofmeister's series.

The substituted ureas increase both Tl and Tu. The effects on Tl are roughly similar for the three tested molecules. On the other hand, the effects on Tu follow the chaotropic characters of the additives as determined by their effect on the nonionic amphiphile cloud point.

4. Effect of the Oil

(a) Oil Molar Volume. The effects of the saturated hydrocarbons and alkylbenzenes on the phase inversion temperature of nonionic amphiphile

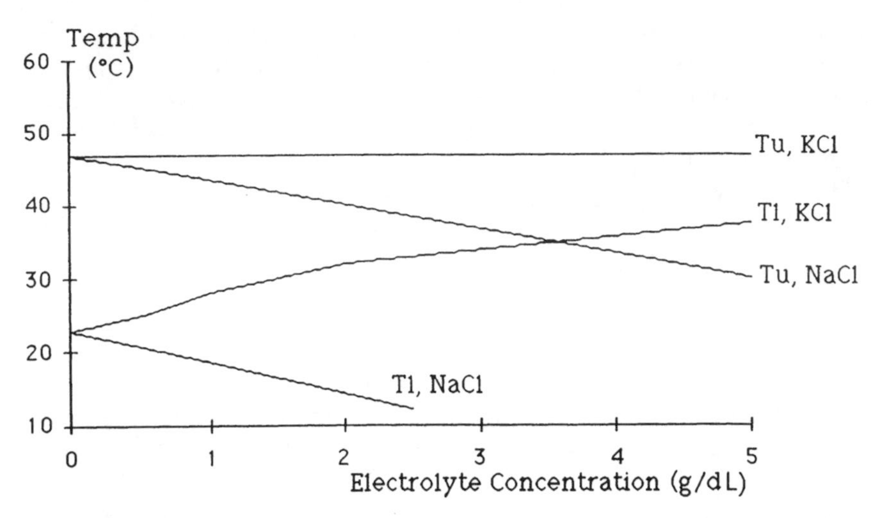

FIG. 6 Effects of electrolyte type and concentration on the phase behavior of ethylene glycol monobutyl ether-decanol acetate-brine system (supertricritical).

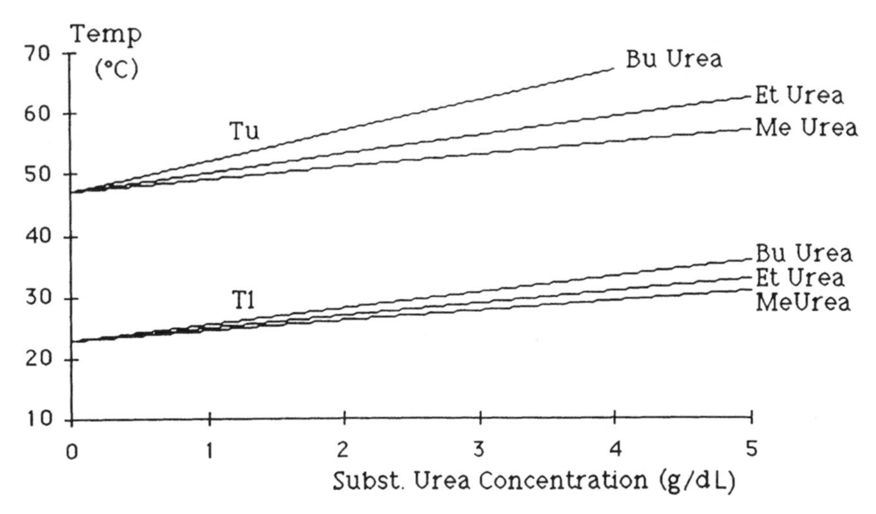

FIG. 7 Effects of alkylated ureas on the phase behavior of ethylene glycol monobutyl ether-decanol acetate-brine system (supertricritical).

containing systems has been disclosed in details by Kahlweit and Strey [9]. Decreasing the oil molar volume results in lower Tl and Tu, and accordingly, the PIT, along with a reduction of the supertricritical character of the system.

With the alkylbenzenes and diethylene glycol monobutyl ether, a tricritical point is observed near hexylbenzene close to 55°C.

The same trends are observed with alkanol acetates, which all give infratricritical systems. Fig. 8 summarizes the phase behavior of the brine-oil-diethylene glycol monobutyl ether systems.

Reducing the oil molar volume (in the same class, either hydrocarbons, alkyl benzenes or alkanol acetates) increases significantly the number of molecules per unit volume; the main reason for PIT reduction is accordingly mixing entropy.

(b) Oil Polarity. The effect of oil polarity is also illustrated by Fig. 8: at constant molar volume (as for instance, 180 mL/mole), increasing the oil polarity from saturated hydrocarbons to alkyl benzenes to alkanol acetates results in a dramatic reduction of the PIT and a trend toward infratricritical states.

Increasing the oil polarity has an effect similar to an oil molar volume reduction. Nevertheless, the thermodynamic reason is different: at constant molar volume, no entropy effect is to be expected; the main reason is a decrease of the oil-water and oil-amphiphile mixing enthalpies. The

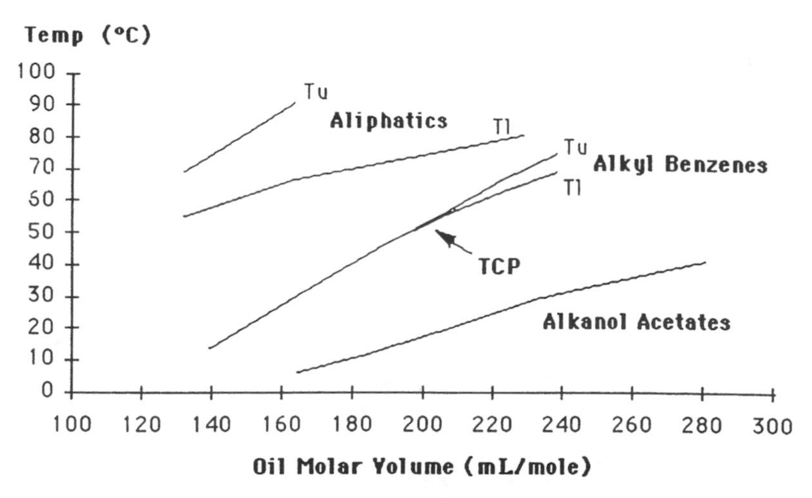

FIG. 8 Effects of oil molar volume and polarity on the phase behavior of diethylene glycol monobutyl ether-hydrocarbon-water system.

presence in the oil structure of a functional group able to establish dipolar interactions and/or hydrogen bonds dramatically reduces the incompatibilities of the oil for water and the hydrophilic part of the amphiphile.

The polarity of the oil can be estimated from Hansen's 3-D solubility parameters. Hansen divided Hildebrand's solubility parameter into three independent components: δd for dispersion contribution, δp for polar contribution, and δh for H-bonding contribution (see Section IV.A. for more information). As an estimation of the oil polarity, we define *Dph* as the square root of the square of the polar component plus the square of the H-bonding component of the solubility parameter, using the table published by Barton [12].

For ternary mixtures of alkanes, alkyl benzenes, or alkanol acetates with water and diethylene glycol monobutyl ether, the phase inversion temperature can be satisfactorily expressed as

$$\text{PIT (°C)} = 1.016V - 0.00121V^2 - 44.72Dph + 4.747Dph^2 - 2.74 \tag{6}$$

V is the oil molar volume in mL/mole and *Dph* is the oil polar character in $\text{MPa}^{1/2}$.

The effects of molar volume and polarity appear to be independent, at least in this specific case.

5. Phase Topography Tailor Making

The phase topography as well as the remoteness from a tricritical thermodynamic point of brine-oil-nonionic amphiphile systems can be established by adjusting the physicochemical parameters of the system.

The qualitative conclusions are summarized in Fig. 9.

Amphiphile

Increasing the HLB increases Tu, Tl, and PIT, due to a direct action on the amphiphile partition coefficient.

Increasing molar volume (at constant HLB) increases the supertricritical character of the system.

Brine

The brine lipophobicity can be reduced by addition of salting-in electrolytes as well as hydrotropic uncharged organic molecules.

The salting-in (chaotropic) character of electrolytes increase with their water solubilization entropy.

The hydrotropic character of uncharged organic molecules increases with the internal "contrast" of the molecule structure, as long as its concentration is significantly below the solubility limit.

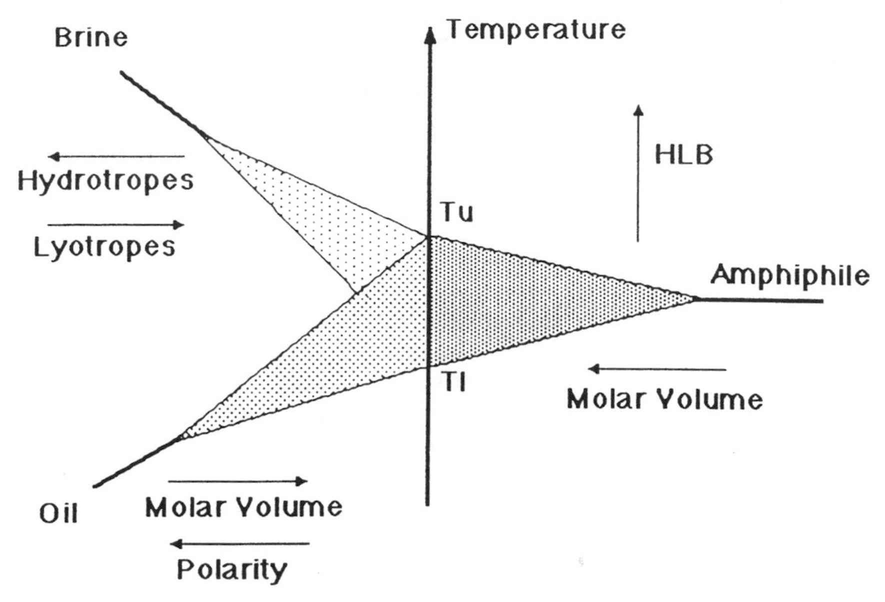

FIG. 9 Tool for phase behavior prediction.

The presence of chaotropic materials (salting-in electrolytes or hydrotropes) increases the PIT and decreases the supertricritical character of the system.

Oil

Increasing the oil molar volume (at constant polarity) increases PIT and the supertricritical character of the system is increased. The reason for this is mainly entropic.

Increasing the oil polarity (at constant molar volume) decreases PIT and reduces the supertricritical character of the system, mainly for enthalpic reasons.

Mastering these basic rules will allow the formulator to predict, at least qualitatively, the effects of the product ingredients, resulting in a significant product development workload reduction.

B. Systems Based on Anionic Surfactants

The ionic surfactants should theoretically show a behavior similar to that of the nonionic surfactants. In practice, the presence of the negative charge on the hydrophilic headgroup makes the molecule highly water soluble. As a consequence, the systems obtained are usually far from the phase inversion condition, in the Winsor I configuration. Accordingly, such systems show poor oil solubilization. Due to ionic repulsions, anionic surfactants show nevertheless an excellent ability to stabilize the oil droplets in water.

The oil uptake capacity of anionic surfactant-based systems can be significantly increased by the addition of an essentially water insoluble "cosurfactant" such as an alcohol containing five or more carbon atoms. The alcohol molecules organize themselves between the molecules of the anionic surfactant and "screen" the electrostatic repulsions to a large extent. This allows much better packing and "corrects" the natural curvature of the interface, to reduce the interfacial tension between an oil and the surfactant aqueous solution to a minimum. As an example, a system containing sodium dodecyl sulfate, pentanol and water can solubilize more than 50% decane.

Anionic surfactant based systems require a lot of optimization as the choice of the cosurfactant is very delicate and depends on the intended task. The domain of existence of a high oil uptake capacity in the water side is usually very narrow and difficult to find.

The addition of a magnesium salt can significantly increase the oil uptake capacity of an anionic surfactant system. This is probably due to the exchange between the magnesium and the sodium, which leads to a magnesium surfactant salt. Magnesium surfactant salts are much less hydrophilic than the sodium analogs.

IV. SOLUBILIZATION IN NONAQUEOUS MEDIA

Another important way of cleaning greasy soil is to dissolve it in an organic solvent. Petroleum ether, ethanol and acetone are currently used solvents. Greasy soils are some times very hard to remove from a cotton fabric because they spread on the surface. The nature of the soil is very complex: it can contain crystallites and dust particles dispersed in an amorphous mass of triglyceride.

The extent of triglyceride crystallization depends on the proportion of unsaturated triglycerides, the average chain length and the thermal history of the grease. It is obvious that the greater the degree of crystallization, the harder the grease is to remove. To maximize the solubilization of any grease, it is desirable to select a solvent or a solvent mixture that exhibits a maximum solvency for tripalmitin. Pure tripalmitin is a linear, fully saturated C16 triglyceride which crystallizes very rapidly to more than 80%.

The organic solvents exhibiting a significant solvency for tripalmitin are very limited and most of them are halogenated hydrocarbons. They are not recommended for use in households, as their handling requires excellent ventilation. According to Barton and Patton [13, 14], it is possible to select blends of two (or more) poor solvents performing not only better than each solvent alone, but at parity with chlorinated hydrocarbons. Such blends can be predicted by using Hansen's 3-D solubility parameters.

A. Hansen's Solubility Parameters

The cohesive energy density (CED) of a compound is the energy required to vaporize one cubic centimeter of that compound. It is the energy required to overcome the cohesion forces that maintain the compound in a condensed phase. The cohesion forces can be divided into three independent classes.

Dispersion forces correspond to the cohesion forces resulting from induced dipole-induced dipole interactions (London forces). These forces are present in any compound. They are the only ones present in hydrocarbons, and are responsible for the liquefaction and solidification of the rare gases.

Polar forces correspond to the cohesion forces resulting from dipole-dipole interactions. These forces occur in any compound having a permanent dipole moment such as ethers, esters, ketones, etc.

H-bonding forces correspond to the cohesion forces resulting from hydrogen bonding interactions. They occur in compounds having a mobile proton (alcohols, carboxylic acids, amines, amids, . . .) or a lone electron pair (ketone, ester, ether, . . .). Compounds exhibiting polar cohesion forces usually exhibit also some degree of H-bonding forces.

To each class of cohesion forces is assigned a *solubility parameter*. δ_d corresponds to the dispersion contribution, δ_p corresponds to the polar contribution, and δ_h corresponds to the H-bonding contribution. The sum of the squares of the solubility parameters of a compound is equal to the cohesive energy density.

$$\text{CED} = \delta_d^2 + \delta_p^2 + \delta_h^2 \qquad (7)$$

Any compound can be represented by a single point in a 3-D space, the three axes of which are δ_d, δ_p, and δ_h (Fig. 10).

According to Barton [12], maximum solubility is obtained when the solvent and the solute exhibit similar solubility parameters. This means that it is not enough to have similar cohesive energy density, the dispersion, polar, and H-bonding components must match each other.

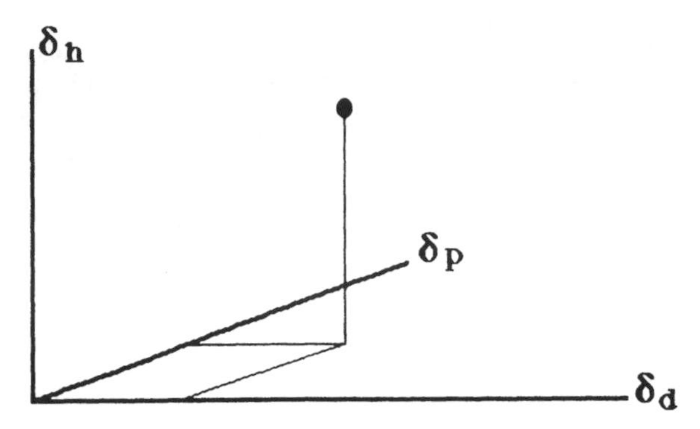

FIG. 10 Representation of compound in the solubility parameter space.

The solubility parameters of a great number of organic compounds are listed in tables such as those published by Barton. When not available, approximations can be calculated according to the addition of the contributions of the different segments of the molecule (group molar attraction contributions) [13].

The calculated solubility parameters of tripalmitin are

$$\delta_d = 16.8\ \text{MPa}^{1/2} \qquad \delta_p = 1.0\ \text{MPa}^{1/2} \qquad \delta_h = 1.6\ \text{MPa}^{1/2}$$

The solubility parameters of a mixture of solvents can be calculated by adding the individual contributions multiplied by the weight fractions. For instance, the polar solubility parameter of a 25% solution of ethanol (15.8, 8.8, 19.4) in hexane (14.9, 0.0, 0.0) is

$$\delta_p = (8.0 \times 0.25) + (0.0 \times 0.75) = 2.0\ \text{MPa}^{1/2} \tag{8}$$

The mixtures of two compounds can be represented on a straight line joining the points corresponding to the two compounds, in the 3-D solubility parameter space. Mixtures of organic solvents can show increased solvency for tripalmitin if their composition has solubility parameters close to tripalmitin. To predict the composition of maximum solvency, one has to minimize the distance from the tripalmitin to the line joining the two solvents.

B. Solvent Molar Volume

Unfortunately, optimizing the solvent blend composition is not enough to guarantee an acceptable performance. We showed that, in addition, the molar volumes of the solvents have to be as small as possible, to provide

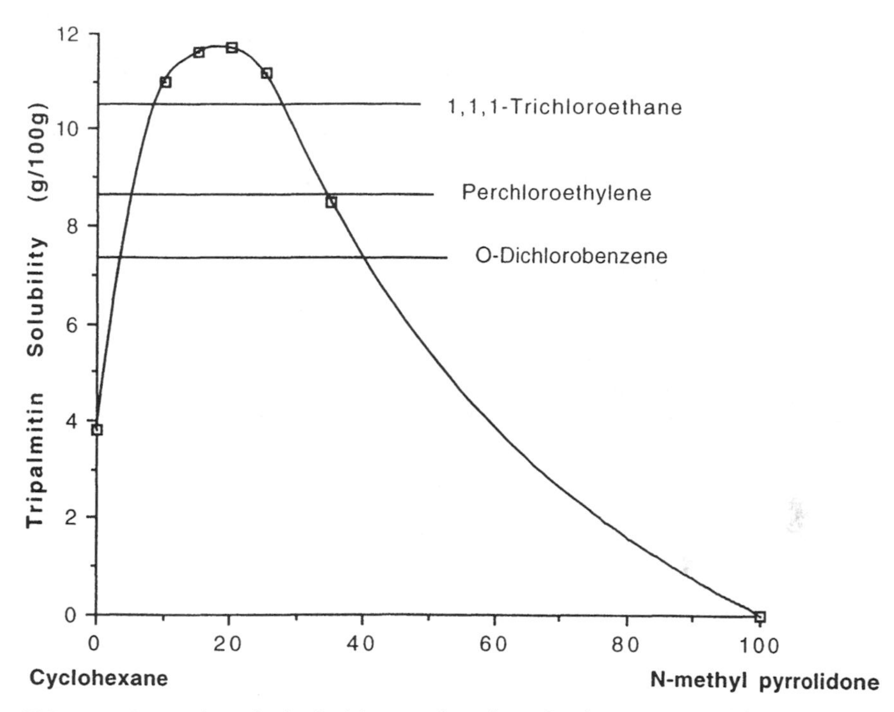

FIG. 11 Solubility of tripalmitin as a function of solvent composition.

enough mixing entropy [15]. Whenever possible, low molar volume solvents should be used. In practice, the molar volume should not exceed 0.2 L $mole^{-1}$.

C. Mixture of Poor Solvents

The following example illustrates the principle of mixing poor solvents to obtain increased solubility. Cyclohexane and *N*-methyl pyrrolidone both exhibit a low solvency for tripalmitin. Fig. 11 shows the solubility of tripalmitin as a function of the composition of the two poor solvents. In a well-chosen composition range, the solvency for tripalmitin exceeds that of many chlorinated solvents.

REFERENCES

1. S. E. Friberg and Q. L. Venable, *Encyclopedia of Emulsion Technology*, Vol. 1 (P. Becher, ed.), Marcel Dekker, New York, 1983, p. 287.

2. L. Gan Zud and S. E. Friberg, *J. Dispersion Sci. Technol. 4*: 19 (1983).
3. C. Solans, E. Garcia Domingues, and S. E. Friberg, *J. Dispersion Sci. Technol. 6*: 523 (1985).
4. F. Candau, Z. Zekhinini and J.-P. Durand, *J. Colloid Interface Sci. 114*: 398 (1986).
5. K. Halmberg and E. Oterberg, *Progr. Colloid Polym. Sci. 74*: 98 (1987).
6. M. Kahlweit, E. Lessner, and R. Strey, *J. Phys. Chem. 87*: 5032 (1983).
7. M. Kahlweit, E. Lessner, and R. Strey, *J. Phys. Chem. 88*: 1937 (1984).
8. M. Kahlweit, R. Strey and D. Haase, *J. Phys. Chem. 89*: 163 (1985).
9. M. Kahlweit and R. Strey, *Angew. Chem. Int. Ed. Engl. 24*: 654 (1985).
10. M. Buzier and J.-C. Ravey, *J. Colloid Interface Sci. 91*: 20, (1983).
11. F. Hofmeister, *Arch. Exp. Pathol. Pharmakol 24*: 247 (1888).
12. A. F. M. Barton, *Chem. Rev. 75*: 731 (1975).
13. A. F. M. Barton, *Handbook of Solubility Parameters and Other Cohesion Parameters*, CRC Press, Boca Raton, FL 1983.
14. T. C. Patton, *Solubility and Interaction Parameters, in Paint Flow and Pigment Dispersion*, 2nd ed., Wiley-Interscience, New York, 1979, pp. 301–334.
15. P. Durbut and G. Broze, *Comm. J. Com. Esp. Deterg., Barcelona, 20*: 383 (1989).

16

Solubilization in Micellar Separations

TIMOTHY J. WARD and KAREN D. WARD Department of Chemistry, Millsaps College, Jackson, Mississippi

SYNOPSIS

When the surfactant concentration increases to a point called the critical micelle concentration, extensive aggregation of the surfactant monomers occurs. These aggregates, called micelles, cause many of the bulk physical properties of the solution to change. Micelles provide a nonhomogeneous environment for the solubilization of solutes and are used to alter the selectivity in many separation techniques. In this work, the structure and properties of micelles, solubilization theory, and applications in various separation techniques were examined.

I. INTRODUCTION

Surfactants, a term derived from surface active agents, are also known as soaps, detergents, and amphiphiles. Surfactant aggregates, called micelles, have been used in many areas of analytical chemistry, including classical titrations, molecular absorption, luminescence spectroscopy, and electrochemistry [1]. The focus of this work is micellar solubilization techniques in liquid chromatography and electrokinetic chromatography. The structure and properties of micelles will be discussed prior to considering the theory of separation and examining selected applications.

II. STRUCTURE AND PROPERTIES OF MICELLES

Surfactants are composed of a hydrophobic tail and a hydrophilic headgroup which can be cationic, anionic, nonionic, or zwitterionic. The hydrophobic tail typically contains from 7 to 21 carbons [2] and may contain single or double hydrocarbon chains and multiple bonds. In very dilute aqueous solutions (less than 10^{-4} *M*), surfactants usually exist as monomers, although dimers, trimers, etc., can exist. At a surfactant concentration known as the critical micelle concentration (cmc), extensive aggregation of the surfactant monomers occurs and many bulk physical properties of the solution change. The aggregate is termed a micelle and contains from 60 to 100 or more surfactant monomers. The number of monomers in a micelle is called the aggregation number. As the surfactant concentration is increased above the cmc, more micelles are formed and the concentration of the nonaggregated surfactant remains nearly constant, approximately equal to the cmc [1]. Aggregation occurs over a narrow range of concentrations, and slightly different cmc values may be obtained when using different measurement methods. Table 1 shows some typical micelle parameters [2].

TABLE 1 Typical Surfactants and Their cmc, Aggregation Number, and Krafft Point Values[a]

Surfactant	cmc (M)	Aggregation number	Krafft point (°C)[b]
Aqueous (normal)			
Anionic			
Sodium dodecyl sulfate (SDS) $CH_3(CH_2)_{11}OSO_3^- Na^+$	0.0081	62	9
Potassium perfluoroheptanoate $C_7F_{15}COO^- K^+$	0.03	[c]	25.6
Sodium polyoxyethylene (12) dodecyl ether $CH_3(CH_2)_{11}(OCH_2CH_2)_{12}OSO_3^- Na^+$ (SDS 12EO)	0.0002	81	<0
Cationic			
Cetylpyridinium chloride[d] $C_{16}H_{33}N^+C_5H_5Cl^-$	0.00012	95	[c]
Cetyltrimethylammonium bromide (CTAB) $CH_3(CH_2)_{15}N^+(CH_3)_3Br^-$	0.0013	78	23
Nonionic			
Polyoxyethylene (6) dodecanol $CH_3(CH_2)_{11}(OCH_2CH_2)_6OH$	0.00009	400	[c]
Polyoxyethylene (23) dodecanol (Brij-35) $CH_3(CH_2)_{11}(OCH_2CH_2)_{23}OH$	0.0001	40	[c]
Zwitterionic			
N-dodecyl-*N*,*N*-dimethylammonium-3-propane-1-sulfonic acid (SB-12) $CH_3(CH_2)_{11}N^+(CH_3)_2(CH_2)_3SO_3^-$	0.003	55	<0
N,*N*-dimethyl-*N*-(carboxymethyl)octylammonium salt $C_8H_{17}N^+(CH_3)_2CH_2COO^-$ (octylbetaine)	0.25	24	<0
Nonaqueous (reversed)			
Bis(2-ethylhexyl) sodium sulfosuccinate (AOT) $NaO_3SCH(CH_2COOC_8H_{17})COOC_8H_{17}$	0.0006	[c]	[c]

[a] Values for aqueous solution at 25°C, except AOT determined in hexane.
[b] Temperature at which the solubility of an ionic surfactant is equal to the cmc.
[c] Not available or not defined.
[d] In 0.0175 *M* NaCl.

Micelle formation is primarily controlled by three forces: the hydrophobic repulsion between the hydrocarbon chains and the aqueous solution, the charge repulsion of ionic headgroups, and van der Waals attraction between the hydrocarbon tails [1–3]. The length of the hydrocarbon tail, size of the headgroup, and interaction of the hydrocarbon tails with one another and with the aqueous solution determines the size and shape of the micelle, the cmc, and the aggregation number. Shapes of micelles can vary from rough spheres to prolate ellipsoids, and the diameter of micelles generally ranges from 3 to 6 nm [2]. This small diameter prevents micelles from being filtered from solution and from appreciably scattering light.

A. Normal Micelles

In aqueous solution the hydrocarbon tails are oriented inward and the headgroups are positioned outward into the bulk solvent [3]. In this model the resulting structure resembles a small hydrocarbon pool or oil droplet surrounded by the polar headgroups, counterions, and water [4–7]. Other micelle models have been proposed in which some water penetrates into the hydrocarbon center or in which some portion of the surfactant tails are exposed to the bulk solvent [2].

Aqueous micelle solutions are often used to solubilize solutes. Solutes can interact with micelles electrostatically with the head groups, hydrophobically in the micelle core, or by a combination of both [8, 9]. Often the solubility of sparingly water-soluble compounds can be dramatically increased by using micellar solutions [10]. For example, 1,2-benzphenanthrene and 2,3-benzphenanthrene have a very low water solubility of less than 9.0×10^{-9} *M*. In the presence of 0.50 *M* potassium dodecanoate, their solubility increases to about 6.4×10^{-4} *M*, an increase of 66,000 times [11].

Micelles are dynamic systems in which the surfactant monomers enter and exit the micelle in microseconds [12–15]. Complete exchange of surfactant monomers in a micelle can occur in milliseconds to seconds. Any solutes associated with the micelle can also be exchanged with the bulk solution or other micelles. A characteristic binding constant can be found for a solute that associates with a micelle. This binding constant is equivalent to the ratio of the entrance and exit rate constants. Typical rates for several solutes have been published [1, 16]. A large entrance-to-exit rate constant ratio indicates that the solute is interacting strongly with the micelle.

Drastic changes in the cmc and aggregation number are observed when additives are added to the micellar solution [10]. The addition of electrolytes generally increases the aggregation number while decreasing the cmc

[10]. Water soluble organic molecules may either enhance or inhibit the formation of micelles. For example, addition of low concentrations of short-chain alcohols (four or fewer carbon atoms) can lower the cmc, and therefore enhance the formation of micelles [10]. Some organic solvents, such as 1,3-propanediol and glycerol, which can have three-dimensional structures in the pure liquid state, have been shown to promote micelle formation when present at low concentrations and allow micelle formation in mixtures with high concentrations of these solvents [17, 18]. Conversely, higher concentrations (10–15%) of short-chain alcohols suppresses micelle formation [17, 19]. Organic solvents that can hydrogen bond with water, such as acetone, dioxane, acetonitrile, and tetrahydrofuran, may also increase the cmc when present at very low concentrations, and suppress micelle formation altogether at higher concentrations (15–20% v/v) [17, 18].

B. Reversed Micelles

Aggregation of monomers can occur when surfactants are dissolved in nonpolar organic solvents. In reversed micelles the hydrocarbon tails are oriented outward into the bulk solvent while the hydrophilic headgroups are oriented inward. These reversed micelles are more complex than normal micelles but can be used to solubilize polar solutes in nonpolar solvents. Information on aggregation forces and aggregation number of reversed micelles can be found in the following references [15, 20–22].

III. SOLUBILIZATION IN MICELLAR LIQUID CHROMATOGRAPHY

Micelles provide a non-homogeneous environment for the solubilization of solutes because they have a polar surface and a nonpolar interior. They are the principal component of the mobile phase in micellar liquid chromatography (MLC), serving as an alternative to the hydro-organic mobile phases used in traditional reversed-phase liquid chromatography. The surfactant monomers present in the micellar solution can adsorb onto alkyl-bonded stationary phases such as C_{18} or C_8 in at least two ways [23]. The first, termed hydrophobic adsorption, involves adsorption of the alkyl tail of the monomer to the stationary phase alkyl group, with the ionic headgroup of the monomer in contact with the polar solution. The second possibility, termed silanophilic adsorption, involves adsorption of the ionic headgroup to free silanol groups on the silica gel, resulting in a more hydrophobic stationary phase. Fig. 1 illustrates these possible adsorption modes. Various studies have shown that the first type of adsorption fre-

Si
Si
SiOH
Si
SiOH
Si
$OSO_3^- Na^+$
$OSO_3^- Na^+$
$OSO_3^- Na^+$

Hydrophobic adsorption

Silanophilic and hydrophobic adsorption

FIG. 1 Possible modes for surfactant adsorption onto alkyl-bonded stationary phases such as a C_{18} bonded reversed-phase column. (Reprinted with permission from Ref. 23. Copyright 1987 American Chemical Society.)

quently occurs on alkyl-bonded stationary phases [23, 24]. Although the amount of surfactant monomer adsorbed to the stationary phase remains constant in most cases, the presence of organic additives can affect the amount of surfactant adsorbed [10, 25–34]. Structural parameters of the stationary phase such as pore diameter and particle size distribution can be altered due to monomer adsorption [10, 35]. For these reasons it is important to properly equilibrate the column with the micellar mobile phase to ensure reproducible results.

There are a number of advantages in using micelles in chromatographic applications, such as enhanced selectivity, low cost, low toxicity, ease of mobile phase disposal, ease of purification of mobile phase (water and surfactant), and ability to separate hydrophilic and hydrophobic solutes simultaneously [2, 25, 36–41]. MLC also has several disadvantages, the most prominent being that separations generally exhibit lower efficiencies than liquid chromatography using hydro-organic mobile phases. The causes and possible solutions to the low efficiency in micellar chromatography are discussed in Refs. 3, 42, and 43.

A. Partitioning Theory

The first effective use of aqueous micellar solutions in reversed-phase liquid chromatography was demonstrated by Armstrong and Henry [37]. They showed that hydrophobic, amphiphilic, and hydrophilic molecules could be separated under isocratic conditions using a micellar mobile phase. The resolution of different types of compounds is possible since the micelle can interact with the solutes via several interactions. The three-phase model for micellar chromatography was first proposed by Armstrong and Nome [44]. This model, shown in Fig. 2, depicts the equilibria involved during separation and allows a theoretical description of micellar LC. First is the partition coefficient of the solute between the bulk solvent and the stationary phase (P_{sw}), and second, a solute partitioning takes place between the bulk solvent and the micelle (P_{mw}). A third partitioning of the solute between the micelle and the stationary phase is also represented (P_{sm}). This model was used by Armstrong and Nome to derive the following equation to account for the reversed-phase retention behavior of uncharged solutes:

$$\frac{V_s}{V_e - V_m} = \frac{\nu(P_{mw} - 1)}{P_{sw}} C + \frac{1}{P_{sw}} \tag{1}$$

where V_s, V_m, and V_e are the volume of the stationary phase, mobile phase, and elution respectively; ν is the partial specific volume of the surfactant in the micelle; and C is the concentration of surfactant present in the micelles in g/mL (found by total surfactant concentration minus the cmc). Equation (1) can be used to describe the retention of nonpolar, polar, and ionic solutes in anionic, cationic and nonionic micellar LC mobile phases [33, 45].

This pseudophase retention equation also can be used for the determination of partition coefficients and binding constants. By plotting $V_s/(V_e - V_m)$ versus C, which should give a straight line of positive slope, P_{sw} and P_{mw} can be calculated from the intercept and the ratio of the slope

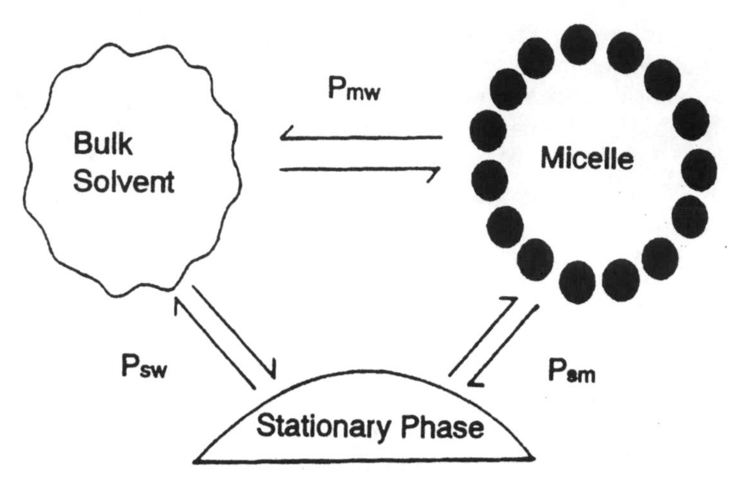

FIG. 2 Representation of the three-phase model which allows a theoretical description of micellar liquid chromatography. *P* is the partition coefficient of a solute between the indicated phases. (Reprinted with permission from Ref. 44. Copyright 1981 American Chemical Society.)

to the intercept, respectively [2, 3]. Other equations have been derived for the calculation of equilibrium constants for the solute between the bulk aqueous phase and micelles [46]. One of these equations does not require knowledge of the values of the partial specific volume of the surfactant or the volume of the stationary phase. Armstrong has listed 12 reexpressions of the pseudophase LC retention equation and discussed statistical and other advantages for the different variations [2].

B. Electrostatic and Hydrophobic Interactions

The separation mechanism in MLC is much more complex than that in traditional LC, owing to the number of potential interactions of the solute with the micellar mobile phase and the modified stationary phase. Solutes experience various microenvironment polarities in a given mobile phase, with solute retention depending on the interaction between the solute, micelle, and surfactant-modified stationary phase. Nonpolar solutes should have only hydrophobic interactions, while electrostatic interactions must be considered with charged solutes. The charged solute can have the same or the opposite charge as that of the surfactant headgroup. An example of solute and surfactant with the same charge is when sodium dodecyl sulfate (SDS) is used to resolve anionic solutes such as undissociated phenol and 2-naphthol, or when cetyltrimethylammonium bromide (CTAB) is used to separate cationic solutes such as amines. In this case, electrostatic repulsion between the micelle and solute will prevent partitioning to the micelle and the solute will stay in the bulk mobile phase. The electrostatic repulsion between the solute and surfactant adsorbed to the stationary phase should decrease retention, since the solute will not adsorb to the stationary phase. However, if any hydrophobic interactions between the stationary phase and solute occur, the solute retention may be affected by the stationary phase [36].

In the case of oppositely charged solute and surfactants, electrostatic attraction will occur between the solute and micelle, and between the solute and surfactant adsorbed to the stationary phase. If the solute has a hydrophobic interaction with the stationary phase, its retention will be increased, while a hydrophobic interaction with the micelle will decrease retention. By choosing an appropriate surfactant, both polar and nonpolar solutes can be resolved simultaneously [47].

C. Binding, Nonbinding, and Antibinding Solutes

Armstrong and Stine [9, 48] investigated the effect of electrostatic interactions between the solute and micelle on selectivity. In this study, compounds are classified as binding, nonbinding, or antibinding based on their retention behavior. Solutes that associate or bind to micelles will demonstrate a decreased retention when the micellar concentration is increased. Solutes whose retention is unaltered by micellar concentration have zero interaction with the micelle, and their interaction is called nonbinding. Highly polar molecules may demonstrate antibonding behavior [9, 48], in which retention increases with increasing micelle concentration. This effect arises when the solute is expelled or excluded from the micelle due to electrostatic repulsion. These effects are only observed on stationary phases that do not appreciably adsorb surfactant monomer, such as CN or C_1 bonded phases [48]. By exploiting the antibinding behavior of some solutes, unusual selectivity in MLC is achieved.

D. Micellar Selectivity in MLC

A comparison of the chromatographic behavior of a homologous series of compounds shows the differences in retention and selectivity between conventional reversed-phase chromatography and MLC [49–51]. In conventional reversed-phase chromatography using nonmicellar mobile phases, the logarithm of the capacity factor of a compound is linearly related to the number of carbons in a homologous series of compounds. However, in micellar chromatography, the hydrophobic selectivity of ionic and nonionic micelles does not follow a linear relationship when ln k' is plotted versus the homologue number of the solute. This is thought to be due to different solute polarities, which lead to different locations in the micelle for different homologues. This behavior was investigated by Hinze and Weber [52], who demonstrated that it was explained by two equilibria in the retention process. The first equilibrium is the partitioning of the solute between the aqueous phase and the stationary phase and the second is the partitioning of the solute between the micelle and stationary phase. An equation for retention was derived that explains the retention data with respect to the equilibria processes.

Functional group selectivity is defined as the ratio of the capacity factor of a substituted compound to that of a nonsubstituted parent compound [36]. Hydrophobic interactions play a large role for many compounds, particularly nonionic solutes. In general, a decrease in functional group selectivity is observed with increasing micelle concentration [36, 53]. Other studies report that ionic, nonionic [54], and zwitterionic [55] micelles have functional group selectivity values that correlate well with octanol-water partition coefficients.

By plotting the logarithm of the capacity factor versus the logarithm of surfactant concentration for different compounds, changes in selectivity become evident [28]. Intersection of the linear plots indicates that the elution order of the compounds has reversed, with an accompanying change in selectivity.

E. Selectivity due to Other Parameters

To utilize the enhanced selectivity, it is necessary to examine the various factors which affect retention in micellar chromatography. Several of these, including pH, ionic strength, and temperature, are listed in Table 2.

Retention of weak acids and bases will be affected by the pH of the micellar phase, since the partition coefficient of the solute into the micelle will be different for the dissociated and undissociated forms of the solute. Small changes in pH can have a great effect on retention, particularly when the pH is close to the pK_a of the solute [56–58]. The effect of pH change depends upon the nature of the surfactant and the solute [10, 59]. The type of stationary phase also influences the effect of pH change. Cyano- and C_{18} columns interact very differently with surfactant monomers, and solutes will demonstrate different elution behavior on these columns with a change in pH. If no electrostatic repulsion occurs between the solute and surfactant, the capacity factor exhibits a sigmoidal dependence on pH when the surfactant concentration remains constant [59].

TABLE 2 General Effect of Parameters in Micellar Liquid Chromatography

Parameter	Effect on retention
Surfactant concentration in the mobile phase	Increasing concentration decreases retention
Addition of short-chain alcohols to mobile phase	Increasing concentration up to 5% decreases retention
pH	Effect varies and depends on solute nature
Temperature	Increasing temperature slightly decreases retention
Ionic strength	Increasing ionic strength decreases retention

Ionic strength of the micellar mobile phase can be changed by the addition of salts to the aqueous phase. If the solute interacts electrostatically with the micelle, retention should be affected by the ionic strength. Generally, retention decreases as ionic strength increases [60–64]; however, there are some exceptions to this trend [59, 60]. An increase in temperature is observed to cause a slight decrease in retention [27].

Addition of organic modifiers also affects solute retention. In general, adding up to 5% (v/v) of short chain alcohols such as *n*-propanol, *n*-butanol, or *n*-pentanol to the mobile phase decreases solute retention while increasing chromatographic efficiency [10].

F. Multiparameter Optimizations in Micellar Chromatography

Khaledi et al. [65, 66] have described a methods development strategy using amino acids and peptides with micellar mobile phases. Three parameters were examined: the concentration of organic modifier, the concentration of surfactant, and a limited range of pH. The amino acids and peptides used in this study showed linear or weakly curved retention behavior when the parameters were varied. This allowed efficient optimization of parameters using a relatively small number of initial experiments. An advantage of a quick and easy methods development protocol is the ability to overcome some disadvantages such as limited efficiency in micellar chromatography [65].

G. Applications Using Micellar Chromatography

To explore the potential and utility of micellar solubilization of solutes in micellar chromatography, a few representative applications are examined.

1. Use of Micellar Gradients

Micellar mobile phases can be used in gradient analyses (with respect to micelle concentration) with a much shorter reequilibration time being required than with traditional mobile phases [67]. In micellar solutions, if additional monomer is added, more micelles are formed and the concentration of surfactant monomer remains nearly constant at approximately the cmc value. Therefore, little column reequilibration time is necessary for repetitive analyses, since the amount of surfactant adsorbed to the stationary phase remains constant [67, 68]. Micellar gradients are also more compatible with electrochemical detectors than are traditional gradients [68, 69]. Another study has shown that the use of reversed micelles in normal phase chromatography reduces the effect of water in the mobile phase on the solute retention [70].

2. Drug and Protein Analyses

An isocratic mixed-micellar HPLC mobile phase can be used for the simultaneous quantitation of active ingredients in commercial cold tablets [71]. Since the active ingredients acetaminophen, chlorpheniramine, and pseudoephedrine have very different functional groups and polarities, they are very difficult to separate using traditional isocratic HPLC methods. A micellar method using a mixture of nonionic and anionic surfactants achieved an isocratic separation of all four components in under 20 minutes. Fig. 3a shows a chromatogram obtained using this method. The addition of 4.5% (v/v) *n*-propanol to the mobile phase improves the efficiency.

Micellar chromatography offers a unique option for the analysis of drugs or endogenous substances in body fluids. Several workers have reported the advantages of using micellar mobile phases for drug monitoring [72–74]. Micellar mobile phases eliminate the necessity of a prechromatographic protein precipitation or sample extraction step since the protein is readily solubilized in the micellar phase. Proteins in biological sample matrices often clog LC columns when traditional hydro-organic mobile phases are employed. In MLC, the protein is solubilized in the micellar mobile phases and eluted with the void volume. An additional reported benefit is that the surfactant monomers appear to displace drugs adsorbed to protein binding sites, thus freeing the drugs for quantitation [3, 73]. Various drugs have been assayed using this technique, such as acetaminophen, acetylsalicylic acid, chloramphenicol, phenobarbital, phenytoin, procainamide, propranolol, morphine, and codeine [72, 73]. Results obtained from the micellar chromatography and EMIT showed good agreement for the serum samples in one study [72]. A micellar chromatography method used amperometric detection to monitor dopamine

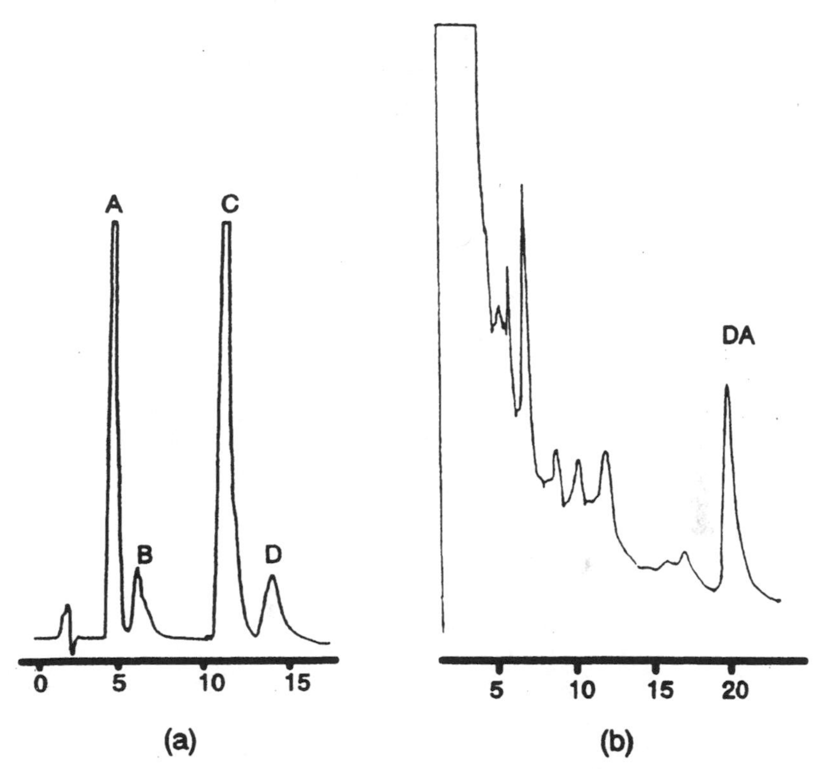

FIG. 3 Representative chromatograms of drugs and endogenous substances. (a) Chromatogram of A = acetaminophen; B = pseudoephedrine; C = chlorpheniramine; D = excipent. Mobile phase: 18.0 g Brij 35 + 3.46 g SDS in 1 L water containing 4.5% (v/v) *n*-propanol. Flowrate: 1.5 mL/min on 250 × 4.2 mm CN column at 65°C. (b) Chromatogram of dopamine = DA in urine. Mobile phase: 0.01 *M* SDS in 1 mM EDTA with 3% (v/v) *n*-propanol at pH 1.5. Flowrate: 0.7 mL/min on 150 × 4.6 mm C_{18} column at 40°C. ((a) reprinted from Ref. 73, by courtesy of Elsevier Science Publishers and (b) reprinted from Ref. 75.

by the direct injection of urine onto the column [75]. The detection limit of dopamine is reported to be 4 pg using this method, and the chromatogram is shown in Fig. 3b.

The separation of proteins can be achieved on a reversed-phase HPLC column using nonionic surfactant micellar phases [76]. The separated proteins were categorized into classes of low, intermediate, and high retention. There was no strict correlation between retention time and molecular weight, isoelectric point, or hydrophobicity. Proteins with high retention usually have low molecular weights and small decreases of surfactant concentration in the mobile phase exponentially increase protein retention.

3. Micelle Exclusion Chromatography

Micellar mobile phases have been used with size exclusion columns for the separation of inorganic anions [77] and heavy-metal cations [78]. Several anions having different hydration energies and molecular shapes were simultaneously separated using micelle exclusion chromatography. Selectivity for the anions is different from conventional ion-exchange or ion-pairing chromatography, due to the partitioning of the solute between the aqueous and micellar phase and the partitioning between the mobile and stationary phase. In order to separate the heavy-metal cations, the complexing agents tartaric acid and citric acid are added to the micellar mobile phase [78].

4. Bile Salt Micellar Mobile Phases

Most micellar mobile phases in HPLC have used linear surfactants such as the quaternary alkylammonium halides, anionic alkylsulfates, and nonionic surfactants like the Triton or Brij series. Recently the use of aqueous solutions of bile salts as the micellar mobile phase has been reported [79]. The bile salts are chiral, natural steroidal surfactants with surface-active properties. The bile salts have different micellar structure and aggregation behavior than the linear surfactants due to the steroidal structure. The bile salts form helical aggregates [80–82] and have been used in the separation of routine compounds, geometric isomers, and structural isomers [79].

IV. ELECTROKINETIC CHROMATOGRAPHY BASED ON MICELLAR SOLUBILIZATION

Micellar electrokinetic chromatography (MEKC), also referred to as micellar electrokinetic capillary chromatography (MECC) and micellar electrokinetic chromatography (MEC) in the literature, is a technique which was first introduced by Terabe et al. in 1984 [83]. MEKC combines many of the operational principles and advantages of micellar liquid chromatography and capillary zone electrophoresis (CZE). In CZE a sample is introduced at one end of a buffer-filled capillary column (ca. 50 μm i.d. $\times$ 100 cm), which is suspended between two reservoirs filled with buffer solution. An electric field is applied and charged solutes are separated by their electrophoretic migration with reported efficiencies often in excess of 100,000 plates/m [84].

MEKC is a relatively new separation technique that utilizes electroosmotically pumped micelles in an aqueous mobile phase to solubilize neutral solutes and affect separation. The aqueous and micellar phases constitute a homogeneous solution; however, these phases migrate at different rates and solutes distribute themselves between these two phases. In this case micelles move at a rate slower than that of the bulk solvent, due to their overall fractional charge, which causes the micelles to act as a moving "stationary" phase. The separation mechanism is similar to that of conventional liquid-liquid partition chromatography in that both ionic and nonionic solutes can partition between the micelles and the surrounding aqueous phase producing a separation [85, 86]. Neutral solutes are eluted by electro-osmotic flow, with any separation being achieved by the differences in their viscous drag. Unfortunately, these differences are usually very small, and neutral solutes generally are not separated by CZE.

A. Theory of Separation

It is known when ions are present at any solid-liquid interface, such as the silica-solution interface in a capillary column, there will be a variation in the ion density near that surface giving rise to an electrical double layer. In this case, the solid surface has an excess of anionic charge resulting from the ionization of surface silanol groups. Counterions surround this anionic surface forming a stagnant double layer adjacent to the capillary walls and a diffuse layer which extends out into the bulk solution. This simplified model is depicted in Fig. 4. The potential created across these layers is termed the zeta potential, ζ. These cations which extend into the diffuse layer are mobile and can migrate under the presence of an applied potential.

Electroosmotic flow in a capillary arises from these hydrated cationic counterions in the diffuse layer as they migrate toward the cathode and drag bulk solvent with them. Equations that describe the various parameters which affect electroosmotic flow and separation have been derived [87, 88]. The strength of the applied potential field, E, and the extent of the potential drop across the double layer governs the electroosmotic velocity, v_{eo}, which can be expressed as

$$v_{eo} = -\frac{\epsilon \zeta_1}{4\pi\eta} E = \frac{l}{t_0} \qquad (2)$$

where ϵ is the dielectric constant, η is the viscosity of the liquid medium, and l is the capillary length to the detector. Experimentally, v_{eo}, can be obtained from the migration time of a neutral species or solvent peak, t_o, as shown in Eq. (2). It is interesting to note from Eq. (2) that the electroosmotic velocity is not constant but rather controlled by a number

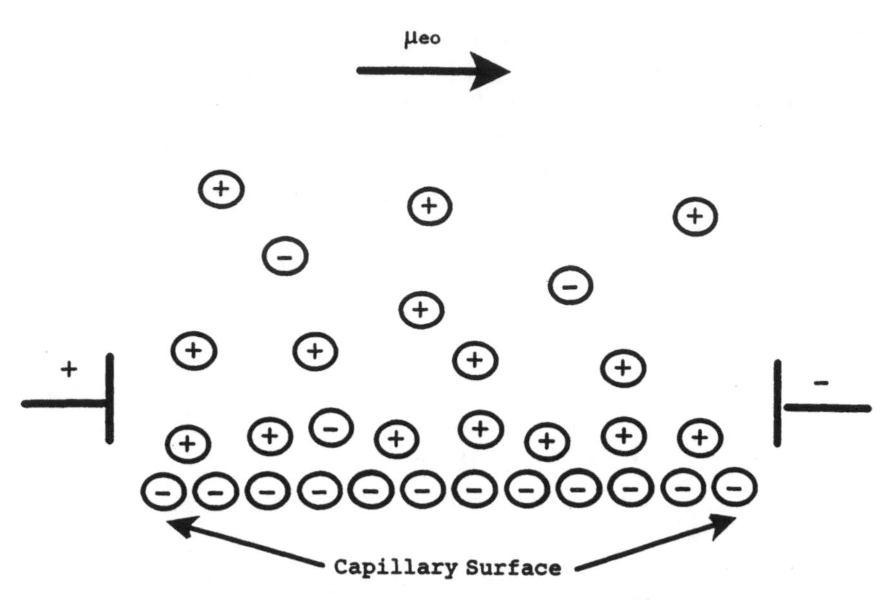

FIG. 4 Schematic of the surface and solvated ions at a silica-solution interface depicting the flow of solvated ions under an applied potential. (Reprinted with permission from *Anal. Chem.*, vol. 61, p. 294A. Copyright 1989 American Chemical Society.)

of variables such as mobile phase composition and the zeta potential of the capillary walls. Since $E = V/L$, where V is the voltage drop across the complete length of capillary, Eq. 2 can be represented as

$$v_{eo} = \frac{\mu_{eo} V}{L} \tag{3}$$

where μ_{eo} is the electroosmotic mobility of the solvent. The electrophoretic velocity ν_{ep} can be written in an analogous form to Eq. (3) as

$$v_{ep} = \frac{\mu_{ep} V}{L} \tag{4}$$

where μ_{ep} is the electrophoretic mobility of a given species. The velocity of a charged solute is given by

$$v_s = v_{eo} + v_{ep} = \frac{(\mu_{eo} + \mu_{ep})V}{L} \tag{5}$$

If the solute migrates electrophoretically in a direction opposite to the bulk flow (for example, a negatively charged species inside a negatively charged fused silica capillary), then v_s will be smaller than v_{eo} (i.e., v_{ep} will be negative).

At this point it is useful to define some of the terms used to describe the separation process and examine the equations that are used to measure the various separation parameters. The migration time of the solute t_s can be obtained from Eq. (5) as

$$t_s = \frac{l}{v_s} = \frac{lL}{(\mu_{eo} + \mu_{ep})\,V} \tag{6}$$

An expression analogous to Eq. (6) can be written for the migration time of a micelle, t_{mc}. When a neutral solute partitions within a micelle, the migration time of that species can be represented as

$$t_s = \left(\frac{1 + \tilde{k}'}{1 + (t_o/t_{mc})\tilde{k}'}\right)t_o \tag{7}$$

where $\tilde{k}'$ is the capacity factor for partition of the solute between the micelle and the aqueous phase, and is expressed as

$$\tilde{k}' = \frac{n_{mc}}{n_{aq}} = K\frac{V_{mc}}{V_{aq}} \tag{8}$$

where n_{mc}/n_{aq} is the mole ratio of solute in the micellar and aqueous phases, respectively, K is the distribution coefficient and V_{mc} and V_{aq} are the volume of the micellar and aqueous phases, respectively. Since the micellar phase is moving toward the detector in MEKC, solutes must elute between the column void time t_o and the micelle elution time t_{mc}. It is expected that hydrophobic solutes would spend most of their time in the micelle and elute near t_{mc}. Therefore the range over which the solute can migrate is between t_o and t_{mc}, and this range, along with the efficiency of the process, determines the peak capacity. Analogously to chromatography, resolution can be expressed as

$$R_s = \frac{N^{1/2}}{4}\left(\frac{\alpha - 1}{\alpha}\right)\left(\frac{\tilde{k}_2'}{1 + \tilde{k}_2'}\right)\left(\frac{1 - t_o/t_{mc}}{1 + (t_o/t_{mc})\tilde{k}_1'}\right) \tag{9}$$

where N is the number of theoretical plates and α is the separation factor k_2'/k_1', where 1 and 2 refer to the first and second components, respectively.

Since resolution in MEKC is generally optimal for capacity factors between approximately 1 to 5, depending on the elution window between t_o and t_{mc}, this technique is usually limited to relatively hydrophilic solutes [89]. The migration behavior of solutes and optimization techniques for MEKC have been examined by several groups [90–92].

B. Micellar Selectivity in MEKC

Factors which affect separation and efficiency in MEKC are discussed in the literature [85, 86, 93, 94]. MEKC can achieve as many as 200,000 to 600,000 theoretical plates, with height equivalent to theoretical plates approximately 1.9 to 3.7 microns [95, 96]. To extend the use of MEKC to more hydrophobic solutes, organic solvents have been added to the mobile phase [97, 98]. The organic solvent interacts with the capillary wall and slows the electroosmotic flow, which lengthens the micelle elution time t_{mc} and widens the elution window [97]. This also affects partitioning since the overall polarity of the mobile phase is decreased, and the capacity factor is reduced for hydrophobic solutes.

Addition of organic solutes can also affect selectivity in MEKC [98]. Higher concentrations of organic solvents can result in poor separations. For example, at levels higher than 15% (v/v), short-chain alcohols dramatically reduce efficiency and the analysis time becomes prohibitively long [99]. Different organic solvents can exhibit different effects on the separation. Methanol and isopropanol significantly alter the electroosmotic flow and widen the elution window, while solvents such as acetonitrile and dioxane reduce sample capacity, but do not dramatically alter the electroosmotic flow [89].

C. Applications of MEKC

Several typical applications are presented to examine the micellar solubilization of solutes in MEKC. Where it is appropriate, MEKC techniques will be contrasted to MLC separations and the potential advantages or disadvantages of each method discussed. In practice, the two techniques often are complementary methods of analysis.

1. Biological and Environmental Applications

A variety of environmental and biological samples can be separated by MEKC. For example, MEKC can be used for the separation of nucleic acid constituents [87], catecholamines [100], and inorganic anions [101]. For oligonucleotides, selectivity is increased by the addition of divalent metals to the SDS micellar phase since the metals are attracted to the surface of negatively charged micelles and the differential complexation of oligonucleotides with the metals affects separation. Inorganic anions are separated due to their differences in their ion association constants with cetyltrimethylammonium chloride (CTAC) and distribution coefficients into the micelle.

Flavonoid drugs, which are plant metabolites, are better resolved, and the separation takes less time using MEKC as compared to reversed-phase HPLC [102]. The level of sensitivity compared to HPLC is also higher, and MEKC has the additional advantage of a very low solvent consumption. The simultaneous separation of water- and fat-soluble vitamins can be achieved by MEKC using a SDS/γ-cyclodextrin mobile phase [103]. Separation parameters such as pH, micelle concentration, and organic modifiers which affect resolution are examined in this work. It was found that a combination of γ-cyclodextrin with sodium dodecyl sulfate provided the best selectivity for separating the two groups of vitamins.

Cyclodextrin-modified MEKC has been used for the solubilization and subsequent separation of polynuclear aromatic hydrocarbon analytes [104–107]. These solutes are not resolved in SDS solutions alone due to their large micellar solubilization. With CD/SDS, the cyclodextrin dissolves the analyte in the mobile phase, while the micelles act as the moving "stationary" phase. The selectivity can be changed with the addition of cyclodextrin, since an additional inclusion mechanism is introduced.

Bile salt surfactants have also been utilized in the separation of polynuclear aromatic hydrocarbons [89]. Use of the bile salts leads to a general decrease in capacity factor due to the increased polarity of the bile salts compared to SDS. The unique structure of the bile salt micelle seems to allow the presence of a higher concentration of organic solvents without degrading the efficiency of the separation or increasing the analysis time.

2. Chiral Separations by Micellar Solubilization

Chiral separations by MEKC are possible using chiral surfactants or mixed micelle-cyclodextrin mobile phases [108–112]. The bile salts mentioned above and L-amino acid derivatives such as sodium *N*-dodecanoyl-L-valinate (SDVal) have been used to achieve chiral separations. For example, chiral separation of naphthalene-2,3-dicarboxaldehyde amino acids is possible using MEKC with cyclodextrin added to the micellar phase [108]. In general, γ-cyclodextrin seemed to give a better separation than β-cyclodextrin. The use of methanol and urea in SDVal/SDS solutions improves peak shape and changes selectivity in the separation of phenylthiohydantoin derivatives of D,L-amino acids [109].

3. Drug Analysis Using MEKC

MEKC also offers a unique option for the analysis of drugs or endogenous substances in body fluids. The advantages of MEKC for drug monitoring are increased efficiency, peak symmetry, and speed when compared to MLC techniques [113–115]. Using MEKC, common drugs of abuse and/or their metabolites, such as opioids, methaqualone, and amphetamines can easily be detected at concentrations as low as 100 ng/mL using low-wavelength UV [114]. Fig. 5 shows an electropherogram of an extracted

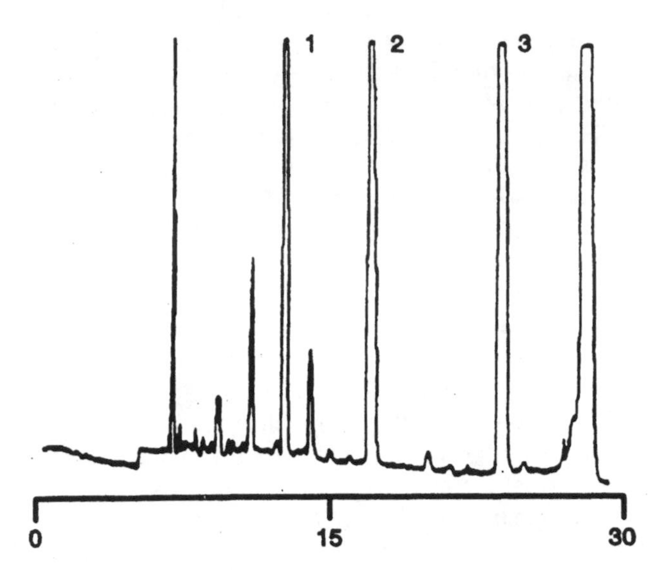

FIG. 5 Chromatogram of 1 = benzoylecgonine, 2 = morphine, 3 = codeine from extracted urine blank. Column: 70 cm × 75 μm i.d. fused-silica capillary. Buffer: 75 mM SDS, 6 mM $Na_2Ba_4O_7$, and 10 mM Na_2HPO_4 at pH 9.1. Voltage: 20 kV. Current: 76–80 μA. (Reprinted with permission from Ref. 114. Copyright 1991 American Chemical Society.)

urine sample spiked with benzoylecgonine, morphine, and codeine at 10 μg/mL each. When compared to the chromatograms in Fig. 3 several observations are apparent: efficiency is much greater, peak asymmetry is excellent, and sensitivity is improved. One disadvantage of MEKC when compared to direct injection MLC techniques, is that a solid-phase extraction must first be performed on the sample before analysis.

For a complex mixture consisting of acidic and neutral impurities present in a seized heroin sample, MEKC resolved twice the number of peaks in approximately one-third the time as an alternate HPLC method [115]. Peaks that can be resolved using this technique include cocaine, opium alkaloids, barbiturates, and cannabinoids. Due to MEKC's rapid analysis time and high resolving power, MEKC is well suited as an alternative technique for general drug screening.

REFERENCES

1. L. J. Cline Love, J. G. Habarta, and J. G. Dorsey, *Anal. Chem. 56*: 1132A (1984).

2. D. W. Armstrong, *Sep. Purif. Methods 14 (2)*: 213 (1985).
3. J. G. Dorsey, *Adv. Chromatogr. 27*: 167 (1987).
4. J. H. Fendler, *Membrane Mimetic Chemistry*, Wiley, New York, 1982.
5. G. S. Hartley, *Trans. Faraday Soc. 31*: 31 (1935).
6. G. S. Hartley, *Quart. Rev. Chem. Soc. 2*: 152 (1948).
7. P. Mukerjee and J. R. Cardinal, *J. Phys. Chem. 82*: 1620 (1978).
8. J. H. Fendler and L. K. Patterson, *J. Phys. Chem. 74*: 4608 (1970).
9. D. W. Armstrong and G. Y. Stine, *J. Am. Chem. Soc. 105*: 2962 (1983).
10. W. L. Hinze, in *Ordered Media in Chemical Separations* (W. L. Hinze and D. W. Armstrong, eds.), American Chemical Society, Washington, DC, 1987.
11. D. Attwood and A. T. Florence, *Surfactant Systems* Chapman and Hall, London, 1983.
12. J. H. Fendler and E. J. Fendler, *Catalysis in Micellar and Macromolecular Systems* Academic Press, New York, 1975.
13. C. A. Bunton, *Pure Appl. Chem. 49*: 969 (1977).
14. H. Wennerstrom and B. Lindman, *Phys. Rep. 52*: 2 (1979).
15. B. Lindman, H. Wennerstrom, and H. F. Eicke, *Micelles: Topics in Current Chemistry*, No. 87, Springer-Verlag, New York, 1980.
16. M. Almgren, F. Grieser, and J. K. Thomas, *J. Am. Chem. Soc. 101*: 279 (1979).
17. L. G. Ionescu, L. S. Romanesco, and F. Nome, *Surfactants in Solution* (K. L. Mittal and B. Lindman, eds.,) Plenum Press, New York, 1984.
18. C. A. Bunton, L. H. Gan, F. H. Hamed, and J. R. Moffatt, *J. Phys. Chem. 87*: 336 (1983).
19. L. Magid, *Solution Chemistry of Surfactants* (K. L. Mittal, ed.,) Plenum Press, New York, 1979.
20. J. H. Fendler, *Acc. Chem. Res. 9*: 153 (1976).
21. H. F. Eicke and A. Denss, *J. Colloid Interface Sci. 64*: 386 (1978).
22. H. F. Eicke, *Top. Curr. Chem. 87*: 85 (1980).
23. A. Berthod, I. Girard, and C. Gonnet, in *Ordered Media in Chemical Separations* (W. L. Hinze and D. W. Armstrong, eds.), American Chemical Society, Washington, DC, 1987.
24. C. H. Giles, in *Anionic Surfactants*, (E. H. Lucassen-Reynders, ed.,) Marcel Dekker, New York, 1981.
25. A. Berthod, I. Girard, and C. Gonnet, *Anal. Chem. 58*: 1362 (1986).
26. R. Weinberger, P. Yarmchuk, and L. J. Cline Love, *Anal. Chem. 54*: 1552 (1982).
27. P. Yarmchuk, R. Weinberger, R. F. Hirsch, and L. J. Cline Love, *J. Chromatogr. 283*: 47 (1984).
28. P. Yarmchuk, R. Weinberger, R. F. Hirsch, and L. J. Cline Love, *Anal. Chem. 54*: 2233 (1982).
29. M. T. Hearn and B. Grego, *J. Chromatogr. 296*: 309 (1984).
30. S. H. Hansen and P. Helboe, *J. Chromatogr. 285*: 53 (1984).
31. W. G. Tramposch and S. G. Weber, *Anal. Chem. 58*: 3006 (1986).
32. A. Berthod, I. Girard, and C. Gonnet, *Anal. Chem. 58*: 1356 (1986).

33. A. Berthod, I. Girard, and C. Gonnet, *Anal. Chem. 58*: 1359 (1986).
34. F. G. P. Mullins and G. F. Kirkbright, *Analyst 111*: 1273 (1986).
35. S. H. Hansen, P. Helboe, and M. Thomsen, *J. Chromatogr. 360*: 53 (1986).
36. M. J. M. Hernandez and M. C. G. Alvarez-Coque, *Analyst 117*: 831 (1992).
37. D. W. Armstrong and S. J. Henry, *J. Liq. Chromatogr. 3 (5)*: 657 (1980).
38. D. W. Armstrong, *Am. Lab. 13*: 14 (1981).
39. D. W. Armstrong, in *Solution Chemistry of Surfactants* (K. L. Mittal and E. J. Fendler, eds.), Plenum Press, New York, 1982.
40. D. W. Armstrong and M. McNeely, *Anal. Lett. 12*: 1285 (1979).
41. D. W. Armstrong, W. L. Hinze, K. H. Bui, and H. N. Singh, *Anal. Lett. 14*: 1659 (1981).
42. D. W. Armstrong, T. J. Ward, and A. Berthod, *Anal. Chem. 58*: 579 (1986).
43. A. Berthod, M. F. Borgerding, and W. L. Hinze, *J. Chromatog. 556*: 263 (1991).
44. D. W. Armstrong and F. Nome, *Anal. Chem. 53*: 1662 (1981).
45. M. F. Borderding and W. L. Hinze, *Anal. Chem. 57*: 2183 (1985).
46. M. Arunyanart and L. J. Cline Love, *Anal. Chem. 56*: 1557 (1984).
47. J. S. Landy and J. G. Dorsey, *Anal. Chim. Acta 178*: 179 (1985).
48. D. W. Armstrong and G. Y. Stine, *Anal. Chem. 55*: 2317 (1983).
49. M. G. Khaledi, E. Peuler, and J. Ngeh-Ngwainbi, *Anal. Chem. 59*: 2738 (1987).
50. M. G. Khaledi, *Anal. Chem. 60*: 876 (1988).
51. M. F. Borgerding, F. J. Quina, W. L. Hinze, J. Bowermaster, and H. M. McNair, *Anal. Chem. 60*: 2520 (1988).
52. W. L. Hinze and S. G. Weber, *Anal. Chem. 63*: 1808 (1991).
53. M. G. Khaledi, J. K. Strasters, A. H. Rodgers, and E. D. Breyer, *Anal. Chem. 62*: 130 (1990).
54. J. P. Berry and S. G. Weber, *J. Chromatogr. Sci. 25*: 307 (1987).
55. F. Gago, J. Alvarez-Guilla, J. Elguers, and J. C. Diez-Masa, *Anal. Chem. 59*: 921 (1987).
56. F. Palmisano, A. Guerrieri, P. G. Zambonin, and T. R. I. Cataidi, *Anal. Chem. 61*: 946 (1989).
57. L. J. Cline Love and J. Fett, *J. Pharm. Biomed. Anal. 9*: 323 (1991).
58. J. Haginaka, J. Wakai, and H. Yasuda, *J. Chromatogr. 488*: 341 (1989).
59. M. Arunyanart and L. J. Cline Love, *Anal. Chem. 57*: 2837 (1985).
60. D. W. Armstrong and G. Y. Stine, *J. Am. Chem. Soc. 105*: 6220 (1983).
61. F. G. P. Mullins and G. F. Kirkbright, *Analyst 109*: 1217 (1984).
62. G. F. Kirkbright and F. G. P. Mullins, *Analyst 109*: 493 (1984).
63. Z. El Rassi and C. Horvath, *J. Chromatogr. 326*: 79 (1985).
64. Z. El Rassi and C. Horvath, *Chromatographia 15*: 75 (1982).
65. J. K. Strasters, S. T. Kim, and M. G. Khaledi, *J. Chromatogr. 586*: 221 (1991).
66. J. K. Strasters, E. D. Breyer, A. H. Rodgers, and M. G. Khaledi, *J. Chromatogr. 511*: 17 (1990).
67. J. S. Landy and J. G. Dorsey, *J. Chromatogr. Sci. 22*: 68 (1984).

68. J. G. Dorsey, M. G. Khaledi, J. S. Landy, and J. L. Lin, *J. Chromatogr. 316*: 183 (1984).
69. M. G. Khaledi and J. G. Dorsey, *Anal. Chem. 57*: 2190 (1985).
70. M. A. Hernandez-Torres, J. S. Landy, and J. G. Dorsey, *Anal. Chem. 58*: 744 (1986).
71. T. A. Biemer, *J. Chromatogr. 410*: 206 (1987).
72. F. J. DeLuccia, M. Arunyanart, and L. J. Cline Love, *Anal. Chem. 57*: 1564 (1985).
73. M. Arunyanart and L. J. Cline Love, *J. Chromatogr. 342*: 293 (1985).
74. F. J. DeLuccia, M. Arunyanart, P. Yarmchuk, R. Weinberger, and L. J. Cline Love, *LC Mag. 3*: 794 (1985).
75. Y. Qu, P. Hu, and P. L. Zhu, *J. Liq. Chromatogr. 14*: 2755 (1991).
76. R. A. Barford and B. J. Sliwinski, *Anal. Chem. 56*: 1554 (1984).
77. T. Okada, *Anal. Chem. 60*: 1511 (1988).
78. T. Okada, *Anal. Chem. 60*: 2116 (1988).
79. R. W. Williams, Z. S. Fu, and W. L. Hinze, *J. Chromatogr. Sci. 28*: 292 (1990).
80. G. Esposito, E. Giglio, N. V. Pavel, and A. Zanobi, *J. Phys. Chem. 91*: 356 (1987).
81. E. Giglio, S. Loreti, and N. V. Pavel, *J. Phys. Chem. 92*: 2858 (1988).
82. A. R. Campanelli, S. C. De Sanctis, E. Giglio, N. V. Pavel, and C. Quagliata, *J. Inclusion Phenomena 6*: 391 (1989).
83. S. Terabe, K. Otsuka, K. Ichikawa, A. Tsuchiya, T. Andro, *Anal. Chem. 56*: 111 (1984).
84. D. Perrett and G. Ross, *Trends Anal. Chem. 11*: 156 (1992).
85. S. Terabe, K. Otsuka, and T. Ando, *Anal. Chem. 57*: 834 (1985).
86. D. E. Burton, M. J. Sepaniak, and M. P. Maskarinec, *Chromatographia 21*: 583 (1986).
87. A. S. Cohen, S. Terabe, J. A. Smith, and B. L. Karger, *Anal. Chem. 59*: 1021 (1987).
88. R. J. Hunter, *Zeta Potential in Colloid Science* Academic, London, 1981.
89. R. O. Cole, M. J. Sepaniak, W. L. Hinze, J. Gorse, and K. Oldiges, *J. Chromatogr. 557*: 113 (1991).
90. J. P. Foley, *Anal. Chem. 62*: 1302 (1990).
91. K. Ghowsi, J. P. Foley, and R. J. Gale, *Anal. Chem. 62*: 2714 (1990).
92. J. K. Strasters and M. G. Khaledi, *Anal. Chem. 63*: 2503 (1991).
93. M. J. Sepaniak and R. O. Cole, *Anal. Chem. 59*: 472 (1987).
94. S. Terabe, H. Utsumi, K. Otsuka, T. Ando, T. Inomata, S. Kuze, and Y. Hanaoka, *JHRC & CC 9*: 666 (1986).
95. S. Terabe, K. Otsuka, K. Ichikawa, A. Tsuchiya, and T. Andro, *Anal. Chem. 56*: 111 (1984).
96. M. J. Sepaniak, D. E. Burton, and M. P. Maskarinec, in *Ordered Media in Chemical Separations* (W. L. Hinze, and D. W. Armstrong, eds.) American Chemical Society, Washington, DC, 1987.
97. A. T. Balchunas and M. J. Sepaniak, *Anal. Chem. 59*: 1466 (1987).

98. J. Gorse, A. T. Balchunas, D. F. Swaile, and M. J. Sepaniak, *JHRC & CC 11*: 554 (1988).
99. A. T. Balchunas and M. J. Sepaniak, *Anal. Chem. 60*: 617 (1988).
100. T. Kaneta, S. Tanaka, and H. Yoshida, *J. Chromatogr. 538*: 385 (1991).
101. T. Kaneta, S. Tanaka, M. Taga, and H. Yoshida, *Anal. Chem. 64*: 798 (1992).
102. P. G. Pietta, P. L. Mauri, A. Rava, and G. Sabbatini, *J. Chromatogr. 549*: 367 (1991).
103. C. P. Ong, C. L. Ng, H. K. Lee, and S. F. Y. Li, *J. Chromatogr. 547*: 419 (1991).
104. T. Imasaka, K. Nishitani, and N. Ishibashi, *Analyst 116*: 1407 (1991).
105. S. Terabe, *Trends Anal. Chem. 8*: 129 (1989).
106. S. Terabe, Y. Miyashita, O. Shibata, E. R. Barnhart, L. R. Alexander, D. G. Patterson, B. L. Karger, K. Hosoya, and N. Tanaka, *J. Chromatogr. 516*: 23 (1990).
107. H. Nishi and M. Matsuo, *J. Liq. Chromatogr. 14*: 973 (1991).
108. T. Ueda, F. Kitamura, R. Mitchell, T. Metcalf, T. Kuwana, and A. Nakamoto, *Anal. Chem. 63*: 2981 (1991).
109. K. Otsuka, J. Kawahara, K. Tatekawa, and S. Terabe, *J. Chromatogr. 559*: 209 (1991).
110. A. Dobashi, T. Ono, S. Hara, and J. Yamaguchi, *Anal. Chem. 61*: 1984 (1989).
111. H. Nishi, T. Fukuyama, M. Matsuo, and S. Terabe, *J. Chromatogr. 1*: 233 (1990).
112. K. Otsuka and S. Terabe, *J. Chromatogr. 515*: 221 (1990).
113. I. Z. Atamna, G. M. Janini, G. M. Muschik, and H. J. Issaq, *J. Liq. Chromatogr. 14*: 427 (1991).
114. P. Wernly and W. Thormann, *Anal. Chem. 63*: 2878 (1991).
115. R. Weinberger and I. S. Lurie, *Anal. Chem. 63*: 823 (1991).

Index